Industrial Biotechnology

Industrial Biotechnology

Dr. Debabrata Das and Dr. Soumya Pandit

CRC Press is an imprint of the
Taylor & Francis Group, an informa business

First edition published 2021
by CRC Press
6000 Broken Sound Parkway NW, Suite 300, Boca Raton, FL 33487-2742

and by CRC Press
2 Park Square, Milton Park, Abingdon, Oxon, OX14 4RN

CRC Press is an imprint of Taylor & Francis Group, LLC

Library of Congress Cataloging-in-Publication Data
Names: Das, Debabrata, 1953– author. | Pandit, Soumya, author.
Title: Industrial Biotechnology / Dr. Debabrata Das, Dr. Soumya Pandit.
Description: First edition. | Boca Raton, FL: CRC Press, 2021. |
Includes bibliographical references and index. |
Summary: "This book offers a comprehensive overview of biochemical processes, technologies, and practical applications of industrial biotechnology. The work comprises chapters that discuss medium preparation, inoculum preparation using industrial strain and upstream processing, various fermentation processes, and physico-chemical separation processes for the purification of products and packaging. It features case studies of a variety of biochemical production processes. The text is aimed at advanced students, industrial practitioners, and researchers in biotechnology, food engineering, chemical engineering, and environmental engineering"—Provided by publisher.
Identifiers: LCCN 2020055187 (print) | LCCN 2020055188 (ebook) |
ISBN 9780367408886 (hardback) | ISBN 9780367822415 (ebook)
Subjects: LCSH: Biotechnology.
Classification: LCC TP248.2 .D37 2021 (print) |
LCC TP248.2 (ebook) | DDC 660.6-dc23
LC record available at https://lccn.loc.gov/2020055187
LC ebook record available at https://lccn.loc.gov/2020055188

ISBN: 978-0-367-40888-6 (hbk)
ISBN: 978-0-367-76956-7 (pbk)
ISBN: 978-0-367-82241-5 (ebk)

Typeset in Times
by Newgen Publishing UK

Contents

Foreword

Bioproducts are becoming increasingly important for our economies and wellbeing. Biological processes are less energy intensive than physical–chemical processes. Our ability to manipulate genetic information allows for the improved production of both existing and novel bioproducts in the fields of chemistry, foods, pharmacy, agriculture and environmental management. For the industrial-scale production of bioproducts, many different processes must be developed.

The purpose of this book is to give an overview of the bioprocessing industry, from upstream, the processing of raw materials, to their downstream processing, through a myriad of transformations, purifications etc. The readers, from students to practitioners, will benefit from detailed information about these bioprocesses, from basic material and energy balances, kinetics and cell growth, to substrate conversion and product formation. This knowledge can be applied in the design of bioreactors, from sterile fermentation to large-scale applications. This text book provides basic and applied information for a variety of large-scale bioproduct formation processes, from basic valves and pumps, to strain development and effluent treatment processes. Mathematical analysis of the processes is presented, derived in simplified form to also allow ready access to those not trained as engineers.

I congratulate the authors, **Debabrata Das** and **Soumya Pandit,** for seeing the need for producing the present volume, *Industrial Biotechnology,* which comprehensively covers most industrial bioprocesses.

I strongly recommend this excellent book to students of biotechnology, biochemical engineering, chemical engineering, food technology etc. and also to researchers and process engineers involved in the production of bioproducts in order to enrich their knowledge of application of biosciences for the benefit of mankind.

Dr. John Benemann

CEO, MicroBio Engineering Inc.

PO Box 15821, San Luis Obispo, CA 93406, USA

Preface

What you learn from a life in science is the vastness of our ignorance.

– David Eagleman

Industrial biotechnology comprises two components: one is industry, and the other is biotechnology. Similarly, biotechnology has two components: one is bio, and the other is technology. Technology essentially deals with science and engineering. Therefore technology is the application of scientific knowledge to develop or create different products on a large scale. Science tells us why things happen, and how they happen. Bio refers to biological systems. Therefore biotechnology is the application of biological sciences to develop various useful products. Industrial biotechnology is the application of biological processes at a commercial level in order to produce products on a large scale for the benefit of mankind. Industrial biotechnology comprises a set of operation processes that use living microorganisms (such as bacteria, yeast, fungi, algae etc.) or components of living organisms such as enzymes, to produce a product of industrial importance which is followed by downstream or purification processes. Biological processes involve different biomolecules which come from living sources. It is now possible to manipulate DNA to get desired changes in biochemical processes. This has influenced our perceptions of the development of products useful for the chemical, food, pharmaceutical, and environmental sectors. New developments in gene expression, protein engineering and cell fusion have made a significant impact on the biotechnology industry regarding new product development. Biochemical reactions are mostly chain and reversible reactions in nature. To understand biochemical engineering it is necessary to know the principles of chemical engineering. The book has been designed accordingly in order to understand the chemical processes followed by biochemical processes such as enzymatic reaction engineering, cell growth kinetics etc. This includes not only mathematical modelling but also designing of bioreactors. Classifications of the bioproducts and several industrial production processes including biological waste treatment processes are discussed in detail. The biochemical industry comprises several processing units which are divided into upstream, fermenter, and downstream processing. Upstream processing mainly involves preparation of the medium, medium sterilization, and air sterilization etc. This is followed by bioreactors where biochemical reactions lead to the desired products. The last crucial step is the purification of the product, known as downstream processing. This is explained with the help of flow diagrams. Material analysis of industrial fermentation processes has been included. Each chapter begins with a fundamental explanation for general readers and ends with in-depth scientific details suitable for expert readers. The book also includes the solution of several industrial problems. It is written in a manner where it can be useful to both senior students and to graduates of biotechnology; it should also be useful in food and biochemical engineering courses. The book is also appropriate for chemical engineering graduates, undergraduates and industrial practitioners. One of the authors has been offering an online 12-week National Programme on Technology Enhanced

Learning (NPTEL) course (equivalent to 3 credit course of IITs in India) on Industrial Biotechnology for the last four years.

We would like to acknowledge the help of Mr. Chandan Mahata at various stages of manuscript preparation. We are also thankful to Mr. Tapas Mohanty for his help in computer drawing of most of the book's illustrations.

We hope this book will be useful to our readers!

Debabrata Das
Soumya Pandit

Author Biographies

Debabrata Das pursued his doctoral studies at the Indian Institute of Technology (IIT) Delhi and post-doctoral research work at the University of Utah (UU), USA. He was a Former Professor and Head at Biotechnology Department, IIT Kharagpur. He was also MNRE Renewable Energy Chair Professor for three years and Professor for more than 32 years at IIT Kharagpur. He has been actively involved in the research of biohydrogen technology for a period of more than twenty-one years. He has made commendable contributions towards the development of a commercially competitive and environmentally benign bioprocess for hydrogen production from organic wastes using both mesophilic and thermophilic microorganisms. His recent work on the biohythane process for the maximization of gaseous energy recovery from organic wastes is worth mentioning. A 10,000 L biohydrogen pilot plant study was successfully carried out; the technology licence agreement of the process has already been transferred to M/s. Dhampur Sugar Mills Ltd. He is the author of five text books: *Biohydrogen Production: Fundamentals and Technology Advances* and *Fundamentals of Biofuel Production Processes* published by M/s. CRC Press, New York and *Biohythane: Fuel for the future*, *Biochemical Engineering: An Introductory Text Book* and *Biochemical Engineering: A Laboratory Manual* published by M/s. Pan/Jenny Stanford Publishing Pte. Ltd., Singapore. He is the Editor of books entitled *Algal Biorefinery: An Integrated Approach* and *A bioelectrochemical system that convert wastes to Watts* published separately by M/s. Springer, Switzerland and M/s. Capital Publishing Company, India. He has three Indian patents and three patents are pending. He has Google *h-index* of 55 for his research work. He has more than 170 research publications in peer-reviewed journals and has contributed more than 38 chapters in books published by international publishers. He offers two NPTEL 12-week courses on Industrial Biotechnology (for the last four years) and Aspects of Biochemical Engineering. He was awarded the IAHE Akira Mitsue award 2008 and BRSI Malaviya Memorial award 2013 for his contributions to hydrogen research. He is a Fellow of International Association for Hydrogen Energy (IAHE), Indian National Academy of Engineering (INAE); Biotechnology Research Society of India (BRSI); West Bengal Academy of Science and Technology (WAScT); and Institution of Engineers (IE). He is a member of the editorial board of several international journals.

Soumya Pandit pursued his doctoral studies at Indian Institute of Technology, Kharagpur and did his postdoctoral research work at the Department of Desalination & Water Treatment, The Zuckerberg Institute for Water Research (ZIWR), Ben-Gurion University, Israel. He is currently working as Senior Assistant Professor at Sharda University, Delhi NRC, India. He is also a recipient of a North West University, South Africa Post Doctoral fellowship. After completion of his PDF, he became Assistant Professor at Amity Institute, Mumbai. He has published various papers at international conferences in the area of environmental biotechnology and bioenergy. He has published 25 research papers in peer-reviewed journals and 25 book chapters. He has three Indian patents. His current research focuses on microbial electrochemical systems for bioenergy harvesting, bacterial biofilm and biofouling study etc. He serves as an editorial board member in several journals.

List of Symbols, Greek letters, Abbreviations

SYMBOLS

A, a	Arrhenius constant, Area
a_v	Surface area per unit volume
b	Effective width of the air flow
C	Concentration, Cunningham's correction factors for slip flow
C_{Dm}, C_D	Drag coefficient
D, D_{max}, $D_{wash\ out}$	Dilution rate, Decimal reduction time, Diameter, Maximum dilution rate,
D_a	Damkohler number
D_{BM}, D_{AB}	Effective diffusivity of substrate
d_p, d_f	Particle diameter, Fibre diameter
D_z	Axial dispersion coefficient
E, E_a, $[E]_o$	Efficiency, Activation energy, Total or initial enzyme concentration
F, F_0, F_a, F_W, F_r, F_G, F_D, F_B	Volumetric flow rate, Mass flow rate, Inflow rate, Overall feed flow rate, Cell mass wasting flow rate, Cell mass recycling flow rate, Gravitational force, Drag force, Buoyant force
G	Free energy
g_c, g	Conversion factor, Acceleration due to gravity
I	Inhibitor concentration
k	Constant, Thermal death rate constant, Rate constant
k_L	Mass transfer coefficient in the liquid phase
K, K_c	Distribution coefficient, Equilibrium constant
K_i,_K_I	Inhibition constant
k_La	Volumetric mass transfer coefficient
K_m	Michaelis Menten constant
k_p	Product inhibition constant
K_S	Saturation constant
k_{-1}	Rate constant of backward reaction
L	Length
m	Maintenance coefficient
N, N_0	Number of the cells at any time t, Initial number of cells
n	Order of reaction
N_{Da}, N_r, N_R	Damkolher number, Reaction number, Geometrical ration
N_s	Rate of mass transfer

P	Pressure
P	Product concentration
P_e	Peclet number
P_o	Initial product concentration
q_P	Specific product formation rate
q_S	Specific substrate consumption rate
r, r_x	Rate of reaction, Radius, Rate of cell mass production
R	Gas constant
Re, N_{Re}	Reynolds number
r_s	Rate of substrate degradation
T	Temperature, Kelvin
$t, t_{gn}, t_d, t_{b,} t_{hd}, t_{total}, t_{down\ time}$	Time, Generation time, Doubling time, Batch time, Holding time, Total time, Down time or Idle time
S, S_0, S_{SS}, S_b, S_S	Substrate concentration, Initial substrate concentration, Steady state substrate concentration, Bulk substrate concentration, Substrate concentration at the surface of the solid matrix
Sc	Scmidth number
u	Fluid velocity
U_0	Relative velocity between the fluid and particle
V, v	Working volume of the reactor, Air velocity, Rate of reaction
v_{max}	Maximum velocity of reaction
V_p ' $V_{R,}$ V_C, V_g	Volume of the particle, Volume of the reactor, Critical air velocity, Settling velocity
$X, x, x_n, x_{max}, x_u, x_e, x_0$	Fraction of the substrate converted, Thickness of the depth filter, Cell mass concentration, cell mass Concentration after the cell division, Maximum cell mass concentration, Settled cell mass concentration, Effluent cell mass concentration, Initial cell mass concentration, Displacement of the particle
X_{90}	Thickness of the depth filter for 90 % removal of contaminants present in air
Y	Mole fraction
Y_C	g C-atom biomass/g C-atom substrate
$Y_{X/S}, Y'_{x/S}, Y$	Overall cell mass yield, True growth yield, Mole fraction
$Y_{P/S}$	Product yield coefficient
$Y_{X/O}$	g biomass/g oxygen as O_2 consumed
z	Length, Height of the column

GREEK LETTERS

μ	Specific growth rate of the cell, Viscosity
μ_{max}	Maximum specific growth rate

μ_d	Specific death rate of the cell
ξ_p	Energetic product yield
σ_b	Weight fraction of carbon in biomass
σ_p	Weight fraction of carbon in product
σ_s	Weight fraction of carbon in substrate
γ_b	Degree of reduction of biomass
γ_p	Degree of reduction of product
γ_s	Degree of reduction of substrate
τ_{CSRT}, τ_{PFR}	Space time in CSTR, Space time in PFR
ε	Void fraction
η, η_0, η_α, $\bar{\eta}$	Energetic growth yield, Effectiveness factor, Overall collection efficiency, Single fibre efficiency, Collection efficiency
α	Growth associated coefficient, Recycle ratio, Volume fraction of the filter
B	Saturation parameter, Non-growth associated coefficient
φ	Thiele modulus, Inertial parameter
A, ρ_p	Density of the fluid, Density of the particle
∇_{total}	Sterilization criterion
θ	Hydraulic retention time
θ_C	Mean cell residence time

ABBREVIATIONS

A	Arrhenius constant
AAB	Acetic acid bacteria
AcCoA	Acetyl co-enzyme
AD	Anaerobic digestion
ADH	Alcohol dehydrogenase
ADP	Adenosine diphosphate
AMP	Adenosine monophosphate
ANN	Artificial neural networks
6-APA	6-aminopenicillanic acid
ASP	Activated sludge process
ATP	Adenosine triphosphate
BFD	Block flow diagram
BOD	Biochemical oxygen demand
BPP	Biohydrogen pilot plant
BSTR	Batch stirred tank reactor
CA	Citric acid
CAA	Citric acid anhydrous
CAM	Citric acid monohydrate
CoA	Coenzyme A
COD	Chemical oxygen demand
CM	Chloramphenicol

CPHE	Continuous plate heat exchanger
CSI	Continuous steam injection
CSTR	Continuous stirred tank reactor
DAP	Dihydroxyacetone phosphate
DHAP	Dihydroxy acetone phosphate
DNA	Deoxyribonucleic acid
DO	Dissolved oxygen
DP	Dynamic pump
DSP	Downstream processing
ED	Entner–Doudoroff
EDTA	Ethylenediaminetetraacetic acid
EMP	Embden–Meyerhof–Parnas
FAD	Flavin adenine dinucleotide (oxidized form)
FADH	Flavin adenine dinucleotide (reduced form)
FDA	Food and Drug Administration
FDP	Positive displacement pump
FFK	Phosphofructokinase
F6P	Fructose-6-phosphate
F1,6 P	Fructose 1,6 diphosphate
HAc	Acetic acid
HEPA	High efficiency particulate air
HFCS	High fructose corn syrup
HMP	Hexose monophosphate
HRT	Hydraulic retention time
IE	Immobilized enzyme
IV	Inoculum vessel
LA, LAB	Lactic acid, Lactic acid bacteria
MLSS	Mixed liquor suspended solids
MLVSS	Mixed liquor volatile suspended solids
MPN	Most probable number
MSW	Municipal solid wastes
NAD^+	Nicotinamide adenine dinucleotide (oxidized form)
NADH	Nicotinamide adenine dinucleotide (reduced form)
$NADP^+$	Nicotinamide adenine dinucleotide phosphate (oxidized form)
NADPH	Nicotinamide adenine dinucleotide phosphate (reduced form)
PEP	Phosphenol pyruvate
PF	Production fermenter
PFD	Process flow diagram
PFFP	Plate and frame filter press
PFR	Plug flow reactor
PID	Piping and instrumentation diagram
PSO	Particle swamp optimization
RNA	Ribonucleic acid
RVF	Rotary vacuum filter
SCP	Single cell protein

SS	Stainless steel
TA	Total acid
TCA	Tricarboxylic acid
TOC	Total organic carbon
UV	Ultra violet
VFA	Volatile fatty acid

1 Introduction

Industrial biotechnology comprise two components; one is industry, and another is biotechnology. Again biotechnology has two components; one is bio, and another is technology. Technology basically deals with science and engineering, and so technology is the application of scientific knowledge to develop or create different products. Science tells us why things happen, and how they happen. Bio is related to biological systems and so biotechnology is the application of biological sciences in order to develop or create different products. Industrial biotechnology is the application of the biological process to the commercial level to produce products for the benefit of mankind. On the other hand industrial biotechnology may be defined as a set of unit operation processes that use living organisms (such as bacteria, yeast, fungi, algae etc.) or components of living organisms like enzymes, to produce a product of industrial importance which is followed by downstream or purification processes.

Chemical engineering course curriculums comprise one important course on chemical process technology. This deals with different chemical processes which are in operation in industry. The purpose of this industrial biotechnology book is to elucidate different industrial biochemical processes which include unit operation processes that may be classified in three broad areas: upstream processing, bioprocess or fermentation process, and downstream processing as shown in Table 1.1.

1.1 UPSTREAM PROCESSING

Upstream processing mainly deals with the processes involved before the operation of the production fermenter as mentioned below:

- Medium preparation
- Inoculum preparation
- Medium sterilization
- Air sterilization

TABLE 1.1
Major Units Present in the Fermentation Industries

Broad areas of fermentation process	Name of the unit
Upstream Processing	➢ Medium preparation ➢ Air and medium sterilization ➢ Inoculum preparation
Bioprocess / fermentation process	➢ Production fermenter
Downstream processing	➢ Product purification units ➢ Packaging ➢ By-products ➢ Effluent treatment process

TABLE 1.2
Medium Composition for the Production of 1 kg of Dry Yeast

Items	Amount (kg)
Cane molasses	4.3
NH_3	0.9
$(NH_4)\ H_2PO_4$	0.3
$(NH_4)_2\ SO_4$	1.1
Air	60

1.1.1 Medium Preparation

Medium composition plays an important role in the microbial fermentation process. Mediums comprise different components such as carbon source, nitrogen source, minerals, and vitamins. Carbon contributes mostly to the formation of product, cell mass and energy molecules like NADH, FADH, ATP etc. Nitrogen mostly contributes to cell mass formation. Minerals and vitamins play important roles in the metabolic pathways as cofactors of different enzymes, e.g. in the case of the citric acid fermentation process by *Aspergillus niger*; the sucrose present in cane molasses contributes to citric acid formation (via EMP and TCA pathways), and cell mass formation as well as to energy generation. The composition of the medium used in the baker's yeast fermentation process is given in Tables 1.2 and 1.3. The raw materials used in the medium should be cheap and easily available. Multiple choices of raw materials influence the cost of the materials for long-time operation of the plant. Both cane and beet molasses are found to be suitable for the yeast fermentation process mainly due to the presence of a good amount of fermentable sugar and trace nutrients like biotin, inositol and pantothenic acid. Cane molasses are largely available in India, Brazil, and USA. Beet molasses are available mainly in western countries.

TABLE 1.3
Typical Composition of Cane Molasses

Composition	Amount (% w/w)
Total solid	78–85
Total sugar	50–58
(contains 30% cane sugar and 28% invert sugar]	
N	0.08–0.5
P_2O_5	0.009–0.07
MgO	0.25–0.8
CaO	0.15–0.8
K_2O	0.8–2.2

TABLE 1.4
Types of Mutagens for Developing Industrial Strains

Classification of mutagens	Name of the technique
Physical mutagens	✓ Non-ionizing radiation ✓ Ionization radiation
Chemical mutagens (can remove, replace or modify DNA bases)	✓ Alkylating reagents ✓ Base analogues ✓ Intercalating agents

1.1.2 Inoculum Preparation

Two types of microbial strains are available: wild and industrial strains. Wild strains mean organisms which are isolated from soil, i.e. available naturally. Industrial strains are produced from wild strains after several treatment processes. Industrial strains are usually produced through a mutation process (Hunter, 2006). Compounds which can induce mutation are called mutagens, and are classified as physical or chemical mutagens as shown in Table 1.4.

The characteristics of wild and industrial strains are summarized in Table 1.5. The major characteristics of industrial strains are higher product titer in the fermentation broth, genetic stability and the capability of using cheaper raw materials. These have significant influence on the cost of product formation. For example, the cost of product purification is inversely proportional to the product concentration in the fermentation broth. The productivity of the microbial strains should not change with the generation of the microbial strain. Therefore the microbial strains should be genetically stable.

Industrial cultures must be preserved and maintained in such a way as to eliminate genetic change, protect against contamination, and retain viability. Based on their

TABLE 1.5
Characteristics of Wild and Industrial Strains

Wild strains	Industrial strains
✓ Poor reproducibility (Growth rate, Product formation rate and titer)	✓ High reproducibility in product formation rate and titer
✓ Susceptible to product inhibition	✓ Highly tolerant to product inhibition
✓ Poor genetic stability	✓ High genetic stability
✓ Ability to use various substrates	✓ Acclimatized to use cheaper and wide variety of substrates
✓ Poor substrate conversion rates	✓ Very high substrate conversion rates etc.

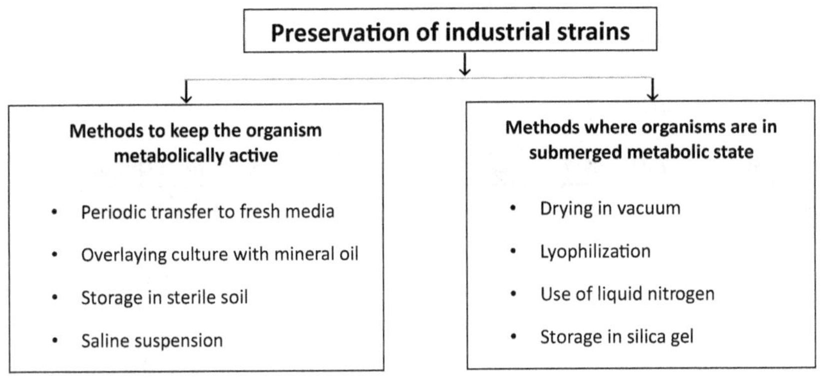

FIGURE 1.1 Techniques of preservation of microbial strains.

properties, techniques for the preservation of microbes are broadly divided into two categories as shown in Figure 1.1. The merits and demerits of these preservation techniques are tabulated in Table 1.6.

Inoculum plays a very important role in the fermentation process. In the life cycle of the microorganism, there are different phases; namely lag, log, stationary and decline phase. Each phase of growth has a different significance. For example the lag phase is considered to be the acclimatization phase, the log phase to be the active growth phase, the stationary phase to be the starvation phase and the decline phase to be the death phase. The microorganisms are active in the log phase. Therefore in the fermentation process, the seed culture is inoculated in between the mid-log phase and the late-log phase to ensure the microorganisms are active. The amount of cell mass present in the culture is very important for any fermentation process. This is determined from the volume of the inoculum and the cell mass concentration. The inoculum volume used for the product fermenter usually varies from 5 to 10% v/v. In the case of unicellular cells, the quantification of the cell can be carried out with respect to number. However, in the case of filamentous cells like *Aspergillus niger*,

TABLE 1.6
Merits and demerits of preservation techniques of industrial strains

Method	Merits and demerits
Periodic transfer	Variables of periodic transfer to new media include frequency, medium used and holding temperature. This can lead to increase mutation rates and production of variants.
Mineral oil slant	A stock culture is grown on a slant and covered with sterilized mineral oil; the slant can be stored at refrigerator temperature.
Minimal medium, distilled water, or water agar	Washed cultures are stored under refrigeration; these cultures can be viable for 3–5 months or longer.
Freezing in growth medium	Not reliable, can result in damage to microbial structures, with some microorganisms; however, this can be a useful means of culture maintenance.
Drying	Cultures are dried on sterile soil (solid stocks), on sterile filter paper discs, or in gelatine drops; these can be stored in a desiccator at refrigeration temperature or frozen to improve viability.
Freeze-drying	Water is removed by sublimation, in the presence of a cryoprotective agent; sealing in an ampoule can lead to long-term viability, with 30 years having been reported.
Ultra-freezing	Liquid nitrogen at -196°C is used, and cultures of fastidious microorganisms have been preserved for more than 15 years.

quantification with respect to number is not possible. Therefore, sporulation of the culture is carried out under moisture stress conditions for the proper quantification of the seed culture.

1.1.3 Air and Medium Sterilization

Air and medium sterilization play a very important role in the case of the microbial fermentation process. This is necessary to allow the desired microorganism to grow in the fermenter to get the targeted product, because other organisms should not interfere in the process. Air and medium sterilization are discussed in detail in Chapter 7.

1.2 FERMENTATION PROCESS

The living organism has some typical characteristics like reproduction, sensitivity to environment, acclimatization etc. Different physico-chemical parameters such as temperature, agitator speed, pH etc. play an important role in the fermentation process. All these parameters should be optimized in order to achieve maximization of product formation. The values of these parameters depend on the microorganism and the product formation. Zymology is a Greek word meaning 'The working of fermentation'. This is an applied science which studies the biochemical process of fermentation and its practical uses.

1.3 DOWNSTREAM PROCESSING

Downstream processing is also known as the purification or product recovery process. The cost of the product depends on its purity level. There are two types of chemical available in the market: commercial and analytical grade. Commercial grade products are usually much lower in price compared with analytical grades which are mostly used as feedstock of different chemical or biochemical processes. Analytical grade chemicals are mostly used for analytical purposes in the laboratory or pharmaceutical or food industries.

Several physico-chemical separation processes are involved in the purification of products such as solid-liquid separators, liquid-liquid extraction, drying, evaporator etc. Examples of solid-liquid separation processes are filtration, centrifugation, sedimentation etc. Filtration processes depends not only on the size of the particles but also their characteristics. For example, centrifuging is considered for the separation of bacterial cells because their size varies from 0.5 to 2 μm; plate and frame filter press is used for the separation of yeast cells of size 3–7 μm; rotary vacuum filter (RVF) for the separation of fungi or moulds with size of several mm. Several other solid-liquid separation processes are in practice in the citric acid producing industry, such as pannevis filter, gypsum filter etc. Different types of evaporator are used in the industries for the enhancement of product concentration such as falling film evaporator, natural/forced circulation evaporator, multiple-effect evaporators etc. Several dryers are largely used by the chemical and biochemical industries, such as rotary drum dryer, spray dryer etc. In addition different types of valves, pumps, heat exchangers and chillers are also required in the fermentation industries.

Different types of packaging techniques are recommended for industries depending on the nature of the product. For example in the brewing industry, the beer may be packed in bottles, cans, or barrels. Penicillin is available in two forms: solid and liquid. Solid is packed in the form of capsules and liquid in injection vials. Citric acid monohydrate is available as solid crystals and is packed in polyethylene pouches. Compressed yeast is packed in paper cartons and marketed under refrigerated conditions (Das and Das, 2019).

1.4 CHARACTERISTICS OF BIOPRODUCTS

The major characteristics of the bioproducts are given below

- Sustainability
- Less carbon and water footprint
- Creates rural employment opportunities
- Fewer emissions/pollutants to the environment
- Biodegradability and recyclability
- High productivity
- Use of raw materials from local sources

Sustainable means that the bioproduct will be available with us for a longer period of time. Lower carbon and water footprints reduce the environmental pollution problem to a great extent. This also creates rural employment and opportunities because some biochemical products can also be produced at small scale. Biodegradability, recyclability, and high productivity and use of raw materials from local sources are important characteristics of the bioproducts (Wittman and Liao, 2017; Thakur, 2013).

1.5 DIFFERENCES BETWEEN CHEMICAL AND BIOCHEMICAL PROCESSES

Differences between the chemical processes and the biochemical processes are shown in Table 1.7. Chemical processes use the synthetic materials for different product formation processes. In chemical processes, if the product is complex in nature then the raw material is also complicated. However, in the case of biochemical processes or biological processes, one substrate such as sucrose can produce different products. For example, sucrose can be used for the production of citric acid, ethanol, lactic acid etc. The biochemical processes are usually operated at ambient temperature and atmospheric pressure whereas chemical processes require mostly high temperature and high pressure. Therefore chemical processes are generally energy intensive as compared to biochemical processes. On the other hand, chemical processes are carried out in unsterile conditions whereas biochemical processes are usually operated under sterile conditions (Bailey and Ollis, 2010; Das and Das, 2019).

TABLE 1.7
Differences between Chemical and Biochemical Processes

Chemical processes	Biochemical processes
➢ Raw materials are changed with respect to the product. ➢ These processes need mostly high temperature and pressure. ➢ Requires unsterile conditions ➢ Expensive catalyst such as Pt required for product formation. ➢ Produces limited variety of products. ➢ Environmental pollution due the presence of toxic and biodegradable compounds.	➢ Same raw material may be used for the formation of different products with the help of microbial cells. ➢ Usually operate at ambient temperature and atmospheric pressure. ➢ Mostly require sterile conditions. ➢ Use living microorganisms and the enzymes for product formation. ➢ Can produce some unique products like interferons, insulin, hepatitis B vaccine etc. ➢ Environmentally friendly because of their biodegradable characteristics.

1.6 CLASSIFICATIONS OF BIOPRODUCTS

Different types of bioproducts are available in the market. These are classified as follows

- High-volume and low-value products
- Medium-volume and medium-value products
- Low-volume and high-value products

The cost of the low-value high-volume products is less than £6/Kg and these are usually required in large quantities. Examples are citric acid, ethanol, lactic acid, cheese, baker's yeast etc. (Prescott and Dunn, 1959). On the other hand the cost of medium volume and medium value products is less than £60/Kg; e.g. antibiotics like penicillin, streptomycin etc. The cost of low-volume and high-value product is around £60/mg; e.g. insulin, interferon etc.

Parameters playing a crucial role in the cost of bioproducts in fermentation industries are

- The cost of raw materials
- Duration of the fermentation process
- Concentration of the product
- Overall cost of the utilities (heating, cooling and air supply etc.)
- Mode of purification

Bioproducts may be broadly classified into three different categories as shown below:

- Bioenergy
- Biochemicals
- Biomaterials

Examples of these categories are illustrated in Figure 1.2. Rapid industrialization and urbanization increase energy demand drastically. This energy demand is mostly fulfilled from fossil fuels sources such as crude petroleum, natural gas and coal. These fossil fuels have limited reserve and will be exhausted very soon. Therefore, bioenergy may play a vital role in overcoming energy demands in the near future. Bioenergy is broadly classified as solid, liquid and gaseous fuels. Transportation problem and poor thermal conversion efficiency are the major issues for utilizing the solid agricultural biomass. These problems can be overcome by using liquid and gaseous fuels. Up to 20% gasoline can be replaced by bioethanol in automobiles (Das et al., 2014; Das and Jhansi, 2019; Duran, 2012, Das, 2015).

Biodiesel and biobutanol are also considered to be good substitutions for gasoline. Gaseous fuels such as hydrogen and methane can also be used in automobiles. Compressed natural gas (CNG) contains mostly methane which is considered to be an effective fuel for removing fog from the atmosphere. Bioproducts have tremendous potentiality both in the chemical and pharmaceutical industries. 70% of citric acid is used in the

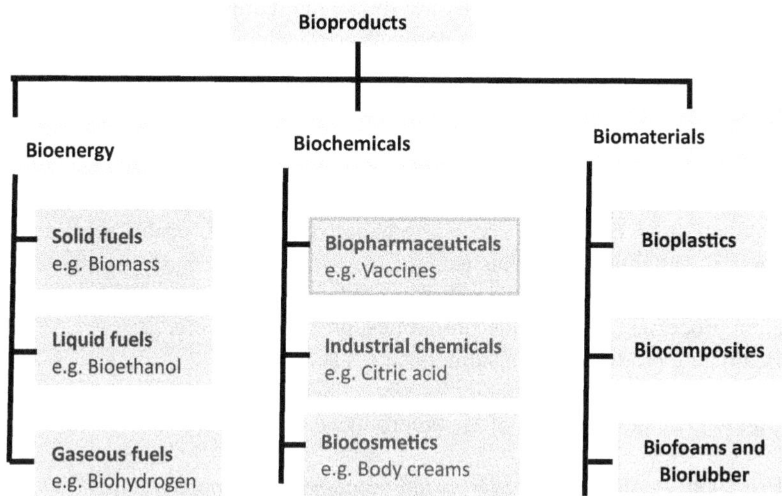

FIGURE 1.2 Broad classification of bioproducts.

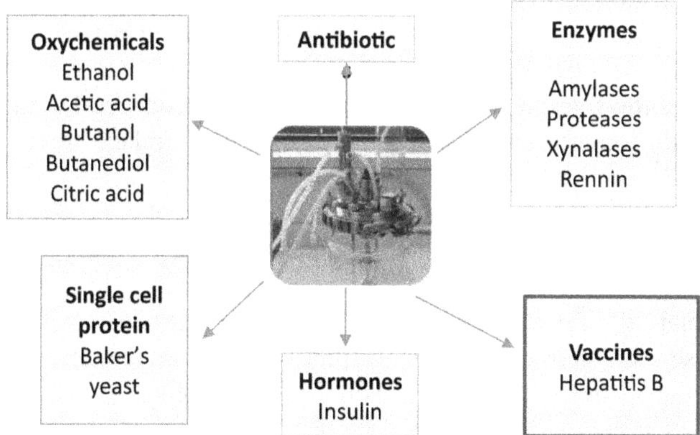

FIGURE 1.3 Bioproducts available in the industries.

food industry. Baker's yeast is largely used in the bread-making industries (Figure 1.3). In addition, bioproducts may be used for human healthcare, such as insulin, hepatitis B etc. Plastics are considered a nuisance because they are mostly non-biodegradable. Bioplastics are environmentally friendly and classified as starch-based bioplastics (derived using corn starch, potato starch, etc.), cellulose-based bioplastics (cellulose acetate), protein-based bioplastics (albumin-glycerol), aliphatic polyesters (polylactic acid, polyhydroxyalkanoate) and biopolymer (bio-polycarbonate, bio-polyamide). Examples of biocomposites are sodium alginate/silk fibroin, starch/lignin etc. Biofoams and biorubbers are used in the production of vehicle door panels or parts.

Bioproducts produced from industry are shown in Figure 1.3. Five major groups of commercially important fermentation products are available:

- Production of microbial cells (or biomass) as the product; e.g. baker's yeast, single cell protein, probiotics etc.
- Production of microbial enzymes; e.g. amylase, protease, catalase, glucose oxidase etc.
- Production of oxychemicals; e.g. ethanol, citric acid, acetic acid, acetone, butanol, glutamic acid, lysine etc.
- Modification of a compound which is produced in the fermentation/transformation processes; e.g. steroids, antibiotics, prostaglandin etc.
- Healthcare products; e.g. insulin, hepatitis B etc.

The chronological development of the bioproducts is given below:

- **Pre 1900**: alcohol and vinegar; batch processes using pure culture
- **1900–1940**: Baker's yeast, glycerol, citric acid, lactic acid, acetone/butanol; fed batch – using pure cultures
- **1940 to date**: penicillin, streptomycin, other antibiotics, gibberellins, amino acid, nucleotides, enzymes, steroids etc.
- **1960 to date**: single cell protein; continuous medium recycle, genetic engineering of production strains
- **1979 to date**: foreign compounds, not normally produced by microbial cells, e.g. insulin, interferon; genetic engineering to introduce foreign genes into microbial host

The present development of industrial fermentation products is shown below:

- Microbial cell of probiotics: capsule, drink/beverages
- Amylase and glucose isomerase for fructose syrup production as diet sweetener
- Colouring agent from microorganism for textile colours
- Biodiesel as energy source to replace petroleum
- Bioinsecticides
- Microbial bioplastics (polyhydroxyalkanoates)
- Isoflavon of soybean
- Lipase for detergent

1.7 MICROORGANISMS USED IN THE FERMENTATION INDUSTRIES

Microorganisms used in the fermentation industries are different according to the desired product formation. Table 1.8 shows examples of microorganisms used in the fermentation industries (Glazer and Nikaido, 2007).

This is the metabolic process occurring in bacteria, yeasts, fungi, or other microorganisms involving the breakdown of organic matter into acids, gases or alcohol. Ethanol fermentation in yeast from sugar is a well-known example of the anaerobic

TABLE 1.8
Microorganisms Used in Biochemical Processes

Product	Microorganism
Citric acid	*Aspergillus niger*
Baker's yeast	*Saccharomyces cerevisiae*
Penicillin	*Penicillium chrysogenum*
Acetic acid	*Acetobacter aceti*
Lactic acid	*Lactobacillus delbrueckii*
Yogurt, kefir, probiotics	Lactic acid producing bacteria
Glutamic acid	*Corynebacterium glutamicum*
Streptomycin	*Streptomyces griseus*
Biopesticide	*Bacillus thuringiensis*

fermentation process. Initially sugar is hydrolysed to glucose and fructose by invertase present in the yeast cells. In the EMP pathways, glucose is converted to pyruvic acid which is converted to acetaldehyde by the decarboxylation reaction. This is followed by the reduction of acetaldehyde to ethanol. This is shown below:

$$\text{Glucose} + 2\ \text{ADP} + 2\ \text{Pi} + 2\text{NAD}^+ \longrightarrow 2\ \text{Pyruvic acid} + 2\ \text{ATP} + 2\text{NADH}$$

$$\downarrow$$

$$\text{Ethanol} + 2\ \text{NAD}^+ \longleftarrow 2\ \text{Acetaldehyde} + 2\ CO_2 + 2\ \text{NADH}$$

1.8 APPLICATIONS OF ENZYMES

Enzymes have tremendous potential in the market for different purposes such as industrial, medicinal, and analytical (Table 1.9). Industrial enzymes are mostly used for the production of industrial products such as processed foods, detergents, cheese production, high fructose corn syrup (HFCS) etc. Medicinal enzymes are marketed by the pharmaceutical industries; e.g. digestive enzymes. Analytical enzymes are mostly used for analytical purposes, e.g. estimation of glucose by glucose oxidase; enzyme sensors etc. (Wittmann and Liao, 2017).

1.9 BOOK OVERVIEW

Bioproducts exhibit enormous potential to meet the demand of several chemicals such as citric acid, penicillin, vaccines, lactic acid etc. in our day to day life; also to overcome the scarcity of fossil fuels by using both liquid and gaseous biofuels which are environmentally friendly; biomaterials such as bioplastics, biomaterials etc. To keep abreast of this rapidly evolving technology, this book is intended to provide a detailed description of fundamental concepts of different commercial bioproduct producing industries including their biochemistry, microbiology,

TABLE 1.9
Industrially Important Enzymes and Their Applications

Enzymes	Applications
Proteases	Food processing, Detergent industry, Health care etc.
Lipases	Dairy and food processing industries
Cellulases	Biofuel industries for the breakdown of cellulose
Isomerases	Used to convert glucose syrup into fructose syrup
Xynalases	Used in the paper processing industry
Ligases and Nucleases	Molecular biology
Rennin	Cheese making
Pectinase	Food processing (fruit pulp processing)
β-glucanase	Brewing industries
Trypsin	Pharmaceutical industry
Tannase	Elimination of tannin

bioreactor engineering, and so on. This textbook consists of twenty chapters that cover the fundamentals of biochemical processes and different types of industrial bioproducts production processes in detail. Chapter 1 includes a basic introduction and history of bioproducts with an insight into the different steps involved in the biochemical fermentation processes. Industrial biotechnology is the application of the biological process at a commercial level in order to produce products for the benefit of mankind. On the other hand industrial biotechnology may be defined as a set of unit operation processes that use living organisms (such as bacteria, yeast, fungi, algae etc.) or components of living organisms such as enzymes, to produce a product of industrial importance which is followed by downstream or purification processes. Microbial strains, medium, and biochemical pathways play a very important role in bioproduct production commercially. Therefore, Chapter 2 deals with the processes involved in developing industrial strains and also in designing the fermentation medium. Chapters 3–5 include reactor design, enzymatic reaction kinetics, microbial growth kinetics etc. in detail. Biochemical industry comprises several unit operations which are shown in Chapter 6. Sterility of the fermentation process is very important to get the desired product via the microbial fermentation process. Chapter 7 highlights the processes involved for maintaining aseptic conditions in the fermentation industry. The processes involved in the purification of bioproducts are included in downstream processing (Chapter 8). Different oxychemicals such as ethanol, citric acid, lactic acid, amino acids, vitamins, penicillin, etc. are produced in the biochemical industry; these are included in Chapters 9–12. Chapter 13 covers the production of vaccines, hepatitis B and insulin. Enzymes can be used for several purposes, such as high-fructose corn syrup production, medicinal purpose etc. which are discussed in Chapter 14. Biochemical industries contribute greatly to marketing different milk products which are included in Chapter 15. Microbial cells are considered to be single-cell proteins which may be implemented to increase the nutritional quality of food (Chapter 16). Bioenergy is considered to be a sustainable energy source for the future because reserves of the fossil fuels will be exhausted very soon. Different bioenergy production processes are included in Chapter 17. Plastics are responsible for environmental pollution problems. These can be overcome by using biopolymers. In addition, chemical pesticides are largely used in agricultural production processes and have carcinogenic properties. Biopesticides are biodegradable in nature. Different biopolymers and biopesticides production processes are discussed in Chapter 18. Chapter 19 covers the microbial metal-leaching processes which are used in the recovery of precious metals present in very small amounts in ores. It has been observed that both chemical and biochemical industries are mostly responsible for water pollution problems. Chapter 20 includes different wastewater treatment processes in detail along with recent developments.

1.10 CONCLUSIONS

Technology has two parts: science and engineering. Scientific knowledge is applied to useful product formation. Biotechnology is technology that utilizes biological systems such as living cells or materials obtained from living systems such as enzymes to get products. Use of baker's yeast in bread-making industries and citric acid in food industries are examples of industrial biotechnological processes since they are produced from *Saccharomyces cerevisiae* and *Aspergillus niger*, respectively. Industrial biotechnology is also known as white biotechnology because it is not only sustainable but also chemicals can be produced from renewable sources using living cells. Industrial biotechnology uses enzymes and microorganisms to make bioproducts such as chemicals, food ingredients, detergents, biofuels etc. These processes are not only environmentally friendly but also less energy intensive as compared to chemical processes. Industrial biotechnology has both small and commercial-scale applications. The limited number of chemicals required for our day-to-day life are produced through industrial biotechnology.

REFERENCES

1. Bailey JE and Ollis DF, Biochemical Engineering Fundamentals, McGraw-Hill Inc., New Delhi, India, 2010.
2. Das D, Algal Biorefinery: An Integrated Approach, Capital Pub. Co, New Delhi and Springer, Switzerland, 2015.
3. Das D and Das D. Biochemical Engineering: An Introductory Text Book, Jenny Stanford Pub., Singapore, 2019.
4. Das D, Khanna N, and Dasgupta CN, Biohydrogen production: Fundamentals and Technology Advances, CRC Press, 2014.
5. Das D and Varanasi JL, Fundamentals of biofuel production processes, CRC Press, 2019.
6. Doran PM, Bioprocess Engineering Principles, Second Edition, Academic press, Waltham, USA, 2012.
7. Glazer AN and Nikaido H, Microbial Biotechnology: Fundamentals of Applied Microbiology, Cambridge University Press, New Delhi, 2007.

8. Hunter SI, Microbial synthesis of Secondary metabolites and Strain Improvement. Fermentation Microbiology and Biotechnology, Second Edition. CRC Press, 2006.
9. Prescott SC and Dunn CG, Industrial Microbiology, McGraw Hill Book Co. Inc., and K O Gakusha Co. Ltd., Tokyo, 1959.
10. Thakur IS, Industrial Biotechnology: Problems and Remedies, I. K. International Publishing House Pvt. Ltd., New Delhi, 2013.
11. Wittmann C and Liao JC (eds.), Industrial Biotechnology: Products and Processes, Wiley-VCH, 2017.

2 Development of Industrial Strain, Medium Characteristics and Biochemical Pathways

The most important contribution of biotechnology in terms of innovation and economy is in the production of industrial products. Fermentation technology has proved to be a major boon to the world to develop products that were otherwise difficult to synthesize chemically. Fermentation is the process by which microbes are grown on a large scale in order to produce commercially essential products and to carry out chemical transformations. The word is derived from the Latin 'fervere' which means 'to boil'.

The major advantage of fermentation is that production is via microbes which are abundant in nature. Cultivation of living forms in controlled conditions to increase productivity is a laborious task in terms of providing optimum conditions and continuous monitoring. These conditions can be well understood if we know the kind of microbe used and its characteristics.

2.1 CLASSIFICATION OF ORGANISMS

Over the years, organisms have been classified into various categories. The most commonly accepted classification of organism is Whittaker's 5-kingdom classification which was based on cell type (prokaryotic or eukaryotic), cell number (single-celled or multi-celled) and nutritional requirements (autotrophy or heterotrophy). This classification includes: i. Monera (bacteria); ii. Protista (algae and protozoa); iii. Plant; iv. Fungi; v. Animal. The latest classification is by Carl Woese which is based on the 16S ribosomal rRNA sequence of prokaryote and the 18S rRNA sequence in eukaryotes. The reasons for choosing this trait are: the rRNA sequence is essential to the normal functioning of the ribosome and is universally distributed in all organisms. It has a single role irrespective of the cell or organism type. It is evolutionarily conserved and has very little mutation over the years. Thus it has a major role in phylogenetic analysis. This classification has helped in classifying the organisms evolutionarily and finding a common ancestor. It has also been useful in finding the genetic distance between two organisms (Song et al., 2015).

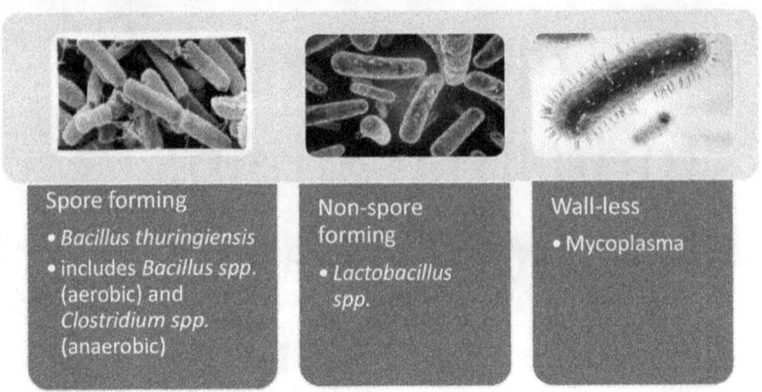

FIGURE 2.1 Different species of industrially important microbes.

2.2 INDUSTRIALLY IMPORTANT MICROORGANISMS

Bacteria and eukaryotes are the most commonly used organisms in industry for commercial production (Figure 2.1). Among these, microbes have various advantages, such as shorter doubling time or rapid growth rate; less space required compared to plant or animal; controlling the physical and chemical conditions essential for growth is easy in microbes; infections caused in microbes can easily be contained (Rowlands, 1984a).

Proteobacteria: This includes diverse group of organisms and thus is named after the Greek God Proteus who had the ability to change shape. The class contains bacteria that are free living; pathogenic; photosynthetic; motile or non-motile. All the organisms are Gram negative. They are facultative or obligate anaerobes.

Firmicutes: They are Gram positive bacteria and also include mycoplasma. These organisms have low G+C content (<50%).

Lactic acid bacteria: Lactic acid bacteria are part of the non-sporulating firmicutes, they have great industrial application. They are rods or cocci from the genera of *Lactococcus, Lactobacillus, Enterococcus, Leuconostoc, Pediococcus and Streptococcus.* The choice of species depends on the type of sugar to be fermented. They are mainly used in the food and pharmaceutical industry. They do not have a cytochrome system and thus cannot carry out electron transport phosphorylation. They are facultative aerobes. They also grow in sugar-rich environments and are fastidious (require amino acids, vitamins and nucleotides). They are classified as homofermentative (only lactic acid production) and heterofermentative (ethanol and CO_2 also produced apart from lactate). They are used because of their short doubling time and growth in cheap sources; in addition, they have relatively little additional growth requirements (N_2 sources). They produce a high yield of lactic acid and they have the ability to grow at low pH and high temperature. Additionally, they also produce low amounts of by-products and produce low cell mass (Stafford and Stephanopoulos, 2001).

Actinobacteria: This type has G+C content >50%. Organisms from the group form hyphae and thus are named accordingly(Greek: actin = rays). Industrially important members includes *Cornybacterium* (amino acid production) and *Actinomycetes.*

Actinomycetes: They have branched hyphae similar to fungi but are classified as bacteria because i) they have a peptidoglycan cell wall; ii) usually 1.0 µm in size which are much smaller than fungi. They are most commonly used in the pharmaceutical industry (e.g. *Streptomyces* for antibiotics). They are mainly soil dwellers and their identification began with finding metabolites in the soil (Stafford and Stephanopoulos, 2001).

Eukaryote: Plant and animal cell culture has advanced to great levels with their use in industry. But fungi are the major organisms amongst the eukaryote that are still used for various purposes in the fermentation industry. Fungi are involved in the production of enzymes (*Rhizopus*), alcoholic beverages (yeasts), pharmaceuticals (*Penicillium*) and in the food industry (*Agaricus* – mushroom) (Song et al., 2015).

2.3 CHARACTERISTICS OF INDUSTRIALLY IMPORTANT MICROORGANISMS

In order for organisms to be used in industry, certain parameters have to be met so as to make the process more profitable and quicker. The organism selected is a compromise between the productivity and the economic constraints of the process. The major characteristics to be noted are: i) The organism should utilize cheap substrates (preferably naturally available) and must not require any additional growth factors other than those already present in the substrate, ii) The organism's doubling time must be short since a slow-growing organism is highly prone to contamination, product turnover rate is very low and thus capital and manpower investment is for a longer time, iii) A fast rate of production should also be maintained in order to increase productivity (Stafford and Stephanopoulos, 2001). End product inhibition should be avoided and toxic intermediates should not be produced. Purification should be easy and cheap. If two downstream processes give the same yield, the cheaper one should be selected. Organisms which have an innate ability to protect themselves against competition should be selected. Organisms which grow in extreme physical conditions (pH or temperature) can be used as they have a natural defence mechanism. Organisms which are less susceptible to contamination should be selected as productivity is highly dependent on contamination. Anaerobic culturing is generally preferred as aeration takes about 20% of the cost of fermentation (Chen and Zeng, 2013).

2.4 BIOCHEMICAL PATHWAYS

The main aim of producing industrial products from microorganisms is to make the process cheaper and fast. It is essential that to optimize production conditions, the metabolic activities of the organisms should be known so as to manipulate them according to our needs. If an organism is supplied with a carbon and nitrogen source, it consumes nutrients in order to divide and grow. The utilization of nutrients and multiplication of the cells is an external indication of the various processes occurring inside a single cell. Several enzymes and factors are produced that are essential in the maintenance of cellular stability (Chen and Zeng, 2013). The series of biochemical reactions that helps in converting a substrate into the final product is known as metabolic reaction. The reaction that degrades complex molecules into smaller ones to produce energy is known as a catabolic reaction or catabolism. Similarly, the reaction that is used in producing complex molecules from simple sources is known as an anabolic reaction or anabolism. The compounds that form a part of the pathway are known as intermediates, while the final product is known as the end product.

2.5 PRIMARY METABOLISM

Any pathway that is associated with the regular growth and maintenance of the organism is known as the primary metabolism. Any disturbance to the primary metabolic pathway will hamper the organisms' survival.

This is mainly concerned with the production of proteins, nucleic acids and similar cellular constituents. Since they are associated with the growth of a microorganism, primary metabolites are mainly produced in the logarithmic phase to ensure maximum cell biomass production (Song et al., 2015) (Figure 2.2).

Secondary Metabolism: Primary metabolites are considered to be essential for the organisms' growth. In contrast, secondary metabolites have no role in the growth of the microorganism. Microorganisms can survive even if secondary metabolite production is stopped. The secondary metabolites are produced understress conditions. After a certain period of growth, the nutrients are exhausted and thus secondary metabolites are produced at the stationary phase (Song et al., 2015). They are restricted to only a certain species and are characteristic of that species. They have unusual chemical structures. Due to strain degeneration, the microorganisms' ability to produce secondary metabolites may be lost. Their production is found to be be controlled by plasmids more than nuclear chromosomes. Inducers of secondary metabolites also trigger morphogenesis.

Trophophase: The feeding phase of growth in which primary metabolites are produced. This coincides with the logarithmic phase of the growth. [Tropho = nutrient]

Idiophase: The phase of secondary metabolite production specific to the organism. Thus secondary metabolites are also termed idiolytes. This coincides with the stationary phase of growth. [idio = peculiar]

Role of secondary metabolites: Secondary metabolites have been highly useful industrial products. Although they have many uses as industrial products, various

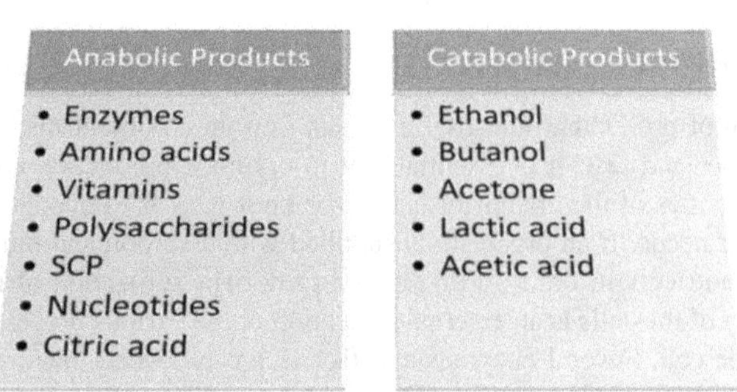

Anabolic Products
• Enzymes
• Amino acids
• Vitamins
• Polysaccharides
• SCP
• Nucleotides
• Citric acid
Catabolic Products
• Ethanol
• Butanol
• Acetone
• Lactic acid
• Acetic acid

FIGURE 2.2 Products of primary metabolism.

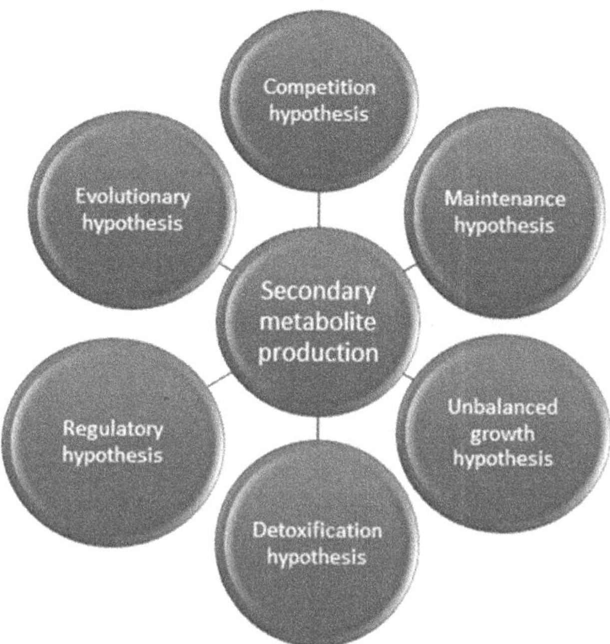

FIGURE 2.3 Hypothesis associated with secondary metabolite production.

theories have been developed to understand the importance of secondary metabolites for the microorganism's own benefit. Some old theories included secondary metabolites being i) Food storage molecules (Figure 2.3), ii) Waste products of cellular metabolism and iii) Breakdown products of macromolecules. These theories are now not accepted and various new hypotheses have been developed (Song et al., 2015).

*Competition hypothesis:*Particularly for antibiotic producing organisms, secondary metabolites are considered to be molecules that help in organisms' survival under low nutrient conditions by preventing the growth of other organisms.

Maintenance hypothesis: Under nutrient deficiency, a secondary metabolite helps to maintain the basic cellular functioning in order for survival of the organism.

Unbalanced growth hypothesis: Some microorganisms cannot control overproduction of primary metabolites which are then converted to secondary metabolites to be excreted from the cell.

Detoxification hypothesis: Molecules accumulated in the cells are detoxified by converting them into antibiotics. This theory developed because the penicillinic acid (precursor of penicillin) is more toxic than benzyl penicillin.

Regulatory hypothesis: Secondary metabolite production has been associated with morphogenesis. In *Neurospora crassa,* carotenoids are produced during sporulation. Similarly, in *Cephalosporium acremonium*, cephalosporin production is associated with arthrospore production. In *Bacillus* spp., peptide antibiotic production is associated with spore formation. Both processes are inhibited by glucose. Similarly, non-sporulating bacilli do not produce antibiotics while reverting to sporulation is accompanied by antibiotic production. Thus morphogenesis is strongly linked to secondary metabolite production.

Evolutionary hypothesis: Put forth by Zahner, this theory states that secondary metabolism is a mixed reaction. Any reaction that affects the cell greatly is converted to primary metabolism over the period of evolution while the other reactions are lost in the evolutionary selection. According to this theory, all organisms must produce secondary metabolites and it is merely due to the lack of an efficient detection system that only some secondary metabolites are identified (Chen and Zeng, 2013).

2.6 INDUSTRIALLY IMPORTANT METABOLIC PATHWAYS

Carbohydrate metabolism is the major set of reactions at industrial level as they are the source of energy for the cells.

Carbohydrate catabolism pathways (Figure 2.4):

2.6.1 Glycolysis or EMP Pathway

Glycolysis or the Embden–Meyerhof–Parnas (EMP) pathway is the first step in respiration in microorganisms. In this pathway, a single 6-carbon compound (glucose) is converted to 2 3-carbon molecules (pyruvate). This process results in the release of energy in the form of ATP which can be used by the cells to carry out metabolic processes. It is divided into an energy investment phase and an energy payoff phase (Stafford and Stephanopoulos, 2001).

Energy investment phase: This phase involves steps of conversion of glucose to glyceraldehyde-3-phosphate (G3P). During this phase, 2 ATP molecules are used for the addition of the phosphate group, which is termed the energy investment phase (Figure 2.5). Most of the glucose is converted to dihydroxyacetone phosphate (DAP). However, for the pathway to progress, formation of G3P is necessary and the continuous utilization of G3P promotes the conversion of DAP to G3P (Stafford and Stephanopoulos, 2001).

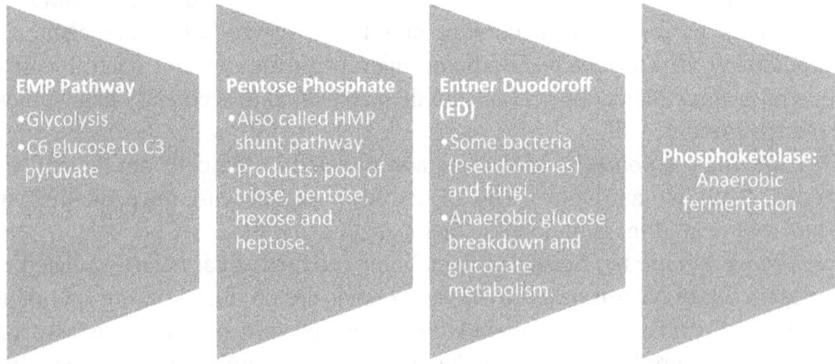

FIGURE 2.4 Common carbohydrate metabolism pathways.

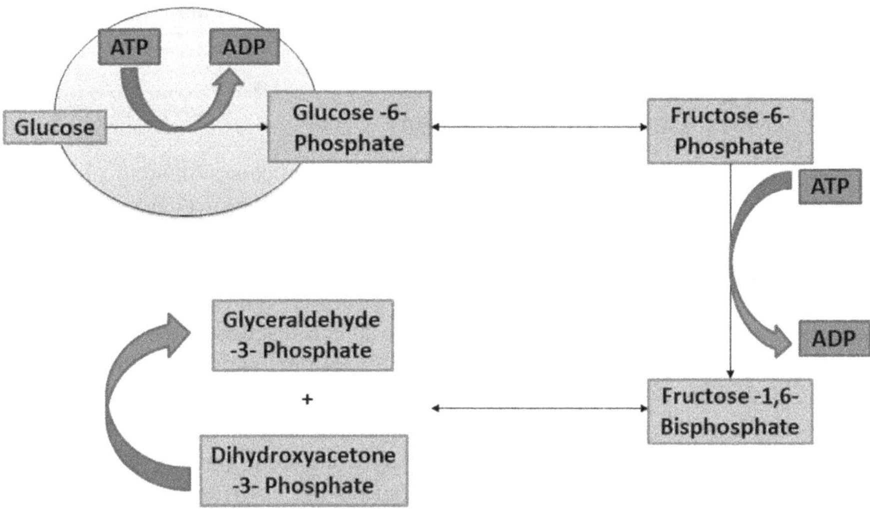

FIGURE 2.5 Energy investment phase of Glycolysis.

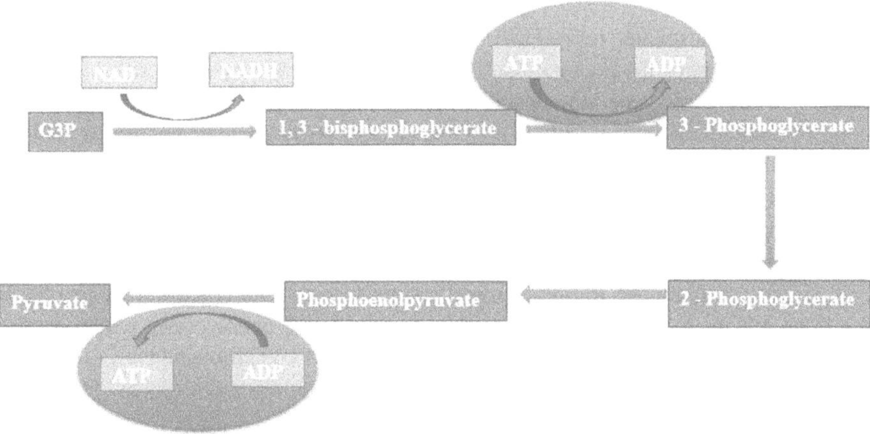

FIGURE 2.6 Energy Payoff phase in Glycolysis.

Energy payoff phase: The 2 glyceraldehyde-3-phosphate molecules are converted to 2 pyruvate molecules releasing 4 ATP molecules in total thus providing cells with energy. In the whole process, 2 molecules of ATP are used and 4 molecules of ATP produced. Thus the net gain of 2 molecules ATP for the conversion of 1 molecule of glucose into 2 molecules of pyruvate (Figure 2.6). Similarly 2 NADH molecules are produced from these reactions (Stafford and Stephanopoulos, 2001).

Regulatory enzymes: Every metabolic pathway has certain regulatory enzymes that control its function. The regulatory enzymes of glycolysis are: hexokinase (conversion of glucose to glucose-6-phosphate), phosphofructokinase (conversion of fructose-6-phosphate to fructose-1, 6-bisphosphate), pyruvate kinase (conversion of phosphoenolpyruvate to pyruvate).

2.6.2 Entner–Doudoroff (ED) Pathway

The Entner–Doudoroff pathway *(ED pathway)* is a process similar to glycolysis that is active in bacterial systems. It involves the conversion of glucose to pyruvic acid through a phosphogluconate intermediate. The phosphogluconate is converted to KDPG, an important intermediate of the reaction. The yield is one molecule of pyruvate and 1 glyceraldehyde-3-phosphate (G3P). The G3P molecule follows the glycolysis pathway to form pyruvate. The site of the reaction is the cytoplasm of the cell. The net yield of the pathway is 1 ATP, 1 NADH and 1 NADPH from one glucose molecule. Distinct features of the Entner–Doudoroff pathway are that it occurs only in prokaryotes and it uses 6-phosphogluconate dehydratase and 2-keto-3-deoxyphosphogluconate aldolase to create pyruvate from glucose. These are the regulatory enzymes of the pathway which controls the progression of the reaction (Stafford and Stephanopoulos, 2001).

2.6.3 Hexose Monophosphate Shunt

Hexose monophosphate shunt (HMP pathway) is also called the pentose phosphate pathway or the phosphogluconate pathway (Figure 2.7). Every living being contains RNA or DNA as its genetic material, which is important for its survival. The backbone of a nucleic acid molecule is a ribose or deoxyribose sugar which is a pentose sugar. Thus it is important that the 6 carbon be converted to a pentose sugar that can be used in the synthesis of the DNA/RNA backbone. The HMP pathway is thus used in this process and produces precursor molecules that can be diverted to various pathways. The two phases of the HMP pathway are:

Oxidative phase: The initial two irreversible steps of the reaction that results in the breakdown of glucose are included in the oxidative phase of the reaction. The first step involves Glucose → phosphogluconate (1 NADPH molecule released); following that is the second step phosphogluconate → ribulose 5 phosphate (1CO2 and 1 NADPH released).

Non oxidative phase: In step three, the conversion of ribulose 5-P to ribose 5-P takes place, after which the 2 R5P molecules are clubbed to yield a 10 C compound. The 10 C molecule is then broken down to 7C and 3 C molecules or 6C and 4C molecules. The 6 C molecule enters the glycolysis cycle while the 4C molecule is utilized in amino acid synthesis.

Net yield of the process:2 NADPH;1 CO_2; 1 H_2O.

Several catabolic reactions start from pyruvic acid as precursor (Figure 2.8). Generation of different type of end product eventually depends on the type of microbes and fermentation conditions or availability of enzymes, electron acceptors etc. Further, ATP generated from glycolysis; TCA cycle is used for anabolic reaction like cell component biosynthesis (Figure 2.9).

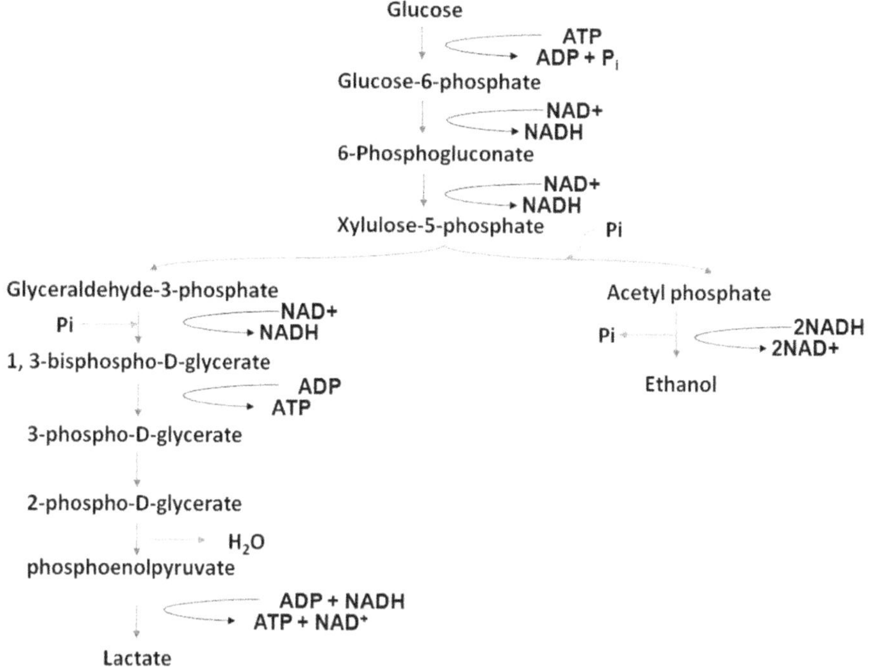

FIGURE 2.7 Phosphoketolase pathway.

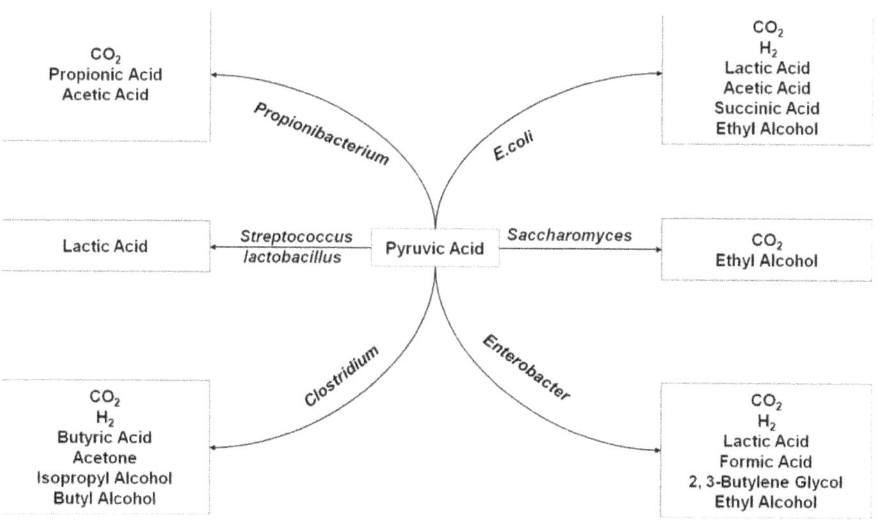

FIGURE 2.8 Products derived from pyruvate.

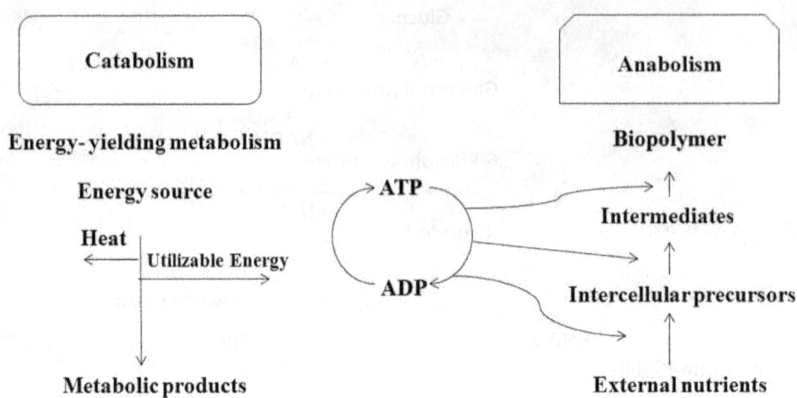

FIGURE 2.9 Coupling reactions of catabolism and anabolism in microbes.

Features of secondary metabolite synthesis: Normal substrates are converted to an important intermediate. Assembly of the intermediates forms complex structures using selected mechanisms. These mechanisms are also an integral part of the primary metabolism. The secondary metabolism is assembled during the cell multiplication stage (Stafford and Stephanopoulos, 2001).

Types of Secondary Metabolite

There are several types of secondary metabolite. Examples are as follows:

i. products with intact glucose skeleton: antibiotics like streptomycin (more common in actinomycetes);
ii. products related to nucleosides: antibiotics like bleomycin (derived from HMP pathway. present in both actinomycetes and fungi);
iii. product derived from Shikimate-Chorismate pathway:amino acids like phenylalanine, tyrosine and tryptophan (most aromatic products fromed by this method is chloramphenicol);
iv. Polyketide pathway: exclusive for fungi. C3 to malonate conversion involved, examples is tetracycline.
v. Terpene and steroids: Acetate to mevalonate conversion. Amino acid derivatives: glucose is a precusrsor for various amino acids. These amino acids produce various compounds like hadacidin, ergot alkaloids and ibotenic acid (Wendisch, 2014). Different types of secondary microbial product are listed in Table 2.1.

2.7 MICROBIAL STRAIN IMPROVEMENT

2.7.1 Steps in Strain Improvement

Selection of microbial strain: The first part of this selection is finding naturally occurring variants that involve improved production of enzymes or those which give a good yield and sort them for improvement. Selection from natural variants is a regular feature of industrial microbiology and biotechnology. For example,

TABLE 2.1
Different Types of Microbial Secondary Metabolites

Sr./No.	Activity	Examples	Producing Organisms
1.	Antibacterials	Penicillin	*Penicillium chrysogenum*
		Cephalosporin	*Acremonium chrysogenum*
		Tetracyclin	*Streptomyces aureofaciens*
		Streptomycin	*Streptomyces griseus*
		Spectinomycin	*Streptomyces spectablis*
		Kenamycin	*Streptomyces kanamyceticus*
		Erythromycin	*Saccharopolyspora erythraea*
		Rifamycin	*Amycolatopsis mediterranei*
		Cephamycin	*Streptomyces clavuligerus*
2.	Antifungals	Aspergillic acid	*Aspergillus flavus*
		Aureofacin	*Streptomyces aureofaciens*
		Candicidin	*Streptomyces griseus*
		Oligomycin	*Streptomyces diastachromogenes*
		Amphotericin	*Streptomyces nodosus*
3.	Enzyme inhibitors	Clavulanic acid	*Streptomyces clavuligerus*
4.	Plant growth regulators	Gibberellin	*Gibberella fujikuroi*
5.	Herbicidals	Bialaphos	*Streptomyces hygroscopicus*
6.	Growth promoters	Tylosin	*Streptomyces fradiae*
		Monensin	*Streptomyces cinnamonensis*
7.	Insecticides	Avermectin	*Streptomyces avermitilis*
	Antiparasites	Milbemycin	*Streptomyces hygroscopicus*
8.	Antitumorals	Actinomycin D	*Streptomyces antibioticus*
		Bleomycin	*Streptomyces verticillus*
		Taxol	*Taxomyces andreanae*
		Mitomycin	*Streptomyces lavendulae*
		Doxorubicin	*Streptomyces peucetius*

penicillin and griseofulvin production by natural variants producing higher yields in submerged rather than in surface culture. The most commonly used microorganisms for strain improvement are *Escherichia coli* and *saccharomyces cerevisiae.* All the strains are selected based on the amount of carbon source they can utilize. The ability of new techniques like CRISPR and Cas9 based systems has allowed gene manipulation of those organisms that were once intractable to it. We have to decide between making an organism more susceptible to gene manipulation or choosing another organism that is easily engineered. Other factors such as cost and ease of upstreaming, mid-streaming and downstreaming processes should be contemplated.

Reconstruction of pathway: Some microbes lack the products of interest and it is important to find the production pathway by finding the candidate enzymes or genes through genome and metagenome analysis. This strategy is not needed in organisms producing the desired product. However, this strategy is of increasing significance because of our interest in the production of products that are not natural or have inefficient production in hosts. Once the pathway has been constructed it can be optimized by expression optimization such as increasing tolerance to product, removing negative regulatory pathways, rerouting fluxes to optimize precursor availability, optimizing metabolic fluxes towards product formation, and optimizing microbial culture conditions. Gene modification to further enhance production and scaling-up fermentation of developed strains are the further steps that need to be taken while improving the strains of microorganisms. Strain improvement techniques play an important role in the strain improvement of organisms for the application step. Mutations in the microorganisms has an important significance in the strain improvement part, which involves changing/mutating a single region or different bases of the organisms. Mutagens (agents that causes mutation) can be distinguished into two categories: physical mutagens (ultraviolet, gamma and X-rays) and chemical mutagens (ethyl methane sulphonate – EMS, nitrosomethyl guanidine – NTG etc.) (Chen et al., 2010). The type of mutation produced is determined by the kind of DNA damage induced by the mutagen as well as the influence of DNA repair pathways on this damage. Spontaneous mutations are those that involve naturally occurring mutations. Mispairing errors, depurination, deletions and insertion sequences and error-prone DNA repair mechanisms are the main reasons for this kind of mutation to occur. They have low frequency of occurrence and usually occur at about 10-10 to 10-6 per generation per gene. A good example of this is the spontaneous mutation of wild as well as mutant strains of *Penicillium chrysogenum* to improve the production of its metabolite.

Classical mutagenesis is a type of mutagenesis that involves the usage of physical and chemical mutagenic agents to manipulate the genetic structure of microbes. This is done to improve the desired characteristic of an organism. The procedure of mutagenesis involves the following steps: (1) exposure of parent strain to a mutagen, (2) random screening of survivors, (3) assay of fermentation media for enhanced formation of product. Every time an improved strain is obtained which is utilized as a parent strain for the next cycle and this process continues until a strain with high throughput is developed (Chen and Zeng, 2013).

Methods to isolate (high yield) mutants: There are four important and most used methods to isolate mutants: i) replica plating; ii) resistance selection method, iii) substrate utilization method, iv) carcinogenicity test.

Replica plating: This method is mainly used to isolate the auxotrophic mutants which help in differentiating between the wild and the mutant strains. This involves the detection of mutants on the basis of strains to grow in the presence of amino acids. This technique was reported by Lederberg in 1952.

Resistance selection method: Generally mutant strains will be resistant to antibiotics and bacteriophages and the wild type strains are not resistant to antibiotics and bacteriophages. In this method the organisms to be found are plated in the media containing antibiotics or bacteriophages and those that can survive are considered mutants and by which the mutant strains are isolated (Chen and Zeng, 2013).

Substrate utilization method: This method mainly depends on the growth of bacteria utilizing the common carbon sources provided. Generally, bacteria utilize the primary carbon source as substrate. In this method the organisms are plated on the media which contains different carbon sources and the mutants are isolated based on utilizing the alternate substrate (carbon source).

Carcinogenicity test: This method is based on detecting the potential carcinogen and testing the mutagenicity in the organisms. Ames testing is the best method to detect the mutagenicity or carcinogens. Strains of *Salmonella typhimurium* are mainly involved in the mutational reversion assay of the Ames test.

Recombination: When two different genetical strains combine to generate a hybrid that is superior and different from either of the parents. Recombination is useful in erasing the neutral and deleterious results which arise during random mutagenesis. Recombinant DNA technology helps in improvement of primary metabolites such as amino acids and extracellular enzymes (Figure 2.10). Recombination techniques involve cloning of genes and expression of the cloned gene into an expression vector and its product formation. Isolation and cloning of genes of interest are the important aspects in rDNA technology which helps in constructing important enzymes and transfers the genes into a suitable host organism. Production of recombinant proteins and metabolic engineering are the two important objectives of recombinant DNA technology (Rowlands, 1984b).

Recombinant proteins: Proteins produced by the transgenes are known as recombinant proteins. They have their own commercial values. Insulin, interferons etc. are the examples of recombinant proteins which are produced in bacteria (Rowlands, 1984b).

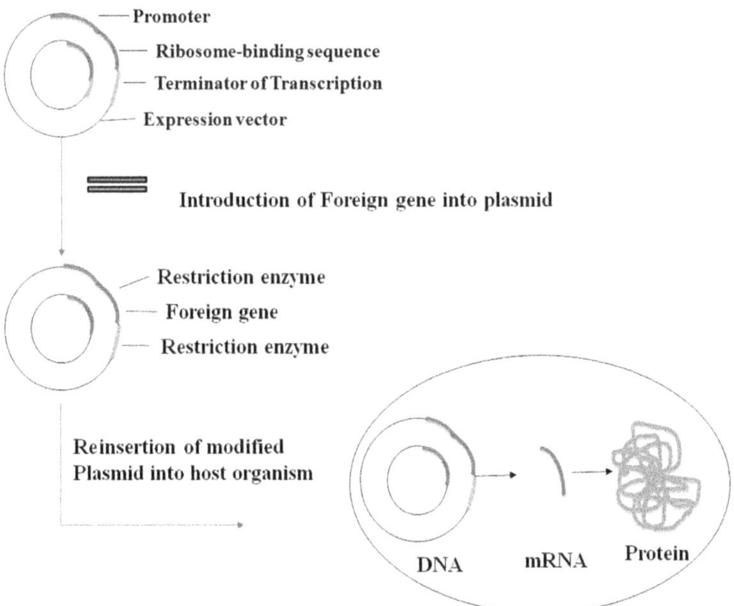

FIGURE 2.10 Process diagram of rDNA technology for protein production.

Metabolic engineering: By introducing a transgene which affects the enzymatic activity, transport and regulatory function of the cell, the metabolic activities of the transgene are activated. Overproduction of ethanol by *E. coli* and amino acid isoleucine in *Corneybacterium sp* are some examples of metabolical engineering of proteins. Product modification, complete metabolite formation, and enhancing growth are the important ways by which the metabolic engineering of proteins is achieved (Stafford and Stephanopoulos, 2001).

Integrated strain improvement: Precision engineering technology is a newly developed strain improvement technology. This technique avoids the previous drawbacks in strain improvement such as formation of undesired products or slow growth, substrate specificity and weak stress tolerance. The integration of classical metabolic engineering and screening methods with profiling technologies gives a clearer understanding of the genetics and physiology of metabolite production in this precision engineering technology. This will have a huge impact in industrial biotechnology (Chen and Zeng, 2013).

Random screening: Surface culture screens can be performed in two ways: the first is based on the diffusion of product from the colony forming a zone which can then be measured (Figure 2.11). The second screening method is based on the elution of

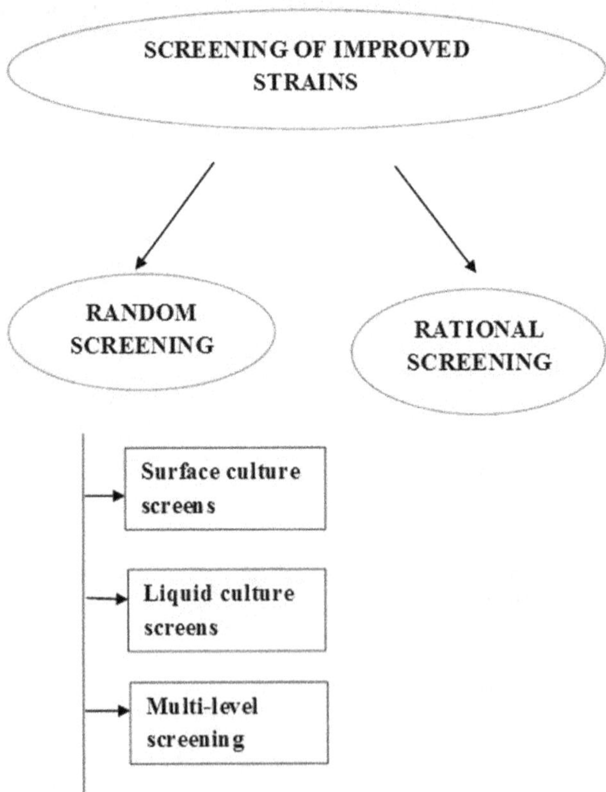

FIGURE 2.11 Screening method for improved strain.

product from growth medium which can then be assayed. Liquid culture screens are based on Erlenmeyer shake flasks. Liquid culture screens are advantageous over surface culture screens because they can mimic production conditions better. Aeration can be controlled in liquid culture screens. Generally, this screening method requires time and a few isolates per unit time are tested. These are, therefore, used for high resolution screens. Multi-level screening is the best when the testing error of screening is high. The isolates pass through different stages of screening where the best strains are selected to go through the next stage. The early stages of multi-level screening have low resolution but the resolution increases as the screening proceeds. The purpose of early stage is to remove the strains that give low titer of product. It is important to ensure that there is no antagonistic reaction between different types of strains used and the selection pressure used should be common to all levels of screening (Rowlands, 1984b).

Rational screening: Some of the methods of selective screening of mutants by rational screening include phenotypic titer depression, reversion of non-producers, resistance/sensitivity to toxic analogues, use of auxotrophs, and selective detoxification (Rowlands, 1984b).

2.7.2 Methods of Strain Improvement

Important methods other than mutations include use of plasmids and protoplast fusion techniques. A mutant requiring oleic acid for neomycin formation by *Streptomyces fradiae* showed a decrease in the intracellular level of neomycin precursors in the mutant. On the other hand, supersensitive mutants of β-lactam antibiotics are another example.

Recent approaches towards strain improvement are discussed below.

Role of Plasmid: Plasmid genes are involved in antibiotic production in *Streptomyces* spp. Plasmids are involved in genetic characteristics in curing experiments. Involvement of plasmids in biosynthesis of aureothricin and kasugamycin in *Str. kasuaensis* was demonstrated some decades ago by Okanishi (1970). The genetic study using *Str. venezuelae* ISP 5230, a chloramphenicol (CM) producer which contains most of the structural genes for the CM biosynthetic steps treated between met and ilu on the chromosome and the plasmid played a role in increasing CM production. A linear plasmid like DNA (pSLA2) of 11.2×10^6 Dalton molecular weight from *Streptomyces* sp. produced antibiotics (Rowlands, 1984b).

Protoplast fusion: Protoplast fusion is one of the useful techniques for obtaining hybrids or recombinants of different microorganism strains. Various studies have been carried out by using protoplast fusion in *Streptomyces, Saccharomyces*, and fungi. Protoplast formation in *Streptomyces* was first reported by Okanishi and his team in the year 1966. Further, they have worked on formation stabilization and regeneration of protoplast of *Str. griseus* and *Str. venezuelae*. Fusion of yeast protoplasts has been reported with *Saccharomyces cerevisiae*. A technique for protoplast fusion in *Brevibacterium flavum* has been used for strain improvement (Song et al., 2015).

2.7.3 Methods of Manipulating the Genetic Basis of Organisms

i) Not involving foreign DNA - conventional mutation
ii) Involving DNA foreign to the organism (i.e. recombination)

Transduction: In this natural process, bacteriophage (transducing particle) as vector, aids in the transfer of genetic apparatus from donor bacterium to recipient bacterium by its infection. It is of two types; generalized transduction and specialized transduction. In generalized transduction, an arbitrary part of the DNA from the host organism is integrated into viral DNA and it is expressed in the recipient organism by infection. The majority of cells infected by generalized transducing phages are abortive in nature. Specialized transduction occurs only in the temperate phage of specialized transducing phages; these rare recombinants incorporate only a specific part of the DNA from the host cell to recipient cells. These phages lacks part of its viral genome and include part of the bacterial genome but mostly these phages cannot complete the infection cycle due to its faulty genome. Therefore, in general, transduction efficiencies are very low and gene transfer is tedious in unrelated strains, limiting the utility of the technique for strain improvement (Chen and Zeng, 2013).

Conjugation: In conjugation, the genetic apparatus from the donor bacterium (F+) is transferred to the recipient bacterium (F-) through direct cell-to-cell contact with the help of sex pilus, trageners also called conjugation genes assist in the formation of a mating pair. Integration of F plasmid to the recipient bacterium occurs through homologous recombination. High-frequency recombination (Hfr) bacteria integrate their sex factor with chromosomal DNA for which the frequency of recombination is very high in case of conjugation between Hfr cells and F- cells. On the other hand, frequency of transfer of the whole F-factor is low. Usually, it takes 100 min for the transfer of entire donor chromosomes to recipient cells, but the mating pair separates spontaneously before the entire chromosomes are transferred, resulting in interrupted mating. Conjugation efficiencies change widely and only a few strains were able to serve as recipients for conjugation (Chen and Zeng, 2013).

Transformation: A short segment of DNA is taken up directly from the extracellular environment by the bacteria without the involvement of a vector. Ability of bacteria to take up extracellular DNA to form transformed cells is known as competence and the cells are called competent cells; homologous recombination occurs between naked DNA and the chromosomes of recipient bacteria. Numerous bacteria are not generally competent in nature but physical or chemical methods via electroporation and calcium shock can be used to make them competent. The drawback of these methods is that they are time consuming; maintaining the environment such as osmotic pressure in a steady state is tedious, time consuming and the process parameters also vary with respect to strain. A limiting factor of this process is that transformation efficiency is highly variable (Rowlands, 1984a).

P*rotoplast fusion:* This is a physical phenomenon which can be used for both prokaryotic and eukaryotic cells. Most of the filamentous fungi are made industrially viable through protoplasmic fusion. Protoplast refers to whole cells except for the cell wall. The cell wall of the bacteria is cleaved by lytic agents such as cellulase, lysozymes or macerozyme and with the presence of fusion agents two or more protoplasts having physical contact fuse together. The surface potential of the membrane is altered when it is brought into close proximity which ultimately results in fusion. Through this process interspecies genes can also be fused together to get the desirable properties of the corresponding strain. Transformation efficiency is also more than 80% (Chen et al., 2010).

Genetic engineering: Genetic engineering allows transfer of a single gene to obtain a specific train in a precise, controllable manner. It appeared after the discovery of molecular scissors known as restriction enzymes and DNA ligase known as molecular glue. DNA cloning is an asexual process in which exotic DNA is introduced into hybrid DNA replicons. Foreign DNA is recognized and cleaved in a sequence-specific manner by class-II restriction enzymes; preferably sticky ends generating restriction enzymes are used because it produces an overhanging sequence. Matching overhangs can stick together by complementary base pairing. Plasmids or phagemids are used for the construction of a DNA library which consists of genes of interest and cloning vectors. The last step in this method of strain improvement is introduction of a cloning vehicle to the target cells followed by screening of recombinants usually based on the phenotype. Due to its robustness and accuracy it is the preferred method for strain improvement. Genetic improvement methods are only used if the process requires only one microbe (Song et al., 2015).

Metabolic engineering: The metabolic architecture of an organism is reprogrammed mostly for the overproduction of metabolites (products) and occasionally for selection of mutant strains. System biology and synthetic biology help in engineering the pathways for redirecting the metabolic flux for the production of native indigenous and extraneous product from the microbial factory (host cell). Though it can be used for the production of any kind of metabolite it is primarily used for the production of secondary metabolites such as nutraceuticals. Regulatory networks can be orchestrated by different approaches such as heterologous expression of gene clusters, gene insertion and deletion, precursor stimulation, redirecting metabolic pathway, quorum sensing and genetic knockout of loci. *Corynebacterium glutamicum* is engineered for the overproduction of isoleucine through altering the regulatory networks (Stafford and Stephanopoulos, 2001).

Site-directed mutation: Alteration of specific oligonucleotide in the DNA sequence of a gene product helps us to analyse the function of a gene or to enhance gene activity; thereby hyper-production of a metabolite can be achieved. Site-specific mutation can be achieved by physical or chemical methods. Target-specific genome editing can be done with high accuracy through a gene editing toolbox (CRISPR/Cas9). It can also be performed in vivo. Using this method, commercial detergent which is prone to chemical oxidation is protected by replacing methionine to alanine; thus its activity is retained. The method is essentially used for improving strains by making them insensitive to product inhibition or substrate inhibition, reducing fermentation time, decreasing or removing byproducts or increasing product concentration (Wendisch, 2014).

2.8 CHARACTERISTICS OF THE MEDIUM FOR INDUSTRIAL FERMENTATION

Choosing the best microorganisms and fermentation materials is a very crucial step for better yield. Estimation of the maturation media is fundamental since it gives supplements vitality for improvement and furthermore gives a substrate in the fermenter for the production of the product. Fermentation or maturation media are composed of two elements, namely the major supplement and the minor supplement components. Major supplements comprise carbon and nitrogen enriched material as these are the building blocks of all the macro and micro molecules. Minor supplements comprise inorganic salts, vitamins as growth factors, growth suppressor, co-factors and biochemical catalysts. Supplements necessary for media preparation rely on fermenting organisms being utilized. Poor combination of substrate or microbial media brings about poor yield. Various supplements found in microbial media determine the yield of the procedure (Stafford and Stephanopoulos, 2001). Media used for fermentation are of two types; namely i) synthetic media or engineered media, ii) industrial media or non-engineered media (Rowlands, 1984b).

Engineered media: Engineered media is helpful in research since every ingredient is synthetically known, along with its composition. This is beneficial, since the ingredient composite amount can be varied and could be specified for each and every process with different organisms. This parameter once optimized correctly would give more yield of product and make the process feasible. An additional benefit of a nicely engineered media is that it does not contain components harmful for the process such as peptidase, RNAase and other hydrolysing catalyst when not required. It also reduces the possibility of contamination. A critical disadvantage of engineered media is the expense. The most significant part of fermentation is that it ought to be economical and beneficial. Since it is not economical, engineered media is not utilized on a large scale. This strategy is suitable for small-scale research centre investigations (Rowlands, 1984b).

Crude media/ industrial media: Crude media or industrial media is utilized in the fermentation process on an industrial scale. It provides all the important development conditions and the micro-cellular environment vital for the growth of the organism and the production of the desired product. It contains a significant amount of supplements, co-factors, proteins, foam reducing agents, and molecular perquisites, whose composition and concentration is unknown. It is vital to ensure that industrial medium must not consist of components altering the developmental and product-producing phases of the organism, that would affect the process output (Stafford and Stephanopoulos, 2001).

2.8.1 Components of Medium

Inorganic components: The medium comprises inorganic salts constituting cations and anions with a carbon source. Sometimes microorganisms used as biocatalysts for fermentation have a particular necessity for ions such as magnesium particles, phosphates, or sulphates. These prerequisites are satisfied by the expansion of these ions to adjust the rough media (Becker and Wittmann, 2012).

Carbon, nitrogen source: Carbon and nitrogen enriched compounds are crucial for microbial mediums for the development of the fermenting organism; referenced below are different carbon and nitrogen rich compounds utilized in maturation mediums.

Saccharine: Sugar cane and beets, molasses, and organic product pulp and juices may be included in this category.

Molasses: This is the result of the sugar cane and beet sugar companies. It can be recovered during the refining procedures. About 95% of the absolute amount of sugar present in cane molasses is utilizable for fermentation and is enriched with thiamine, biotin, phosphorus, sulfur and pantothenic acid.

Beet molasses: Beet molasses are delivered by a similar procedure utilized for sugar cane molasses. Nutrients, for example, biotin, pyridoxine, thiamine, pantothenic acid, and inositol, are available in beet molasses too. Beet molasses have constrained biotin. In this manner, in fermentation, including yeast culture, a limited quantity of cane blackstrap molasses or other biotin providing material ought to be included in the production medium since yeasts require biotin for their development. The biggest usage of cane black strap molasses in India is in the liquor business, which uses it for the production of spirits, alcohols, rum, gin, and whisky.

Fruit juices: Fruit juices comprise easily utilizable sugars. Berry juices constitute glucose and fructose. Consequently, organic juices are utilized as a carbon source in maturation media.

Starch: Major sources of commercially available starches are i) oats constituting rice, maize & wheat, ii) Roots such as beetroot, potato, radish and tubers such as turnip, horseradish.

Starch requires some treatment to be able to achieve transformation to utilizable sugars for the maturation process. This is achieved by enzymatic or chemical treatment.

Cellulose enriched compounds: These complex carbohydrates are comprised of recurring units of β-glucose. The development of β-cellobiose needs two particles of β-glucose, which are connected through α-1, 4-linkage. Units of cellobiose are linked throughout via 1, 4-β-glucosidic linkages. Thus, cellulose needs pretreatment.

Sulfite waste liquor: Sulfite waste liquor is a byproduct of the manufacturing process of wood pulp. It is generated after the digestion procedure in which the pulp undergoes bisulfite treatment; it contains sugars such as hexose and pentose and a fraction of hemi-cellulose is also present. It is currently used in the synthesis of ethyl alcohol by yeast, before it needs to undergo pretreatment procedures for the removal for sulfur dioxide and certain acids that can have a harmful impact on the product.

Wood molasses: This is obtained by acidic breakdown of wood-sourced cellulose. It produces 65–85% of utilizable sugars for maturation process. 0.5% of sulfuric acid is utilized between 150 to 185°C. Utilized in continual mode, a syrup might be synthesized from sawdust. This syrup may contain 4 to 5% of reducible sugar.

Rice straw: Rice straw is used as modest cellulose source which is a low-quality domestic supplement in its normal state on the basis of its bulkiness, poor palatability, high protein unavailability and inedibility. Various microbes are fit for utilizing cellulose for development. Rice straw is used as a maturation medium in the synthesis of single cellular protein (SCP) and silage.

Hydrocarbons and vegetable oils: Hydrocarbons utilized as maturation media component are typically blends of different hydrocarbon segments. These crude materials for fermentation are feasible even though purified hydrocarbon mixtures are progressively costly.

Vegetable oils: When exposed to air, vegetable oils undergo the oxidation of non-saturated structure. Vegetable oils such as maize oil are used in the presence of a surface active agent as antifoams. Vegetable oils are defined as the oil produced by the de-oiling process of vegetable seeds. Depending upon the level of unsaturation they can be classified as i. Oleic: Incorporating groundnut and olive oils. ii. Linoleic: oils included in this are maize, sunflower and cotton seed having a higher degree of unsaturated fat with double bonds. iii. Linolenic acid: including compounds having unsaturated fats with triple bonds e.g. soya bean and linseed oil.

Nitrogen enriched compound: Corn steep liquor (CSL): CSL is a byproduct in the manufacturing process of gluten, starch and various corn products. Corn steep liquor was initially seen as helpful for penicillin generation. It is now perceived as significant in numerous parasitic anti-toxin maturation media. Moreover, it is similarly utilized in the synthesis of foodstuffs. Soya bean oil: soya bean meal is the waste or the leftover once the soya bean oil has been extracted. The slurry is rich in nitrogen constituting about 9%w/w; it can also be used as animal feed once it has been processed and meets the specification. It is frequently selected over CSL since it is easily available and is richer in nitrogen when compared. It is already utilized in the production process of antibiotics (Chen et al., 2010).

Pharmamedia: Used in tetracycline synthesis. Pharmamedia is finely ground cotton seed embryo, constituting 24%, 56% and 5% starch, protein and oil, respectively (Chen and Zeng, 2013).

Distillers soluble: In the production of liquor utilizing grain or maize, liquor is refined from aged grain or maize, leaving the residue, containing 6 to 8%w/v complete solids. The suspended solids from the residue are wiped out by screening, leaving the effluent. From that point the effluent is exposed to concentration, until the solid content reaches 35%w/v giving 'evaporator syrup'. This syrup is then drum dried to yield 'distillers soluble'. This might be utilized as a generation medium segment, since it supplies nitrogen (Chen and Zeng, 2013).

Precursors and inducers: Precursors and inducers are a chemical moiety that enhances the output of the process and can also alter the biochemical pathway to take a specific route for production of desired products. For instance, CSL is used in the production of various type of penicillin, and addition of phenyl acetic acid to the same process results in the synthesis of penicillin G. Other examples of precursors and inducers include cobalt (used in the synthesis of vitamin B12), magnesium ion, etc. (Chen et al., 2010).

Repressors: Repressors are class of compounds added to industrial media to slow down the maturation process in order to obtain high quality products (Stafford and Stephanopoulos, 2001).These compounds are added in the maturation process so as to increase the product yield and keep the procedure contaminant free by suppressing the growth of other microbes. Repressors also help the microbes to slow down the utilization of media components so as to increase the yield of the product. For instance, soya meal along with paraline is used as the nitrogen source for the concoction of streptomycin since it is slowly utilized by the microbe to yield a quality product. By using a media in which repressors are present, the microbe shifts its aim from growth to survival, i.e. it uses the media for product synthesis rather than development.

Antifoams: Antifoams are surface-active compounds that destabilize the hydrophobic interaction between two different surfaces, which results in the disruption of protein films. This helps to increase the mass transfer rate of the process since the media can readily be accessible to the active sites of the enzymes. Some of the characteristic of antifoams are as follows: i. non-toxic to the procedure and unavailable to the microbe; ii. should be economic and easily sterilizable; iii. active even at low concentration. Examples include esters, stearyl alcohol, siliconates, sulphonates etc. Other alternatives for chemical antifoams are mechanical impellors that could be continuously used to reduce the froth in the maturation procedure (Rowlands, 1984b).

2.8.2 Characteristics of an Ideal Fermentation Medium

Buffering limit: This is the ability of the media to resist any pH change caused during the fermentation procedure. pH is a vital physicochemical parameter that should be regulated in order to achieve the desired aim. Buffers can also be added to the medium separately in order to increase the buffering capacity; automatic pH probes are being launched in the market that are used to monitor the pH of the bioreactor and manage it accordingly. $CaCO_3$ and phosphates are added to the media as buffers (Rowlands, 1984a).

Avoidance of foaming: Oil blends and octadecanol are used in penicillin production to avoid foam formation, since foaming can cause media contamination and also affects the product synthesis. Anti-foams have already been used previously in different fermentation media. Froth formation not only affects the media condition chemically but it also changes the pressure that was initially applied in the reactor. This could end issues in which bioreactor damage is also included (Wendisch, 2014).

Toxicity: The industrial medium should not contain any harmful component that could have an undesirable impact on the microbe or on the product like furfural (Wendisch, 2014).

Consistency: Viscosity of the media plays an indirect role in the generation of a non-homogenous medium in which mechanical stirrers are not used. Oxygen needs to be distributed throughout the media to maintain homogeneity, so that oxygen is available to be utilized by each and every microbe for product synthesis to which viscosity might act as a hurdle (Song et al., 2015).

Contamination: Change in the physical characteristics of the media during contamination is the most prominent technique to detect contamination, but it is not very helpful in avoiding contamination. Certain medium stages could be used as contamination indicators; for instance, sudden change in pH. This could act as the trigger in avoiding contamination in citrus extract production (Rowlands, 1984b).

Recovery: The major cost of the procedure involved is its recovery and purity. The components of the medium should not form any complex compound with the desired product which could affect its quality and the recuperation process. Once a media-product complex is formed it will result in the addition of one more step in the recovery procedure. In order to make the product economical such mistakes should be avoided and optimization should be carried out in order to make the recovery process less expensive (Chen and Zeng, 2013).

Availability of raw materials: should be readily available at a feasible rate, so as to continue the maturation process irrespective of the environment.

2.8.3 Media Design and Process Optimization

Various methods are used to design and optimize the media and the fermentation process.

Classical approaches to media design. The classical approach has been in use for centuries. It is designed to mimic the elemental composition of the cells in question. The selection of the sources is based on trial-and-error or prior knowledge of the metabolic pathway of the organism. The salt components need to be balanced as well (Chen and Zeng, 2013).

Statistical experimental design: A set of data is analysed using commercial software and an empirical model in the form of a second order polynomial is constructed. This model is used to predict the concentrations of the additives so as to obtain the desired result (Rowlands, 1984b).

Evolutionary computational methods: This method has developed due to the recent advances in computational intelligence. It is iterative and does not require manual initial inputs. Genetic algorithms are implemented to search the multi-factorial design space with a population of individual media types. Their fitness is calculated based on various factors and then new individuals are generated based on these calculations, followed by the start of the next iteration. Portions of media attributed assigned to the parent individuals are exchanged during crossovers based on the fitness calculations. Another example is the particle swamp optimization (PSO) (Stafford and Stephanopoulos, 2001).

Artificial neural networks (ANN): These methods are used more often for fermentation process control than optimization. ANN requires minimal knowledge of the biological system and is capable of generalizing beyond the training examples. It can deal with non-linear responses. ANN needs huge amounts of process data and does not provide information on the biology of the focus of the model (Stafford and Stephanopoulos, 2001).

2.9 CONCLUSIONS

Yield and productivity are the two critical parameters involved in the economics and viability of a bioprocess which can be achieved through industrial strain development and metabolic engineering. Upstream process plays a significant role in industrial biotechnology. Different technologies adopted to improve the microbial strain, media formulation and optimization have been summarized in this chapter.

REFERENCES

Becker J, Wittmann C, Systems and synthetic metabolic engineering for amino acid production – the heartbeat of industrial strain development. Curr. Opin. Biotechnol. Tissue Cell Pathway Eng 23, 718–726, 2012.

Chen Z, Wilmanns M, Zeng A-P, Structural synthetic biotechnology: from molecular structure to predictable design for industrial strain development. Trends Biotechnol 28, 534–542, 2010.

Chen Z, Zeng A-P, Protein design in systems metabolic engineering for industrial strain development. Biotechnol. J. 8, 523–533, 2013.

Rowlands RT, Industrial strain improvement: Mutagenesis and random screening procedures. Enzyme Microb. Technol. 6, 3–10, 1984a.

Rowlands RT, Industrial strain improvement: rational screens and genetic recombination techniques. Enzyme Microb. Technol. 6, 290–300, 1984b.

Song CW, Lee J, Lee SY, Genome engineering and gene expression control for bacterial strain development. Biotechnol. J. 10, 56–68, 2015.

Stafford DE, StephanopoulosG, Metabolic engineering as an integrating platform for strain development. Curr. Opin. Microbiol. 4, 336–340, 2001.

Wendisch VF, Microbial production of amino acids and derived chemicals: Synthetic biology approaches to strain development. Curr. Opin. Biotechnol., Chem. Biotechnol. Phar. Biotechnol. 30, 51–58, 2014.

3 Chemical Reaction Kinetics, Reactor/Bioreactor Analysis and Stoichiometry of Bioprocesses

All engineering disciplines have been developed from the basic sciences. For example, electrical engineering and electronic engineering developed from physics; chemical engineering comes from chemistry etc. To understand biochemical engineering, it is essential to know chemical engineering. Therefore, the present chapter comprises chemical reaction kinetics and chemical reactor analysis. Stoichiometry of the chemical and biochemical processes are essential to carry out the mass and energy analyses of the processes. This is very useful for industrial fermentation processes (Das and Das, 2019; Bhatt and Thakore, 2010).

3.1 CHEMICAL REACTION KINETICS

In the fermentation or biochemical industry, one of the major units is the bioprocess or bioreactor where product formation takes place. Now, the questions arise as to how the reactions take place, what are the modes of reaction and how to study the kinetics of these reactions. The reactor is the vessel in which reactions take place. Chemical reactions may be broadly classified as homogeneous and heterogeneous reactions. In homogenous reactions, reactions take place in only one phase. These reactions mostly take place either in a gaseous or liquid phase e.g. ammonia formation takes place with the help of nitrogen and hydrogen where all the reactants and products are in a gas phase. Therefore this is an example of a homogeneous reaction. Similarly, in the biochemical process, isomerization of glucose to fructose takes place in the presence of glucose isomerase where all reactants and product are in the liquid phase. Therefore this is also considered as a homogeneous reaction.

Heterogeneous reactions take place in more than one phase. There are three phases: gas, liquid and solid. In heterogeneous reactions a minimum of two phases should be present in the reaction mixture e.g. burning of wood is an example of heterogeneous reaction. Wood burns in the presence of oxygen and produces carbon dioxide and water. Here the product is gas (carbon dioxide and water) and the reactants are wood and oxygen which are solid and gas. Therefore this is an example of a heterogeneous reaction. Similarly, in the microbial fermentation process, microorganisms are an insoluble mass; they use soluble materials present in the fermentation medium for their growth and metabolism purposes to produce the desired product. Therefore this is also an example of heterogeneous reaction due to the presence of solid and liquid phases.

In the case of heterogeneous reaction, one phase should encounter the other phase; only then will the reaction take place. This means one phase should come into contact with the other phase before the reaction, which is known as the diffusion process. Diffusion phenomena play a very important role in the heterogeneous reaction process. Therefore, two things simultaneously take place; the diffusion as well as the heterogeneous reaction. An interesting point is that if the rate of diffusion is greater compared to the rate of reaction then reaction is the controlling factor. To improve the rate of product formation, it is necessary to improve the rate of reaction. Similarly in the case where the rate of diffusion is less than the rate of reaction then one should increase the rate of diffusion to increase the rate of product formation (Levenspiel, 2010; Ghasem and Henda, 2012).

3.1.1 Rate Equations

The typical homogeneous irreversible reaction may be represented as

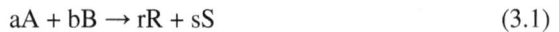

$$\mathrm{aA + bB \rightarrow rR + sS} \tag{3.1}$$

$$\text{Rate of degradation of A} = -r_A = -\frac{dA}{dt} \tag{3.2}$$

$$\text{Similarly, rate of R formation} = r_R = \frac{dR}{dt} \tag{3.3}$$

Minus signs indicate degradation of A and positive quantities signify rate of formation of R. In the case of a reaction, the reactant concentration decreases and that of the product increases with respect to time. The rate of reaction of all components present in the reaction mixture may be expressed as follows

$$-\frac{r_A}{a} = -\frac{r_B}{b} = \frac{r_R}{r} = \frac{r_S}{s} \tag{3.4}$$

Composition and the energy of material influence the rate of reaction. Energy means temperature, light intensity, magnetic field intensity etc. The concentration of reactant or product may be expressed as moles per unit volume or mass per unit volume. Usually S.I. units are taken into consideration. In addition, time can be expressed in seconds (s), minutes (min), hours (h). The rate of the reaction of the following reaction is expressed in Eq. 3.5.

$$A \xrightarrow{k} B$$

$$-r_A = k\, C_A^{\,n} \tag{3.5}$$

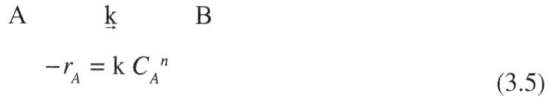

Where k is the rate constant and n is the order of reaction. Both are constants and experimental parameters. Eq. 3.5 may be modified as

$$\ln(-r_A) = \ln k + n \ln C_A \tag{3.6}$$

The plot of ln $(-r_A)$ vs ln C_A gives a straight line curve. The slope will give the value of n and the intercept of the Y-axis will give the value of k. Order of reaction (n) does not have any unit. It can be integral or fractional. If n = 0, the reaction is a zero-order reaction; in the case n = 1, the reaction is first order. In case of second-order reaction the value of n is equal to 2. The unit of k depends on the value of n, e.g. if n = 0, the unit of k is moles L^{-1} s^{-1} assuming unit of C_A as moles L^{-1} and reaction time as seconds (s). In the case n = 1, the unit of k is s^{-1}.

Again k depends on temperature which is expressed by the Arrhenius equation as follows

$$k = A\, e^{\frac{-E_a}{RT}} \tag{3.7}$$

where A is the frequency or pre-exponential factor, E_a is the activation energy, R is the gas constant and T is the absolute temperature. The natural log of Eq. 3.7 may be written as

$$\ln k = \ln A - \frac{E_a}{RT} \tag{3.8}$$

The plot of ln k vs $\frac{1}{T}$ gives a straight line curve where the slope is equal to $\frac{E_a}{R}$ and the intercept of the Y-axis will give the value of A. The value of the gas constant R is known. Therefore, the activation energy of a particular reaction can be calculated (Levenspiel, 2010).

The biological reactions are usually reversible in nature where there are both forward reaction and backward reaction. The first-order reversible reaction may be written as

$$A \underset{k_2}{\overset{k_1}{\rightleftharpoons}} \mathbf{B}$$

Overall reaction = Rate of forward reaction – Rate of backward reaction

$$= k_1\, C_A - K_2\, C_B \tag{3.9}$$

k_1 is the rate constant for the forward reaction, k_2 is the rate constant for the backward reaction.

At equilibrium condition, Rate of forward reaction = Rate of backward reaction

$$\text{Therefore, } k_1 C_A = k_2 C_B$$

$$\frac{k_1}{k_2} = \frac{C_B}{C_A} = K_c = \text{Equilibrium constant} \tag{3.10}$$

The equilibrium constant (K_c) is the ratio of the product concentration and the substrate concentration. Biochemical reactions are mostly reversible in nature. The following two strategies might be taken into consideration for the increase of product concentration because K_c remains constant.

- Removing the product from the reaction mixture
- Increasing the substrate concentration

For example, in the case of ethanol production, ethanol can be evaporated out from the reaction mixture with the help of vacuum because ethanol has a lower boiling point compared with water. This strategy increases the conversion efficiency of sugar to ethanol.

If N denotes the ratio of C_{B0}/C_{A0}, the rate equation can be written as

$$\begin{aligned} -r_A = r_B = -\frac{dC_A}{dt} = \frac{dC_B}{dt} &= k_1 C_A - k_2 C_B \\ &= k_1\left(C_{A0} - C_{A0}X_A\right) - k_2\left(NC_{A0} + C_{A0}X_A\right) \end{aligned} \tag{3.11}$$

Where $X_A = \frac{\left(C_{A0} - C_A\right)}{C_{A0}}$ substrate conversion efficiency. The value of dC_A/dt corresponds to 0 at equilibrium. One gets the following expression:

$$K_c = \frac{C_{Be}}{C_{Ae}} = \frac{N + X_{Ae}}{1 - X_{Ae}} \tag{3.12}$$

In Eq. 3.12, K_c may be represented as the ratio of $\frac{k_1}{k_2}$. Also, X_{Ae} is the fraction of substrate conversion at equilibrium. Putting these into Eq. 3.11 and integrating, we get

$$-\ln\left(1 - \frac{X_A}{X_{Ae}}\right) = \frac{N+1}{N+X_{Ae}} k_1 t \tag{3.13}$$

Eq. 3.13 shows that if $\ln\left(1-\frac{X_A}{X_{Ae}}\right)$ is plotted for various t, then a straight line will be obtained with slope of $\frac{N+1}{N+X_{Ae}}k_1$.

3.1.2 Autocatalytic Reaction

Another very important reaction is autocatalytic reaction. In autocatalytic reaction, the product acts as a catalyst. This can be expressed as follows

$$A + B \rightarrow B + B$$

An example is the baker's yeast fermentation process where sugar is used for the formation of new cell mass and again the new cell mass will be involved in the utilization of sugar. Therefore this is an example of autocatalytic reaction.

3.1.3 Chain Reactions

A chain reaction is defined as reaction in series. This usually takes place in the microbial metabolic pathways where a series of reactions are carried out by the different enzymes and substrates. In these reactions, the initial substrate concentration will decrease and the final product concentration increase with respect to time whereas the concentration of the intermediate products initially increases and then decreases. In the simplest first order chain reactions, the reactant A forms an intermediate product B which further reacts to form the final product C as shown in Eq. 3.14.

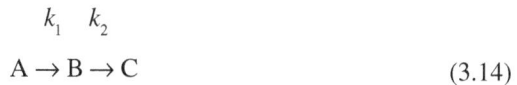

$$A \xrightarrow{k_1} B \xrightarrow{k_2} C \tag{3.14}$$

The rate equations for A, B and C can be written as follows considering first-order kinetics

$$-r_A = -\frac{dC_A}{dt} = k_1 C_A \tag{3.15}$$

$$r_B = \frac{dC_B}{dt} = k_1 C_A - k_2 C_B \tag{3.16}$$

$$r_S = \frac{dC_C}{dt} = k_2 C_B \tag{3.17}$$

Eq. 3.15 is considered as a first order homogenous differential equation and it can be integrated as shown in Eq. 3.18.

$$C_A = C_{A0}e^{-kt} \tag{3.18}$$

Eq. 3.18 represents the exponential decay of C_A with time. Putting Eq. 3.18 in Eq. 3.16, we get a first order differential equation which on integration gives Eq. 3.19 (Das and Das, 2019).

$$C_B = C_{A0}k_1\left(\frac{e^{-k_1t}}{k_2 - k_1} + \frac{e^{-k_2t}}{k_1 - k_2}\right) \tag{3.19}$$

Assuming equimolecular conversion in the stoichiometry of the chain reactions, one can write,

$$C_{A0} = C_A + C_B + C_C \tag{3.20}$$

Putting Eqs. 3.18 and 3.19 in Eq. 3.20 and further integrating the modified differential equation, we get,

$$C_C = C_{A0}\left(1 + \frac{k_2e^{-k_1t}}{k_1 - k_2} + \frac{k_1e^{-k_2t}}{k_2 - k_1}\right) \tag{3.21}$$

Assuming the rate of formation of C is higher than that of B, then we can write $k_2 >> k_1$, and Eq. 3.20 reduces to

$$C_C = C_{A0}\left(1 - e^{-k_1t}\right) \tag{3.22}$$

The rate of the reaction is determined by the first step.

Again, in the case of $k_1 >> k_2$, then

$$C_C = C_{A0}\left(1 - e^{-k_2t}\right) \tag{3.23}$$

Hence, the second step of the chemical reaction is the rate determining and k_2 governs the reaction.

In a series reaction, the product 'B' increases to a maximum and then it decreases as it further reacts to give the final product C. This can be easily found out from Eq. 3.21. In order to find out the time requires to get the maximum concentration of B, Eq. 3.19 can be differentiated with respect to time (t) and equated to 0 ($dC_B/dt = 0$). The time requires to get maximum concentration of B $\left(t_{max}\right)$ can be expressed as

$$t_{max} = \frac{\ln\left(\frac{k_2}{k_1}\right)}{k_2 - k_1} \tag{3.24}$$

Now, Eq. 3.19 and 3.24 can be combined and that will further yield the maximum concentration of the intermediate product, B.

$$\frac{C_{B,max}}{C_{A0}} = \left(\frac{k_1}{k_2}\right)^{k_2/k_2 - k_1} \quad (3.25)$$

Eq. 3.19 is required to find the correlation between the intermediate substrate concentration and reaction time. This equation determines the percentage of intermediate substrate conversion after any time, t. In industry, higher substrate conversion efficiency and higher product concentration are the major criteria. Eq. 3.24 determines the time requires for the maximum intermediate product formation. Eq. 3.25 indicates maximum concentration of the intermediate product.

3.2 DIFFERENT REACTORS/BIOREACTORS

Reactor means the vessels in which the reaction takes place. In the reactor, chemical transformations take place to generate the desired product. On the other hand, bioreactor is defined as the vessel in which the reactions take place either in the presence of living organisms or reactions in the presence of materials from different biological sources such as enzymes. The reactor or bioreactor is connected to other unit processes such as the separator, heat exchanger, utility systems and the process control.

The reactor may be classified into three categories on the basis of the mode of operations, e.g. batch, fed-batch and continuous process. In the batch process, the materials are taken one at a time inside the reactor and the reaction takes place; after the reaction is over, the materials are taken out. In fed-batch, the raw materials are added in batch mode at a fixed interval of time by maintaining the concentration of the substrate below the inhibition level without any output from the system. After the desired volume is reached and the reaction completed, the product is taken out from the reactor. The mode of feeding may be changed in the fed-batch reactor like continuous intermittent feeding by maintaining concentration of the substrate below the inhibitor level. In continuous processes, there is a continuous in-flow and out-flow in the reactor. All continuous processes are usually operated in batch mode initially for the maximization of the rate of reaction followed by continuous feeding and output. This is considered a very efficient process.

Geometrical configuration of the reactor is mostly tubular. This comprises a mechanical agitator for maintaining the homogeneity of the reaction mixture. Other type of reactors are based on contact patterns between the phases, such as packed bed reactor, expanded bed reactor, fluidized bed reactor, bubble column reactor and the airlift reactor. In packed bed reactors, the solid particles touch each other due to the gravitational force of attraction. In expanded bed reactors, the particles will be detached from each other due to higher axial force as compared with the gravitational force. In fluidized bed reactors, particles are more detached from each other due to higher axial force. A bubble column reactor is a vertical tubular reactor which is aerated by sparging air from the bottom through the liquid, whereas in air-lift reactor there is an inner draft tube in addition to air sparging (Nauman, 1987; Sinclair and Kristiansen, 1987).

Operation of the batch reactor is very simple. Usually this is called a batch stirred tank reactor (BSTR). This is the oldest type of reactor. The raw materials are added in the reactor at the same time and the reactions take place; after the reactions are over, the material is taken out. India has more than 400 distilleries which produce ethanol from cane molasses mostly in a batch fermentation process. These have been converted to the continuous process nowadays for better productivity. Chemical industries such as the dye-producing industry use the batch process and particularly wastewater treatment processes are mostly operated in batch mode.

Inhibition is a common phenomenon, particularly in the biological process. Both substrate and product inhibitions take place. The fed-batch reactor is mostly used in case of substrate inhibition. No product is removed during the fed-batch process; once the desired volume is reached and the reaction completes then the product is taken out. This is applied in the baker's yeast fermentation process where the sugar can be converted into yeast cells. Usually one gram of sugar can produce approximately 0.5 g of dry yeast cells. Therefore use of more substrate leads to a higher amount of yeast cell production. In the pharmaceutical industry, particularly for the production of secondary metabolites, a fed-batch process is used; e.g. penicillin.

In the continuous flow reactor, the feed is continuous inside the reactor and products are withdrawn continuously from it. Most of the continuous reactors are usually operated in batch mode first and after attainment of maximum reaction rate, feeding is done from one end and the product is taken out from other end. There are various applications of continuous systems such as single cell protein production, high-fructose corn syrup production etc. High-fructose corn syrup production usually takes place by using immobilized glucose isomerase enzyme where glucose is converted to fructose. Another application is the activated sludge process, which is nothing but a continuous stirred tank reactor (CSTR) with cell recycling.

Reactor classification is based on geometrical configuration, e.g. tubular reactors such as plug flow reactors. Tubular reactors look like a tube where the liquid feeding is done through one end and the product exits from the other end. These reactors consist of a hollow pipe or tube through which the reactant flows and where the reactant moves as a plug along the pipe. Plug flow in the chemical engineering language is known as piston flow. Piston flow means there is no velocity gradient across the cross-section of the tube due to the friction effect between the fluid and the tube or reactor wall while the liquid flows through the pipeline, and so the piston flow is usually considered an ideal flow where there is no friction. In normal cases there will be a velocity gradient due to friction. In the case of plug flow reactors (PFR), there is no axial mixing taken place across the reactor, but there may be radial mixing. In the case of product inhibition PFR is recommended, e.g. the organic acid fermentation process. In the acetic acid fermentation process, high citric acid concentration has some inhibitory effect on the organism. Other examples are citric acid, lactic acid, ethanol fermentation processes etc. However, CSTR is very easy to operate as compared to PFR. The reactants are continuously added and the product is withdrawn while the contents within the vessels are vigorously stirred using internal mechanical agitation.

The main purpose of the mechanical stirrer is to maintain the homogeneity of the reaction mixture. PFR is very difficult to operate but the operation of CSTR is very simple. PFR can be replaced by a series of CSTR known as a cascade.

In a pack-bed reactor, the catalytic pellets, e.g. immobilized enzyme, immobilized microbial cell, are packed in the reactor and then reacting fluid passes through the void space. Pumps can be used to make the fluid move through the packed bed. This has application in wastewater treatment processes, particularly by using immobilized whole cells. Immobilized enzymes are used for the production of different products, for example high-fructose corn syrup by using immobilized glucose isomerase.

The trickle-bed reactor is largely used in the wastewater treatment process. A major problem with this process is channelling due to growth of microorganisms. Since the particles are in contact with each other, as the time increases the microorganisms grow on the surface of the solid matrix leading to loss of porosity of the bed. As the porosity of the bed is lost then there will be a channelling effect. The performance of the reactor will be drastically reduced due to the channelling effect. This is the major problem with the packed-bed reactor.

Another very efficient reactor is the fluidized bed reactor. It is efficient, because the particles are not in contact with each other, i.e. they are separated from each other. Here the fluid can pass through at very high velocity so that the axial force is more than the gravitational force which will lead to detaching the particles from each other. At high fluid flow rate, the hydraulic retention time (HRT) is reduced. HRT may be represented as shown in Eq. 3.26.

$$\text{HRT} = \frac{\textit{Working volume or void volume of the reactor}}{\textit{Volumetric flow rate of the fluid}} \tag{3.26}$$

HRT is the time that liquid resides in the reactor and that is actually the reaction time; and since hydraulic retention time decreases, the reaction time also decreases and one will get a much lower amount of product formation. Recycling of the fluid can increase the reaction time. The typical fluid velocity varies from 6 to 20 m/h and the recycle ratio is very high at 5 to 500. Bed expansion takes place at 30 to 100%. One can easily find how much expansion of the bed takes place.

Bubble column reactors are also used in biochemical processes. In algal fermentation processes, algae are taken in media to grow and air is supplied to increase the dissolved carbon dioxide concentration in the medium. This carbon dioxide is utilized by the algal cells and fixed in the form of cell mass. Here the outgoing gas mostly contains oxygen and unconverted carbon dioxide and nitrogen that are present in the air. It is largely used for single-cell protein production from cheese whey and for the algal cultivation process.

In the airlift reactor, this is similar to the bubble column reactor, only the difference is that there will be an inner draft tube present in the reactor. This is the inner draft tube through which air is passed, but there will be some kind of dragging force here, and some of the air bubble will recycle back to the system (Figure 3.1). These are the two patterns of aeration of systems such as internal or external circulation systems where modes of feeding of air bubbles are different so that some air bubbles

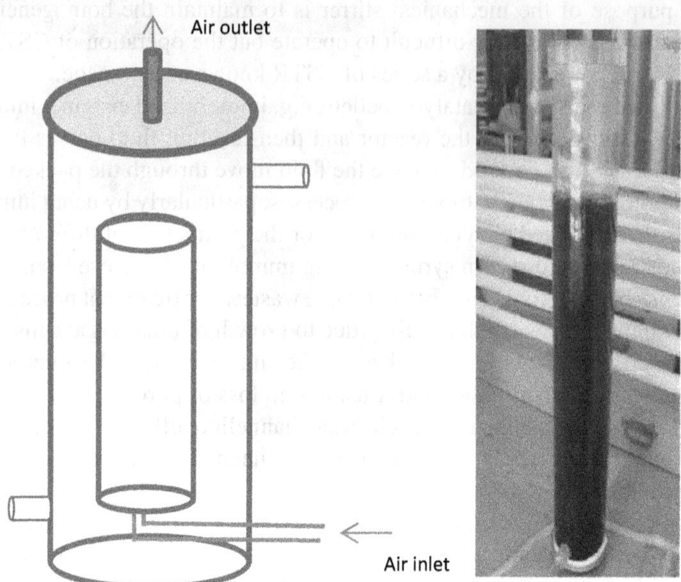

FIGURE 3.1 Experimental set up of the air-lift reactor for algal cultivation.

can be recycled back to the liquid again. This will increase the mass transfer in the system. The draft tube also equalizes the shear stress, because some living cells are very sensitive to shear force. If the shear force is very high, that will hinder the growth of the organism. Therefore, if the shear force is less then it will be good for some organisms to grow, e.g. plant cells and animal cells. For plant cell growth, mostly airlift fermenters are recommended (Doran, 2012).

There are different types of miscellaneous reactors such as membrane reactors. In the membrane reactor, enzyme or microbial cells are retained by a membrane. When substrate is passed through the membrane, it will be exposed to these enzymes and cells and then one can get the desired product at the end. This is largely used in monoclonal antibody production.

The material of construction of the reactor in case of biochemical industrials is mostly stainless steel. There are several characteristics desirable for the material of construction of the reactor. It should be flexible and durable, because the reactor's life should be a minimum of 10 years to 20 years. Again it should be non-toxic to the reactants and the product. It should be resistant to the chemicals and metabolic products created by the microorganism. Cost is the very important factor for the industries concerned. Within most of the biochemical industry, construction of material is mainly made of stainless steel and the stainless steel characteristics depend on the composition of the alloy; e.g. if the alloy contains 12% of chromium that prevents corrosion and 8% of nickel gives the austenitic structure which is known as the smoothness structure of the stainless steel. If we use a small amount of molybdenum, that further increases the acid resistance property of the stainless steel. Therefore, different types of stainless steel have been used as materials of construction in the chemical industry and biochemical industries. In the pilot and commercial scale, bioreactors are usually made of stainless steel because in the microbial fermentation process different types of organic acid formation take place. These organic acids have corrosive property, and due to this corrosive property stainless steel is recommended as the material of construction. Different types of stainless steel are available such as SS304, SS314, SS316, SS317 etc. In the citric acid industry, SS317 is recommended. For anaerobic fermentation processes like ethanol, mild steel coated with epoxy resin is used as the material of construction. Epoxy resin coating on the surface of the mild steel makes a barrier between the liquid and the material of construction. It has the corrosive resistance property, particularly against hydrogen sulphide in the case of the anaerobic digestion process for methane production. In addition, in the presence of acidic media, the epoxy is a very good coating material, which can give protection to the mild steel material. However, in the case of photobiological processes, highly transparent material of construction is required in order to utilize light energy. In the case of polycarbonates the transparency is 100% and so light can penetrate very easily. This is very much required for the growth of the algal cell (Sinclair and Kristiansen, 1987; Blanch and Clark, 1997).

In the case of lab fermenters usually transparent reactors are recommended so that from the outside one can see what is going on inside. One can have visual observation. Usually glass is considered as a material of construction for small-size laboratory fermenters. In stainless fermenters, watch glasses are located at the top cover plate; at one end can be a light so that at the other one can watch the inside of the fermenter.

Inside the reactor/bioreactor, there will be a mechanical stirrer to maintain the homogeneity of the reaction mixture. In the case of microbial cells which are insoluble solids these will settle down without a stirring system. Therefore, for uniform mixing one should use a stirrer. Foam formation is the major problem in the fermentation industries. Biochemical systems are operated either aerobically or anaerobically. In the case of aerobic systems, air is sparged through the reactor to increase the dissolved oxygen concentration in the medium because microorganisms can take the oxygen which is dissolved in the medium for their growth and metabolism. During the fermentation process protein formation takes place in the fermentation broth. When protein and air are mixed with each other, they form foams. If the foam accumulation takes place at the top of the fermenter, it will slowly go up and touch the mechanical seal of the stirrer which will rupture the seal. Then an air contamination problem in the reactor will take place. This mechanical seal is very crucial for any biochemical industry. This can be overcome by using anti-foam oil or a mechanical foam breaker. Usually sterilized air should not contain any kind of contaminant. It passes through the air filter to get the sterile air. In the case of power shut down, the air compressor will be stopped. Since there is no airflow then there is a possibility that the back suction of air in the reactor will cause contamination. Therefore, there should be an air filter at the exit of the air stream of the reactor (Das and Das, 2020).

There are two types of sterilization: one is *in situ* sterilization and another is *in vitro* sterilization. The basic difference between chemical and biochemical processes is that biochemical processes are operated under sterile conditions so that desired organisms can grow inside the reactor. Therefore, the reactor should be perfectly airtight. Hot and cold liquid is passed through the jacket in the reactor to maintain the temperature of the bioreactor. In case of small reactors and big reactors, the modes of sterilization are *in vitro* and *in situ*, respectively. Wet steam is found most effective for the medium sterilization. Different monitoring devices are used in the fermentation processes; e.g. temperature monitoring systems with the help of thermostats, pH probes, dissolved oxygen probes, antifoam sensors, agitator speeds etc. (Das and Das, 2019; Shular and Kargi, 2002)

3.3 REACTOR/BIOREACTOR ANALYSIS

Reactor/bioreactor analysis is usually done with the help of material analysis of the process as follows

Rate of input = Rate of appearance = Rate of output
+ Rate of disappearance + Rate of accumulation (3.27)

In Eq. 3.27, one can consider one particular parameter to write this balance equation; e.g. substrate balance, product balance, cell mass balance, energy balance etc. One can write the balance equation with respect to any such parameters. For example if one wants to write the balance equation with respect to substrate, the equation is to be developed with respect to substrate balance only.

3.3.1 Analysis of Batch Process

Figure 3.2 shows the batch stirred tank reactor (BSTR). In the batch reactor, the reactants are allowed to react for a particular time. A mechanical agitator/impeller is used for maintaining the homogeneity of the reaction mixture for a specific period of time. After the specified time, the mixture (products + unreacted reactants) is then taken out of the reactor. Therefore the batch process is considered an unsteady state because reactant concentration changes with respect to time and the rate of reaction depends

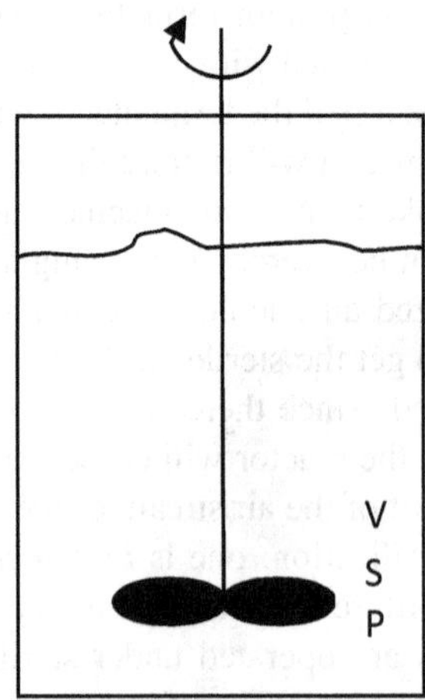

FIGURE 3.2 Batch stirred tank reactor (BSTR).

on the concentration of the reactant. In the batch reactor, there is no continuous inflow and outflow of the reactants or products. There are two types of continuous process: continuous stirred tank reactor (CSTR) and plug flow reactor (PFR). In contrast to a batch reactor, the CSTR always operates at steady state and the components are thoroughly mixed and stirred in the reactor. Therefore, there is a uniform distribution of the contents throughout the reactor. It is important to note that the concentration of the outlet stream and the contents inside a reactor at any point of time is supposed to be identical. A CSTR is also known as the mixed flow reactor (MFR) since the contents are uniformly mixed throughout the reactor at any instant of time. The other ideal steady-state flow reactor is known as the plug flow reactor (PFR). This has already been discussed in the previous section.

Let us assume the following irreversible reaction in the batch reactor:

$$A \rightarrow B$$

Different terms shown in Eq. 3.27 may be written as follows:

Rate of input = 0, Rate of output = 0, in case of substrate balance rate of appearance = 0

Therefore, Eq. 3.27 is modified as

$$-\text{Rate of disappearance of A} = \text{Rate of accumulation of A} \tag{3.28}$$

$$\text{Rate of disappearance of A by reaction } (\text{moles/time}) = -r_A V \tag{3.29}$$

$$\text{Rate of accumulation of A } (\text{moles/time}) = \frac{dN_A}{dt} \tag{3.30}$$

Hence, Eq. 3.28 can be written mathematically as follows:

$$-(-r_A)V = \frac{dN_A}{dt_b} \tag{3.31}$$

Where N_A = no. of moles of A/volume, t_b = time of the batch process.

$$-r_A = -\frac{1}{V}\frac{dN_A}{dt_b} = -\frac{dC_A}{dt_b} \tag{3.32}$$

Integrating Eq. 3.32, we get

$$t_b = -\int_{C_{A0}}^{C_A} \frac{dC_A}{-r_A} \tag{3.33}$$

For a constant volume batch reactor,

$$C_A = C_{A0}\left(1 - X_A\right) \tag{3.34}$$

Putting Eq. 3.34 back to Eq. 3.33, one will get,

$$t_b = C_{A0}\int_0^{X_A} \frac{dX_A}{-r_A} \tag{3.35}$$

Eq. 3.35 represents the time required to achieve a conversion X_A for either isothermal or non-isothermal operation. The rate of reaction remains under the integral sign because it changes as the reaction proceeds. In Eq. 3.35 't_b' represents the reaction time in a batch reactor. After the batch reaction is over, the materials have to be taken out followed by cleaning of the vessel and refilling the vessel with fresh reactant. The time requires for this operation is known as down time or idle time of the reactor. Therefore, the total time of the batch process can written as

$$t_{total} = t_b + t_{\text{down time}} \tag{3.36}$$

Eq. 3.36 is used for finding the volume of the batch reactor.

Problem 3.1 In a batch reactor A is converted into B. This is a liquid phase reaction. The stoichiometry of the reaction is

$$A \rightarrow B$$

The rate of reaction at different concentrations of reactant A is given in the table below. How long must we react each batch for the concentration to drop from C_{A0} = 1.3 mol/L to C_A = 0.3 mol/L and volume of the reactor to get 100 moles of B/d assuming 2 h is the down time?

C_A mol/L	$-r_A$ mol/L min
0.1	0.1
0.2	0.3
0.3	0.5
0.4	0.6
0.5	0.5
0.6	0.25
0.7	0.10
0.80	0.06
1.0	0.05
1.3	0.045
2.0	0.042

SOLUTION

Given data,

$$C_{A0} = 1.3 \frac{mol}{L}$$

$$C_A = 0.3 \frac{mol}{L}$$

$$\text{We know, } t_{batch} = -\int_{C_{A0}}^{C_A} \frac{dC_A}{-r_A}$$

The integration can be evaluated by a graphical method or by a numerical method (area under the curve of C_A vs $\frac{1}{-r_A}$ between C_{A0} and C_A).

C_A mol/L	$-r_A$ mol/L min	$1/(-r_A)$ L min/mol
0.1	0.1	10.00
0.2	0.3	3.33
0.3	0.5	2.00
0.4	0.6	1.67
0.5	0.5	2.00
0.6	0.25	4.00
0.7	0.10	10.00
0.80	0.06	16.67
1.0	0.05	20.00
1.3	0.045	22.22
2.0	0.042	23.81

Area under the curve (using trapezoidal rule) (Figure 3.3) = t_{batch} = 12.70 min
Therefore the total time required for the batch process = 12.70 min + 2 × 60 min = 132.7 min

$$\text{Total number of batches per day} = \frac{24 \times 60}{132.7} = 10.85$$

$$\text{Therefore, the amount of B to be produced per batch} = \frac{100 \text{ } moles}{10.85} = 9.22 \text{ moles}$$

From the stoichiometry, we get 1 mole A converted to 1 mole B.
Hence, the amount of A requires per batch = 9.22 moles

$$\text{Conversion efficiency of the process} = \frac{1.3 - 0.3}{1.3} = 0.77$$

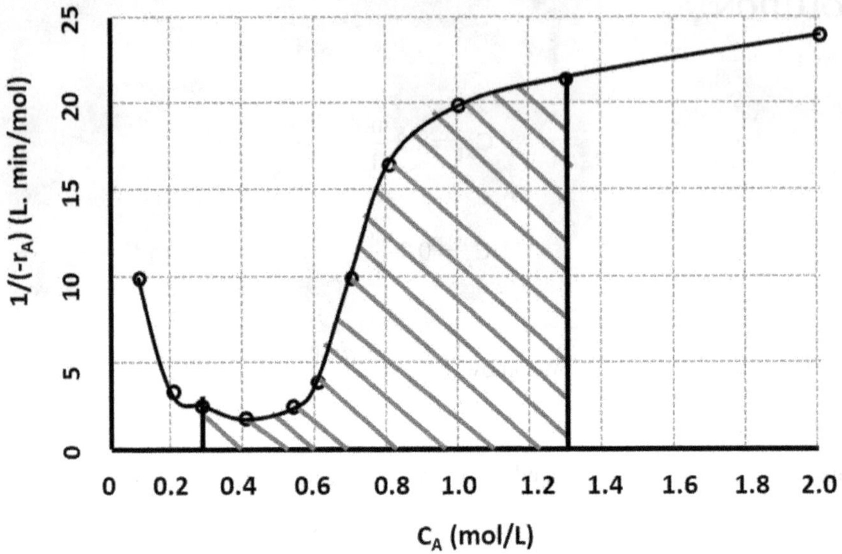

FIGURE 3.3 Plot of $\frac{1}{-r_A}$ vs. C_A.

$$\text{Actual amount of A requires per batch} = \frac{9.22}{0.77} = 11.97 \text{ moles}$$

$$\text{Volume of the batch reactor} = \frac{\text{Actual amount of A requires per batch}}{\textit{Initial concentration of A}}$$

$$= \frac{11.97 \textit{ moles}}{1.3 \textit{ moles/L}} = 9.2 \text{ L}$$

3.3.2 Analysis of Continuous Stirred Tank Reactor (CSTR)

A continuous stirred tank reactor (CSTR) is also named a mixed flow reactor (MFR). In this reactor, there is continuous input and output along with uniform agitation in the reactor (Figure 3.4). In such processes, the rate of accumulation of the substrate is equal to zero under steady state conditions. Also, component A reacts to give product B. Therefore, the rate of accumulation of A is also zero. Hence, Eq. 3.27 may be modified as

$$\text{Rate of substrate input} - \text{Rate of substrate output} - \text{Rate of substrate disappearance} = 0 \quad (3.37)$$

Let the input to the reactor be 'F_{A0}' moles/time and the corresponding output be 'F_A' moles/time. The rate of disappearance of the component A from the reactor is equal to $-r_A V$. Hence, we can write,

$$F_{A0} - F_A = (-r_A)V \quad (3.38)$$

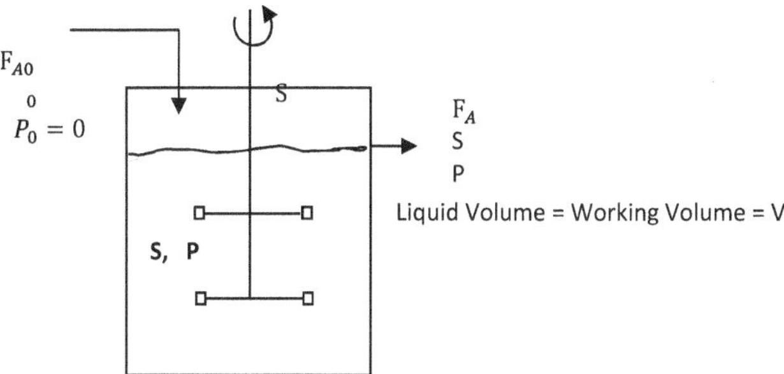

FIGURE 3.4 Continuous stirred tank reactor (CSTR).

For a constant volume system, 'F_A' in Eq. 3.38 can be written as,

$$F_A = F_{A0}\left(1 - X_A\right) \tag{3.39}$$

Putting Eq. 3.39 in Eq. 3.38, we get

$$F_{A0} - F_{A0} + F_{A0}X_A = \left(-r_A\right)V$$

$$F_{A0}X_A = \left(-r_A\right)V \tag{3.40}$$

The above Eq. 3.40 on rearrangement gives us

$$\frac{V}{F_{A0}} = \frac{X_A}{-r_A} = \frac{\tau_{CSTR}}{C_{A0}} \tag{3.41}$$

In Eq. 3.41, τ_{CSTR} represents the space time or the hydraulic retention time (HRT) of CSTR. A HRT of 3 min means that every 3 min one reactor volume of feed at specified conditions is being treated by the reactor. If we further rearrange Eq. 3.41, we get

$$\tau_{CSTR} = \frac{V}{F} = \frac{C_{A0}\, X_A}{-r_A} \tag{3.42}$$

In Eq. 3.42, the volumetric flow rate is 'F'. Eq. 3.42 shows the co-relation of five parameters: X_A, $-r_A$, V, F and C_{A0}. Therefore, in the case where four parameters are known, the fifth parameter can be found directly. In design, Eq. 3.42 is used to find out the volume of reactor needed for a given conversion directly. In kinetic studies each steady-state run gives the same rate of reaction for the conditions within the reactor. The ease of interpretation of data from an MFR or CSTR makes its use very attractive in kinetic studies, in particular with messy reactions (e.g. multiple reactions and solid catalysed reactions).

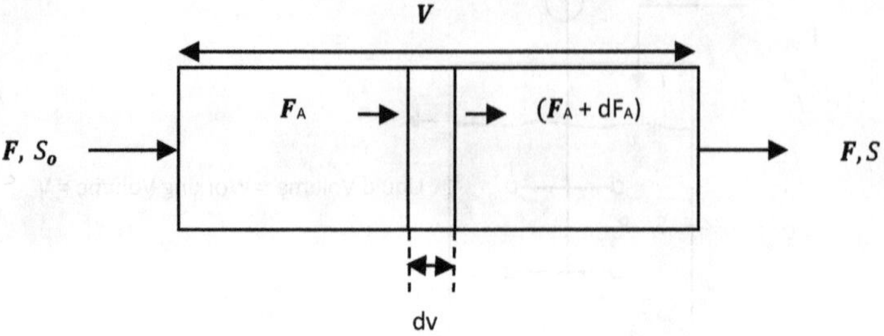

FIGURE 3.5 Plug flow reactor (PFR).

3.3.3 Analysis of Plug Flow Reactor (PFR)

In a plug flow reactor (PFR), the composition of the fluid changes from point to point along the flow path (Figure 3.5). Therefore, the material balance for a reaction component A must be made for a differential element of volume 'dv'. Similarly to the CSTR, the rate of accumulation of substrate at a particular point is zero under steady-state conditions and the reactant A undergoes degradation or decreases rather than formation. Therefore, Eq. 3.27 is also applicable for designing the PFR (Kelly et al., 1994; Katoh and Yoshida, 2010; Ghasem and Henda, 2012).

At each small volume 'dv', the input of the reactant A is F_A moles/time whereas the output of the reactant A from that small volume is $(F_A + dF_A)$ moles/time.

The disappearance of the component A can be written as $(-r_A)$ dv

Putting the above information in the material balance Eq. 3.37, we get

$$F_A - F_A - dF_A = (-r_A)dv \tag{3.43}$$

$$-dF_A = (-r_A)dv \tag{3.44}$$

$$-d\left[F_{A0}(1 - X_A)\right] = (-r_A)dv \tag{3.45}$$

$$F_{A0}dX_A = (-r_A)dv \tag{3.46}$$

Eq. 3.46 accounts for A in the differential section of reactor of volume dv. For the reactor as a whole the expression must be integrated. Now F_A, the feed rate, is constant, but r_A is certainly dependent on the concentration of substrate. Eq. 3.46 can be written as

$$\int_0^V \frac{dv}{F_{A0}} = \int_0^{X_A} \frac{dX_A}{-r_A} \tag{3.47}$$

$$\frac{V}{F_{A0}} = \frac{\tau_{PFR}}{C_{A0}} = \int_0^{X_A} \frac{dX_A}{-r_A} \tag{3.48}$$

Further, Eq. 3.48 can be modified as follows:

$$\tau_{PFR} = C_{A0} \int_0^{X_A} \frac{dX_A}{-r_A} \tag{3.49}$$

$$\tau_{PFR} = -\int_{C_{A0}}^{C_A} \frac{dC_A}{-r_A} \tag{3.50}$$

Eqs. 3.48–3.50 can be written either in terms of concentrations or conversions. For systems of changing density it is more convenient to use conversions. However, there is no particular preference for constant density systems. Whatever their form, the performance equations interrelate the rate of reaction, the extent of reaction, the reactor volume, and the feed rate, and if any one of these quantities is unknown it can be found from the other three (Das and Das, 2019).

Problem 3.4 For the reaction of the previous problem (**PROBLEM 3.3**), find out the volume of the reactor required for the CSTR and PFR for the production of 100 moles of B/ d.

SOLUTION

From Eq. 3.42, we get

$$\tau_{CSTR} = \frac{C_{A0}X_A}{-r_A} = \frac{1.3 \times 0.77}{0.5} = 2\ min = \frac{V}{F}$$

F = volumetric flow rate

$$= \frac{Moles\ of\ A\ requires\ per\ day}{Initial\ concentration\ of\ A} = \frac{100\frac{moles}{d}}{1.3\frac{moles}{L}} = 76.92\frac{L}{d}$$

$$= \frac{76.92\ \text{L}}{24 \times 60\ \text{min}} = 0.0534\frac{L}{min}$$

Therefore, the reactor volume required (V) = 2 min × 0.0534 $\frac{L}{min}$ = 0.1067 L

Again in the case of PFR:

The time required for the batch process = Space time of PFR

From the solution of Problem 3.1, we get

$$t_{batch} = 12.70\,\text{min}$$

Therefore the volume of the required for the PFR = 12.70 min × 0.0534 $\frac{L}{min}$ = 0.68 L

3.4 STOICHIOMETRY OF BIOPROCESSES

Stoichiometry of the bioprocess gives the following information: firstly, it gives the information to find out the quantitative relationship between the amount of the reactant used and the amount of product formed by a biochemical reaction or mutual relationship and internal limitations within the biochemical system. Therefore, the stoichiometry indicates how much substrate reacts to give the desired amount of products. Secondly, this also gives us the information about the validity of the experimental results. This is very useful to determine the accuracy of the experimental data. Thirdly, this determines the amount of heat evolved in the aerobic fermentation process because most of the biochemical reactions are exergonic in natures which are exothermic. Most of the microorganism grows close to the ambient temperature and atmospheric pressure. However, the temperature shoots up particularly during the summer in countries like India (which is a tropical country) to as high as 40–50 °C. Therefore, additional cooling arrangements are required to maintain the temperature of the fermenter in the range 30–35 °C. Thus, if you want to calculate how much heat is evolved during the bioprocess then we can easily find it out.

Stoichiometry is based on the law of conservation of mass. This states that mass can neither be created nor destroyed in a chemical/biochemical reaction. Thus, the amount of matter cannot change. Figure 3.6 visualizes the things that will be very clear where the mass enters through the system boundary and comes out through the system boundary. Therefore some mass enters the system and subsequently comes out plus mass generated within the system minus the mass consumed within the system that is equal to mass accumulated within the system. The mass balance equation for this particular system is shown in Eq. 3.51.

$$\text{Mass inflow} - \text{Mass outflow} + \text{Mass generated} - \text{Mass consumed} - \text{Mass accumulation} = 0 \tag{3.51}$$

The elemental balance of biological reactions can be written on the basis of the law of conservation of mass when the composition of the substrate, product and cellular materials are known. In the case where glucose is used as a carbon source,

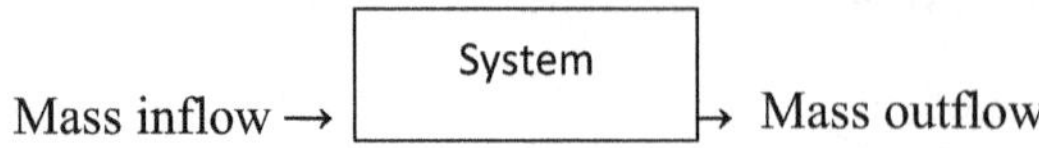

FIGURE 3.6 Mass flow in the system.

one portion of glucose will be used for cell mass production, and another portion for ethanol and carbon dioxide by *Saccharomyces cerevisiae* under anaerobic conditions. The stoichiometry will differ based on the target of the bioprocess or from reaction to reaction.

Usually electron–proton balances are required in addition to elemental balances to determine the stoichiometric coefficient of biochemical reactions. Accurate determination of the composition of cellular material is a major problem. Typical cell mass or biomass composition is represented by an empirical formula: $CH_{1.8}O_{0.5}N_{0.5}$. This biomass composition is approximate with variation of within the acceptable range of ± 5 %. One mole of biological material is defined as the amount containing one-gram atom carbons such as $CH_{\alpha}O_{\beta}N_{\delta}$. The microscopic mass balance of the microbial system concerning the biomass production and another product can be written in its original form. This is based on per carbon atom substrate, per carbon atom biomass, per carbon atom product etc.

The formation of the cell mass requires a nitrogen source in addition to a carbon source and most of the biochemical process are aerobic in nature which requires oxygen. Therefore, it gives the biomass, product, carbon dioxide and water. The composition of substrate, biomass and the product in the equation are expressed by the elemental chemical analysis.

Degree of reduction is used for the analysis of the stoichiometry of the biochemical reaction. Degree of reduction means the number of free electrons present in one-gram atom carbon. For example biomass ($CH_{1.8}O_{0.5}N_{0.5}$) comprises carbon, hydrogen, oxygen and nitrogen. The number of free electrons requirement depends on the electronic structure of the atom. The free electrons of carbon and hydrogen are +4 and +1, respectively due to the donation of the electrons. However, in the case of oxygen and nitrogen, the free electrons are –2, and –3 respectively because oxygen and nitrogen can take 2 and 3 electrons to complete their outermost orbit.

The aerobic fermentation process may be represented as follows:

$$\text{Cells} + \text{Medium} + O_2 \rightarrow \text{More cells} + \text{Product} + CO_2 + H_2O \tag{3.52}$$

The medium usually comprises a carbon source such as sugars, a nitrogen source such as $(NH_4)_2SO_4$, and metal ions and vitamins as cofactors.

The macroscopic mass balance of the microbial system concerning the biomass production and another product can be written in its original form:

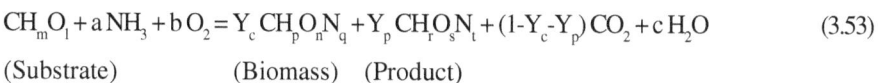

$$\underset{\text{(Substrate)}}{CH_mO_l} + a\,NH_3 + b\,O_2 = Y_c\,\underset{\text{(Biomass)}}{CH_pO_nN_q} + Y_p\,\underset{\text{(Product)}}{CH_rO_sN_t} + (1\text{-}Y_c\text{-}Y_p)\,CO_2 + c\,H_2O \tag{3.53}$$

The degree of reduction of substrate, product and biomass are written as follows:

$$\text{Biomass: } \gamma_b = 4 + p - 2n - 3q \tag{3.54}$$

$$\text{Product: } \gamma_p = 4 + r - 2s - 3t \tag{3.55}$$

$$\text{Substrate: } \gamma_s = 4 + m - 21 \tag{3.56}$$

This is how one can write free electron present per gram atom of biomass, product and substrate. However, there is no free electron of the metabolic products such as water, carbon dioxide, ammonia. This can be justified as follows considering ammonia:

$$\text{Degree of reduction of ammonia} = 1 \times (-3) + 3 \times (1) = 0$$

On the other hand, oxygen in the form of O_2 accepts four electrons.

It has been found that the degree of reduction of the biomasses is different; although they are very close to each other they are different from each other. The standard deviation is 3%.

The empirical formula of the biomass of *Escherichia coli* is $CH_{1.77}O_{0.49}N_{0.24}$. The degree of reduction *E. coli* may be calculated as follows:

$$4 + 1.77 + (-2 \times 0.449)\,(-3 \times 0.24) = 4.07$$

The oxygen requirement is directly related to the free electron available for the transfer to oxygen. This is very important for finding out the heat evolved in the fermentation process.

In Eq. 3.53, the free electron balance can be written with respect to the available electrons in the one gram atom substrate plus the number in oxygen equal to the number of available electrons in the biomass and product. This can be represented as follows:

$$\gamma_s + b\,(-4) = Y_c\,\gamma_b + Y_p\,\gamma_p \tag{3.57}$$

$$b = \frac{\left(\gamma_s - Y_c\,\gamma_b - Y_p\,\gamma_p\right)}{4} = \text{oxygen demand} \tag{3.58}$$

The oxygen requirement of the fermentation process can be calculated with the help of Eq. 3.58. Eq. 3.58 can be modified as follows:

$$1 = \frac{4b}{\gamma_S} + \frac{Y_c\,\gamma_b}{\gamma_S} + \frac{Y_p\,\gamma_P}{\gamma_S} \tag{3.59}$$

Three fractions of Eq. 3.59 indicate available electrons transferred to oxygen, biomass and product from the substrate. In Eq. 3.59 the energetic growth yield (η) is expressed as

$$\eta = \frac{Y_c\,\gamma_b}{\gamma_s} \tag{3.60}$$

Similarly, the energetic product yield (ξ_p) may be represented as follows:

$$\xi_p = \frac{Y_p \gamma_p}{\gamma_s} \tag{3.61}$$

The thermodynamic efficiency of any biological process depends on the values of η and ξ_p which are considered for finding out the validity of the experimental results.

Another way of characterizing the compounds participating in the microbial process is carried out by using the weight fraction of carbon in the organic matter. The weight fraction of carbon present in 1 gram atom carbon biomass, product and substrate may be expressed as follows:

$$\text{Biomass}: \sigma_b = \frac{12}{12+p+16n+14q} \tag{3.62}$$

$$\text{Product}: \sigma_p = \frac{12}{12+r+16s+14t} \tag{3.63}$$

$$\text{Substrate}: \sigma_s = \frac{12}{12+m+16\,l} \tag{3.64}$$

Cell mass yield coefficient ($Y_{x/S}$) can be determined experimentally. In the fermentation process, the substrate concentration decreases and cell mass concentration increases with respect to time. The ratio of cell mass produced and substrate consumed is expressed as $Y_{x/S}$. The unit of this parameter is g of cell mass produced per g substrate consumed. The biomass yield is greater in the aerobic process than that of the anaerobic process. Therefore, the nutritional requirement for the anaerobic fermentation process is much less compared with the aerobic fermentation process. Y_c, yield coefficient with respect to g atom biomass per g atom substrate can be calculated from $Y_{x/S}$ as follows

$$Y_c = \frac{g\,atom\,biomass}{g\,atom\,substrate} = \frac{\text{g biomass produced}}{\text{g substrate consumed}}$$

$$\times \frac{\text{molecular weight (MW) of 1 g atom substrate}}{\text{MW of1 g atom biomass}}$$

$$= Y_{x/S} \times \frac{\text{MW of 1 g atom substrate}}{\text{MW of 1 g atom biomass}} \tag{3.65}$$

Again from Eq. 3.60, we can write

$$Y_c = \frac{\eta \gamma_s}{\gamma_b} = Y_{x/S} \times \frac{\text{MW of1 g atom substrate}}{\text{MW of1 g atom biomass}} \tag{3.66}$$

$$Y_{x/S} = \frac{\eta\,\gamma_s}{\gamma_b} \times \frac{\text{MW of1 g atom biomass}}{\text{MW of1 g atom substrate}}$$

$$= \frac{\eta\gamma_s}{\gamma_b} \times \frac{\text{atomic weight of carbon}}{\text{MW of substrate}} \Bigg/ \frac{\text{atomic weight of carbon}}{\text{MW of biomass}}$$

$$Y_{x/S} = \eta \frac{\gamma_{S}\,\sigma_S}{\gamma_{b}\,\sigma_b} \tag{3.67}$$

where, $\eta \rightarrow$ energetic biomass yield

Similarly, the product yield coefficient can be expressed as follows

$$Y_{P/S} = \varepsilon_P \frac{\sigma_S\,\gamma_S}{\sigma_P\,\gamma_P} \tag{3.68}$$

The question arises as to how the stoichiometry of the bioprocess gives the information on the validity of the experimental results. Experimentally the cell mass yield may be determined from the ratio of (final cell mass concentration – the initial cell mass concentration) and (initial substrate concentration – final substrate concentration). This ratio is known as the yield coefficient. Now from the empirical formula of the substrate and the biomass, the values of σ_S, σ_b, γ_S, and γ_b can be calculated. The values of η and ξ_p can be estimated with the help of Eqs. 3.60 and 3.61. The thermodynamic yield coefficient is equal to ($\eta + \xi_p$). The thermodynamic yield coefficient for the aerobic process varies from 0.5 to 0.6 and but in the case of the anaerobic process this is equal to 0.7. If the thermodynamic yield coefficient values come within this range then the experimental data is acceptable. However, in the case where this value deviates from the expected value, then the experimental data has some errors. This problem can be sorted out on the basis of less accuracy of the experimental data.

The cell mass yield with respect to biomass can be expressed as (Das and Das)

$$Y_{x/o} = \frac{3\eta}{2\sigma_b\gamma_b\left(1-\eta-\xi_p\right)} \tag{3.69}$$

Eq. 3.69 directly calculates how much cell mass is produced per gram of oxygen consumed.

It has already been mentioned that the range of the thermodynamic coefficient for the aerobic fermentation process is from 0.5 to 0.6 and approximately 0.7 in the case of the anaerobic process. For the determination of the thermodynamic coefficient of the bioprocess, the concept of process is important; e.g. in the anaerobic digestion process, methane and carbon dioxide are mainly produced from organic wastes.

Since it is anaerobic fermentation process, one can ignore the amount of cell mass produced because the cell mass growth in aerobic process is usually 10 times more as compared with that of the anaerobic fermentation process. Therefore the value of η can be neglected. The thermodynamics coefficient of the anaerobic process is 0.7 which will be equal to the value of ξ_p. Now if there is a variation of the value from 0.7 then one should check that experimental data. Therefore, it is a very easy way to detect the error of your experimental results (Bhatt and Thakore, 2010; Ghasem and Henda, 2012).

Stoichiometry of the bioprocess also determines the heat evolved in the case of the aerobic fermentation process. Most bioproducts are produced through the aerobic fermentation process. In the fermentation process, the amount of oxygen required for both the aerobic and anaerobic process comes from the different sources; e.g. in the aerobic process the organism requires the molecular oxygen but in the anaerobic process the oxygen is utilized from compounds such as nitrate, sulphate, and nitrite. Heat evolved in the aerobic fermentation process is determined from molecular oxygen consumption. Heat evolved in the aerobic fermentation process can be calculated with the help of the equation:

$$Q = 4\,Q_0\,b \text{ [kJ/g atom of substrate consumed]} \tag{3.70}$$

Where Q_0 is approximately 133 kJ/equivalent of free electron transferred from the substrate to O_2. The invariant Q_0 directly links the mass balance of the process with its energy balance (Volesky and Votruba, 1992).

Heat produced per unit mass of substrate consumption and cell mass formation are expressed as

$$\text{Heat evolved/ g of substrate consumed} = \frac{4Q_0\,b}{M_{substrate}}\frac{kJ}{g} \tag{3.71}$$

$$\text{Heat evolved/ g of biomass produced} = \frac{4Q_0\,b}{M_{substrate}\,Y_{x/S}}\frac{kJ}{g} \tag{3.72}$$

To find out the heat evolved to produce 10 kg of baker's yeast in the baker's fermentation process, initially the stoichiometry of the process is to be determined. Then the yield of biomass per gram of substrate can be calculated. This value we can use in Eq. 3.72 to find out the heat evolved for the production of 10 kg baker's yeast.

Problem 3.3: *Candida utilis* is used for the production of fodder yeast by the aerobic fermentation process considering ethanol as a major carbon source. Find out the stoichiometry of the process and heat evolved during the fermentation process (Volesky and Votruba, 1992).

Solution:

Here *Candida utilis* is considered to be the fodder yeast and this is used as animal feed. The speciality of this yeast is that it can use both hexose and pentose sugar.

In the present problem *Candida utilis* utilizes ethanol as a substrate for the production of the biomass. Basically microorganisms require carbon sources for cell mass formation, and nitrogen sources also contribute for cell mass formation. Molecular oxygen is required for this fermentation process because cell mass grows aerobically. The empirical formula of ethanol can be written as C_2H_6O and we can write this as $CH_3O_{0.5}$ on the basis of per gram atom carbon.

The stoichiometric equation is given:

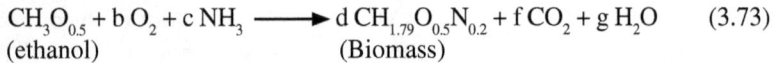

$$\underset{\text{(ethanol)}}{CH_3O_{0.5}} + b\,O_2 + c\,NH_3 \longrightarrow d\,\underset{\text{(Biomass)}}{CH_{1.79}O_{0.5}N_{0.2}} + f\,CO_2 + g\,H_2O \qquad (3.73)$$

Where 1 gram atom of ethanol, b moles of oxygen and c moles of ammonia combine to form d moles of biomass, f moles of carbon dioxide and g moles of water.

Calculation of $\gamma_s, \gamma_b, \sigma_s$ and σ_b:

$$\gamma_s = 4 + 3 - 2\times0.5 = 6$$

$$\gamma_b = 4 + 1.79 - 2\times0.5 - 3\times0.2 = 4.19$$

$$\sigma_s = \frac{12}{12+3+16\times0.5} = \frac{12}{23} = 0.522$$

$$\sigma_b = \frac{12}{12+1.79+16\times0.5+14\times0.2} = \frac{12}{24.59} = 0.488$$

For the aerobic process the value of the thermodynamic coefficient is in range of 0.5–0.6.

Assuming 0.6 is the value of the thermodynamic coefficient

$$\eta = 0.6$$

$$Y_{x/s} = \eta\frac{\sigma_s\gamma_s}{\sigma_b\gamma_b} = 0.6\frac{(0.523)(6)}{(0.488)(4.19)} = 0.92\,\frac{\text{g of biomass}}{\text{g of substrate}}$$

O_2 consumption:

$$Y_{x/o} = \frac{3\eta}{2\sigma_b\gamma_b(1-\eta-\varepsilon_p)} \qquad (\text{assuming, } \varepsilon_p = 0)$$

$$Y_{x/o} = \frac{3(0.6)}{2(0.488)4.19(1-0.6)} = \frac{1.8}{1.63} = 1.1\,\frac{\text{g of biomass}}{\text{g of } O_2 \text{ use}}$$

Oxygen demand as O_2:

$$b = \frac{\left(\gamma_s - \gamma_c\gamma_b - Y_p\gamma_p\right)}{4}$$

$$\eta = \frac{Y_c\gamma_b}{\gamma_s}, \quad Y_c = \frac{\eta\gamma_s}{\gamma_b} = \frac{0.6\times 6}{4.19} = 0.859$$

$$Y_c = d = 0.859$$

$$b = \frac{6 - 0.859\times 4.19}{4} = 0.6 \quad \frac{\text{g mole of } O_2}{\text{g atom of carbon in substrate}}\left(Y_p\gamma_p = 0\right)$$

Calculation of stoichiometric coefficient

$$C{:}\ 1 = d + f, \quad f = 1 - d = 1 - 0.859 = 0.141$$

$$O{:}\ 0.5 + 2b = 0.5d + 2f + g = 0.5 + 2(0.6) = 0.5(0.859) + 2(0.141 + g)$$

$$1.7 - 0.4295 - 0.282 = g = 0.988$$

$$H{:}\ 3 + 3C = 1.79d + 2g$$

$$3 + 3C = 1.79(0.859) + 2(0.98), \qquad C = 0.171$$

$$N{:}\ C = 0.2d$$

Stoichiometric equation,

$$CH_3O_{0.5} + 0.60_2 + 0.171NH_3$$

$$\rightarrow 0.859CH_{1.79}O_{0.5}N_{0.2} + 0.141CO_2 + 0.989H_2O \tag{3.74}$$

Heat evolved during process

$$Q = 4\times Q_o \times b$$

$$= 4\times 133\times 0.6 = 319.2\,\frac{\text{kJ}}{\text{gram atom C}} \qquad \left(\text{MW of } CH_3O_{0.5} = 23\right)$$

$$= 319.2\times\frac{1}{23} = 13.8\,\frac{\text{kJ}}{\text{g of substrate}}$$

Eq. 3.74 gives us an idea regarding the contribution of different components for the product and byproduct formation. This is very useful for the industry to find out the stoichiometry of this particular process.

Problem 2: Cane molasses is used as a major raw material for baker's yeast production. Cane molasses contents 50%w/w of sugar. Find out the stoichiometry of the process and determine the amount of cane molasses required for 1 kg dry baker's yeast production.

Solution: The general stoichiometry equation for baker's yeast production may be written as follows, assuming sucrose as a carbon source and no ethanol formation:

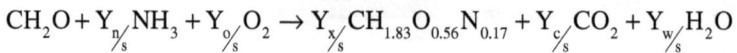

$$CH_2O + Y_{n/s}NH_3 + Y_{o/s}O_2 \rightarrow Y_{x/s}CH_{1.83}O_{0.56}N_{0.17} + Y_{c/s}CO_2 + Y_{w/s}H_2O$$

$$\gamma_s = 4 + 2 - 2 = 4$$

$$\gamma_b = 4 + 1.83 - 0.56 \times 2 - 0.17 \times 3 = 4.2$$

$$\sigma_s = \frac{12}{12 + 2 + 16} = \frac{12}{30} = 0.4$$

$$\sigma_b = \frac{12}{12 + 1.83 + 0.56 \times 16 + 0.17 \times 14} = 0.476$$

Assuming thermodynamic efficiency in aerobic process

$$= 0.55 = \eta = \frac{Y_{x/s}\gamma_b}{\gamma_s}$$

$$Y_{x/s} = \frac{\eta\gamma_s}{\gamma_b} = \frac{0.55 \times 4}{4.2} = 0.52$$

From C-balance, $Y_{c/s} = 1 - Y_{x/s} = 1 - 0.52 = 0.48$

From N-balance, $Y_{n/s} = 0.52 \times 0.17 = 0.0884$

From H-balance, $2 + 0.0884 \times 3 = 0.5 \times 1.83 + Y_{w/s} \times 2$

$$Y_{w/s} = \frac{2.2652 - 0.915}{2} = 0.6751$$

The stoichiometry of baker's yeast production

$$CH_2O + 0.0884NH_3 + 0.4685O_2 \rightarrow 0.52CH_{1.83}O_{0.56}N_{0.17} + 0.48CO_2 + 0.6751H_2O$$

30 17 32 25.17 44 18

(3.75)

$$b = \frac{\gamma_s - Y_{x/s}\gamma_b}{4} = \frac{4 - 0.52 \times 4.2}{4} = 0.4685$$

Typical material analysis

From Eq. 3.75, we can write

1 g-atom substrate produced 0.52×25.17 g of cell mass

Therefore, sugar required for 1 kg baker's yeast production

$$= \frac{1000 \text{ g}}{0.52 \times 25.17 \text{ g}} \text{ g-atom} = 76.4 \text{ g} = \text{atom}$$

$$= 76.4 \times 30 = 1,292 \text{ g}$$

Cane molasses is usually content 50 %w/w sugar.

Therefore, amount of cane molasses required = $\frac{1,292 \text{ g}}{0.5} = 2,584$ g

3.5 CONCLUSIONS

The bioreactor is the heart of a bioprocess which is responsible for product formation. Biochemical processes are different from those of chemical processes with respect to aseptic condition, temperature etc. Chemical reaction kinetics and chemical reactor analysis help to understand the kinetics of the biochemical processes. Biochemical processes are broadly classified as batch, fed-batch, and continuous. Continuous processes are found to be efficient as compared to batch processes with respect to product formation. Stoichiometry of the chemical and biochemical processes play an important role for the materials and energy analysis of the processes. In addition to the intermolecular relationship of the reactants and product, the stoichiometry of the bioprocesses provides information on the heat generated during the aerobic fermentation process and the validity of the experimental results.

REFERENCES

Bailey JE, and Ollis DF, Biochemical Engineering Fundamentals, McGraw-Hill Inc., New Delhi, 2010.

Bhatt BI, and Thakore SB, Stoichiometry, Tata McGraw-Hill Education, 2010.

Blanch HW and Clark DS, Biochemical Engineering, Marcel Dekker, New York, 1997.

Das D and Das D, Biochemical Engineering: An Introductory Text Book, Jenny Stanford, Singapore, 2019.

Das D and Das D, Biochemical Engineering: A laboratory manual, Jenny Stanford, Singapore, 2020.

Doran PM, Bioprocess Engineering Principles, Second Edition, Academic Press, Waltham, 2012.

Gavhane KA, Introduction to Process Calculations Stoichiometry, Nirali Prakashan, 2012.

Ghasem N, and Henda R, Principles of Chemical Engineering Processes, CRC Press, 2012.

Katoh S, Yoshida F, Biochemical Engineering: A Textbook for Engineers, Chemists and Biologists, Wiley-VCH, 2010.

Kelly RM, Wittrup KD, and Karkare S (Eds.), Biochemical Engineering VIII, Vol. 745, Annals of the New York Academy of Sciences, 1994.

Levenspiel O. Chemical reaction Engineering, Third Edition, Wiley-India, 2010.

Nauman EB, Chemical Reactor Design, John Wiley, 1987.

Shuler ML and Kargi F, Bioprocess Engineering: Basic Concepts, Second Edition, Prentice-Hall, New Delhi, 2002.

Sinclair CG, and Kristiansen B, Fermentation kinetics and Modelling, Open University Press, 1987.

Volesky B and Votruba J, Modeling and Optimization of fermentation processes, Elsevier, Netherlands, 1992.

4 Enzymatic Reaction Kinetics and Immobilization of Enzymes

Enzymes are mostly globular proteins and act as catalysts. A catalyst is defined as a substance which increases the rate of a chemical reaction without undergoing a permanent chemical change. The activation energy of the reactant molecules is lowered to increase the rate of a catalytic chemical reaction but it does not influence the equilibrium of the reaction (Figure 4.1). Enzymes are very specific to one substrate, because one enzyme usually does not participate in different reactions. For example glucose isomerase acts on glucose only and converts glucose to fructose.

Different metabolic pathways are present in the microorganism. Each metabolic pathway comprises a series of reactions. Each step of the reaction requires one specific enzyme. Therefore the microorganism is considered as a multi-enzyme system. The differences between the enzymatic reaction and microbial fermentation process are Table 4.1.

4.1 DIFFERENCES BETWEEN ENZYMATIC REACTION AND MICROBIAL FERMENTATION

Enzymatic reaction requires a particular pH, temperature and specific substrate but microbial fermentation processes are mostly not substrate specific for microbial growth and metabolic reactions. For example, glucose can produce several products such as ethanol, acetic acid, citric acid etc. However, the pH and temperature required for different living systems may vary. Carbon sources present in the medium contribute to the growth of cells, as a source of energy and product formation. Nitrogen sources are mostly used for the growth of cells. Minerals and vitamins are considered as cofactors in the different enzymatic reactions.

Enzymes are mostly protein molecules with active sites. They have a globular structure. On the other hand, synthetic or artificial enzymes are non-protein in nature; e.g. some RNA molecules can act as enzymes. Catalytic RNA molecules are called ribozymes. In any case, most enzymes are proteins in nature and they have a globular structure. A globular structure is a kind of folded structure which

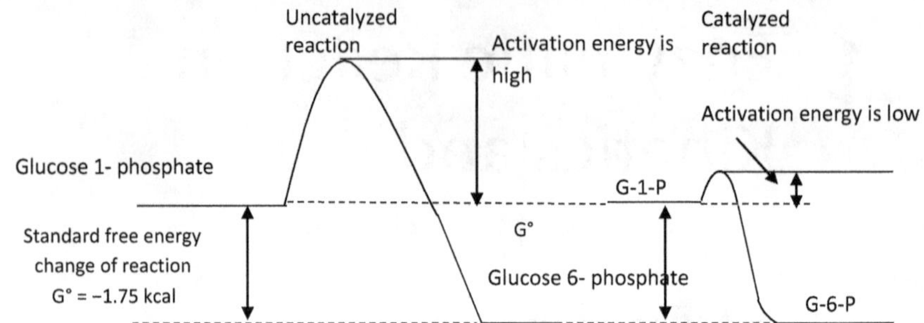

FIGURE 4.1 Change of free energy during conversion of G-1-P to G-6-P by phosphorgluco mutase.

TABLE 4.1
Differences between Enzymatic Reaction and Microbial Fermentation

Enzymatic Reaction	Microbial Fermentation
Enzymes are mostly globular proteins and very specific to the substrate	Not that much substrate specification for microbial growth
Required particular pH and temperature	Required pH and temperature would be in optimum range
Specific substrate to be added	Medium is to be added. Medium contains Carbon source- for energy, body building and product formation Nitrogen source- for body building, product formation Minerals and vitamins act as a cofactor

is formed due to the hydrogen bonding between -NH_2 and -COOH functional groups with the R group of the different amino acids. During this folding they form a particular site which is very specific to the configuration of a particular substrate; this is known as an active site of the enzymes. Proteins without active sites are not considered to be enzymes.

4.2 ROLES OF COFACTORS IN ENZYMATIC REACTIONS

The three-dimensional configuration of the protein molecule is responsible for the formation of the active site which is in turn responsible for the specificity as well as the catalytic ability of the enzyme. Most enzymes require cofactors in order to make them active. A cofactor is a nonprotein compound which in combination with an inactive protein (known as apoenzyme) becomes a catalytically active protein (known as holoenzyme). Cofactors may be classified as shown in Figure 4.2.

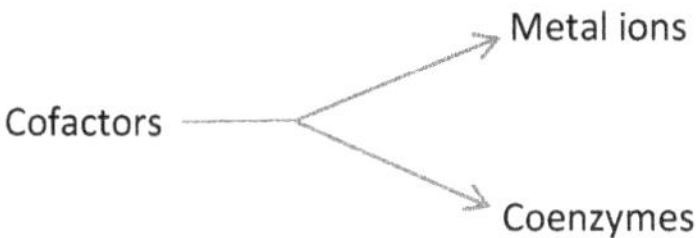

FIGURE 4.2 Classification of the cofactors present in the enzymatic reactions.

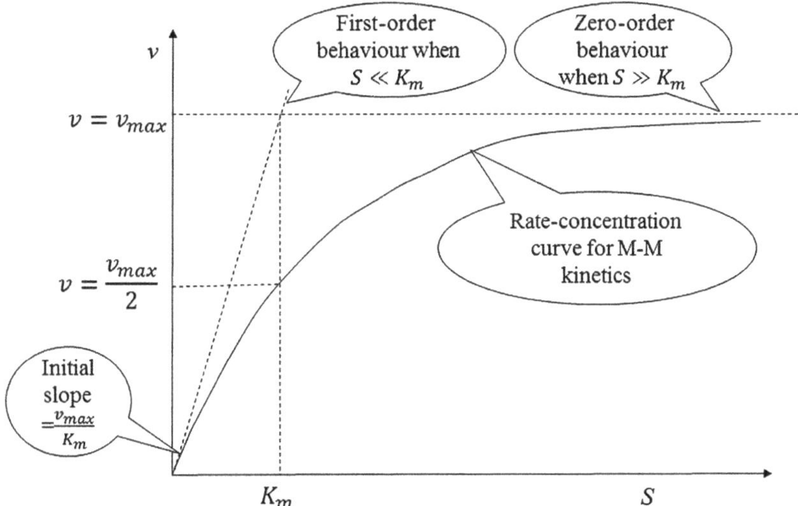

FIGURE 4.3 Plot of velocity of reaction vs. substrate concentration in an enzymatic reaction.

Different properties and catalytic activities of the same enzyme obtained from different organisms are due to the different amino acid sequences. For example, glucose isomerase obtained from *B. coagulans* requires Co^{++} as a cofactor whereas glucose isomerase from mutant of *B. coagulans* does not require Co^{++} at pH greater than 8.

Enzymatic activity is expressed via the international unit (I.U.). 1 I.U. is defined as the amount of the enzyme that catalyses the conversion of one micromole of substrate per min under specified conditions. Specific enzymatic activity is expressed as micromole of substrate converted per min/mg of protein. Enzymatic activity may also be expressed as a turnover number where protein molecules have more than one active site. This is possible in the case of larger protein molecules. The turnover number is defined as the number of substrate molecules reacted per catalyst site per unit time.

4.3 MICHAELIS–MENTEN EQUATION

The kinetics of the enzymatic reaction are expressed with the help of the Michaelis–Menten (M–M) equation. The Michaelis–Menten equation represents a graphical correlation between v and S as shown in Figure 4.3.

M–M proposes a correlation between the velocity of the enzymatic reaction (v) and substrate concentration (S) on the basis of above co-relation of v and S as follows

$$v = \frac{v_{max}\, S}{K_m + S} \tag{4.1}$$

where v = velocity of the reaction, S = substrate concentration, v_{max} = maximum velocity of reaction and K_m = M–M constant.

This is known as the M–M equation for enzymatic reaction kinetics.

Special features of the M–M equation

- At low substrate concentration, $S \ll K_m$, the reaction follows first-order kinetics, $v \approx \frac{v_{max}}{K_m} S$
- At high substrate concentration, $S \gg K_m$, the reaction follows zero-order kinetics. $v \approx v_{max}$
- The maximum velocity of reaction is directly proportional to the total enzyme concentration (E_0).

Again, when $v \approx \frac{v_{max}}{2}$, from Eq. 4.1, $K_m = S$

Briggs and Haldane justified the M–M equation on the basis of reaction kinetics of the enzymatic reaction assuming quasi-steady state condition. This quasi-steady state condition is valid when the $\frac{[E]_0}{[S]_0}$ value is very small as shown in Figure 4.4.

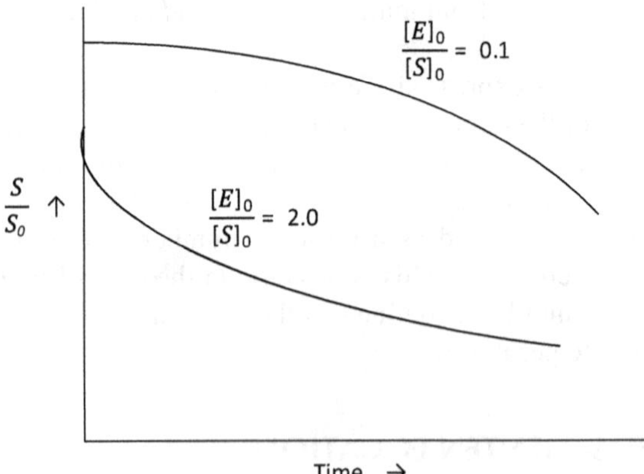

FIGURE 4.4 Substrate degradation profiles at different $\frac{[E]_0}{[S]_0}$ ratio.

Briggs and Haldane assumed the substrate initially tries to sit at the active site of the enzyme molecule. However, when it tries to sit at the active site, only a portion of the substrate can sit properly. Briggs and Haldane considered this sitting phenomenon to be a reversible reaction. The substrate which sits properly at the active site forms an ES complex which converts to a product which is an irreversible reaction. The reactions may be written as follows:

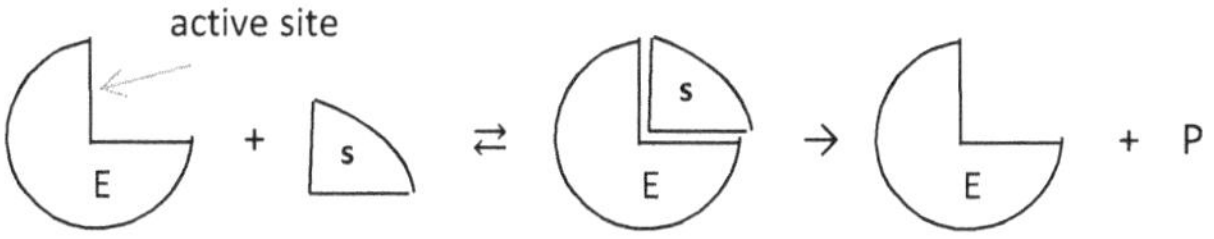

active site

E + S ⇄ E S → E + P

$$E+S \underset{k_{-1}}{\overset{k_1}{\rightleftarrows}} ES \overset{k_2}{\rightarrow} P+E \tag{4.2}$$

The velocity of the above reaction may be represented as follows

$$v=-\frac{dS}{dt}=k_1[S][E]-k_{-1}[ES] \tag{4.3}$$

$$\text{and } \frac{d[ES]}{dt}=k_1[S][E]-(k_{-1}+k_2)[ES] \tag{4.4}$$

$$\text{At quasi-steady state condition,} \frac{d[ES]}{dt}=0 \tag{4.5}$$

$$\text{Again, } [E]_0=[E]+[ES] \tag{4.6}$$

$$[E]=[E]_0-[ES] \tag{4.7}$$

$$\frac{d[ES]}{dt}=k_1[S]([E]_0-[ES])(k_{-1}+k_2)[ES]=0 \tag{4.8}$$

$$[ES]=\frac{k_1[S][E]_0}{k_1[S]-(k_{-1}+k_2)} \tag{4.9}$$

$$v=k_1[S]([E]_0-[ES])\text{-}k_{-1}[ES] \tag{4.10}$$

$$v=k_1[S][E]_0-[ES](k_1[S]-k_{-1})=k_1[S][E]_0-\frac{k_1[S][E]_0}{k_1[S]-(k_{-1}+k_2)}(k_1[S]-k_{-1}) \tag{4.11}$$

$$v = \frac{k_1 k_2 [S][E]_0}{k_1 [S] - (k_{-1} + k_2)} \tag{4.12}$$

$$v = \frac{k_2 [E]_0 [S]}{\frac{k_{-1} + k_2}{k_1} + [S]} = \frac{v_{max} [S]}{K_m + [S]} \tag{4.13}$$

Where $v_{max} = k_2 [E]_0$ and $K_m = \frac{k_{-1} + k_2}{k_1}$

This is how Briggs and Haldane justified the M–M (Eq. 4.13).

4.4 DETERMINATION OF THE SUBSTRATE CONCENTRATION PROFILE

The M–M equation (Eq. 4.1) may be written as

$$v = \frac{v_{max} [S]}{K_m + [S]} = -\frac{dS}{dt}$$

Eq. 4.1 may be expressed in the following form (Eq. 4.14) by integrating from 0 to time, t

$$-\left(K_m \int_{S_o}^{S} d\ln + \int_{S_o}^{S} ds \right) = v_{max} \int_{o}^{t} dt \tag{4.14}$$

Assuming at t = 0, substrate concentration = S_0 and at time t, substrate concentration = S
Eq. 4.14 may be written as

$$K_m \ln \frac{S_0}{S} + (S_o - S) = v_{max}\, t \tag{4.15}$$

Eq. 4.15 may be used to find out the substrate concentration after any time t to find the substrate concentration profile in an enzymatic reaction.

4.5 DETERMINATION OF THE KINETIC CONSTANTS

The kinetic constants of the enzymatic reactions v_{max} and K_m can be determined by using different plots developed from the M–M equation. These plots are known as the Lineweaver–Burk (Eq. 4.16), Hanes–Woolfe (4.17) and Eadie–Hofstee plots (4.18).

Eq. 4.1 may be written as

$$\frac{1}{v} = \frac{1}{v_{max}} + \frac{K_m}{v_{max}} \frac{1}{S} \tag{4.16}$$

$$\frac{S}{v} = \frac{K_m}{v_{max}} + \frac{S}{v_{max}} \tag{4.17}$$

$$v = v_{max} + K_m \frac{v}{S} \tag{4.18}$$

Eq. 4.16 is used to draw the plot $\frac{1}{v}$ vs. $\frac{1}{S}$ which is known as the Lineweaver–Burk plot. This is shown in the Figure 4.5. The slope of the straight line curve gives the value of $\frac{K_m}{v_{max}}$ and the intercept gives the values of $\frac{1}{v_{max}}$. as shown in the figure.

Similarly, $\frac{S}{v}$ vs. S and v vs. $\frac{v}{S}$ plots are known as Hanes–Woolfe and Eadie–Hofstee plots, respectively as shown in Figures 4.6 and 4.7.

Both the kinetic constants v_{max} and K_m can be determined by the above plots as shown in Figures 4.6 and 4.7.

4.6 INHIBITION OF ENZYMATIC REACTIONS

An inhibitor is responsible for the retardation of the enzyme-catalysed reaction when it is present in the reaction mixture. Inhibition may be due to the interaction between

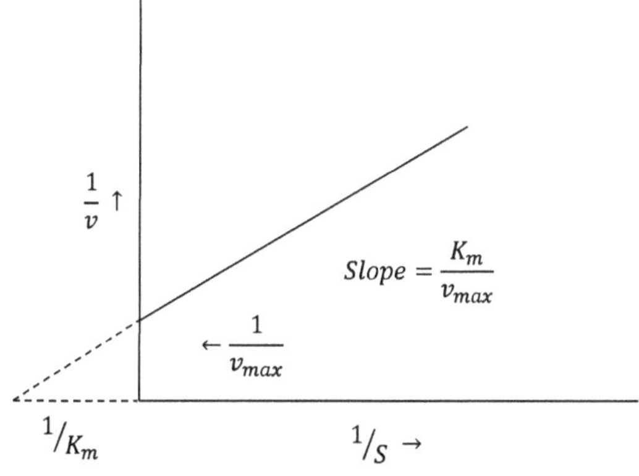

FIGURE 4.5 Lineweaver–Burk plot.

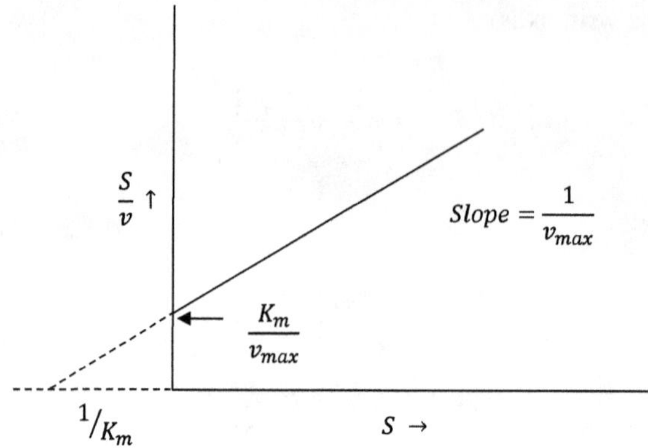

FIGURE 4.6 Hanes–Woolfe plot.

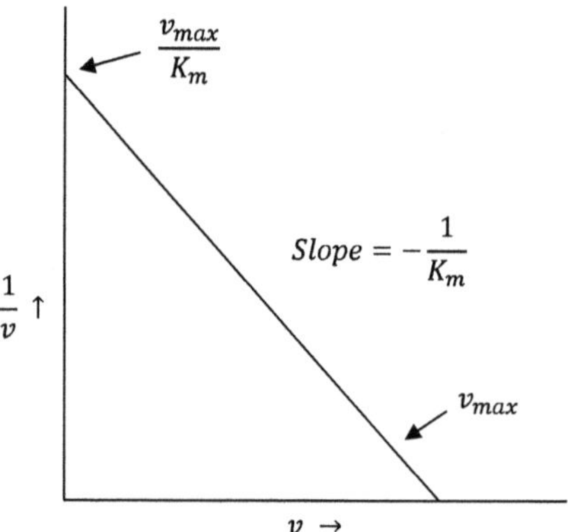

FIGURE 4.7 Eadie–Hofstee plot.

inhibitor and enzyme, or between inhibitor and substrate, and it is mostly reversible in nature. Reversible enzyme inhibition can be classified into three types:

- Competitive
- Non-competitive
- Uncompetitive

4.6.1 Competitive Inhibition

Competitive inhibition can be represented as shown below:

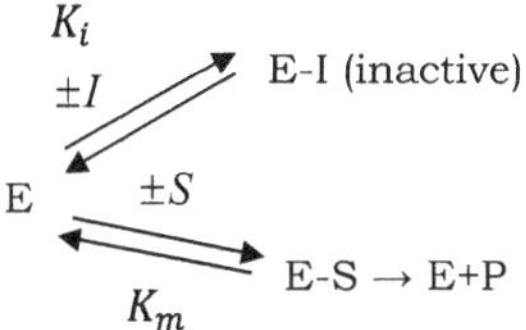

An example of competitive inhibition reaction is the conversion of succinic acid to fumaric acid.

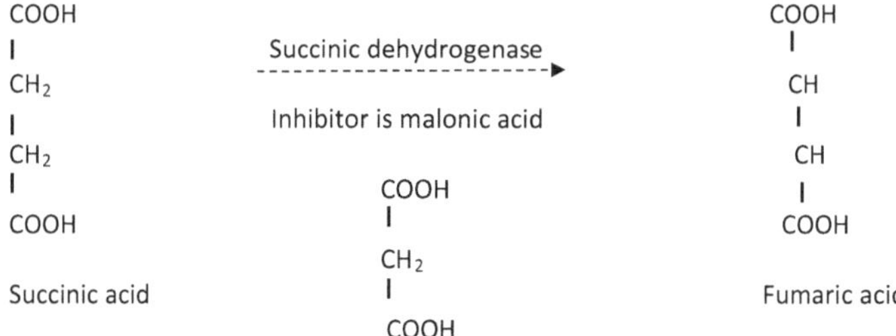

Another example is methotrexate whose structure has a resemblance to that of vitamin folic acid. It acts as an inhibitor for the dihydrofolate reductase which prevents the formation of dihydrofolate from tetrahydrofolate.

The velocity of competitive reaction may be expressed as

$$v = \frac{v_{max} S}{S + K_m \left(1 + \frac{I}{K_i}\right)} \tag{4.19}$$

Where K_i is ithe equilibrium constant for the formation of the enzyme inhibitor constant and I is the inhibitor constant. The Lineweaver–Burk plot in the case of competitive inhibition is shown below.

In competitive inhibition, v_{max} remains constant and K_m increases (Figure 4.8).

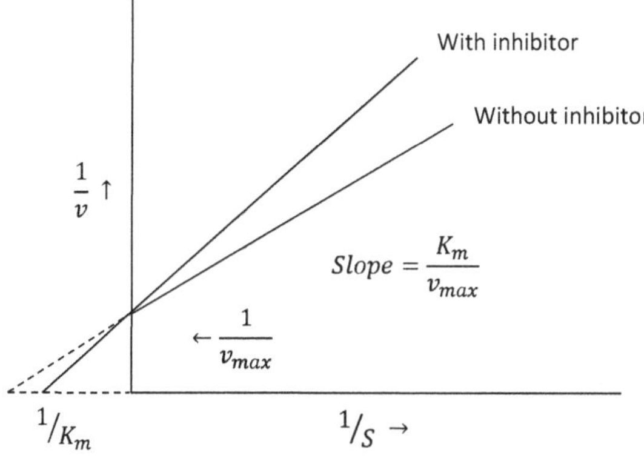

FIGURE 4.8 Lineweaver–Burk plot with competitive inhibition.

4.6.2 Non-Competitive Inhibitor

In non-competitive inhibition, the inhibitor can bind to an enzyme somewhere other than the active site. In non-competitive inhibitors there is a formation of both di-molecular and tri-molecular complexes as shown below.

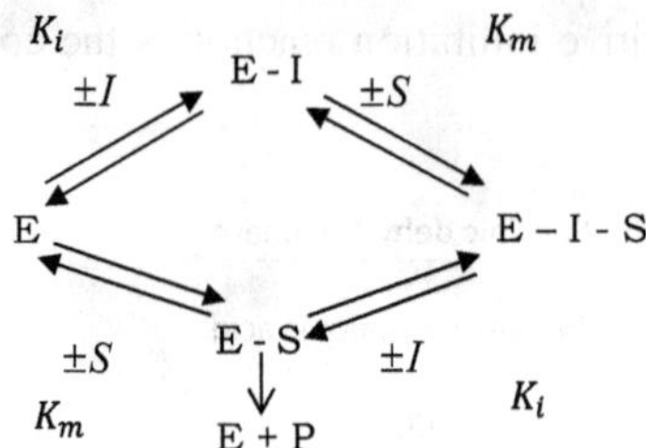

The velocity of the non-competitive reaction is given below:

$$v = \frac{v_{max}}{\left(1+\frac{I}{K_i}\right)\left(1+\frac{K_m}{s}\right)} \tag{4.20}$$

For example pyruvate kinase is inhibited by alanine (synthesized from pyruvate) in the glycolysis pathway.

The Lineweaver–Burk plot for the non-competitive inhibition reaction is shown in Figure 4.9.

In non-competitive inhibition, v_{max} decreases and K_m remains constant.

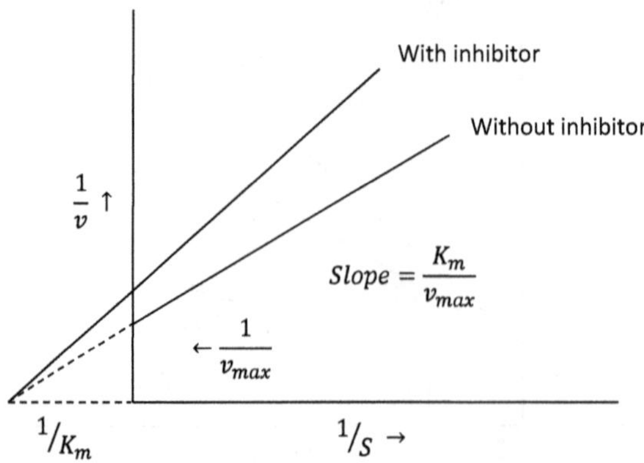

FIGURE 4.9 Lineweaver–Burk plot with non-competitive inhibition.

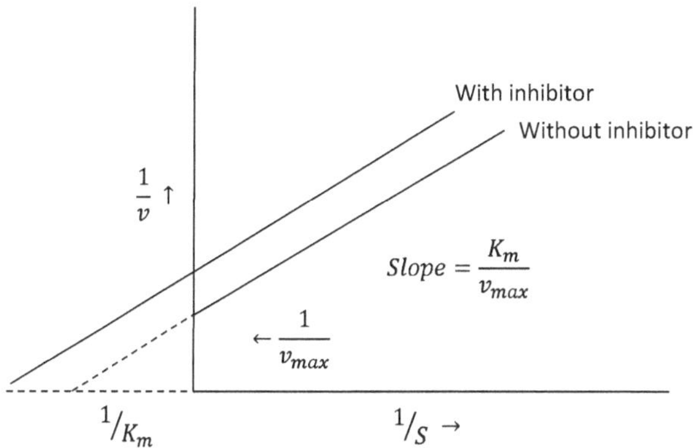

FIGURE 4.10 Lineweaver–Burk plot with uncompetitive inhibition.

4.6.3 Uncompetitive Reactions

Uncompetitive inhibitors do not bind to free enzymes but effect enzymatic reaction by binding to the ES complex at locations away from the active site. The reaction scheme is shown below.

$$\begin{array}{c} \mathbf{E + S \leftrightarrow ES \rightarrow E + P} \\ + \\ | \\ \updownarrow \\ \mathbf{ESI} \end{array}$$

The velocity of reaction of uncompetitive reactions is given below:

$$v = \frac{v_{max}}{\left(1+\frac{I}{K_i}\right)} \frac{S}{\left[\frac{K_m}{1+\frac{I}{K_i}} + S\right]} \tag{4.21}$$

The Lineweaver–Burk plot of uncompetitive reactions is shown in Figure 4.10.

In the case of uncompetitive inhibition reactions, both v_{max} and K_m values are changed.

4.7 ENZYMATIC REACTIONS IN BATCH PROCESSES

In batch processes, the materials are taken once at a time, and after the reaction is over the product is taken out (Figure 4.11). In this process the reaction time is consider as

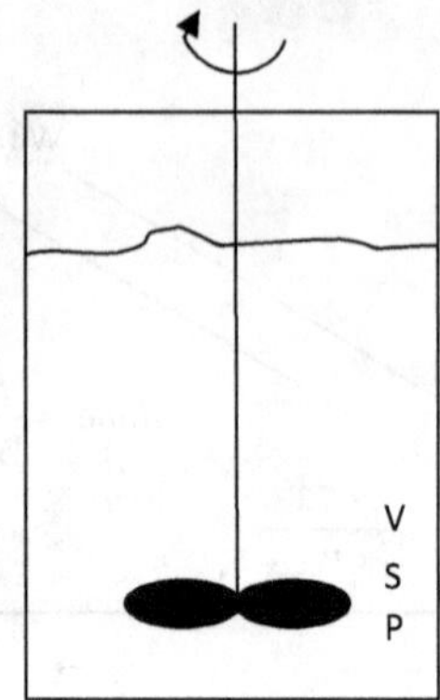

FIGURE 4.11 Schematic diagram of the batch process.

t_b which can be calculated by Eq. 4.15. In practice, batch processes involve unproductive time such as harvesting, cleaning and refilling of the reaction mixture in the reactor, which is known as down time, $t_{down\ time}$.

$$\text{Total batch time, } t_{total} = t_b + t_{down\ time} \tag{4.22}$$

The volume of the batch reactor is calculated on the basis of t_{total}.

4.8 ENZYMATIC REACTIONS IN CONTINUOUS-STIRRED TANK REACTOR (CSTR)

In the case of freely suspended enzymes in CSTR, the enzymes are continuously withdrawn from the reactor in the outflow (in the product stream) (Figure 4.12). Enzymes are not reproduced during the reaction, and so in CSTR the enzyme is to be added continuously to keep the enzyme concentration constant. The following assumptions are made in the operation of CSTR:

- Steady-state operation where rate accumulation of the substrate is zero
- Working volume (liquid volume) (V) of the reactor is kept constant

Substrate balance of the process:

Rate of input of substrate + Rate of substrate generation =
Rate of output of the substrate + Rate of substrate consumption
+ Rate of substrate accumulation (4.23)

$$F S_0 + 0 = F S + (-r_S)V + 0 \tag{4.24}$$

$$F(S_0 - S) = (-r_S)V \tag{4.25}$$

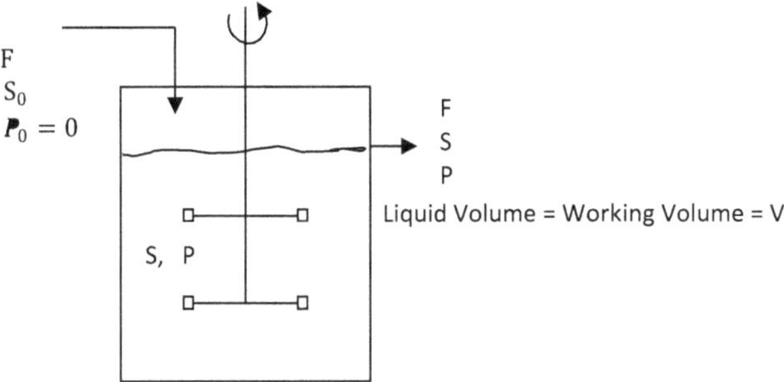

FIGURE 4.12 Schematic diagram of CSTR.

$$\frac{F}{V}(S_0 - S) = (-r_S) = \frac{v_{max}\ S}{K_m + S} \tag{4.26}$$

$$D(S_0 - S) = \frac{v_{max}\ S}{K_m + S} \tag{4.27}$$

$$\frac{1}{D} = \frac{(S_0 - S)}{\frac{v_{max}\ S}{K_m + S}} = \tau_{\text{CSTR}} \tag{4.28}$$

Eq. 4.27 is used to calculate the dilution rate to get a desired amount of substrate conversion.

4.9 ENZYMATIC REACTIONS IN PLUG FLOW REACTOR (PFR)

Enzymatic reactions in plug flow reactors (Figure 4.13) have the following characteristics

- Plug flow reactor (PFR) is an alternative to CSTR for continuous operation
- Liquid passes through the reactor as discrete 'plug' and does not interact with neighbouring fluid elements
- There is no axial mixing; only radial mixing
- Composition varies along with the flow path

$$F\,S\,|_z = F\,S\,|_{z+\Delta z} + (-r_S)A\,\Delta z \tag{4.29}$$

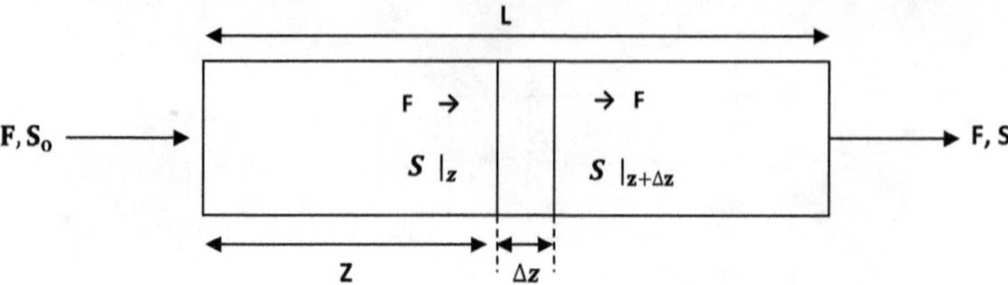

FIGURE 4.13 Schematic diagram of continuous plug flow reactor.

Rearranging the above equation, we get

$$\frac{F\left(S|_{z} - S|_{z+\Delta z}\right)}{A\,\Delta z} = \left(-r_S\right) \tag{4.30}$$

$$\frac{u\left(S|_{z} - S|_{z+\Delta z}\right)}{\Delta z} = \left(-r_S\right) \tag{4.31}$$

By definition

$$\frac{u\,ds}{dz} = \left(-r_S\right) = \frac{V_{max}\,S}{K_m + S} \tag{4.32}$$

$$\frac{1}{u}\int_0^L dz = \int_{S_0}^{S} \frac{K_m + S}{V_{max}\,S}\,dS \tag{4.33}$$

$$\frac{L}{u} = \tau_{PFR} = \frac{K_m}{v_{max}} \ln\frac{S_0}{S} = \frac{1}{v_{max}}(S_0 - S) \tag{4.34}$$

The space time required for the PFR reactor is same as the batch time for a particular reaction (Eq. 4.15). Therefore, it may be written as

$$\tau_{PFR} = t_b \tag{4.35}$$

Problem 4.1 An enzymatic reaction follows M–M kinetics. The kinetic constants of the enzyme are as follows: $V_{max} = 2.5$ mmol m^{-3} s^{-1} and $K_m = 5$ mM. The reaction is carried out in a batch-stirred tank reactor. Determine the time required for 50% conversion of the substrate if the initial substrate concentration is 0.2 M?

SOLUTION

The given data: $S_0 = 0.2$ M, $V_{max} = 2.5$ mmol m^{-3} s^{-1}, $K_m = 5$ mM

Now, $S = (1 - 0.5)\, S_0 = 0.5\,(S_0) = 0.1$ M

$$V_{max} = 2.5 \text{ mmol m}^{-3}\text{s}^{-1} = 2.5 \text{ mmol m}^{-3}\text{s}^{-1} \times \frac{1\,\text{mol}}{1000\,\text{mmole}} \times \frac{1\,\text{m}^3}{1000\text{ L}}$$

$$= 2.5 \times 10^{-6}\,\text{M s}^{-1}$$

$$K_m = 5 \text{ mM x} \frac{1\,\text{mol}}{1000\,\text{mmole}} = 0.005 \text{ M}$$

$$\text{For a batch reactor, } t_{batch} = \frac{1}{V_{max}} K_m \ln\frac{S_0}{S} + (S_0 - S)$$

$$= \frac{1}{2.5 \times 10^{-6}}\left[0.005 \ln\frac{0.2}{0.1} + (0.2 - 0.1)\right] = \mathbf{11.5\ h}$$

Problem 4.2 A series of experiments were performed for determining the K_I values for two competitive inhibitors. K_I values for the inhibitors A, B, and C are 5, 1, and 0.2, respectively. Find the following:

(a) Higher affinity of the inhibitor to the free enzyme?

(b) If the same concentration of inhibitors were considered in each experiment, which inhibitor would give the smallest value of K_m?

(c) What is the ratio of K_I / K_m for each inhibitor in case K_m is 1 μM? How this is the related to the competing equilibria for binding of the substrate vs. the inhibitor to the enzyme?

SOLUTION

(a) The lowest K_I (dissociation constant) value of inhibitor binds with the highest affinity, so inhibitor C possesses the higher affinity to the enzyme.

(b) In the case where K_I value is higher, K_m value will be smallest. Therefore, inhibitor A would give the lowest value for K_m.

(c) The ratio of K_I/K_m indicates the binding constant for both the inhibitor and the substrate.

If $K_I/K_m > 1$, the substrate is bound more tightly, **so the inhibitor will not have that much effect** on kinetics.

If $K_I/K_m = 1$, both substrate and inhibitor **bind with the same affinity.**

If $K_I/K_m < 1$, the inhibitor binds with greater affinity, and therefore will have a 'stronger' inhibition effect on the kinetics.

In this case, $K_I/K_m = 5$, 1, and 0.2 for A, B and C, respectively. Thus, as was the case in (a), **Inhibitor C** binds with greatest affinity and will have a greater effect on the enzyme kinetics than A or B.

Problem 4.3 D-(-)-hydroxphenylglycin is the optically active intermediate in the synthesis of the broad-spectrum antibiotics amoxicillin. This intermediate is among others produced from a hydantoin derivative which is poorly soluble in water (about 1 kg m^{-3}). The price of the substrate is cost determining and the degree of conversion should therefore be very high; at least 99%.

Calculate the volume needed to produced 100 kg of product per day by hydantoinase in

(a) Batch
(b) CSTR and
(c) Plug flow reactor

The following data is available: $v_{max} = 1.5$ kg $min^{-1} m^{-3}$; $K_m = 5$ kg m^{-3}; $Y_{p/s} = 1$ kg kg^{-1} degree of conversion = 99%, Down time = 6 h; The enzyme activity is assumed to be constant with respect to time.

SOLUTION

The given data:

$$v_{max} = 1.5 \text{ kg min}^{-1}\text{m}^{-3}$$

$$K_m = 5 \text{ kg m}^{-3}$$

$$Y_{P/S} = 1 \text{ kg kg}^{-1}$$

$$S_0 = 1\, kg\, m^{-3}$$

Degree of conversion = 99%

Down time = 6 h

(a) Batch Process

From Eq. 3.57, we can write

$$t_b = -\int_{S_0}^{S} \frac{dS}{-r_S} = -\int_{S_0}^{S} \frac{dS}{\frac{v_{\max} S}{K_m + S}} = -\int_{S_0}^{S} \frac{K_m}{v_{max}} \frac{dS}{S} - \int_{S_0}^{S} \frac{1}{v_{max}} dS = \frac{K_m}{v_{max}} \ln \frac{S_0}{S} + \frac{1}{v_{max}}(S_0 - S) \tag{4.36}$$

In the given problem, S = (1– 0.99) 1 $kg\ m^{-3}$ = 0.01 $kg\ m^{-3}$

$$\text{Therefore, } t_b = \frac{5\ \text{kg m}^{-3}}{1.5\ \text{kg min}^{-1}\text{m}^{-3}} \ln \frac{1\ kg\,m^{-3}}{0.01\ kg\,m^{-3}} = 3.33 \times 4.6\ \text{s} = 15.33\ \text{min}$$

Total time required for the batch process = $t_b + t_{down\ time}$

$$= 15.33\ \text{min} + 6 \times 60\ \text{min} = 375.33\ \text{min}$$

$$\text{Number of batches per day} = \frac{24 \times 60}{375.33} = 3.84$$

$$\text{Therefore, amount of product produces/ batch} = \frac{100\,kg}{3.84} = 26.04\ \text{kg}$$

Since $Y_{P/S}$ = kg kg^{-1}
Amount of substrate requires/ batch = 26.04 kg
∴ Volume of the batch reactor = 0.01 $kg\,m^{-3}$ = **26.04 m^3**

(b) CSTR

$$\text{From Eq. 3.66, we get } \tau_{\text{CSTR}} = \frac{\text{V}}{\text{v}} = \frac{(S_0 - S)}{-\text{r}_\text{S}} = \frac{(S_0 - S)}{\text{v}} = \frac{(S_0 - S)}{\dfrac{v_{\max S}}{K_m + S}}$$

Under steady state conditions, S = 0.01 $kg\ m^{-3}$

$$\tau_{\text{CSTR}} = \frac{(1 - 0.01)\ \text{kg m}^{-3}}{\dfrac{1.5\ \text{kg min}^{-1}\text{m}^{-3} \times 0.01\ \text{kg} + \text{m}^{-3}}{(5\ \text{kg m}^{-3} + 0.01\ \text{kg m}^{-3})}} = 330\ \text{min}$$

Amount of product formation $= 100\,kg\,d^{-1}$
Since $Y_{P/S} = 1$ kg kg^{-1}
Amount of substrate required ≈ $100\,kg\,d^{-1}$

$$\text{Volumetric flow rate} = \frac{\textit{total amount of substrate requires per day}}{\textit{initial concentration of substrate}}$$

$$= \frac{100\,kg\,d^{-1}}{1\,kg\,m^{-3}} = \frac{100\,m^3}{24 \times 60\ \text{min}} = 0.069\,m^3\,min^{-1}$$

Volume of the reactor = 0.069 $m^3\ min^{-1} \times 330$ min = **22.77m^3**

(c) Plug Flow Reactor (PFR)

From Eqs. 3.73 and 4.34, we get

$$\tau_{PFR} = \int_{S_0}^{S} \frac{dS}{-r_s} = \frac{K_m}{v_{max}} \ln \frac{S_0}{S} + \frac{1}{v_{max}} (S_0 - S) = 15.33 \text{ min}$$

Volume of the reactor = 0.069 $m^3\ min^{-1}$ × 15.33 min = **1.06 m^3**

∴ Volume of the PFR reactor is less compared to CSTR and batch reactor.

4.10 IMMOBILIZED ENZYMES

Immobilization of an enzyme means that it has been confined or localized in a solid matrix so that it can be reused continuously. Immobilization increases stabilization of the enzymes. Classification of the immobilization of enzymes is based on the following parameters:

- The type of interaction between enzyme and solid matrix
- The characteristics of the solid matrix
- The nature of the enzyme and solid matrix complex

In 1969, the first commercial application of immobilized enzyme technology took place in Japan by using amino acylase obtained from *Aspergillus oryzae* for the industrial production of L-amino acids. Consequently, immobilized penicillin acylase (PA) was used for the conversion of penicillin G and cephalosporin to 6-aminopenicillanic acid (6-APA) and 7-aminocephalosporanic acid (7-ACA), respectively. Another example is the conversion of glucose to fructose by immobilized glucose isomerase.

Advantages of immobilized enzyme as compared to the free enzyme are given below.

- Stable and more efficient in function.
- Can be reused again and again.
- Products are enzyme free.
- Control of enzyme function is easy.
- Suitable for industrial and medical use.
- Minimize effluent disposal problems.

4.10.1 Activity of the Immobilized Enzymes

Activity of the immobilized enzymes is expressed in international units (IU). IU defines as micromoles substrate degraded per min per g immobilized enzyme or per unit surface area of the solid matrices. The immobilized enzymes activities are sensitive to environmental conditions such as

- Initial substrate concentration,
- Concentration immobilized enzymes,

- Temperature,
- pH,
- Reaction time,
- Ionic strength,
- Buffer concentration,
- Agitation or flow rate, and
- Physical dimensions of the carriers.

Bound protein indicates the amount of protein fixed or immobilized on the solid matrix. This is also important for studying the kinetic behaviour of the complex. The bound protein unit is expressed as mg protein per g of carrier.

This specific activity of bound protein is expressed as micromole substrates degraded per min per mg bound protein. The effectiveness of the immobilization technique or coupling efficiency is determined by comparing overall activity of the immobilized enzymes and that of the soluble enzyme before immobilization (Messing, 1975; Pitcher, 1980; Bernath et al., 2014).

Coupling efficiency is defined as follows

$$Coupling\,efficiency(\%) = \frac{Overall\,activity\,of\,the\,immobilized\,enzyme}{Overall\,activity\,of\,the\,initial\,enzyme\,solution} \times 100 \quad (4.37)$$

Solid matrices play a very important role in the preparation of immobilized enzymes. Properties of solid matrices are as follows:

- Inert
- Physically strong and stable
- Should be cheap enough to discard
- Regeneration characteristics reduce the cost of the process
- Higher surface area

4.10.2 Immobilization Techniques

There are several immobilization methods available. The classification of the immobilization methods is shown in Figure 4.14.

Different immobilization techniques have already been reported (Das & Das, 2019). The selection of the immobilization technique for the enzymes depends on selection criteria such as binding force, cost, mode of preparation, enzyme leakage, applicability, operation problems, microbial protection, diffusional problems etc. (Das and Das, 2019).

4.10.3 Immobilized Enzymatic Reaction Kinetics

Immobilized enzymatic reaction is heterogeneous where mass transfer resistance is to be introduced which is absent in a free solution enzymes system. Mass transfer

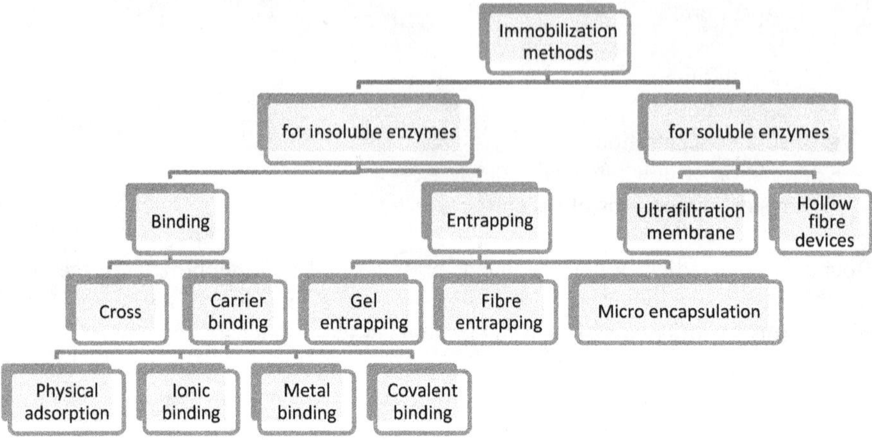

FIGURE 4.14 Immobilization techniques classification for enzymes (Das & Das, 2019).

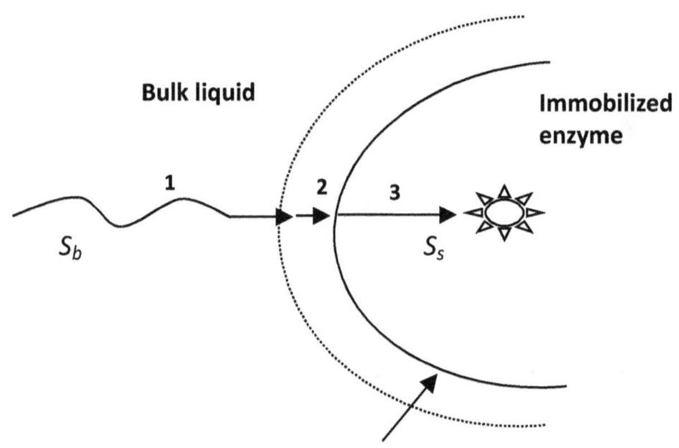

FIGURE 4.15 Diffusional path of substrate from bulk solution to immobilized enzyme surface. (Steps 1 and 2 are the external mass transfer resistance. Step 3 is the intra-particle mass transfer resistance. S_b and S_s are the substrate concentration at bulk and surface of the solid matrix.)

resistance occurs due to the large particle size of immobilized enzymes or due to the inclusion of enzymes in the polymeric solid matrix.

The substrate diffusion path from liquid to the reaction site in an immobilized enzyme is shown in Figure 4.15.

- Diffusion of substrate present in the liquid from the bulk (S_b) to a relatively unmixed liquid layer surrounding the immobilized enzymes
- Diffusion of substrate through the relatively unmixed liquid layer
- Diffusion of substrate from the surface of the particle to the active site of the enzyme present in the inner layer

The rate of mass transfer can be expressed by N_s equal to $k_L a(S_b - S_s)$. This $k_L a$ is taken as a volumetric mass transfer coefficient. Moreover, S_b is the concentration of substrate in the bulk and S_s is the concentration at the surface of the solid metric. The dragging force is $(S_b - S_s)$. Under steady state conditions

Rate of mass transfer = Rate of reaction

$$\therefore\ k_L a(S_b - S_s) = (-r_S) = \frac{v_{max} S_s}{K_m + S_s} \tag{4.38}$$

The above equation can be expressed in dimensionless form:

$$\frac{(S_b - S_s)}{\dfrac{v_{max}}{k_L a}} = \frac{S_s}{K_m + S_s} \tag{4.39}$$

Dividing by S_b, we get

$$\frac{\left(\dfrac{S_b}{S_b} - \dfrac{S_s}{S_b}\right)}{\dfrac{v_{max} S_b}{k_L a}} = \frac{\dfrac{S_s}{S_b}}{\dfrac{K_m}{S_b} + \dfrac{S_s}{S_b}} \tag{4.40}$$

Assuming $\dfrac{S_s}{S_b} = x_S$ and $\dfrac{S_b}{K_m} = \beta$

$$\frac{(1 - x_S)}{N_{Da}} = \frac{\beta x_S}{1 + \beta x_S} \tag{4.41}$$

N_{Da} is known as the Damkohler number. It is equal to the maximum velocity of reaction divided by the maximum mass transfer. If S_s. is equal to 0, the dragging force or mass transfer will be maximum.

If the Damkohler number is lesser than 1, the overall reaction will be controlled by the enzymatic reaction because the rate of mass transfer is much greater than the rate of reaction

$$(-r_S) = \frac{v_{max} S_b}{K_m + S_b} \tag{4.42}$$

On the other hand, if the Damkohler number is greater than 1, the overall reaction will be controlled by the rate of mass transfer because the rate of reaction is much greater than the rate of mass transfer.

$$(-r_S) = k_L a \; S_b \tag{4.43}$$

There is a parameter called the effectiveness factor (η) for the measurement of the extent to which the reaction rate is lowered because of mass transfer resistance. The effectiveness factor of an immobilized enzyme can be represented as

$$\eta = \frac{\text{actual velocity of reaction}}{\text{velocity of the reaction when there is no mass transfer limitation i.e. } S_s = S_b} \tag{4.44}$$

$$\eta = \frac{\dfrac{v_{max} S_s}{K_m + S_s}}{\dfrac{v_{max} S_b}{K_m + S_b}} = \frac{\dfrac{\beta x_S}{1+\beta x_S}}{\dfrac{\beta}{1+\beta}} \tag{4.45}$$

If $x_S = 1$, the concentration of substrate at the surface of the solid matrix is equal to the bulk substrate concentration. This indicates that there is no mass transfer limitation. However, if $x_S << 1$, then mass transfer is very slow as compared to the reaction rate (Bailey and Ollis, 2010; Doran, 2012; Das and Das, 2019; 2020; Kennedy et al. 1985).

4.10.4 Applications of Immobilized Enzymes

Immobilized enzyme systems are largely used in industry for the production of several products as mentioned below:

- Industrial applications: antibiotics, beverages, amino acids etc.
- Biomedical applications: treatment, diagnosis and drug delivery.
- Food industries: production of high-fructose corn syrups (HFCS).
- Textile industries: scouring, bio-polishing.
- Detergent industries: immobilization of lipase/protease for effective dirt removal.

For example large-volume high-fructose corn syrup (HFCS) production from glucose using immobilized glucose isomerase.

$$\text{Corn starch} \xrightarrow{\text{Amylases}} \text{Glucose} \xrightarrow{\text{Immobilized glucose isomerase}} \text{Fructose}$$

Fructose is 10 times sweeter than glucose. In 1976, USA alone produced more than a billion pounds of HFCS (dry basis, containing approximately 42% fructose).

Immobilized penicillin acylase is used for the hydrolysis of penicillin G to 6-aminopenicillanic acid (6-APA) which is used commercially in Europe. Other examples of the application of immobilized enzymes are given below.

1. Use of fibre-entrapped lactase for the hydrolysis of whole milk lactose.
2. Saccharification of soluble starch by immobilized glucoamylase.
3. Immobilized β-galactosidase for the hydrolysis of cheese whey lactose.
4. Production of chill proof beer.
5. Steroid transformation.
6. Residual oxygen removal from various food products.

Immobilized enzymes can be used for analytical purposes, e.g. glucose sensors for the detection of glucose; alcohol sensors for the detection of alcohol etc.

Immobilized enzymes are widely used in the pharmaceutical industries; for examples for clinical analysis, therapeutic use, preventative and environmental medicine and biochemical and biophysical research, e.g. treatment of enzyme-sensitive disorders including hereditary enzyme defects can be overcome by using immobilized enzymes (Messing, 1975).

Problem 4.4 Immobilized glucose isomerase is used for the conversion of glucose to fructose. What is the height of the immobilized enzymes column? The following data is available:

Diameter of the column (D_T) = 5 *cm*
Particle size 30/40 mesh (about 0.71 mm average diameter, d_p),
Feed rate (F)= 500 mL/h
Glucose concentration in feed at 60°C = 500 g/L,
Glucose conversion efficiency = 60%,
Viscosity of the liquid (μ) = 3.6 c.p. at 60°C
Density of the liquid (ρ) = 1.23 g/mL at 60°C,
Substrate diffusivity (D) = 0.21 x 10^{-5} cm^2/ sec at 60°C
Void fraction (ε) = 0.35

SOLUTION

Satterfield has suggested an expression for column height as follows

$$Z = \frac{\varepsilon(Re)^{\frac{2}{3}}(Sc)^{\frac{2}{3}}}{1.09a_v}\ln\left(\frac{Y_1}{Y_2}\right)$$

Where Z = height of the column, ε = void fraction, a_v = ratio of the particle surface area to volume, Y_2 = mole fraction of substrate in the outflow stream, Y_1 = mole fraction of substrate feed, *Re*= Reynolds number.

Again

$$Re = \frac{D_T u\rho}{\mu}$$

where

$$u = \frac{Volumetric\ Feed\ flow\ rate}{crosssectional\ Area\ of\ the\ column} = \text{Velocity of the fluid,}$$

D_T = Diameter of the column, ρ = Density of the liquid,
μ = Viscosity of the liquid.

$$Volumetric\ liquid\ \ flow\ rate = 500\,\frac{\text{mL}}{\text{h}} = 0.139\,\frac{\text{mL}}{\text{S}}$$

$$Crosssectional\ area\ of\ the\ column = \pi\left(\frac{D_T}{2}\right)^2 = \pi(2.5)^2\ cm^2$$

$$u = \frac{0.139\frac{mL}{S}}{\pi(2.5)^2\ cm^2} = 7.079\times10^{-3}\ {cm}/{s}$$

$$Re = \frac{(5)(7.079\times10^{-3})(1.23)}{3.6} = 0.0121$$

$$Sc = Schmidt\ No = \frac{\mu}{D\rho} = \frac{3.6}{0.21\times10^{-5}\times1.23} = 1.3937\times10^{6}$$

$$a_v = \frac{surface\ area\ of\ paricle}{volume\ of\ particle} = \frac{4\pi\left(\frac{d_p}{2}\right)^2}{\frac{4}{3}\pi\left(\frac{d_p}{2}\right)^3} = \frac{6}{d_p} = \frac{6}{0.71\ mm} = \frac{6}{0.071\ cm} = 84.50\ cm^{-1}$$

$$Y_1 = \frac{mole\ of\ glucose}{total\ moles\ of\ all\ constituents} = \frac{\frac{500}{180}}{\frac{500}{180}} = 1$$

$$Y_2 = \frac{mole\,of\,glucose}{total\,moles\,of\,all\,constituents} = \frac{\frac{500\times(1-0.60)}{180}}{\frac{500}{180}} = \frac{\frac{500\times0.40}{180}}{\frac{500}{180}} = 0.40$$

Putting all the known values into the proposed equation

$$Z = \frac{0.35(0.0121)^{\frac{2}{3}}(1.3937\times10^{6})^{\frac{2}{3}}}{1.09\times84.50\,cm^{-1}}\ln\left(\frac{1}{0.40}\right) = 2.29\,\boldsymbol{cm}$$

∴ Height of the column is 2.29 cm.

4.11 CONCLUSIONS

Enzymes are considered to be biocatalysts which are very specific with respect to the substrate. Enzymes are mostly proteins with active sites. The kinetics of the enzymatic reaction play an important role in biochemical processes. Most of the biochemical reactions are reversible in nature. The values of the kinetic constants are useful for designing the biochemical reactor. A major drawback of the enzymatic reaction process is the removal of enzymes after the completion of the process for the purification of the product. Enzymes are very costly. Therefore, it is necessary to find suitable devices such as immobilized enzymes in order to reuse the enzyme again. Immobilized enzymes are largely used in industry for the production of useful products, for analytical purposes and for medicinal uses. A major bottleneck of immobilized enzymatic reactions is the diffusion problem. This problem must be overcome for the maximization of product formation.

REFERENCES

Bailey JE and Ollis DF, Biochemical Engineering Fundamentals (2nd Edition), McGraw Hill Book Company, 2010.

Bernath FR, Venkat Subramanian K and Veith WL, "Immolozed Enzymes", Annual reports on Fermentation Processes, Vol. 1, (Ed. Perlman D), Academic Press, 2014.

Das D and Das D, Biochemical Engineering: An Introductory Text Book, Jenney Stanford, Singapore, 2019

Das D and Das D, Biochemical Engineering: A Laboratory Manual, Jenney Stanford, Singapore, 2020

Doran PM, Bioprocess Engineering Principles, Second Edition, Academic press, Waltham, 2012.

Kennedy JF et al., "Principles of Immobilized Enzymes", Handbook of Enzyme Biotechnology, Wiseman A (Ed.), 1985.

Messing RA, Immobilized Enzymes for Industrial reactors, Academic Press, New York, 1975.

Pitcher WH, Immobilized Enzymes for Food Processing, CRC Press, 1980.

5 Microbial Growth Kinetics, Product Formation and Substrate Degradation

Microbial reaction is an example of autocatalytic reaction because microbial cells grow in the medium using substrate to produce cell mass and products. Therefore, the overall reaction may written as follows

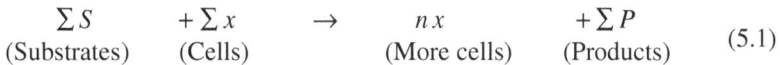

$$\underset{\text{(Substrates)}}{\sum S} \quad \underset{\text{(Cells)}}{+\sum x} \quad \rightarrow \quad \underset{\text{(More cells)}}{n x} \quad \underset{\text{(Products)}}{+\sum P} \tag{5.1}$$

Microbial cells are a living system. The living system has several special characteristics as shown below:

- Reproduction
- Sensitive to the environment
- Acclimatization

Reproduction of the cells is expressed in terms of either doubling time or generation time. The specific growth rate of the cell (μ) is represented by Eq. 5.1.

$$\mu = \frac{1}{x}\frac{dx}{dt} \tag{5.2}$$

Where x is the cell concentration and t denotes the time of growth of the cells. The concentration of cells may be expressed either as mass per unit volume or number of cells per unit volume in the case of unicellular cells like *E. coli*. However, in the case of filamentous cells such as *A. niger*, the cell concentration is expressed only in mass per unit volume.

$$\text{Again,} \int_{x_0}^{x} \frac{dx}{x} = \mu \int_0^t dt \tag{5.3}$$

Where x_0 = initial cell concentration and x = cell mass concentration after time t.

The doubling time (t_d) is defined as the time required for getting double the cell mass in the growth medium. Therefore, this can be mathematically represented as Eq. 5.5.

$$\int_{x_0}^{2x_0} \frac{dx}{x} = \mu \int_0^{t_d} dt \tag{5.4}$$

$$t_d = \frac{\ln 2}{\mu} \tag{5.5}$$

Generation time is the time required for cell division. If x_0 cell divided into x_n number of cells, then the generation time may be expressed as Eq. 5.7.

$$\int_{x_0}^{x_n} \frac{dx}{x} = \mu \int_0^{t_{gn}} dt \tag{5.6}$$

$$t_{gn} = \frac{\ln \frac{x_n}{x_0}}{\mu} \tag{5.7}$$

In each cell division if each cell divides into two, then

$$t_{gn} = t_d \tag{5.8}$$

The living cells are very sensitive to the environmental conditions such as medium composition, pH, temperature, agitator speed etc. The medium of the cells mainly comprises carbon sources, nitrogen sources, minerals and vitamins. Carbon sources are mostly contributed by the cells, products and energy molecules (ATP, NADH, FADH) formation. Nitrogen sources mostly contribute to cell formation and minerals and vitamins are the cofactors of the enzymatic reactions in the metabolic pathways.

Another important property of living cells is acclimatization. *Pseudomonas putida* is suitable for the degradation of toxic chemicals like phenol, hydrocarbons etc. present in wastewater. Phenol has germicidal properties. Therefore phenol at higher concentration will inhibit the growth of *P. putida*. Now, if the cell is allowed to grow at low phenol concentration followed by slowly increasing the concentration, then the cells will acquire the ability to withstand higher concentrations of phenol (Pelczar et al., 1998).

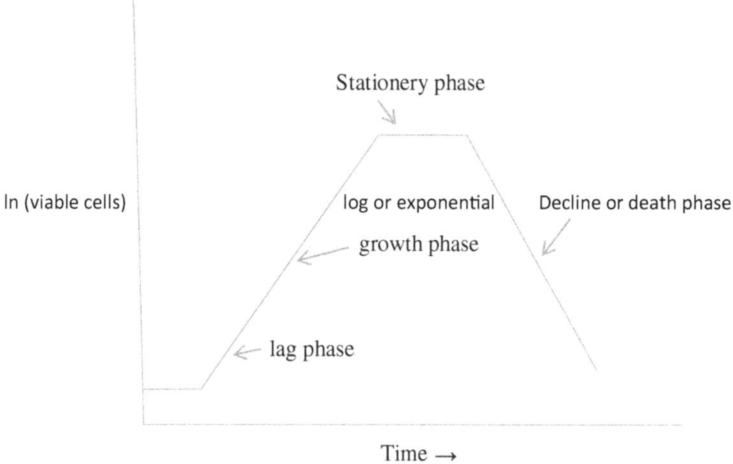

FIGURE 5.1 Cell growth cycle in a batch process.

5.1 LIFE CYCLE OF THE MICROBIAL CELL

To handle the new microorganism it is very much essential to know the growth cycle of the cell to find out the status of the microorganism (Figure 5.1). For example, the growth cycle of the cell comprises four distinct phases; lag, log, stationery and decline phases. Therefore we should know their status: then and only then we can handle the microorganism in a proper way. In the life cycle of the cell, the lag phase is considered to be an acclimatization phase because every organism requires some time to adjust to the new environment. Therefore this is usually considered as a non-productive phase. The length of this phase should be as small as possible. In the case of *Saccharomyces cerevisiae* the lag phase will be reduced significantly in the presence of biotin in the medium, which is a very important characteristic of the microbial cell. If the lag time is greater, then the total fermentation time will increase which is undesirable in the industrial fermentation process. The log phase or exponential growth phase plays a very important role in the preparation of the inoculum because cells remain active at this phase. Primary metabolites formation takes place in this phase; e.g. the ethanol fermentation process.

Usually the inoculation of the organism is carried out in between the mid log phase and late log phase. The stationary phase signifies the starvation phase. Secondary metabolite formation takes place in this phase, e.g. the penicillin fermentation process. In this phase the rate of growth of the cell is equal to the rate of death of the cells. The cell growth cycle of a microbial cell can be prepared only when the cells grow in a batch process. If we extend the stationary phase then we will get more secondary metabolites. This is done by adding a little nutrient to the fermentation media so that the stationary phase may be extended. In the case of secondary metabolite formation like penicillin, this phase is extended by feeding a small amount of nutrient into the medium in the stationery phase (Doran, 2012; Shular and Kargi, 2002).

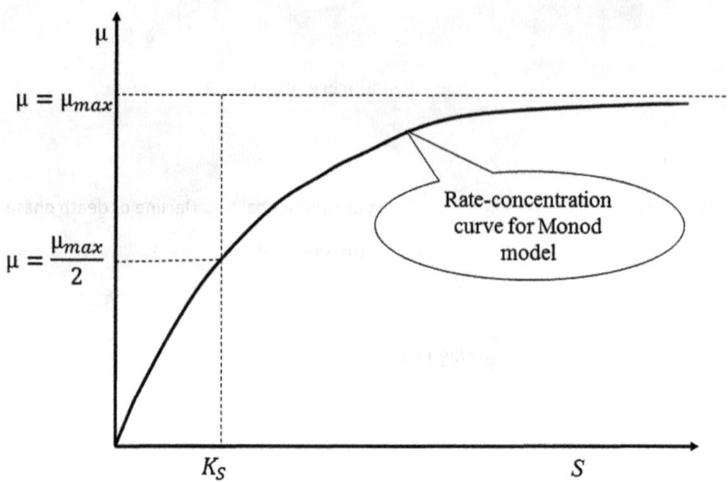

FIGURE 5.2 Correlation between specific growth rate of the cell and concentration of the growth limiting substrate.

5.2 CELL GROWTH KINETICS

Monod proposed an equation (Eq. 5.9) to correlate between specific growth rate of the cell and concentration of the growth limiting substrate homologous to the M–M equation. A graphical representation of this correlation is shown in Figure 5.2.

$$\mu = \frac{\mu_{max} S}{K_S + S} \tag{5.9}$$

where μ_{max} = maximum specific growth rate, K_S = saturation constant, S = concentration of growth limiting substrate, and μ = specific growth rate.

The M-M equation deals with the correlation between the velocity of reaction and the substrate concentration. In the case of enzymatic reactions, the substrate is very specific, e.g. glucose isomerase acts only on glucose to produce fructose. In the case of cell growth medium, there are several components present (Bailey and Ollis, 2010; Aiba et al., 1973; Harzevili and Chen, 2015). If the specific cell growth rate changes with respect to any component present in the medium keeping the concentration of the other component above the desired level, it is known as a growth-limiting substrate. Higher values of K_S indicate a high amount of substrate to get the desired amount of cells. However, the Monod equation has the following limitations

- If S is finite then μ is finite
- When $S \rightarrow \infty$, $\mu \rightarrow \mu_{max}$
- It does not explain what will happen in case of $S \rightarrow 0$
- Does not take account of the death of the cell
- Does not apply in the case of substrate and product inhibitions

5.2.1 Batch Cell Growth Kinetics

Batch processes are considered to be an unsteady state operation where composition of the medium changes with time. Cell mass balance may be expressed as follows:

Rate of cell input + Rate cell generation = Rate of cell output + Rate of cell death + Rate of cell accumulation (5.10)

$$0 + r_x V = 0 + r_d V + \frac{d(xV)}{dt} \tag{5.11}$$

Where r_x = rate of cell generation, r_d = rate of cell death, V = volume of the medium, x = cell mass concentration. When V is constant, then we can write Eq. 5.12.

$$\frac{dx}{dt} = (\mu - \mu_d)x \tag{5.12}$$

Where μ_d = specific death rate of the cell.

If the specific death rate of the cell is negligible as compared to cell growth, then we can write

$$\frac{dx}{dt} = \mu x \tag{5.13}$$

$$\frac{dx}{x} = \mu\, dt \tag{5.14}$$

Assuming μ remains constant

$$\int_{x_0}^{x} \frac{dx}{x} = \mu \int_0^t dt \tag{5.15}$$

$$\ln \frac{x}{x_0} = \mu t \tag{5.16}$$

$$t = \frac{\ln \frac{x}{x_0}}{\mu} \tag{5.17}$$

Eq. 5.17 finds the time required to increase the cell mass concentration from x_0 to x. Again, Eq. 5.17 may be modified to find the cell concentration after time t as shown in Eq. 5.18.

$$x = x_0\, e^{\mu t} \tag{5.18}$$

Specific substrate consumption rate may be defined as the rate of substrate consumption per unit cell mass concentration as shown in Eq. 5.19.

$$q_S = \frac{1}{x}\frac{ds}{dt} \tag{5.19}$$

Similarly, specific product formation rate can be expressed as shown in Eq. 5. 20.

$$q_P = \frac{1}{x}\frac{dP}{dt} \tag{5.20}$$

The cell yield coefficient is defined as the amount of cells produced per unit mass of substrate consumed (Eq. 5.21).

$$Y_{x/S} = \frac{mass\ of\ cell\ produced}{mass\ of\ substrate\ consumed} = \frac{(x - x_0)}{(S_0 - S)} \tag{5.21}$$

Similarly, product yield coefficient may be expressed as shown in Eq. 5.22.

$$Y_{P/S} = \frac{mass\ of\ product\ produced}{mass\ of\ substrate\ consumed} = \frac{(P - P_0)}{(S_0 - S)} \tag{5.22}$$

Eq. 5.13 may modified by using the Monod equation as shown in Eq. 5.23.

$$\frac{dx}{dt} = \mu x \tag{5.23}$$

$$\text{Again,} -\frac{dS}{dt} = -\frac{dS}{dx}\frac{dx}{dt} = -\frac{1}{Y_{x/S}}\frac{\mu_{max} S}{K_S + S}x = -r_S \tag{5.24}$$

Where $-r_S$ is the rate of substrate degradation.

5.2.2 Determination of the Kinetic Constants

Eq. 5.9 may be written as follows

$$\frac{1}{\mu} = \frac{K_S}{\mu_{max}}\frac{1}{S} + \frac{1}{\mu_{max}} \tag{5.25}$$

From the plot $\frac{1}{\mu}$ vs. $\frac{1}{S}$, the cell growth kinetic constants can be found as shown in Figure 5.3.

Unit of μ_{max} ans K_S are $time^{-1}$ and $mass\ volume^{-1}$, respectively.

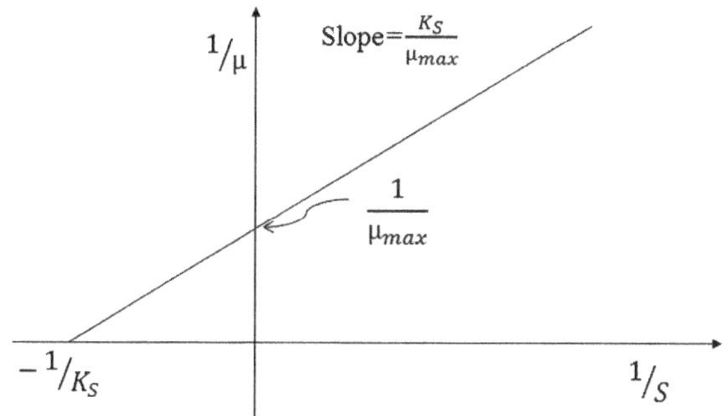

FIGURE 5.3 Plot of $\frac{1}{\mu}$ vs. $\frac{1}{S}$

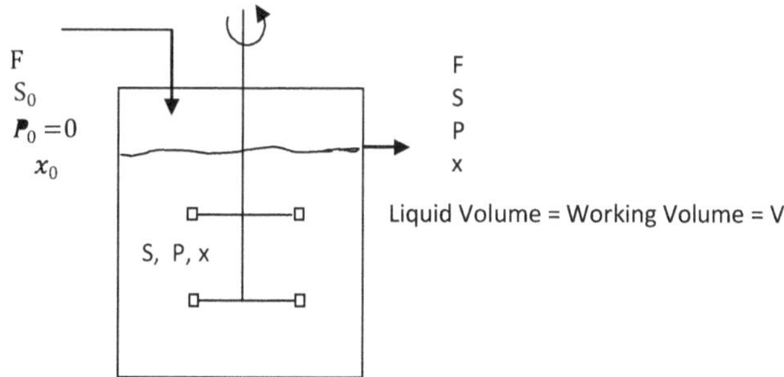

FIGURE 5.4 Schematic diagram of a chemostat.

5.3 CHEMOSTAT

A chemostat is similar to a continuous stirred tank reactor (CSTR) where the chemical composition of the living microbial system in a bioreactor is kept constant under steady state conditions (Figure 5.4). Fresh medium is continuously added to the bioreactor, while culture liquids containing left-over nutrients, metabolic end products and microorganisms are continuously removed at the same rate to keep the culture volume constant. Chemostats have a number of advantages as compared to batch cell growth and metabolism (Lydersen et al., 1994; Lim and Shin, 2013; Blanch and Clark, 1997):

- It is possible to maintain cells at steady conditions for infinite period of time, where cell growth, the pattern of metabolism and the medium composition remain invariant with time. This enables detailed study of the nature of cell growth and metabolism at specific growth rates which would not otherwise be possible under the transient conditions of batch culture.
- The effect of a limiting nutrient on the cell growth can be studied.
- To find out the effect of growth limiting substrate on the cell morphology and metabolism of the living cells.

- The chemostat can be used to study the period of unbalanced growth which occurs during the transition period between steady states at different specific growth rates. It has been found that some plant cells can produce some special metabolites during the transient phases of the cell. Therefore, chemostats are found suitable for this purpose.

Cell balance of the chemostat can be achieved with the help of Eq. 5.10 as follows.

$$F x_0 + r_x V = F x + r_d V + \frac{d(xV)}{dt} \tag{5.26}$$

Under steady-state condition, the rate of cell accumulation = $\frac{d(xV)}{dt} = 0$

Again, in the case of sterile feed, $x_0 = 0$ and assuming no death of the cell, Eq. 5.26 may be modified as

$$F x = r_x V \tag{5.27}$$

$$\frac{F}{D} = \frac{1}{x} r_x = \mu \tag{5.28}$$

$$D = \mu = \frac{\mu_{max} S}{K_S + S}. \tag{5.29}$$

Where D is known as the dilution rate. Therefore, under steady-state conditions, sterile feed and no death of the cell, the dilution rate is equal to the specific growth rate. The main drawback of the batch process is the change of the cell growth phases with respect to time. Maximum growth rate of the cells takes place at the log phase. Therefore, this phase cannot be operated for longer periods of time in a batch process. However, in the case of chemostats, this lag phase can be operated for an infinite period of time by controlling the dilution rate (Volesky and Votruba, 1992; Liu, 2013; Villadsen et al., 2009).

From Eq. 5.29

$$S = \frac{D K_S}{(\mu_{max} - D)} = Steady\ state\ substrate\ concentration \tag{5.30}$$

Again, from Eq. 5.21

$$x = x_0 + Y_{x/S}\left(S_0 - S\right) = x_0 + Y_{x/S}\left(S_0 - \frac{D K_S}{(\mu_{max} - D)}\right) \tag{5.31}$$

In Eq. 5.31, x is the steady-state cell mass concentration.

Again the substrate balance equation of the chemostat can be written as

Rate of substrate input + Rate substrate generation = Rate of substrate output + Rate of substrate disappearance + Rate of substrate accumulation (5.32)

$$F\,S_0 + 0 = FS + (-r_s)V + \frac{d(VS)}{dt} \tag{5.33}$$

Under steady-state conditions, the rate of consumption of the substrate = 0

$$\therefore F(S_0 - S) = (-r_s)V \tag{5.34}$$

$$\frac{F}{V}(S_0 - S) = (-r_s) = \frac{dS}{dx}\,\frac{dx}{dt} = \frac{1}{Y_{x/S}}\,\mu\,x \tag{5.35}$$

$$D(S_0 - S) = \frac{1}{Y_{x/S}}\mu x \tag{5.36}$$

Eq. 5.30 is known as the chemostat model. Again Eq. 5.30 can be modified as

$$D(S_0 - S) = \frac{1}{Y_{x/S}}\,\frac{\mu_{max}\,S}{K_S + S}\,x \tag{5.37}$$

Eq. 5.31 is known as the Monod chemostat model.

Again, the cell productivity = $\dfrac{cell\,mass\;produced}{time} = \dfrac{dx}{dt} = \mu\,x = D\,x$

$$= D\,Y_{x/S}\left(S_0 - \frac{D\,K_S}{(\mu_{max} - D)}\right) = Y_{x/S}\left(D\,S_0 - \frac{D^2\,K_S}{(\mu_{max} - D)}\right) \tag{5.38}$$

At maximum dilution rate, D_{max} the rate of cell mass production will be maximum i.e.

$$\frac{d(D\,x)}{dD} = 0 \tag{5.39}$$

Now, by differentiating Eq. 5.38 with respect to D, we get

$$\frac{d(D\,x)}{dD} = Y_{x/S}\left(S_0 - \frac{2D\;K_S}{(\mu_{max} - D)} - \frac{D^2\,K_S}{(\mu_{max} - D)^2}\right) = 0 \tag{5.40}$$

$$\frac{S_0}{K_S} = \left(\frac{2D}{(\mu_{max} - D)} + \frac{D^2}{(\mu_{max} - D)^2} \right) \tag{5.41}$$

$$\frac{S_0}{K_S} + 1 = \left(1 + \frac{2D}{(\mu_{max} - D)} + \frac{D^2}{(\mu_{max} - D)^2} \right) \tag{5.42}$$

$$\frac{S_0 + K_S}{K_S} = \left(1 + \frac{D}{(\mu_{max} - D)} \right)^2 \tag{5.43}$$

$$\therefore D_{max} = \mu_{max} \; (1 \pm \frac{K_S}{K_S + S_0}) \tag{5.44}$$

The variations of the steady state cell mass productivity, substrate and cell mass concentration in a chemostat with the increase of the dilution rate are shown in Figure 5.5. $D_{wash\ out}$ is the dilution rate when the bioreactor does not contain any cells ($x = 0$). The corresponding steady-state substrate concentration will be S_0 because no reaction will take place in the bioreactor. Therefore, $D_{wash\ out}$ may be expressed as follows.

$$D_{wash\ out} = \frac{\mu_{max} \; S_0}{K_S + S_0} \tag{5.45}$$

$$\text{Again, Hydraulic retention time} = \frac{1}{D} \tag{5.46}$$

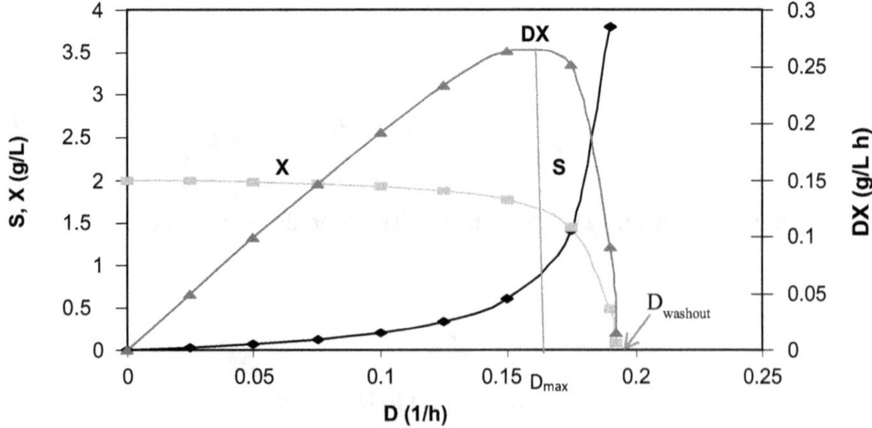

FIGURE 5.5 Effect of dilution rates on the cell mass concentration (x), substrate concentration (S) and cell mass productivity (Dx).

The reasons for the $D_{wash\ out}$ condition are given below

- If the hydraulic retention time < generation time of the cell.
- A major drawback of the chemostat is cells wasting from the bioreactor. Therefore, the cell wash out condition arises in the case where rate of cell wasting is more than cell growth in the bioreactor at infinite time.

The relationship among D_{max}, $D_{wash\ out}$, and μ_{max} in a chemostat for the cultivation of a cell is given below:

$$D_{max} < D_{wash\ out} < \mu_{max}$$

Therefore, Eq. 4.44 may be written as

$$D_{max} = \mu_{max}\left(1 - \frac{K_S}{K_S + S_0}\right) \tag{5.47}$$

From Eq. 5.38, we can find the expression for maximum cell mass productivity (Eq. 5.47).

$$\text{Maximum cell mass productivity} = D_{max}\, x = Y_{x/S}\left(D_{max}\, S_0 - \frac{D_{max}{}^2\, K_S}{(\mu_{max} - D_{max})}\right) \tag{5.48}$$

From Eq. 5.31, the maximum cell mass concentration using sterile feed may be expressed as follows

$$x_{max} = Y_{x/S}\left(S_0 - \frac{D_{max}\, K_S}{(\mu_{max} - D_{max})}\right) \tag{5.49}$$

$$x_{max} = Y_{x/S}\left(S_0 - \frac{K_S}{\left(\frac{\mu_{max}}{D_{max}} - 1\right)}\right) \tag{5.50}$$

$$x_{max} = Y_{x/S}\left(S_0 - \frac{K_S}{\frac{\mu_{max}}{\mu_{max}\left(1 - \frac{K_S}{K_S + S_0}\right)} - 1)}\right) \tag{5.51}$$

$$x_{max} = Y_{x/S}\left(S_0 - \sqrt{K_S(K_S + S_0)} + K_S\right) \tag{5.52}$$

Eq. 5.52 is considered as the determination of maximum cell mass concentration in a chemostat. A drawback of the chemostat is usually that the D_{max} value is very close to $D_{wash\ out}$, so it is not practicable to operate the chemostat at D_{max}. Small variations of dilution rate in this region may cause large fluctuations of the values of x and S, unless the flow rate of the feed to the bioreactor is controlled precisely. The $D_{wash\ out}$ problem of a chemostat can be avoided in two ways:

- Chemostat with cell recycling
- Using an immobilized whole cell system

It is possible to maintain the same concentration of cells in the bioreactor by recycling excess cells which are exiting the bioreactor. On the other hand, in the case of immobilization of whole cells, it is easily possible to avoid the cell wash out problem because the cells are fixed to the solid matrix (Das and Das, 2019).

5.3.1 Determination of Cell Growth Characteristics in Chemostat

From Eq. 5.29, we get

$$D = \frac{\mu_{max}\, S}{K_S + S}$$

$$\frac{1}{D} = \frac{K_S}{\mu_{max}}\frac{1}{S} + \frac{1}{\mu_{max}} \tag{5.53}$$

From the plot $\frac{1}{D}$ vs. $\frac{1}{S}$, we get a straight line plot as shown in Figure 5.6. The slope gives the values of $\frac{K_S}{\mu_{max}}$ and the intercept of the Y-axis gives the values of $\frac{1}{\mu_{max}}$.

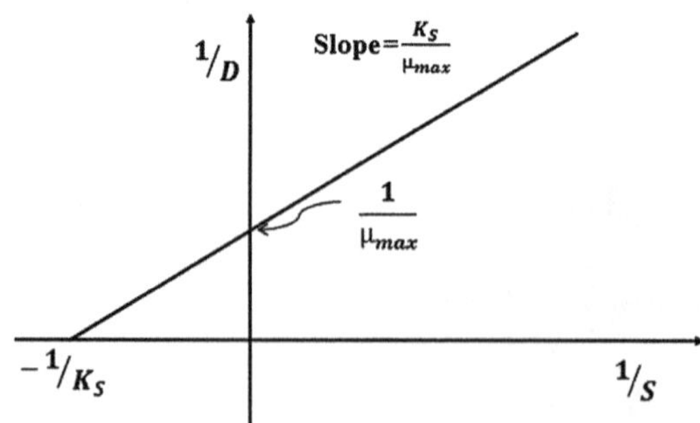

FIGURE 5.6 Plot $\frac{1}{D}$ vs. $\frac{1}{S}$ of a chemostat for the determination of kinetic parameters.

5.3.2 Plug Flow Reactor

Plug flow reactors (PFR) are found to be suitable for processes where products act as an inhibitor. Plug flow is nothing but piston flow, which is known as ideal flow. Therefore, a major drawback of PFR is that it is very difficult to operate because there is no back-mixing in the bioreactor. It has been found that the number of chemostat (CSTR) in series (known as a cascade) is equivalent to PFR. Cascades can be operated easily as compared to PFR. (Das and Das, 2019)

5.3.3 Multistage Chemostat in Series

For secondary metabolites production, the growth and product-formation steps need to be separated, since optimal conditions (temperature, pH and limiting nutrients) for each step are different; e.g. citric acid, penicillin etc. To analyse the multistage chemostat, the following assumptions are made (Figure 5.7):

- Steady-state operation
- No substrate addition in the consecutive stages
- Balanced growth at all stages
- Sterile feed ($x_0 = 0$)
- Negligible endogenous respiration of the cells ($r_d = 0$)

Cell mass balance in the first reactor from Eq. 5.26, we get

$$F\,x_0 + r_x\,V_1 = F\,x_1 + r_d\,V_1 + \frac{d\left(x_1 V_1\right)}{dt} \tag{5.54}$$

At steady state, the rate of cell accumulation, $\dfrac{d\left(x_1 V_1\right)}{dt} = 0$

Eq. 5.54 may be written as

$$0 + r_x V_1 = F x_1 \tag{5.55}$$

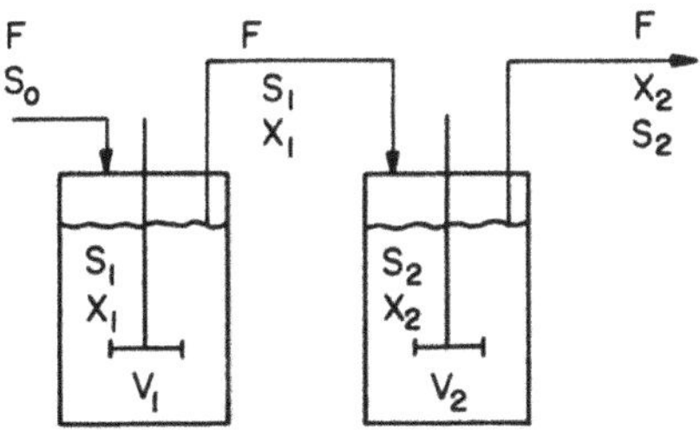

FIGURE 5.7 Schematic diagram of multistage chemostat.

$$\frac{F}{V_1} = \frac{1}{x_1} r_x \tag{5.56}$$

$$D_1 = \mu_1 \tag{5.57}$$

Again, $D_1 = \mu_1 = \dfrac{\mu_{max}\, S_1}{K_S + S_1}$

$$S_1 = \frac{K_s\, D_1}{\left(\mu_{max} - D_1\right)} \tag{5.58}$$

From Eq. 5.21, we get

$$x_1 = Y_{x/S}\left(S_0 - \frac{K_s\, D_1}{\left(\mu_{max} - D_1\right)}\right) \tag{5.59}$$

Cell mass balance in the second reactor from Eq. 5.26 under steady state conditions, we get

$$F\, x_1 + \mu_2\, x_2\, V_2 = F\, x_2 + 0 + 0 \tag{5.60}$$

$$\mu_2\, x_2\, V_2 = F\, x_2 - F\, x_1 = F\left(x_2 - x_1\right) \tag{5.61}$$

$$\mu_2 = \frac{F}{V_2}\left(1 - \frac{x_1}{x_2}\right) = D_2\left(1 - \frac{x_1}{x_2}\right) \tag{5.62}$$

$$\text{Where } x_1 < x_2, \text{ so } D_2 > \mu_2 \tag{5.63}$$

Similarly, from the substrate balance in the second reactor

$$F\, S_1 + 0 = F\, S_2 + \frac{\mu_2\, x_2}{Y_{x/S}} V_2 \tag{5.64}$$

$$F\, S_2 = F\, S_1 - \frac{\mu_2\, x_2}{Y_{x/S}} V_2 \tag{5.65}$$

$$S_2 = (S_1 - \frac{\mu_2\, x_2}{Y_{x/S}} \frac{1}{D_2}) \tag{5.66}$$

$$\text{Therefore from Eq. 5.21,} x_2 = Y_{x/S}\left(S_1 - \frac{K_s D_2}{(\mu_{max} - D_2)}\right) \tag{5.67}$$

Thus one can find the steady state concentration of substrate and cell mass in different chemostats connected in series (Blanch and Clark, 1997).

5.3.4 Chemostat with Cell Recycling

Chemostats with cell recycling can be operated at D_{max} without the cells wash out problem because the excess cells which are wasting from the bioreactor are recycled back. Several biochemical industries have already adopted this process; e.g. Biostill for the ethanol production, activated sludge process for the wastewater treatment etc. Cell recycling increases the cell productivity of the process.

Figure 5.8 shows the schematic diagram of an activated sludge process (ASP) which is similar to the CSTR system with cell recycling. This is used for the stabilization of biodegradable liquid organic wastes. To correlate different parameters mathematically, the following assumptions are made:

1. The influent and effluent cell mass concentrations are negligible.
2. So becomes S in the bioreactor due to complete mix regime
3. All reactions occur only in the bioreactor

The cell mass balance of the ASP on the basis of input and output is as follows.

$$F_0 x_0 + V\left(\frac{\mu_{max} S x}{K_S + S} - \mu_d x\right) = (F_0 - F_w) x_e + F_w x_u \tag{5.68}$$

Where F_0 = Feed flow rate, x_0 = cell mass concentration in the feed, S = steady-state substrate concentration, x = steady-state cell mass concentration, μ_d = specific death rate of the cell, μ_{max} = maximum specific cell growth rate, K_S = saturation constant,

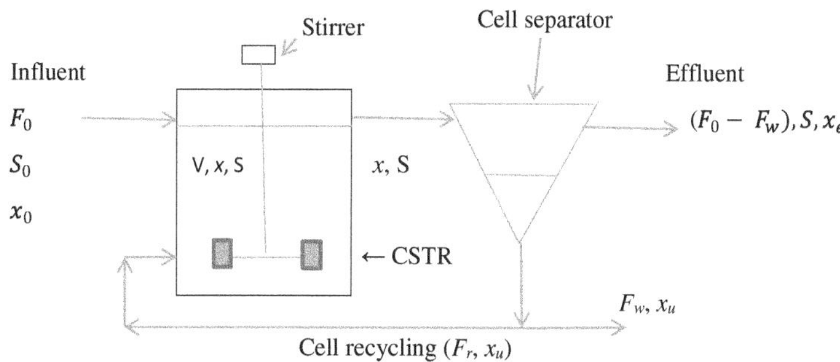

FIGURE 5.8 Chemostat with cell recycling.

F_w = cell wasting flow rate from the cell separator, x_e = cell mass concentration in the effluent, x_u = settled cell mass concentration from the cell separator.

It is assumed that x_0 x_e, are negligible. Then Eq. 5.68 becomes

$$V\left(\frac{\mu_{max} S x}{K_S + S} - \mu_d x\right) = F_w x_u \tag{5.69}$$

$$\left(\frac{\mu_{max} S}{K_S + S}\right) = \frac{F_w x_u}{V x} + \mu_d \tag{5.70}$$

Similarly, under steady-state conditions from the substrate balance we can write

$$F_0 S_0 - \frac{\mu_{max} S x V}{(K_S + S) Y_{x/S}} = (F_0 - F_w) S + F_w S = F_0 S \tag{5.71}$$

$$\left(\frac{\mu_{max} S}{K_S + S}\right) = \frac{Y_{x/S}}{V x} F_0 (S_0 - S) \tag{5.72}$$

From Eqs. 5.70 and 5.72, we get

$$\frac{F_w x_u}{V x} + \mu_d = \frac{Y_{x/S}}{V x} F_0 (S_0 - S) \tag{5.73}$$

Again, hydraulic retention time (HRT) (θ) = $\dfrac{V}{F_0}$ (5.74)

and mean cell residence time = $\theta_c = \dfrac{V x}{F_w x_u}$ (5.75)

In the case of CSTR, $\theta = \theta_c$, but in the case of CSTR with cell recycling i.e. chemostat, $\theta_c > \theta$.

Putting θ and θ_c in Eq. 5.73, we get

$$\frac{1}{\theta_c} + \mu_d = \frac{Y_{x/S}}{\theta x} (S_0 - S) \tag{5.76}$$

$$x = \frac{Y_{x/S} (S_0 - S) \theta_c}{\theta (\theta_c \mu_d + 1)} \tag{5.77}$$

Eq. 5.77 is used to find the cell mass concentration in the bioreactor under steady-state conditions. On the other hand, if x is known, then Eq. 5.77 can be used to find the bioreactor volume as follows:

$$V = \frac{Y_{x/S}\left(S_0 - S\right)\theta_c F_0}{x\left(\theta_c \mu_d + 1\right)} \tag{5.78}$$

Therefore the reactor volume of the chemostat with cell recycling can be found by using Eq. 5.80 (Metcalf and Eddy, 1991).

5.3.5 Separation of Slow-Growing Cells from Fast-Growing Cells in the Chemostat

The anaerobic digestion process is carried out by two groups of microflora: acidogens and methanogens. Acidogens are fast-growing microorganisms as compared to methanogens. Therefore, the $D_{wash\ out}$ values of methanogens are smaller as compared to those of acidogens (Figure 5.9). Thus, if the dilution rate of the chemostat is maintained just above the $D_{wash\ out}$ value of methanogens, then it is possible to separate the acidogens from the anaerobic digester very easily. These acidogens have the potential to produce hydrogen from organic wastes. In addition, the operating conditions of the acidogens and methanogens are different with respect to substrate requirement, temperature and pH. Hence, the two-stage anaerobic digestion process is very easy to control and also improves the methane production to a great extent (Das and Das, 2019; Bailey and Ollis, 2010).

5.4 LUEDEKING–PIRET MODEL FOR PRODUCT FORMATION

Product formation kinetics of the microbial system are determined by using the Luedeking–Piret model as given in Eq. 5.79.

$$\frac{dP}{dt} = \alpha\frac{dx}{dt} + \beta x \tag{5.79}$$

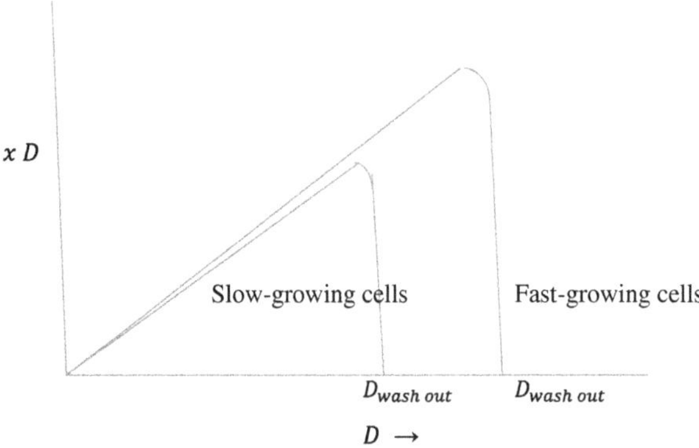

FIGURE 5.9 Variation of rate of cell mass formation with respect to the dilution rate in a chemostat for the slow- and fast-growing microbial cells.

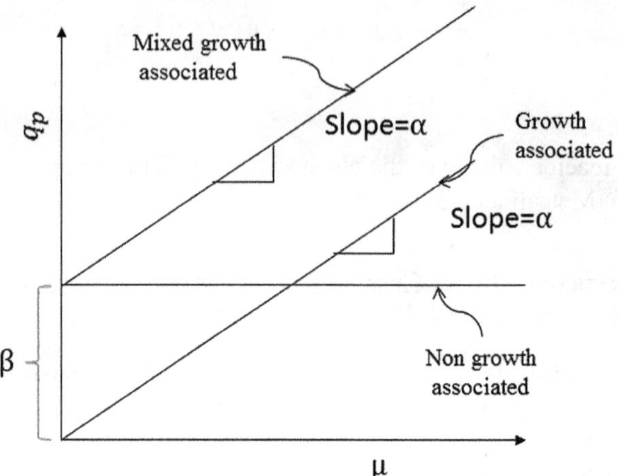

FIGURE 5.10 Plots of q_P vs. μ for different types of fermentation processes.

Where $\frac{dP}{dt}$ = rate of product formation, $\frac{dx}{dt}$ = cell growth rate, α = growth associated coefficient, β = non-growth associated coefficient.

The Luedeking–Piret model indicates the product formation kinetics that combine both the growth associated and non-growth associated contributions.

Eq. 5.79 can be written as

$$\frac{1}{x}\frac{dP}{dt} = \alpha\frac{1}{x}\frac{dx}{dt} + \beta \tag{5.80}$$

$$q_P = \frac{1}{x}\frac{dx}{dt} + \beta = \alpha\mu + \beta \tag{5.81}$$

Where q_p is the specific rate of product formation. Plots of q_p vs. μ for different types of fermentation processes are shown in Figure 5.10.

In the case of mixed growth associated products the values of both α and β will be significant. However, in the case of growth associated product formation like ethanol fermentation process, $\beta \approx 0$ as shown in Figure 5.10. In this case the rate of product formation is directly proportional to the rate of cell growth which takes place in the log phase. On the other hand, in the case where non-growth associated product $\alpha = 0$, the rate of product formation is proportional to the cell mass concentration only which is taken place in the stationery phase, e.g. all the antibiotic fermentation processes (Das and Das, 2020),

5.5 PIRT MODEL FOR THE MAINTENANCE OF CELLS

The Pirt model deals with the maintenance of cells. Therefore the rate of substrate consumption contributes to both cell formation and the maintenance of the cells as shown in Eq. 5.82.

$$\left(\frac{dS}{dt}\right)_{overall} = \left(\frac{dS}{dt}\right)_{growth} + \left(\frac{dS}{dt}\right)_{maintenance} \tag{5.82}$$

$$\frac{\mu x}{Y_{x/S_{overall}}} = \frac{\mu x}{Y'_{x/S_{growth}}} + m x \tag{5.83}$$

Where $Y_{x/S_{overall}}$ = overall yield coefficient, $Y_{x/S_{growth}}$ = true growth yield coefficient, m = maintenance coefficient, x = concentration of viable cells.

Again from Eq. 5.85, we get

$$\frac{1}{Y_{x/S_{overall}}} = \frac{1}{Y'_{x/S_{growth}}} + \frac{m}{\mu} \tag{5.84}$$

From the plot $\frac{1}{Y_{x/S_{overall}}}$ vs. $\frac{1}{\mu}$, the slope will give the value of m and intercept of the Y-axis will give the value of $\frac{1}{Y'_{x/S_{growth}}}$, as shown in Figure 5.11.

PROBLEM 5.1 The characteristics of microorganisms used in the chemostat are as follows: $\mu_{max} = 0.5\ h^{-1}$ and $K_S = 2$ g/L.

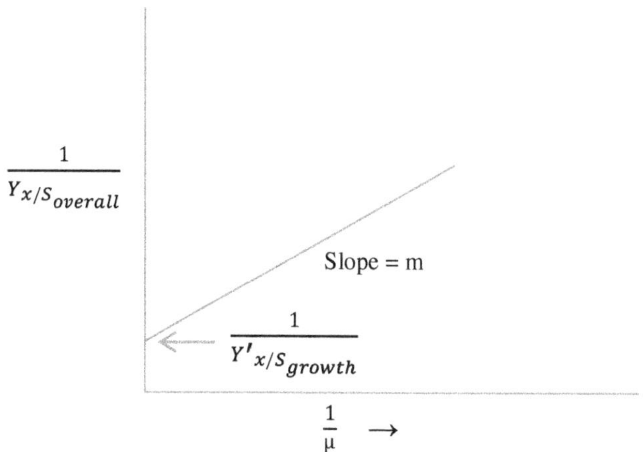

FIGURE 5.11 Plot of $\frac{1}{Y_{x/S_{overall}}}$ vs. $\frac{1}{\mu}$

(a) Under steady-state conditions with no cell death, if S_0 = 50 g/L and $Y_{x/S}$ = 1, what dilution rate D will give the maximum rate of cell production?
(b) For the same value of D using reactors of the same size in series, how many reactors will be required to reduce the substrate concentration to 1 g/L?

SOLUTION

(a) We know that

$$D_{max} = \mu_{max}[1-\sqrt{\frac{K_s}{(K_s+S_o)}}]$$

$$= 0.5[1-\sqrt{\frac{2}{2+50}}]$$

$$= 0.402\,h^{-1}$$

(b) Again, under steady conditions and sterile feed, the steady-state substrate concentration (S_1) is written as

$$S_1 = \frac{\boldsymbol{D K_s}}{\boldsymbol{\mu_{max} - D}} = \frac{\mathbf{(0.402)(2)}}{\mathbf{(0.5)-(0.402)}} = 8.2 \text{ g/L}$$

Again concentration of cell mass at the outgoing stream under steady-state conditions

$$x_1 = Y_{x/s}\left[S_o - \frac{\boldsymbol{D K_s}}{\boldsymbol{\mu_{max} - D}_s}\right] = 1\ [50-8.2] = 41.8 \text{ g/L}$$

Substrate balance under steady-state conditions in the second reaction from Eq. 5.69 we get

$$\frac{F}{V}(S_1 - S_2) - \frac{x_2\,\mu_{max}\,S_2}{Y_{x/s}(S_2+K_s)} = o \tag{5.85}$$

$$D(S_1 - S_2) - \frac{x_2\,\mu_{max}\,S_2}{Y_{x/s}(S_2+K_s)} = o \tag{5.86}$$

$$\text{Yield coefficient, } Y_{x/s} = \frac{(x_2 - x_1)}{(S_1 - S_2)} \tag{5.87}$$

where x_1, S_1 are the concentrations of biomass and substrates in the inlet of the second reaction respectively and x_2, S_2 are the same for the outlet of the second reactor under steady-state conditions.

$$x_2 = x_1 + Y_{x/S}(S_1 - S_2) \tag{5.88}$$

Substituting the x_1 value in Eq. 5.86, we get

$$D(S_1 - S_2) - \frac{[x_1 + Y_{x/S}(S_1 - S_2)]\,\mu_{max}\,S_2}{Y_{x/s}(S_2 + K_s)} = o \tag{5.89}$$

Putting the values of D, S_1, μ_{max}, $Y_{x/S}$, K_s in Eq. 5.89, we get

$$S_2 = 0.293\frac{g}{L} < 1\,g/L$$

Therefore, two reactors will be enough for the above substrate conversion by the microbial cells.

PROBLEM 5.2 Baker's yeast production takes place in a chemostat of a 1 m^3 bioreactor. The initial substrate concentrations in the feed and rate of cell mass (yeast) production are $0.5\,kg\,m^{-3}$ and $0.1\,kg\,h^{-1}$, respectively. The steady-state substrate concentrations at a feed flow of $0.5\,m^3\,h^{-1}$ and $1.0\,m^3\,h^{-1}$ are $0.1\,kg\,m^{-3}$ and $0.3\,kg\,m^{-3}$, respectively. Compute the following:

a) The overall yield of yeast
b) The kinetic equation for yeast formation
c) The flow rate for maximum yeast production
d) The maximum production rate of yeast

SOLUTION

(a) We know that

Feed flow rate ($m^3\,h^{-1}$) x concentration of cells ($kg\,m^{-3}$) = Rate of cell mass production ($kg\,h^{-1}$)

Therefore, concentration of yeast = $\frac{0.1 kg\, h^{-1}}{0.5 m^3\, h^{-1}} = 0.2\ kg/m^3$

In the case of sterile feed, $x_{ss} = Y_{x/S}\left(S_0 - S_{ss}\right)$

Therefore, $Y_{x/S} = \frac{0.2\, kg/m^3}{\left(0.5-0.1\right) g/m^3} = 0.5$ = Overall yield of yeast cells

(b) From the Monod model under steady-state and sterile conditions

$$\mu = \frac{\mu_{max} S}{K_S + S} = D$$

Again, in the first case, D = $\frac{F}{V} = \frac{0.5 m^3/h}{1 m^3} = 0.51/h$

In the second case, D = $\frac{1m^3/h}{1m^3} = 11/h$

Therefore, from the Monod equation, in the first case, $0.5 = \frac{\mu_{max} 0.1}{K_S + 0.1}$ (5.90)

Similarly in the second case, $1 = \frac{0.3}{K_S + 0.3}$ (5.91)

Dividing Eq. 5.90 by Eq. 5.91, we get $\frac{0.5}{1} = \frac{(K_S + 0.3)\mu_{max} 0.1}{(K_S + 0.1)\mu_{max} 0.3}$

$$\therefore K_S = 0.3\ kg/m^3$$

Putting the value of K_S in Eq. 5.90, we get

$$\mu_{max} = 2\frac{1}{h}$$

Therefore, the kinetic equation is $\mu = \frac{2S}{0.3 + S}$ (5.92)

(c) We know that $D_{max} = \mu_{max}\left(1 - \sqrt{\frac{K_S}{\left(K_S + S_0\right)}}\right)$

Putting the values of μ_{max}, K_S and S_0 in the above equation, we get

$D_{max} = 0.776\ 1/h = \frac{F}{V}$

Flow rate for the maximum yeast production = $0.776 \times 1 = 0.776\, m^3/h$

(d)

$$S = \frac{K_S\, S}{(\mu_{max} - D)} = \frac{0.3 \times 0.776}{(2 - 0.776)} = 0.19\ kg/m^3$$

$$X = Y_{X/S}\left(S_0 - S\right) = 0.5\left(0.5 - 0.19\right) = 0.155\ kg/m^3$$

Maximum cell mass productivity = $0.776\ m^3/h \times 0.155\ kg/m^3 = 0.12$ kg/h m^3

PROBLEM 5.3 100 m^3 rectified spirit (containing 90%v/v ethanol) is produced by one distillery industry in a chemostat from cane molasses (containing 50%w/w sugar) using *S. cerevisiae*. The characteristics of the yeast are as follows

$$\mu_{max} = 0.05 h^{-1},\ K_S = 2\, kg/m^3,\ Y_{X/S} = 0.05,\ Y_{P/S} = 0.5, \text{ and } S_0 = 300\, kg/m^3$$

Determine the volume of the bioreactor and amount of cane molasses required per day.

SOLUTION

For sterile media, $X_0 = 0$

Steady-state biomass concentration,

$$X = X_0 + Y_{X/S}\left(S_0 - S\right)$$

$$\text{Now, } D_{max} = \mu_{max}\left(1 - \sqrt{\frac{K_S}{(K_S + S_0)}}\right)$$

$$= 0.05\left(1 - \sqrt{\frac{2}{(2+300)}}\right) = 0.046\, h^{-1}$$

Also, steady-state substrate concentration

$$S_{ss} = \frac{K_S\, D_{max}}{\mu_{max} - D_{max}} = \frac{2 \times 0.046}{0.05 - 0.046} = 23\, kg/m^3$$

In the case of sterile feed, $x_{ss} = Y_{x/S}\left(S_0 - S_{ss}\right)$

$$\therefore x_{ss} = 0.05\ \left(300 - 23\right) = 13.85\, kg/m^3$$

Basis: 100 m^3 rectified spirit ≡ 90 m^3 ethanol production per day

Density of ethanol = 780 kg/m^3

Amount of ethanol = 780 $kg/m^3 \times 90\ m^3/d = 70200\ kg/d$

$$\text{Substrate required} = \frac{70200\,kg/d}{Y_{P/S}} = \frac{70200\,kg/d}{0.5} = 140400\,kg/d$$

$$\text{Actual amount of sugar required} = \frac{140400}{0.98} \approx 143000\ kg/d$$

(Assuming sugar conversion efficiency 98 %)

Volumetric feed flow rate =

$$F = \frac{substrate\,required}{initial\,substrate\,conc} = \frac{143000\,kg/d}{300\,kg/m^3} = 476\frac{m^3}{d} = 19.86\frac{m^3}{h}$$

$$\text{Now, } \tau_{CSTR} = \frac{S_0 - S}{(-r_S)}$$

$$(-r_S) = \frac{1}{Y_{X/S}}\mu X$$

We substitute D_{max} for μ in the case of sterile feed and under steady-state conditions,

$$(-r_S) = \frac{1}{Y_{\frac{X}{S}}} D_{max}\ x = \frac{1}{0.05} \times 0.046 \times 13.85\frac{kg}{m^3\,h} = 12.74\frac{g}{m^3\,h}$$

$$\tau_{CSTR} = \frac{S_0 - S}{(-r_S)} = \frac{V}{F}$$

$$\frac{V}{F} = \frac{(300-23)\,kg/m^3}{12.74\frac{g}{m^3\ h}} = 21.74\ h$$

$$\therefore V = 19.86\frac{m^3}{h} \times 21.74\,h \approx 432\,m^3$$

Volume of the reactor = 432 m^3

Substrate required per day = 143000 kg/d

Cane molasses is 50%w/w sugar

$$\text{Therefore, cane molasses required} = \frac{143000\,kg/d}{0.5} = 286000\frac{kg}{d} = 286\frac{MT}{d}$$

5.6 CONCLUSIONS

Microorganisms remain active in the log phase of the cell growth cycle. Therefore, the inoculum added to the production fermenter is usually between the mid-log phase and late-log phase to ensure the microbial strain is active. The productivity of the batch fermentation process is much less as compared to the chemostat mainly due to the absence of down time. In addition, the chemostat process can be operated in the log phase for an infinite period of time by controlling the dilution rate whereas that of the batch process is for a very short time. The major drawback of the chemostat is the problem of wash out of the cell mass. This will take place when the hydraulic retention time is less than the generation time of the cell. This problem can be overcome by using either cell mass recycling or using immobilized whole cell. Slow-growing microbial cells can be separated from fast-growing cells by controlling the dilution rate. The strategy of operation of the fermentation process may be planned on the basis of the nature of the product formation (growth associated or non-growth associated or mixed growth associated). Substrate consumption for maintenance of the cells can also be determined.

REFERENCES

Aiba S, Humphrey AE and Millis NF, Biochemical Engineering, Academic Press, 1973.

Bailey JE and Ollis DF, Biochemical Engineering Fundamentals, McGraw-Hill, New Delhi, 2010.

Blanch HW and Clark DS, Biochemical Engineering, Marcel Dekker, New York, 1997.

Das D and Das D, Biochemical Engineering: An Introductory Text Book, Jenny Stanford, Singapore, 2019.

Das D and Das D, Biochemical Engineering: A Laboratory Manual, Jenny Stanford, Singapore, 2020.

Doran PM, Bioprocess Engineering Principles, Second Edition, Academic press, Waltham, 2012.

Glazer AN and Kikaido H, Microbial Biotechnology (2nd Edition), Cambridge University Press, 2007.

Harzevili FD and Chen H, Microbial Biotechnology: Progress and Trends, CRC Press, 2015.

Lydersen BK, D'elia NA, and Nelseon KL, Bioprocess Engineering: Systems, Equipment and Facilities, Wiley India, New Delhi, 1994.

Levenspiel O, Chemical reaction Engineering, Third Edition, Wiley-India, 2010.

Lim HC and Shin HS, Fed-Batch Culture: Principles and Applications of Semi-Batch Bioreactors, Cambridge University Press, 2013.

Liu S, Bioprocess Engineering: Kinetics, Biosystems, Sustainability, and Reactor Design, Elsevier, Amsterdam, 2013.

Metcalf and Eddy, Wastewater Engineering: Treatment, Disposal and Reuse (Revised by Tchobamoglous G and Burton F.L.), TATA McGraw-Hill, 1991.

Pelczar MJ, Chan ECS, Kreig NR, Microbiology (5th Ed.), Tata McGraw-Hill, New Delhi, 1998.

Shuler ML and Kargi F, Bioprocess Engineering: Basic Concepts, Second Edition, Prentice-Hall, New Delhi, 2002.

Sinclair CG and Kristiansen B, Fermentation Kinetics and Modelling, Open University Press, 1987.

Villadsen J, Nielsen J and Lida G, Bioreaction Engineering Principles, Springer, 2009.

Volesky B and Votruba J, Modeling and Optimization of Fermentation Processes, Vol. 1, Elsevier, 1992.

6 Overview of the Fermentation Industry

The flow diagram of the process is the most effective way of communicating information about the different units present in the biochemical process. Flow diagrams are classified as shown in Figure 6.1.

A biochemical industry does not only have fermenters; it also has different upstream processes and downstream processes. Upstream processes involve pretreatment of the raw materials, air sterilization, medium sterilization and inoculum preparation for the production fermenter. This is followed by production fermentation where product formation in the bioreactor takes place. Downstream processing deals with the purification of products. Different steps are involved for the purification of products. It is well known that products are usually marketed in purified form. There are two types of chemical available in the market: analytical grade and commercial grade. Commercial grade chemicals are less purified because these are mostly used as chemical feedstock for the chemical industries. Analytical grade chemicals are of about 99.99% purity and are used for analytical purposes in the laboratory or the food and pharmaceutical industries. The cost difference between analytical grade and commercial grade chemical is quite significant. Therefore purification is a very important part of any chemical and biochemical industry (Prescott and Dunn, 1959; Moo-Young, 2019; Wittmann and Liao, 2017).

Thus the main purpose of the flow diagram is to give information about different units involved in the industry to get the product. These can be represented in three ways: block flow diagrams, process flow diagrams and piping and instrumentation diagrams (Figure 6.1).

6.1 BLOCK FLOW DIAGRAM

A block flow diagram (BFD) is the simplest form of flow diagram used in industry. One block in a BFD can represent a single piece of equipment e.g. medium preparation, air sterilization process, medium sterilization process, inoculum vessel, production fermenter, cell separator etc. A flow diagram of the downstream processing of citric acid fermentation is shown in Figure 6.2 (Das and Das, 2019).

There are several units are included, such as fermentation liquor, calcium citrate precipitation, pannevis filter, evaporator etc. In the citric acid industry, after citric acid

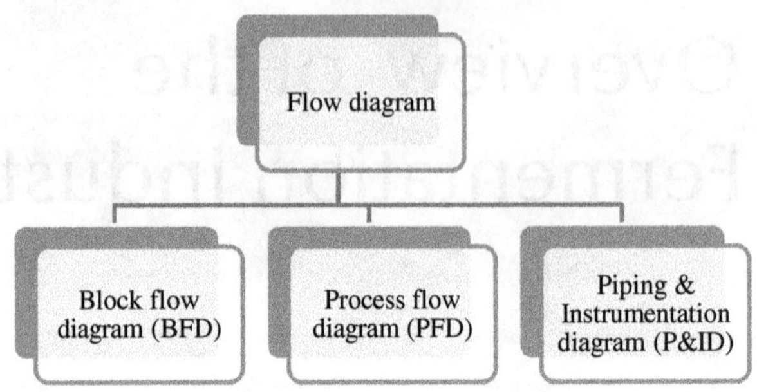

FIGURE 6.1 Classifications of the flow diagrams.

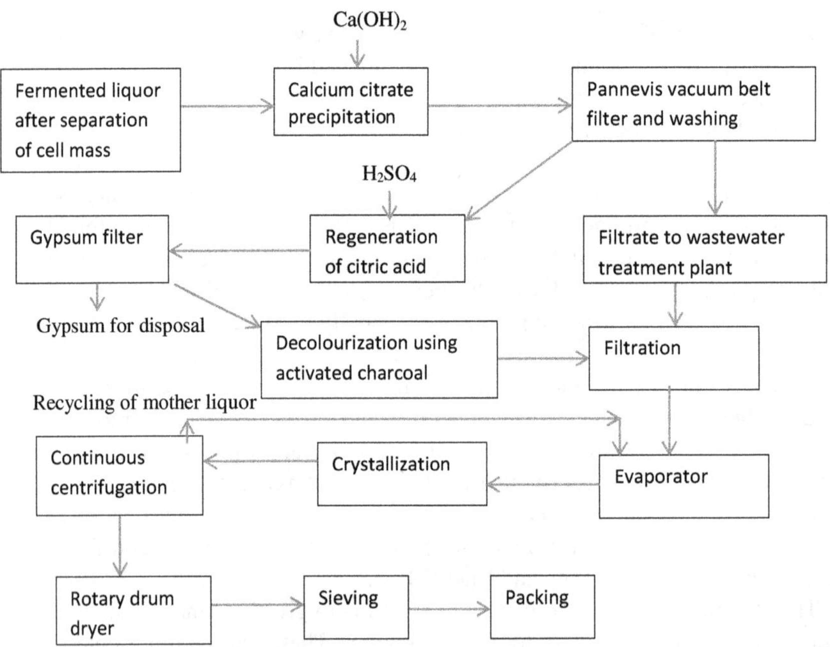

FIGURE 6.2 Block flow diagram of the downstream processing of the citric acid fermentation process.

production takes place in the fermenter, it passes through different processing units such as calcium citrate precipitation, filtration and washing, regeneration of citric acid, decolourizing treatment, evaporator, crystallization, centrifugation, drying, sieving, packing etc. Thus there are many steps involved before the final product formation. Every step is represented with the help of a block. The fermented liquor collected after separation of the microbial cells is treated with lime to precipitate out the citric acid in the form of calcium citrate. Calcium citrate is an insoluble mass and has to be separated with the help of a filtration process. The calcium citrate is washed with water, so that the colour present is partly removed. Then it is hydrolysed in the presence of concentrated H_2SO_4. Calcium citrate is converted to citric acid and calcium sulphate. Calcium sulphate is removed by gypsum filter. This is disposed of as gypsum which is a by-product of the cement industry. The citric acid is passed through decolourizing methods such as treatment with activated carbon to remove the colour because if colour is present in the product then it will be very difficult to market it. Again filtration is carried out to remove the activated carbon. The citric acid is concentrated with the help of an evaporator. After concentration, it is cooled down and then passed through a crystallization process to form crystals of citric acid monohydrate (CAM). This is followed by centrifugation to separate the crystals, which are dried and sieved to get different sizes of crystals. Finally packaging is done in a polythene pouch. Thus, this overall block flow diagram explains the citric acid production process (Prescott and Dunn, 1959; Das and Das, 2019).

6.2 PROCESS FLOW DIAGRAM

The process flow diagram provides a visual representation of the different units in a biochemical industry. Flow diagrams are also referred to as Process Mapping. This has the following benefits

- To get a clear understanding of the process
- To find out non-value-added operations
- Facilitates teamwork and communication
- Keeps every processing unit on the same page

The process flow diagram gives a clear picture about the whole plant. A process flow diagram (PFD) represents the general flow of plant processes and equipment in the fermentation industry. The PFD includes the relationships among the major equipment of a plant facility. However, it does not give minor details such as piping details and designations.

Different utilities and units are represented by different configurations as shown in Figures 6.3 and 6.4.

There are different valves that are essential for the operation of the industry. In PFD these valves are also represented through different symbols as shown in Figure 6.5.

Ball valves are largely used in the biochemical industry to draw off samples. They can be opened and closed instantly. In the wash basin, a globe valve is used in the water pipe line. It can be opened and closed slowly.

PFD of the ethanol fermentation process from corn stover is shown in Figure 6.6. Corn stover usually comes by truck and is unloaded. The unloading is followed by grinding and then there is a conveyor belt along which it passes to the pretreatment unit. This unit comprises steaming and acid treatments for the hydrolysis of corn stover. Then it passes through the solid–liquid separation unit. After separation the liquor is taken out and its pH adjusted with lime. This is followed by saccharification and fermentation processes in the presence of enzymes and microorganisms, respectively for ethanol production. The ethanol is separated out from the distillation column and some leftover solid materials (lignin) are taken out for running the boilers. The

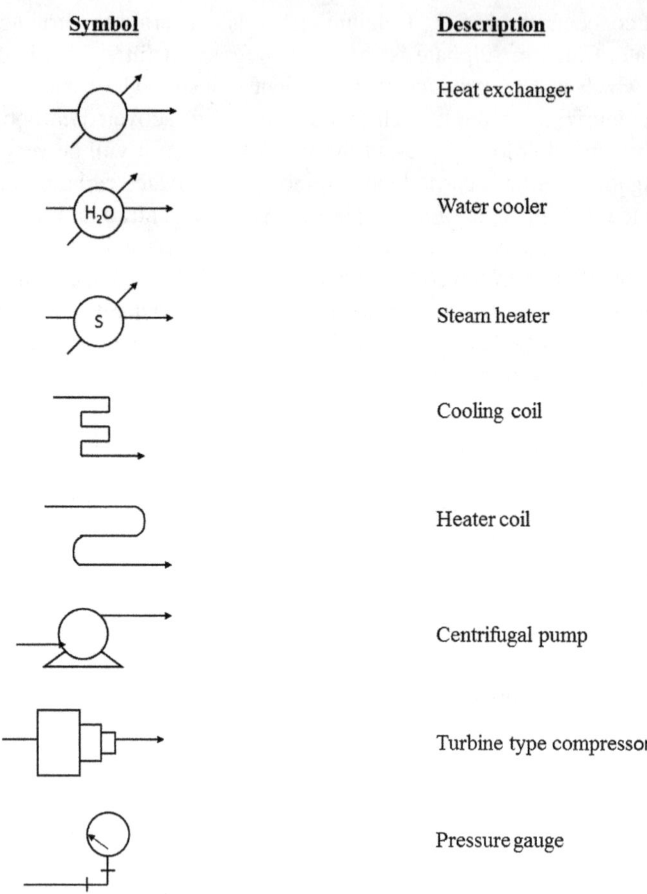

FIGURE 6.3 Different utilities of the biochemical industry.

wastewater comes from the bottom portion of the distillation column and passes through the wastewater treatment process for the reduction of its pollution load. This is one typical example of a process flow diagram to convert corn stover into ethanol (Wittmann and Liao, 2017; Thakur, 2013; Das, 2015).

6.3 PIPING AND INSTRUMENTATION DIAGRAM/DRAWING (PID)

A piping and instrumentation diagram/drawing (PID) displays the piping and vessels in the process flow together with the instrumentation and control devices (Figure 6.7). Therefore, a piping and instrumentation diagram is usually located in the control room of any fermentation industry. From P & ID, one can find out how the plant is running, how the liquid is flowing from one end to other, which valves are open/ closed, and which parts are functioning. All this information is available from PID. This information is very essential for running an industry.

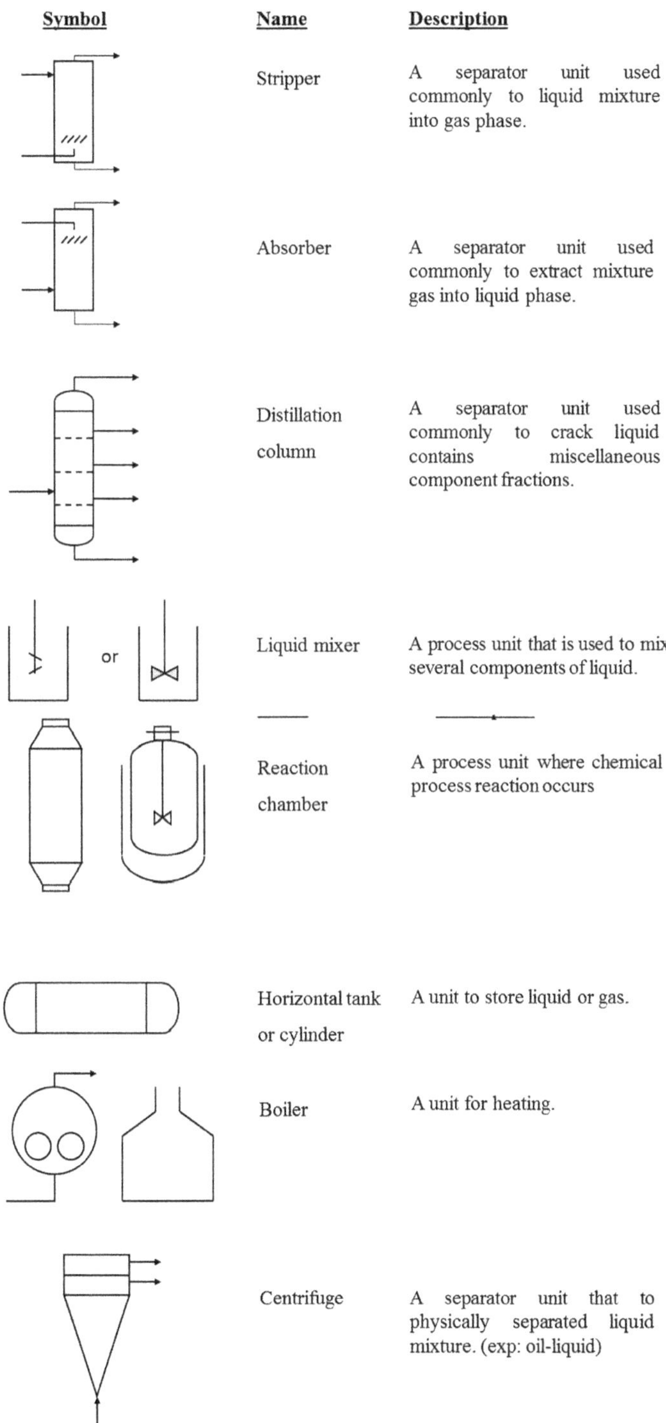

Symbol
Name
Description
Stripper
A separator unit used commonly to liquid mixture into gas phase.
Absorber
A separator unit used commonly to extract mixture gas into liquid phase.
Distillation column
A separator unit used commonly to crack liquid contains miscellaneous component fractions.
or
Liquid mixer
A process unit that is used to mix several components of liquid.
Reaction chamber
A process unit where chemical process reaction occurs
Horizontal tank or cylinder
A unit to store liquid or gas.
Boiler
A unit for heating.
Centrifuge
A separator unit that to physically separated liquid mixture. (exp: oil-liquid)

FIGURE 6.4 Configurations of different units.

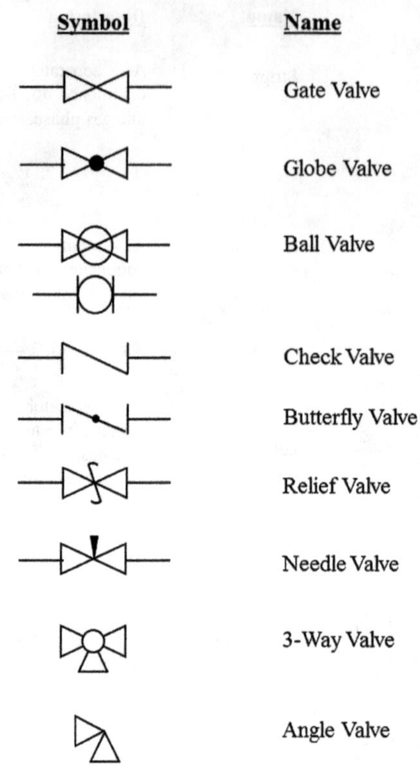

FIGURE 6.5 PFD of different valves used in the industry.

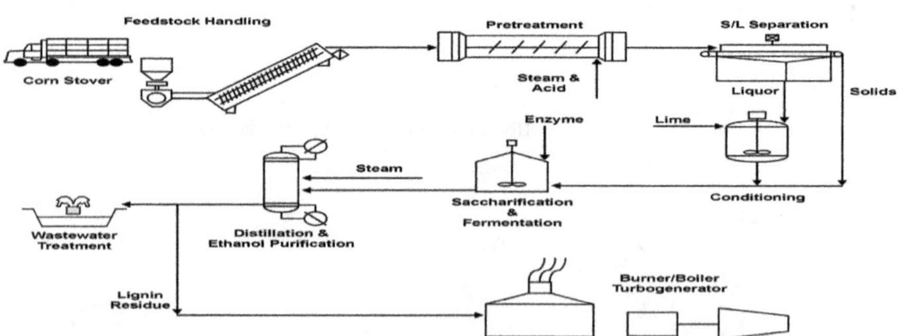

FIGURE 6.6 Process flow diagram for the production of ethanol from corn stover.

(Source: http://re.emsd.gov.hk/english/other/biofuel/bio_tech.html)

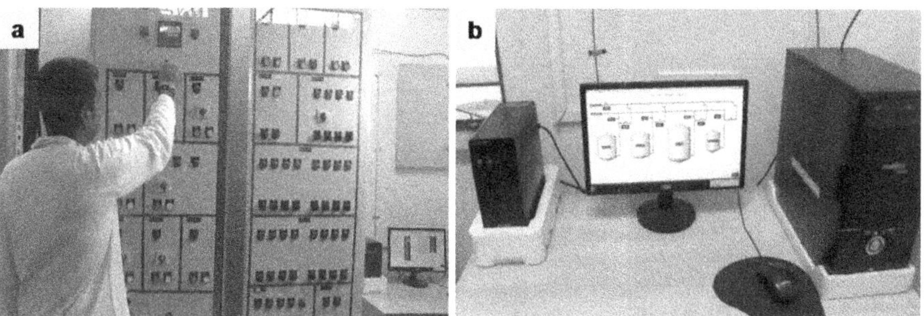

FIGURE 6.7 Piping or instrumentation diagram/drawing (PID) of 10^3 biohydrogen pilot plant at the Indian Institute of Technology Kharagpur.

In industry usually there is a lot of noise due to the functioning of the stirrer, several pumps, compressors etc. The author has experience of working in the citric acid production industry. The capacity of the two production fermenters is 200 m^3. A 200 hp motor was used for the operation of the mechanical agitator. It gave tremendous noise. P & ID is usually located in the control room where one can do the process analysis very nicely. Display of process control installation in the schematic drawing is very important for better understanding of the operation of the process.

6.4 PUMPS AND VALVES

Different pumps and valves are used in the biochemical industry.

6.4.1 Pumps

A pump is a device that moves fluids (liquid or gas), sometimes slurries, by mechanical action. It is used for domestic, commercial, industrial, agricultural service, municipal water and wastewater services. Pumps can be classified into two major types according to the principles by which the energy is added to the fluid; one is the positive displacement pump and the other is the non-positive displacement pump (Figure 6.8).

Positive displacement pumps (PDP) use a backward and forward movement to move a fluid. They comprise a piston, plunger, and diaphragm. The advantage of a diaphragm is that liquid being transferred is not in direct contact with the pump due to the presence of a membrane. The reciprocating pump is one in which the liquid follows the movement of a piston during suction and delivery strokes. Here a moving fluid is captured in a cavity which then discharges that fixed amount of fluid. Some of these pumps have an expanding cavity at the suction side and a decreasing cavity at the discharge side. In these pumps no fluid comes back to its casing during pumping out. Examples are progressive cavity pumps, peristaltic pumps, gear pumps, screw pumps, rotary gear pumps, lobe pumps etc. In these pumps pressure is applied directly to the liquid by a reciprocating piston or by rotating members. They are used for shear-sensitive liquid, high-pressure application and variable viscosity applications. Types include reciprocating and rotary pumps.

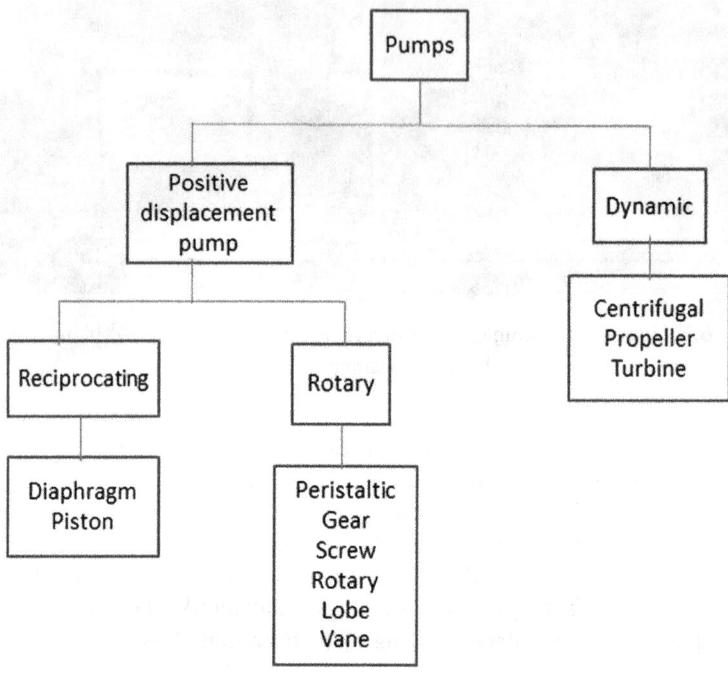

FIGURE 6.8 Classification of the pumps.

Cane molasses is an important raw material in both the citric acid and ethanol making industries. Molasses is a viscous liquid, so an ordinary pump cannot transfer it. Gear pumps are used for transferring the cane molasses from the molasses storage tank (located outside the fermentation plant) to the molasses measuring tank which is located in the fermentation plant. Progressive cavity pumps are used to transfer the thick slurry. Peristaltic pumps are used for the aseptic transfer of the liquid, and are essential in most biochemical industries. Silicon tubes (flexible) are used for transferring liquid. It wraps around the rollers. When the roller rotates it drags the liquid from one end to other. Here all the materials remain inside the tube. Therefore, the liquid does not come into contact with the outside atmosphere, and thus sterility of this system is maintained (Das and Das, 2019; Doran, 2012).

A non-positive-displacement pump produces a continuous flow. However, its output varies significantly due to variation of pressure because it does not provide a positive internal seal against slippage. An example is the centrifugal pump largely used in the fermentation industry. Centrifugal pumps are an example of dynamic pumps where kinetic energy is added to the fluid by increasing the flow velocity. Centrifugal pumps are also largely used in our day-to-day requirements in our homes. In a multi-storey building a centrifugal pump is required to drag water from ground level to the water tank present at the top of the building. Centrifugal pumps generate high rotational velocity that converts the resulting kinetic energy of the liquid into pressure energy. These pumps are generally used where high flow rates and moderate head increases are required. They cannot handle fluids containing suspended solids (Das and Das, 2019).

6.4.2 Valves

Beside pumps, there are different valves used in the fermentation industry. The valve is a device that regulates and controls the fluid flow (gases, liquids, fluidized solids or slurries) by opening, closing or partially obstructing different passageways. The valves can be operated manually, by pneumatic pressure and electrically. Basic valve components are the body, bonnet, trim, packing and actuator.

According to the motion of the fluid, valves can be classified into two types: control valves with linear motion and control valves with rotary motion. Types of linear motion valve are globe valve, diaphragm valve and gate valve. Globe valves are further divided into four types: single port globe, double port globe, angle and 3-way. The globe valve restricts the flow of liquid by altering the distance between a movable plug and a stationary seat. The globe valve is used in our homes particularly in the wash basin for controlling the flow rate of water. Gate valves are largely used for harvesting purposes and are located at the bottom of the fermenter, after the fermentation is over. Examples of rotary motion valves are ball valve, butterfly valve and disc valve. Ball valves are largely used in the biochemical industry for drawing samples from the fermenter.

The different valves used as utilities of the reactor are listed below:

Gate valve: This valve is used in the harvesting lines of reactors. The gate valve functions by inserting a gate into the path of the flow to restrict it, in a manner similar to the action of a sliding door. Gate valves are more often used for on/off control (e.g. drain outlet of the fermentation broth) than for throttling.

Ball valve: This is a spherical ball with a passageway cut through the centre; it rotates to allow the fluid more or less access to the passageway. It can be used for sample collection. The ball is a circular device, so that when the handle moves, the ball rotates. This ball has an opening/passageway. By moving the handle in one way the passageway of the ball opens and the liquid flows through the pipeline and by moving the handle the other way, the passageway closes and the flow gets arrested. Therefore this valve can be instantaneously opened or closed. Similarly, butterfly valve can also be opened and closed instantaneously. Disc valves are also of the same nature.

Needle valve: This is a type of valve having a small port and a threaded, needle-shaped plunger. It allows precise regulation of flow, although it is generally only capable of relatively low flow rates.

Safety valve: This is used to control the pressure (it can withstand certain pressure) within set limits. Above the set pressure, the spring will go up and pressure will be released. Therefore, safety valves open slowly as the pressure increases above the set point and only opens as necessary.

Solenoid valve: This is an electromechanically operated valve. It is electrically controlled. The solenoid valve is used in different control systems e.g. temperature control of the fermenter, connected to anti-foam sensor etc. Anti-foam sensors are connected to pumps through solenoid valves and the valve is opened so that anti-foam oil is drawn inside the fermenter. Therefore, solenoid valves are automatically controlled. In the case of a two-port solenoid valve, the flow is switched on or off. In the case of a three-port valve, the outflow is switched between the two outlet ports. Multiple solenoid valves can be placed together on a manifold. Manual valves cannot be operated electrically; that is why solenoid valves are largely used by the fermentation industry.

Pneumatic valve: This valve is controlled with the help of pressure. The term pneumatic is derived from Greek word 'pneuma' meaning the wind or breath. Pneumatic power is controlled by compressed air. Thus, one can regulate this valve on the basis of air pressure.

6.5 FERMENTER OR BIOREACTOR

The reactor is a vessel in which the reaction takes place. A bioreactor is a vessel in which the reaction takes place with the help of some living cells or biomolecules. In the fermentation process, reactions are mostly carried out in a bioreactor by living cells. A schematic diagram of a laboratory fermentation process is shown in Figure 6.6 (Das and Das, 2020). The major difference between the chemical reactor and the fermenter is to maintain the sterile conditions for the growth of the desired living cell. Industrial fermentation plant mostly comprises the following parts:

- Bioreactor
- Mechanical stirrer, motor, mechanical seal and baffles
- pH probe, dissolved oxygen (D.O.) probe, thermistor, and antifoam oil sensor
- Air filter
- Medium sterilizer
- Different valves, heating and cooling systems
- Air sparger

6.5.1 Schematic Diagram of Bioreactor or Fermenter

Fermenters or bioreactors are usually a cylindrical vessel (Figures 6.9 and 6.10). Laboratory bioreactors are made of either glass or stainless steel whereas industrial bioreactors are mainly made of stainless steel. Glass vessels are usually smooth which makes them nontoxic and corrosion proof. It is easy to examine the interior of the vessel. These vessels require autoclaved sterilization. However, in the case of stainless vessels, usually there are two watch glass windows in the lid of the fermenter. A light source is applied at one glass window so that one can see the medium from the other window. The type of stainless steel used in the bioreactor of the fermentation industries is the nonmagnetic 300-series. Several varieties of the 300-series exist. Those most common to brewing are 314 and 316 stainless steel. For the citric acid fermenter, SS 317 is used. The quality of the stainless steel depends on the composition of the metals.

- If stainless steel (SS) contains 12% Cr, it prevents surface corrosion of the tank by acid.

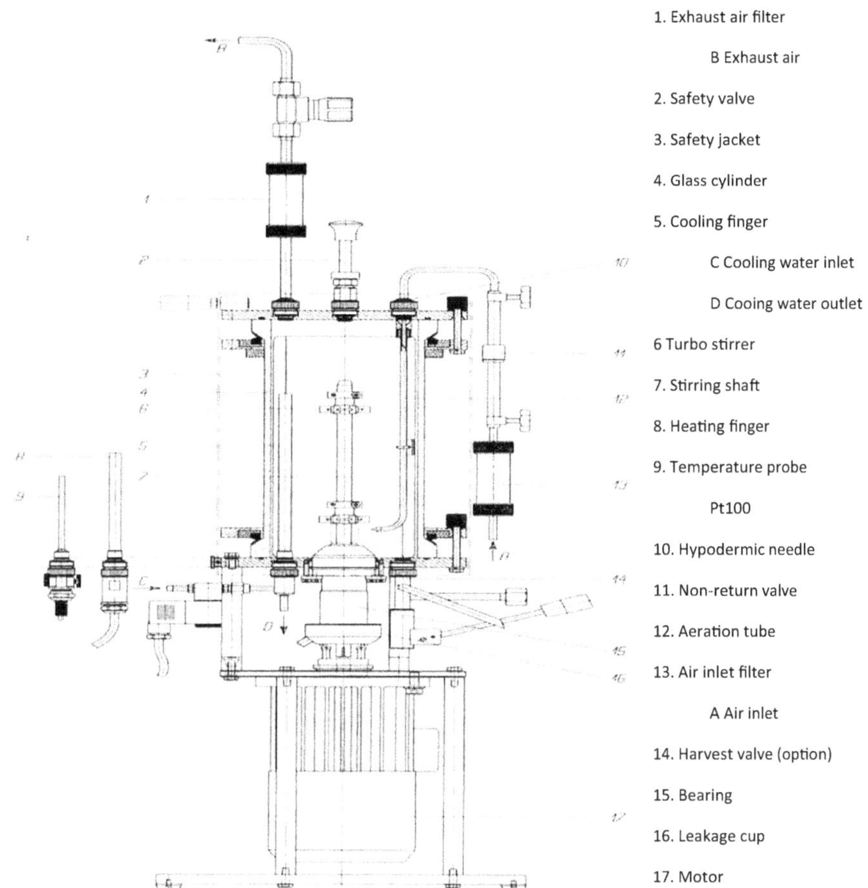

FIGURE 6.9 Schematic diagram of stirred tank bioreactor. (Model: 3.7 KLF 2000, BioEngineering AG, Switzerland.)

- 8% Ni contents in SS gives an austenitic or smooth structure.
- 2–5% Mo in SS can increase the resistance power of steel to acid because in the fermentation process acids are usually formed.

6.5.2 Mechanical Stirrer, Mechanical Seal and Baffles

The purpose of the mechanical stirrer is twofold: i) It mixes the gas bubbles through the liquid culture medium and ii) it keeps the insoluble cells in suspension so that they can freely interact with the different components present in the medium for their growth and metabolism. A Rushton impeller connected with the shaft is used in the fermentation vessel which is shown in Figure 6.11.

A mechanical stirrer causes a vortex in the liquid medium which is responsible for improper mixing and lower mass transfer. This problem can be overcome by

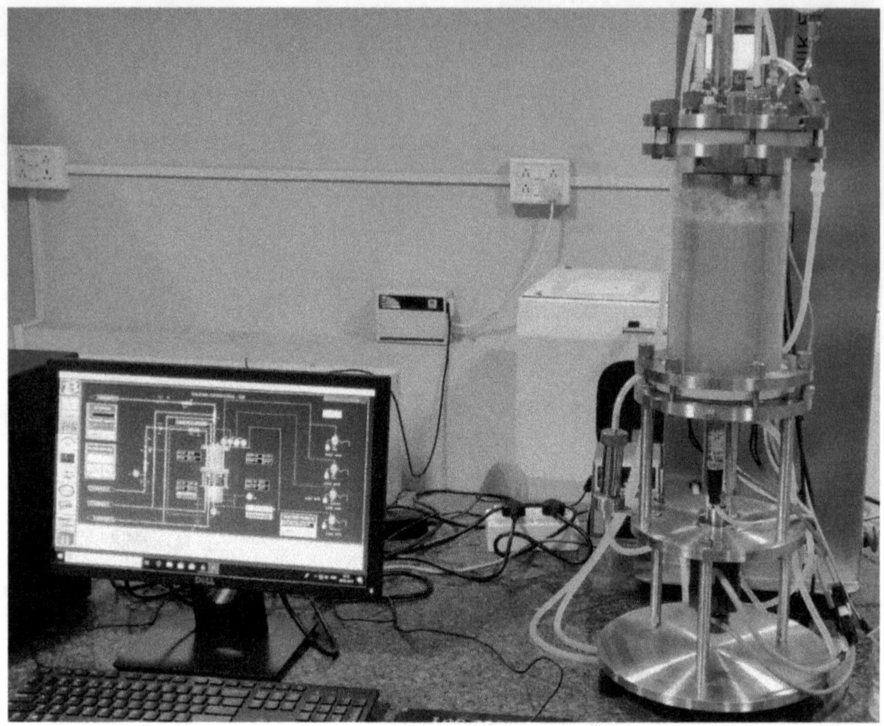

FIGURE 6.10 Photograph of laboratory fermenter, Biojenik Engineering, India.

FIGURE 6.11 Rushton impeller.

using baffles. Baffles are rectangular blades placed inside the fermenter. In the industrial fermentation process hollow rectangular baffles are used. These are intended for cooling purposes in tropical countries such as India during summer. Baffles improve mass transfer to increase the DO concentration of the fermentation medium. In addition, the scouring action due to agitation reduces microbial growth on the walls of fermenter. The shaft of the impeller is connected to the motor which is passed through the mechanical seal (MS) placed at the top of the cover lid or at the bottom of the fermentation vessel. The MS consists of two parts: the stationary part is the bearing and the other part is the rotation shaft.

The two components are pressed together by springs and packaging materials. In the industrial fermentation process, sterilized antifoam oil is used to lubricate the shaft. Antifoam is passed through the chiller for cooling purposes. MS is very important in the case of fermentation industries because there is every possibility of leakage of air through it. Therefore, MS is carefully designed and sterilized antifoam oil is passed to the surface of the shaft in the MS for lubrication. This will reduce friction of the shaft with the MS significantly.

6.5.3 pH Probe, Dissolved Oxygen (DO) Probe, Thermistor and Foam Sensor

A sterilizable pH probe is used in the fermenter to monitor the pH profile of the fermentation medium. This differs from the normal pH probe because it can withstand the high steam pressure (15 psi) of the fermenter. This pressure is maintained on the surface of the electrolyte present in the pH probe with the help of a hand pump. These are manufactured by Mettler Toledo, Ingold etc. In the case of maintaining a particular pH of the fermentation medium, this is connected to the acid and alkali tank with the help of pumps. The DO probe is used for monitoring the dissolved oxygen concentration of the medium because living cells take only the DO for their growth and metabolisms. These are also manufactured by Mettler Toledo. A thermistor is a biometallic device to monitor the temperature in the fermenter. This is connected to the pumps. These pumps are connected to the hot and cold water tanks. Living cells are very sensitive to temperature. Therefore, if the temperature rises, the pump connected to the cold water tank will be energized for reducing the temperature. Similarly the pump connected to the hot water tank will be energized when the temperature decreases. This is how the temperature of the fermenter is maintained. All the monitoring devices are connected to the PID controller of the fermentation plant. The foam sensor has a sharp metallic head and is connected to the pump. This pump is connected to the sterilized anti-foam oil tank. The same anti-foam oil may not be suitable for all fermentation processes. During the fermentation process, foam formation takes place. If the foams are allowed to build up this can damage the mechanical seal which is usually located at the cover lid of the fermenter (Bailey and Ollis, 2010).

6.5.4 Air Filter and Medium Sterilizer

Air and liquid are the major sources of contaminations within the fermentation industries. There are two types of air: stagnant air and moving air for sparging in the fermenter. Stagnant air can be sterilized mostly by using either UV rays or germicidal spray. Different oxidizing gases such as chlorine, iodine, hypochlorites etc. are considered for the sterilization of stagnant air. Sterilization of moving air is done in industry by using glass wool filter. Contaminants present in the air are removed by a physical separation technique, mostly by using glass wool fibres. However, membrane filters may be used for small-scale applications. Every fermentation industry has two air filters or sterilizers: one is used for the inoculum vessel (IV) and another is used for the production fermenter (PF). The volume of IV is 5–10% of that of PF. Therefore the capacity of the air filter for PF is much higher compared to that of IV (Doran, 2012).

Wet steam heat is found to be very effective for the sterilization of the medium. Medium sterilization is done in two modes: batch and continuous. Batch sterilization process is mostly used for small-scale application. Continuous sterilization processes are found most effective both with respect to the amount of steam consumption and the time of sterilization. Air and medium sterilization processes are discussed in detail in Chapter 7.

6.5.5 Air Sparger

Air supply to sparger is carried out through air filters or sterilizers. There are three types of sparger; namely, porous sparger, orifice sparger and nozzle sparger. The size of the air bubbles influences the mass transfer of the fermenter to a great extent. The surface area of the bubbles should be large because this contains a larger volume of air. Efficiency of aeration increases with the size of the holes in the sparger. The size of the air bubbles is between 0.0001 and 0.1 inch diameter. A round pipe provides a number of holes. The diameter of the holes is 1/64 to 1/32 inch. Air is sent by a sparger which contain branches of pipes, known as distributor pipe which contain several holes. The diameter of the pipe hole is as mention above. This distribution pipe cover is the interior part of the base. The optimum rate of aeration is 1 VVM (volume of air/liquid volume/min). Air flow rate in the fermentation industry is controlled by rotameters. Different capacity rotameters are available. Rotameters have a vertical linear scale which is to be calibrated with the help of a gas flow meter.

The purposes of different accessories used in the fermenter are discussed below:

Aeration: The purpose of aeration is to provide sufficient oxygen in the submerged fermentation process for the growth and metabolism of microorganisms.

Compressor: This is used to supply oxygen (air). It is a mechanical device that increases the pressure of the air by reducing its volume and forces more and more air into the storage tank. The compressed air passes through the air filter, then sterilized air is passed to the fermenter.

Rotameter: Flow rate of the air is determined with the help of a rotameter. In the rotameter there is a linear scale and a float (a small ball). When air comes in, the flow rate increases and this ball will float/move up. It is a device that measures the flow rate of fluid in a closed tube. The rotameter must be calibrated before use. The rotameter is calibrated with the help of a gas flow meter. With gas flow meters one can find the flow rate of air. The calibration curve of flow rate vs. linear scale is prepared. At different flow rates, the values at linear scale will also change. Therefore from the calibration curve, one can easily find the actual flow rate of the air. Rotameters are largely used in fermentation industries to find out the flow rate of air. On the other hand, in the chemical industry, the flow rate of the gas is very high. Therefore different types of flow meters (e.g. venture meter and orifice meter etc.) are used.

Air filter: This is a device composed of fibrous material which removes solid particulars such as dust pollen, moulds and bacteria from air. Air filters are very important in fermenters because we know that air contains a lot of particulate matter and particulate matter that are the carrier of microorganisms. If we do not take out this particulate matter from the air, then it will easily enter into the liquid media and will contaminate the system. To remove this particulate matter we use fibrous filters; in particular, glass wool fibres are largely used.

Non-return valve: This is also called a one-way valve. Non-return valves normally allow fluid to flow through themselves in only one direction. The purpose of the check valve is to prevent accidental reversal of the flow of liquid or gas in a pipe. The non-return valve is usually placed in the air inlet line of the fermenter to prevent the back flow of the media.

Mechanical seal: Mechanical seals play a very important role in fermenters. At the connection between the motor and shaft, a mechanical seal (M-seal) is present. The seal consists of two parts; the bearing is the stationary part and the shaft is the rotating part. The two components are pressed together by springs. Sterilized antifoam oils are used to lubricate and cool seals during operation. In industry anti-foam oil is passed at high pressure to maintain positive pressure inside the mechanical seal. If there is positive pressure inside the mechanical seal then there will be less possibility of air entering into the system, and sterility of the process can be maintained. However, the main purpose of passing the oil is to lubricate the shaft.

Feed port: These are silicon tubes connected to the nutrient reservoir. They are used to add nutrients and other important substituents into the fermenter vessel. In lab fermenters the purpose of silicon tubing is for transferring the liquid into the fermenter whereas in industrial fermenters, a stainless steel pipe line is present through which media is transferred into the fermenter.

Baffles: These are metal strips attached radially to the wall of the fermenter. They are used to prevent vortex and to improve aeration capacity. Baffles maintain a gap with the vessel wall to enable scouring action thus minimizing microbial growth on the walls of fermenter. When the agitator rotates, in absence of baffles vortex formation takes place. If a vortex is formed the retention time of the air bubble will vary across the cross-section of the fermenter. Therefore metal strips on the radial wall of the vessel will prevent vortex formation.

Impellers: This is very important part. These are used for agitation purposes, which are required to ensure that a uniform suspension of microbial cells is achieved in a homogeneous nutrient medium.

One example of an impeller is the Rushton impeller. This has a disc which is connected to many blades. The number of blades can vary (2, 3, 4, 5 or 6). The number of blades plays a very important role because it can cause tremendous vibration of the fermenter. Therefore in industry the combination of the blades is arranged in such a way as to minimize vibration in the fermenter. Higher vibration of the fermenter is undesirable because some of the connections will be loosened and contamination problems will occur.

Sparger: This is used for aeration. The purpose of aeration is to provide sufficient oxygen to the microorganism for metabolic requirements; the sparger introduces air into liquid in the vessel. Spargers can be of three types: porous, orifice, and nozzle.

Reflux cooler: This is actually fitted in the outgoing airstream. The air flowing out of the fermenter has the same temperature as during cultivation and is also correspondingly saturated. The moisture is condensed out with the reflux cooler and then is returned to the fermenter, so that the volume of liquid will remain constant.

Exhaust air filter: This is used to stop contamination problems.

Safety valve: This is very much required. If the pressure inside the reactor increases for some reason then there is a possibility of a hazard taking place. In pressure cooker vessels the safety valve is used to control the steam pressure. Similarly in fermenters a safety valve is also required. This valve ensures that the pressure never exceeds the safe upper limit of the specified value.

Temperature loop: This comprises a heater and a cooling finger. In the fermenter a stainless steel jacket is usually used to avoid any explosion during *in situ* sterilization. The surface of the fermenter is wrapped with a pipe line and then the water is passed through it, so that water flows uniformly through the surface of the fermenter and maintains the temperature. Both cold water and hot water can be passed through this pipe according to the requirement.

pH probe: The pH probe used in the fermenter is different from a normal pH probe. The pH probe attached to the fermenter should withstand steam sterilization pressure of 15 psi. The speciality of this pH probe is that one can maintain a pressure of 15 psi inside with the help of a hand pump. Otherwise, in the case of ordinary pH probes, liquid medium can enter into the probe through the porous glass and affect the electrolyte and damage the probe. Therefore, above the electrolyte, 15 psi pressure is maintained to avoid damage to the probe (Das and Das, 2020).

6.6 OPERATION OF BIOHYDROGEN PILOT PLANT

A 10 m^3 biohydrogen pilot plant (BPP) has been designed, fabricated and commissioned at Indian Institute of Technology Kharagpur (Figure 6.12). The details of the plant accessories and the operation are discussed here in order to provide some practical exposure. This may give some idea of the operation of industrial fermentation plants. The BPP has different utilities, such as a steam generator, hot water bath, chiller and air compressor. It comprises several vessels as shown in Table 6.1. The block flow diagram of the process is shown in Figure 6.10. Initially inoculum of *Enterobacter cloacae* IIT-BT08 is prepared in a 5 L conical flask using a shaker. This is followed by the inoculation of the culture to the 50 L fermenter in between mid-log and late-log phase to ensure that the inoculum has active microbial cells. This is transferred to the 50 L fermenter under sterile conditions. Anaerobic conditions are maintained. This is followed by transferring of the culture to a 500 L vessel when the rate of hydrogen production is maximum. 4,500 L feed is taken in the 10 m^3 reactor and anaerobic conditions are maintained by sparging nitrogen gas. When the hydrogen production in the 500 L reactor is maximum, then the culture is transferred to the 10 m^3 fermenter containing 4.5 m^3 medium. The volume is increased to 10 m^3 when the rate of hydrogen production is at a maximum. The cumulative rate of hydrogen production is about 76 m^3 / d. This is shown in the flow diagram Figure 6.13 (Balachandar et al., 2020; Das et al., 2014).

The carbon dioxide present in the gas is removed by passing the gas through the absorber containing 50%w/v KOH solution. The amount of gas production is monitored with the help of a hydrogen flow meter. Hydrogen can be used as a fuel or may

FIGURE 6.12 10 m^3 biohydrogen pilot plant at Indian Institute of Technology Kharagpur.

TABLE 6.1
Different Vessels Used in the Pilot Plant

Name of the vessels, utilities and accessories	Photograph
10 m^3 Biohydrogen reactor	 
Two 3 m^3 feed tanks	
50 L and 500 L inoculum vessels	

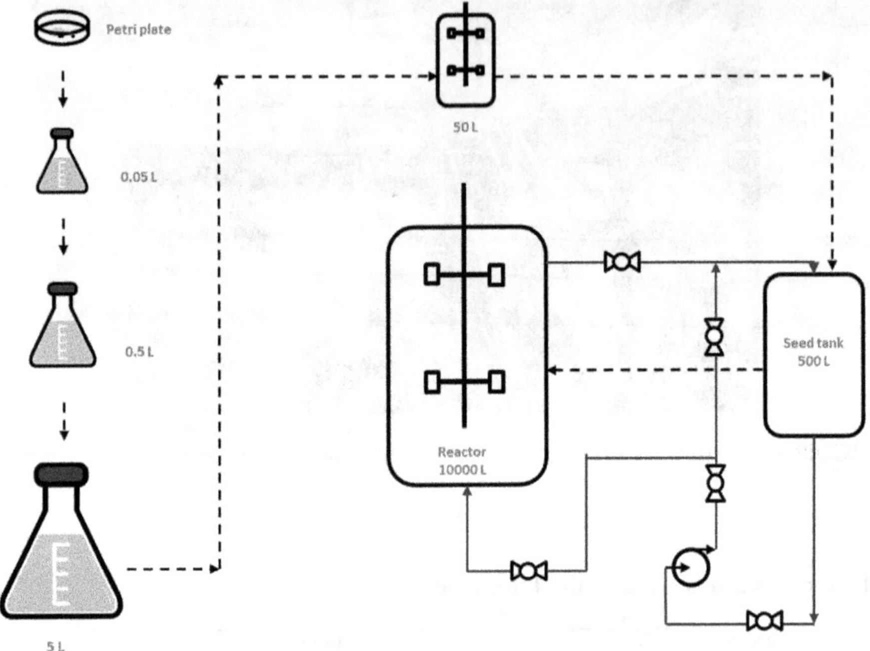

FIGURE 6.13 Flow diagram for the inoculum preparation and biohydrogen production.

be used for the generation of electricity by passing it through a fuel cell. Excess hydrogen gas is passed through the exit pipe line containing a fire arrestor to avoid fire hazard. The composition of the gas is determined by using an online gas analyser from Gas Chromatograph. The overall flow diagram of the plant is shown in Figure 6.14. The live demonstration of the process is shown on the website: www.bioh2iitkgp.in.

The PI & D diagram of the biohydrogen pilot plant is shown in Figure 6.15. This depicts the use of several pneumatic control valves, a centrifugal pump and a hydrogen flow meter. Pneumatic control valves have off and on devices with the help of air pressure through the PI & D controller located in the control room of the plant as shown in Figure 6.4 (Balachandar, 2019).

6.7 CONCLUSIONS

The industrial fermentation process comprises several units besides the fermenter. This is explained with the help of either block flow diagrams (BFD) or process flow diagrams (PFD). Individual blocks in the BFD indicate one unit whereas symbols / configurations for the individual units are depicted in the PFD. The PID controller deals with various connections; pipe lines, valves, pumps etc. This is located in the control room. The most important part of the fermenter is the mechanical seal because this may cause contamination problems. Ball valves are used in the fermentation industry for drawing the sample from the fermenter. Peristaltic pumps are used for

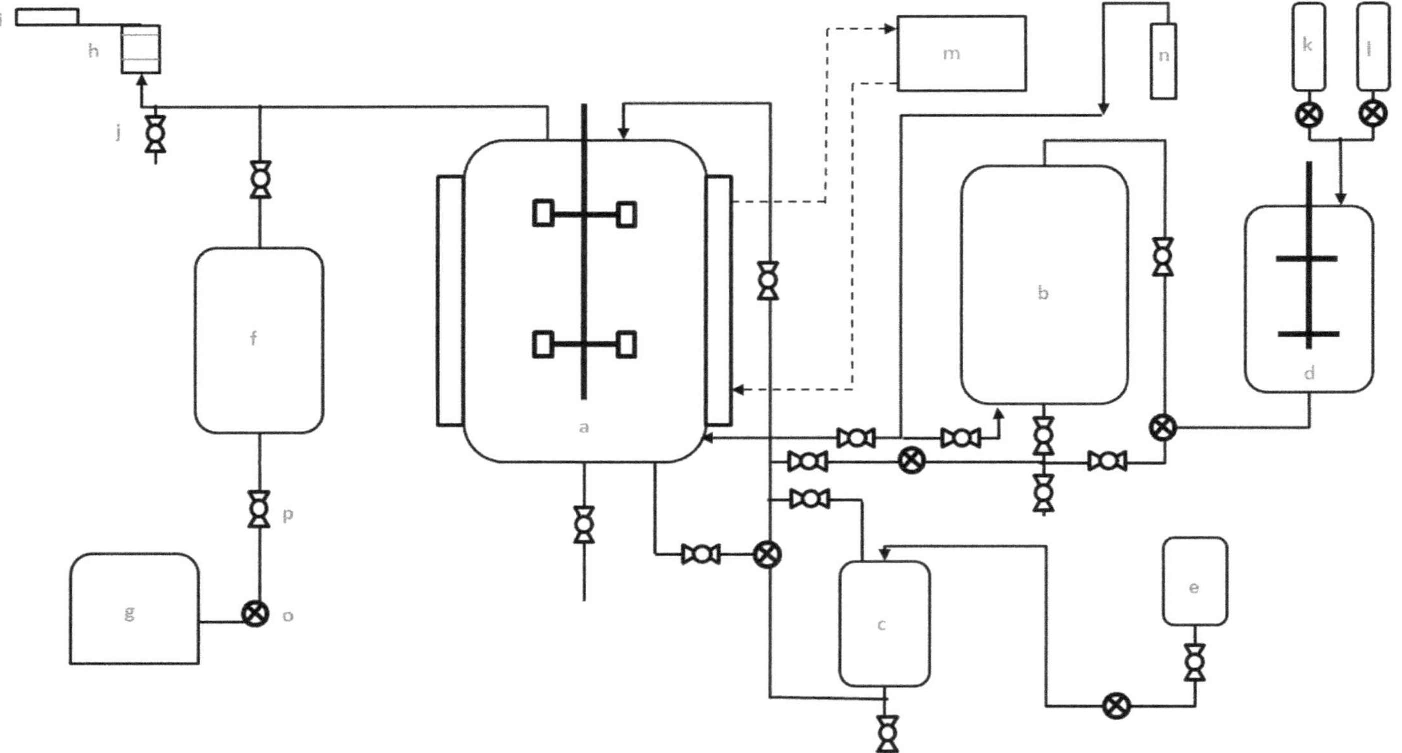

FIGURE 6.14 Block flow diagram of the biohydrogen pilot plant.

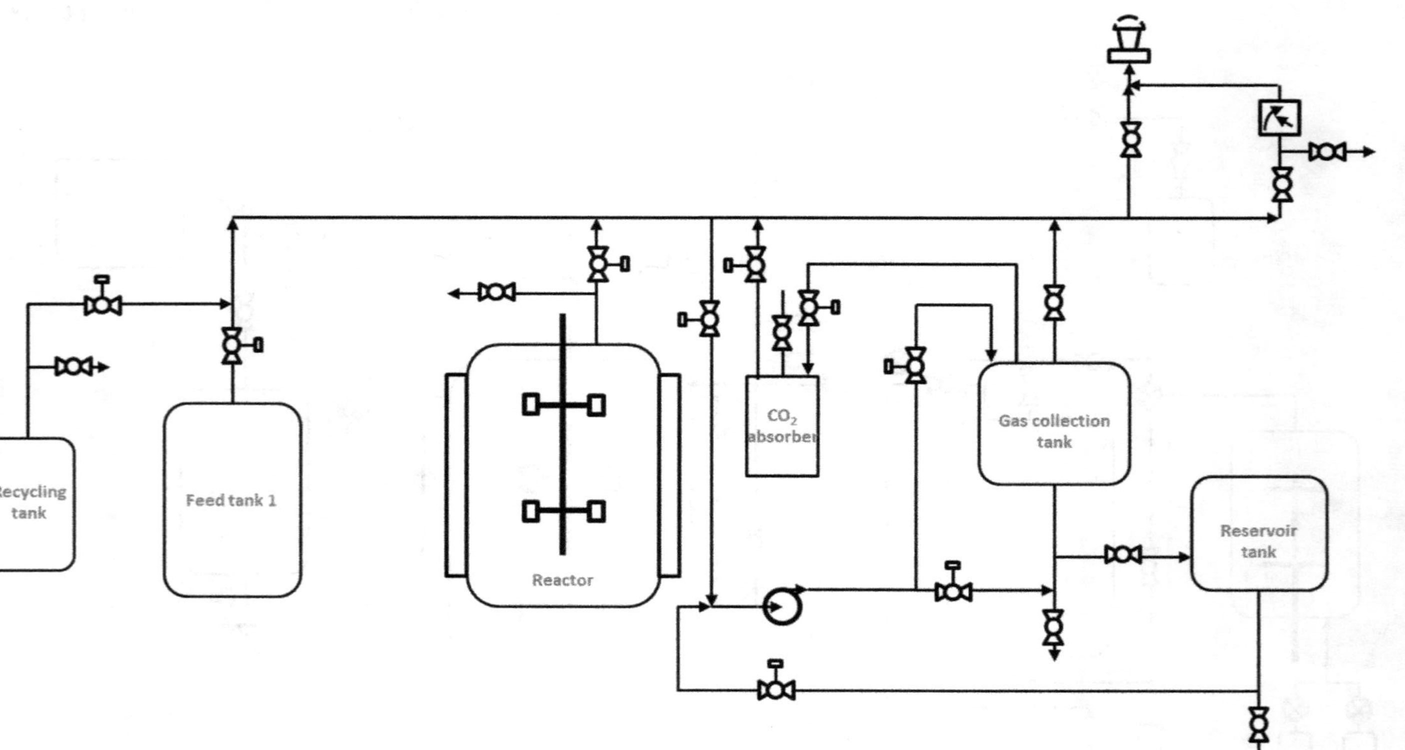

FIGURE 6.15 PID diagram of the biohydrogen pilot plant.

the aseptic transfer of the medium/acid/alkali/anti-foam in the fermenter. Fermenters have various online monitoring systems, such as pH, DO, temperature, rpm of the stirrer etc. The 10 m^3 biohydrogen pilot plant has been discussed in detail to give practical exposure to the fermentation industry.

REFERENCES

Bailey JE and Ollis DF, Biochemical Engineering Fundamentals, McGraw-Hill, New Delhi, 2010.

Balachandar G, Biohydrogen production from organic wastes and residues by dark fermentation, Ph.D. thesis, Indian Institute of Technology Kharagpur, 2019.

Balachandar G, Varanasi Jhansi L, Singh Vaishali, Singh Harshita, Das Debabrata, Biological hydrogen production via Dark fermentation: A holistic approach from Lab-scale to Pilot-scale, *International Journal of Hydrogen Energy*, 45: 5202–5215, 2020.

Das D, Algal Biorefinery: An Integrated Approach, Capital Publications, New Delhi and Springer, Switzerland, 2015.

Das D and Das D, Biochemical Engineering: A Laboratory Manual, Jenny Stanford, Singapore, 2020

Das D and Das D. Biochemical Engineering: An Introductory Text book, Jenny Stanford, Singapore, 2019

Das D, Khanna N, and Dasgupta CN, Biohydrogen production: Fundamentals and Technology Advances, CRC Press, 2014.

Doran PM, Bioprocess Engineering Principles, Second Edition, Academic press, Waltham, 2012.

Moo-Young M (Editor-in-Chief), Comprehensive Biotechnology (3rd edition), Academic Press, 2019

Prescott SC and Dunn CG, Industrial Microbiology, McGraw Hill Book, and K O Gakusha, Tokyo, 1959.

Thakur IS Industrial Biotechnology: Problems and Remedies, I. K. International Publishing House, New Delhi, 2013.

Wittmann C and Liao JC (eds.), Industrial Biotechnology: Products and Processes, Wiley-VCH, 2017.

7 Sterility in the Biochemical Industries

In the fermentation industry one important aspect is sterility because in most cases one desired microorganism should be allowed to grow in a controlled environment. There are three major sources of contamination:

- Air
- Water
- Fermentation vessel

Fermentation processes are broadly classified as aerobic and anaerobic. Most bioproducts are produced through the aerobic fermentation processes where dissolve oxygen (DO) is required for cell growth and metabolism, e.g. penicillin, streptomycin, citric acid etc. However, anaerobic fermentation processes are also used for a few biochemical industries, e.g. ethanol, butanol, methane, hydrogen etc. These processes are operated in the absence of oxygen.

7.1 AIR STERILIZATION

In the aerobic fermentation process, microorganisms take the oxygen which is dissolved in the fermentation medium. Oxygen content in the air is about 21%v/v. Air is sparged in the fermentation medium to increase its DO concentration. Oxygen is sparingly soluble in water. Therefore the air distribution in the fermentation medium plays a very important role.

In aerobic fermentation, the air should be sterilized, and free from suspended particles. The quality of outdoor air depends on the location of the fermentation industry. The maximum average particulate matter concentration is 279 μg/ m^3. Higher amount of contaminants present in the sparging air may increase contamination problems. Therefore, proper air sterilization is very important to the fermentation industry. The amount of particulate matter present in the air depends on the location of the industry e.g. particulate matter present in the air of metropolitan cities such as New Delhi or New York is more compared with cities located at higher altitude such as Dehradun. Microorganisms are present in the air in the range of 80 to 2000/ m^3. Air comprises different microorganisms such as 50% fungal spores, 40% Gram negative bacteria and small trace amount of viruses. The size of these organisms varies from 0.5 μm to several hundred μm. Normally airborne particles considered for destruction or removal during air sterilization are about the size of small bacteria, namely 0.5 to 1 μm. However, this depends on the fermentation process; e.g. in the case of the citric acid fermentation process, the major contaminant that influences the fermentation process is yeast whose size varies from 3 to 7 μm.

Air can be sterilized by several methods, namely

- Filtration
- Heat
- UV rays
- Chemical agents

There are two types of air, namely moving air used in the fermenter and stagnant air used in the culture preparation laboratory, operation theatre etc. Air used in the fermenter is mostly sterilized by filtration. Filtration is a physical separation technique. There are two types of filter

- Membrane filter
- Depth filter

Membrane filters are defined as a filter – especially membrane such as cellulose acetate – that has pore size less than the size of the microorganism to be removed e.g. viruses or bacteria or yeast. This is mostly used in the laboratory. Depth filters use a porous filtration medium such as compact glass wool fibres to retain particles throughout the medium, rather than just on the surface of the medium. Heat is not suitable for sterilizing the air because air is a non-conductor of heat so a higher amount of heat is required for removing the contaminants present in the air. UV rays and chemical agents (known as germicides) such as chlorine, formaldehyde etc. are used for the sterilization of stagnant air.

7.1.1 Depth Filter

Depth filters are mostly used for the industrial fermentation process for the sterilization of air. Depth filters comprise fibrous materials whose typical pore size is greater than the minimum size of the particle to be removed. The basis of the particle removal is based on the probability that a particle will be retained in the filter. Glass wool filter is considered for the sterilization of air. The particles are trapped and removed when air is passed through a depth filter containing glass wool fibres. The principles for the physical removal of the particles in the filtration technique are inertia impaction, interception, gravitational force, electrostatic force of attraction and diffusion. Microbial cell present in the air are very small in size so particle separation by gravitational force may be neglected. The charge distribution in the microorganisms present in the air is different. Therefore particle separation by electrostatic attraction may be neglected. Thus the particle separation by inertia impaction, interception, and diffusion are taken into consideration (Figure 7.1).

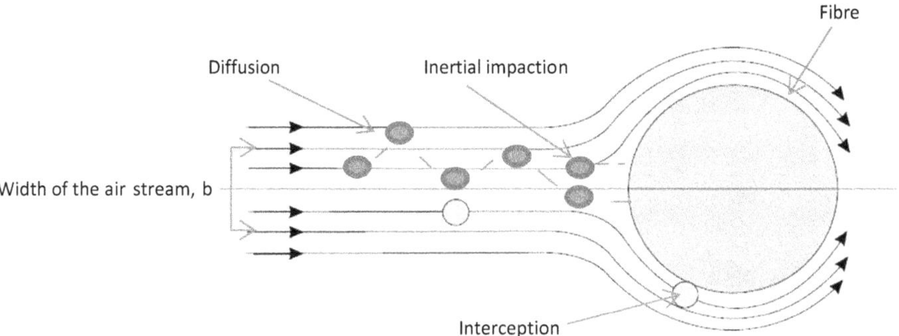

FIGURE 7.1 Mode of air flow is perpendicular against the cross-section of the fibre in the depth filter (Das and Das, 2019).

The following assumptions are made:

- Single fibres are considered to be cylindrical and are placed perpendicularly to the air flow; the air flow around the cylinder is laminar with no vortices.
- 2-dimensional system.

In the case of inertial impaction, due to the inertia of the particles the flow pattern of the particles deviates from that of the air flow as they approach the surface of the fibre (cylinder) (Figure 7.1). This takes place at high air velocity. According to Newton's first law of motion, the particles tend to travel in a straight line due to their momentum. This law says that every particle tries to move in the same direction because of the very high air velocity due to inertial force. Inertia may be defined as the property of matter which indicates that a particle remains at rest or in motion in the same straight line unless some external force is applied. The suspended particles present in the air flow have momentum. The resistance of the particles present in the air by the filter is less. However, the particles tend to travel in straight lines due to their momentum (Figure 7.1). In Figure 7.1, b denotes the width of the upstream air flow. The particles that move beyond 'b' in the flow line air will not touch the cylindrical surface of the fibre even after they deviate from the air flow line near the cylinder. Eq. 7.1 represents the collection efficiency of a single fibre (η_o') due to the inertial effect of the particles.

$$\eta_o' = \frac{b}{d_f} \tag{7.1}$$

Where b = effective width of the air flow for the separation of the particles due to inertial impact, d_f = fibre diameter.

$$\eta_o' = 0, \text{at } \varphi = \frac{1}{16} \tag{7.2}$$

$$\text{Where inertial parameter} = \varphi = \frac{C\rho_P d_p^2 V_o}{18\mu d_f} \tag{7.3}$$

$$\text{Critical air velocity} = V_c = (1.125)\frac{\mu d_f}{C\rho_P d_p^2} \tag{7.4}$$

Where ρ_P = particle density, g/cm^3; d_P = diameter of the particle, μm or cm; C = Cunningham's correction factors for slip flow; d_f = diameter of the fibre, μm or cm; μ = viscosity of air, g/cm s.

At the velocity of air lower than V_c, the inertial impaction of particle may be neglected.

In the case of interception, particles are collected by contact with the fibres during the entrainment in streamlines of air. The particles are considered to be intercepted. The limited condition for the deposition of entrained particles as they pass a cylindrical fibre streamline of air flow which is $\frac{d_P}{2}$ from the fibre surface at a location $\theta = \frac{\pi}{2}$. Eq. 7.5 shows the collection efficiency due to interception.

$$\eta_o'' = \frac{1}{2(2-\ln N_{Re})}\left\{2(1+N_R)\ln(1+N_R)-(1+N_R)+\frac{1}{(1+N_R)}\right\} \tag{7.5}$$

$$\text{Where } N_R = \frac{d_p}{d_f} = \text{Geometrical ratio, Reynolds no. } (N_{Re}) = \frac{d_f u\rho}{\mu} \tag{7.6}$$

Where ρ = density of air, g/cm^3, u = linear air velocity, cm/s.

A filter comprises fibres which are compressed to get openings of various pore sizes. Particles larger than the size of the pores of the filter are removed by interception. However, particles which are smaller than the filter pore sizes, are also retained by interception.

In the case of diffusion, small particles may be collected on the surface of the fibres due to their Brownian motion. The collection efficiency due to diffusion may be expressed as

$$\eta_o''' = \frac{1}{2(2-\ln N_{Re})}\left[2\left(1+\frac{2X_o}{d_f}\right)\ln\left(1+\frac{2X_o}{d_f}\right)-\left(1+\frac{2X_o}{d_f}\right)+\frac{1}{1+2X_o/d_f}\right] \tag{7.7}$$

$$\text{Where } \frac{2X_o}{d_f} = 1.12\frac{2(2-\ln N_{Re})D_{BM}}{Vd_f} \tag{7.8}$$

Where 2 x_0 = displacement of the particle, $D_{BM} = \frac{CKT}{3\pi\mu d_p}$ = Diffusivity of the particle, V = velocity of air.

Therefore, overall collection efficiency,

$$\text{Overall collection efficiency } (\eta_o) = (\eta_o' + \eta_o'' + \eta_o''') \tag{7.9}$$

7.1.2 Design of the Air Filter

Practically, in case of air sterilization at higher air velocity, the collection efficiency due to inertial impact is usually dominant. Therefore η_0' can be neglected in finding η_0 at low velocity (less than 10 cm/s by using glass wool fibre). Eq. 7.10 represents the overall collection efficiency of a fibrous filter.

$$\text{Collection efficiency} = \overline{\eta} = \left(1 - \frac{N}{N_0}\right) \tag{7.10}$$

Where N_0= number of contaminants in the inflow air, and N = number of contaminants in the outflow air.

$$\text{Again, single fibre efficiency } (\eta_\infty) = \pi d_f \frac{(1-\alpha)}{4L\alpha} \ln \frac{N_0}{N} \tag{7.11}$$

$$= \pi\, d_f \frac{(1-\alpha)}{4\,L\,\alpha} \ln \frac{1}{(1-\overline{\eta})} \tag{7.12}$$

Where L = thickness of filter bed, α = volume fraction of the filter

Eq. 7.12 represents a log-penetration relation where a fraction of particles are collected within the thickness of the air filter, L. The thickness of the air filter is usually less than 4 cm. The correlation between η_∞ and η_0 is expressed as

$$\eta_\infty = \eta_0 (1 + 4.5\alpha), \qquad 0 < \alpha < 0.10 \tag{7.13}$$

The plot $\eta_0\, N_R\, N_{Pe}$ ($= \eta_0\, N_R\, N_{Sc}\, N_{Re}$) vs. $N_R\, N_{Pe}^{1/3}\, N_{Re}^{\frac{1}{18}}$ is shown in Figure 7.2. The η_0 (overall collection efficiency) is estimated from the Figure 7.2 at a specific set of operating conditions. Using the η_0 value in Eq. 7.13, η_∞ value can be determined. Eq. 7.12 is used to find out the depth of the air filter (L) using this η_∞ value.

The exponential relationship between $\overline{\eta}$ and L is shown in Figure 7.3. This indicates that at infinite depth of the air filter one can achieve 100% collection efficiency of the air filter. Therefore 100% collection efficiency is not possible. For designing the air filter, one can consider one in 1 million or one in 10 million according to the requirement of the sterility level of the fermentation process.

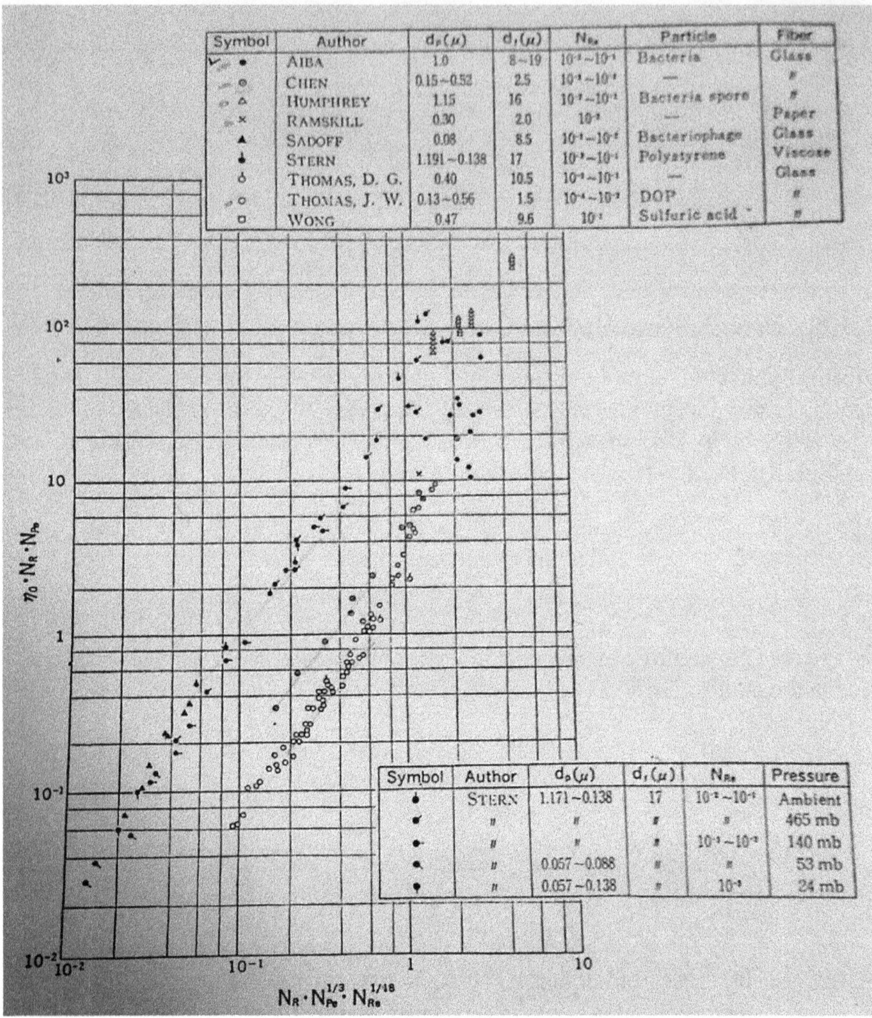

FIGURE 7.2 Plot of $\eta_0 \; N_R \; N_{Pe}$ vs. $N_R \; N_{Pe}{}^{1/3} \; N_{Re}{}^{\frac{1}{18}}$.

7.1.3 Theory of Depth Filtration

In the case of depth filters, it is assumed that the particles remain attached to the surface of the fibre, and are uniformly distributed in the filter at any depth (X), then the reduction of the particles in each layer of a unit thickness of the filter by the same proportion which may be mathematically expressed as

$$\frac{dN}{dX} = -kN \tag{7.14}$$

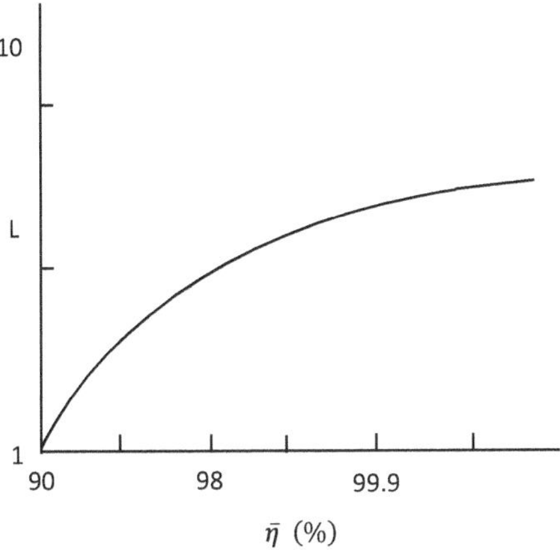

FIGURE 7.3 Correlation between thickness of the air filter and the collection efficiency.

where N is the concentration of particles in the air outflow stream of the filter of depth X, k is a constant.

$$\ln\frac{N}{N_0} = -k\,X \tag{7.15}$$

On integrating the above equation we get

$$\frac{N}{N_0} = e^{-kX} \tag{7.16}$$

where N_0 is the number of particles in the inflow air flow of the filter, and N is the number of particles in the outflow of the filter.

This equation is known as the log penetration relationship.

From Eq. 7.15, a plot of the logarithm of (N/No) vs. X will be a straight line where the slope is equal to k.

The efficiency of the filter may be expressed mathematically as given in Eq. 7.17.

$$E = \frac{(N_0 - N)}{N_0} \tag{7.17}$$

where E is the filtration efficiency of the air.

$$\text{But } \frac{(N_0 - N)}{N_0} = 1 - \frac{N}{N_0} \tag{7.18}$$

$$\text{Substituting, } \frac{N}{N_0} = e^{-kX}$$

$$\text{Thus } \frac{(N_0 - N)}{N_0} = 1 - \frac{N}{N_0} = 1 - e^{-kX} \qquad (7.19)$$

Air filters may be designed on the basis of the log penetration relationship. X_{90} is assumed as the depth of filter required for the removal of 90% of the total number of particles entering the filter.

If $N_0 = 10$ and N = 1, Eq. 7.15 may be written as

$$\ln \frac{1}{10} = -k X_{90}$$

$$-\ln 10 = -k X_{90}$$

$$X_{90} = \frac{2.303}{k} \qquad (7.20)$$

Where X_{90} is the depth filter thickness for the removal of 90% of the air contaminants.

The nature of the filter material and the linear air velocity through the filter are influenced by the value of k. The value of k increases with the air velocity and then reaches a maximum value which is known as the optimum value because the value of k decreases during the further increase in air velocity (Figure 7.4).

Increase of impaction plays a very important role during the increase of the k value of the removal of the particles with the increase of air velocity. This is the principle behind the removal of the contaminants present in the air. However, at high air velocities, k values decrease, probably due to the formation of channels in the filter and also vibration of the fibres. These are responsible for the removal of previously captured microorganisms.

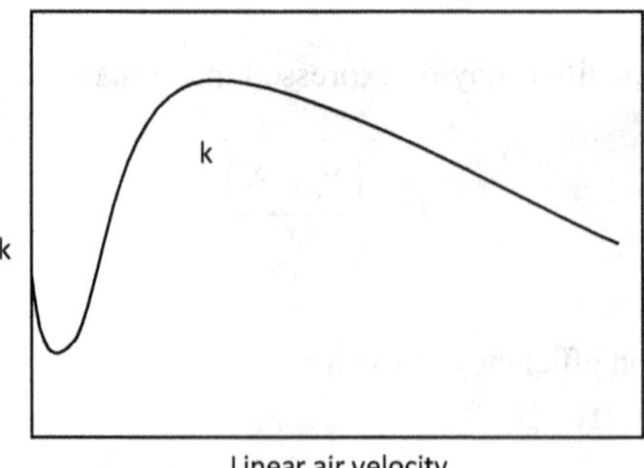

FIGURE 7.4 Plot of linear velocity of air vs. k.

7.1.4 Selection Criteria of the Air Filter

For the selection of air filter, the following criteria are taken into consideration

- Filter retention efficiency
- Economy of operation
- Ease of filter use
- Service provided by the manufacturer

Filter efficiency and reliability of organism retention are the most important selection criteria. The highest level of filtration efficiency may be achieved at a fixed submicron pore size of membrane filters. Wetting of the filter materials due to air moisture content depends on their hydrophobic character (Richard, 1967; Whittet et al., 1965).

7.1.5 Suitability of Glass Wool Fibres for the Air Filter

Kimura et. al. proposed a mathematical correlation among the modified drag coefficient (C_{Dm}) and the pressure drop and other variables present in the air filter as shown in Eq. 7.21 (Das and Bhattacharyya, 1989).

$$C_{Dm} = \frac{\pi g_c d_f \Delta P}{2\rho L V^2 (1-\varepsilon)^m} \tag{7.21}$$

Where ΔP = pressure drop of air flow, g_c = conversion factor, ρ = density of air, ε = void fraction (1-α), V = air velocity, m = empirical component.

The value of empirical component (m) depends on the filter depth (L) as shown in Figure 7.5. This indicates that the value of m depends on not only the length of the filter but also on the diameter of the fibre. Figure 7.6 represents the correlation between the drag coefficient and the Reynolds number by using glass wool and a cotton filter. For a particular air flow, glass wool filters gives less pressure drop as

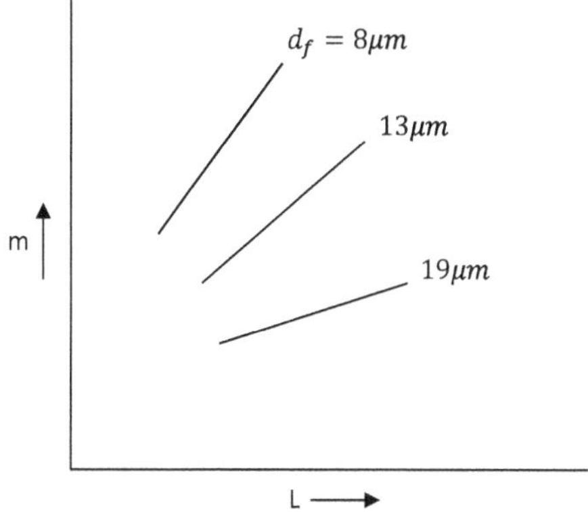

FIGURE 7.5 Correlation between m vs. L using different diameter glass wool fibres.

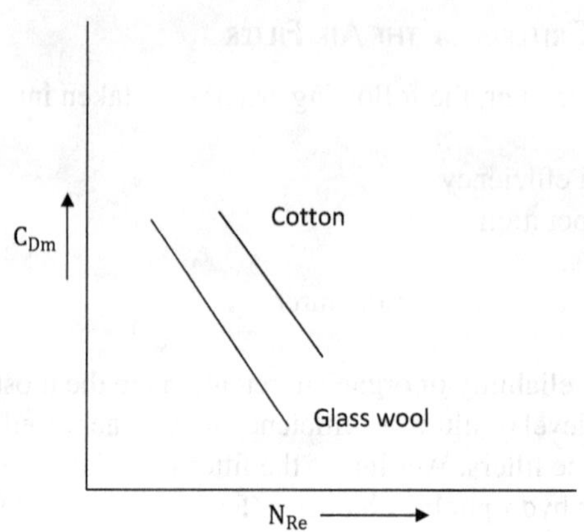

FIGURE 7.6 Correlation between drag coefficient and Reynolds number.

compared to that of cotton filters. Again, pressure drop of the air filter is directly proportional to the drag cofficient. Therefore, it is observed that the drag coefficient in the case of cotton fibre is more as compared to glass wool fibres for a particular air flow with a specific Reynolds number. Thus, cotton fibres cause more heat development. Cotton fibres can also catch fire very easily at lower temperature as compared to glass wool fibres. In addition, glass fibres can be regenerated very easily. Therefore, glass wool filters is recommended for air sterilization in the industrial fermentation process.

7.1.6 Membrane Filters

Membrane filters have a fixed pore. These are largely used in the fermentation industry for the sterilization of air in the case where there is a small amount of air requirement. These filters are usually made from different polymeric materials e.g. cellulose diacetate, cellulose nitrate, polycarbonate and polyester. The pore size of the membranes varies from 0.015 μm to 12 μm. Autoclave is suitable for the sterilization of these filters. There are two processes considered for membrane filter production: the pores of capillary pore membranes are produced by radiation and forced evaporation of solvents from cellulose esters and are considered for the production of labyrinthine pore membranes. The membrane filter has a very high pressure drop. The maintenance cost of the membrane is directly proportional to the pressure drop. In addition, the life of the membrane depends on the pressure drop. Thus the major bottlenecks for the use of the membrane filter are the operation cost and life of the membrane.

7.2 MEDIUM STERILIZATION

Medium sterilization is one of the first and most critical unit operations required to make the fermentation process a success. The main objective is to stop the growth of contaminants or undesired microorganisms during the fermentation process, bioconversion, enzymatic reaction, or storage of the medium. Sterilization means destruction or removal of all viable organisms from an object or from a particular environment. Methods of sterilization are classified as follows:

A. **Physical Method**
   - Thermal (Heat) methods
     - Dry heat
     - Moist heat
   - Radiation method
   - Filtration method

B. **Chemical Method**
   - Disinfectants
   - Alcohols
   - Oxidizing agents
   - Salts
   - Surface active agents

Physical methods of sterilization deal with the destruction or removal of all viable microorganisms and their spores using heat, radiation and filtration. Dry heat requires higher temperatures (160–180 °C) and requires exposure times of 2 h, depending upon the operating temperature. Moist heat sterilization is found to be more effective as compared to dry heat which involves the use of wet steam in the range of 121–140°C. High steam pressure is used to generate the high temperature needed for sterilization. Sterilization can also be done using electromagnetic radiation such as electron beams, X-rays, gamma rays, or irradiation by subatomic particles. Electromagnetic or particulate radiation is of two types: energetic enough to ionize atoms or molecules (ionizing radiation), and less energetic (non-ionizing radiation). Ultraviolet (UV) light irradiation is used for the non-ionic radiation for sterilization of solid surfaces and also some transparent objects. UV irradiation is routinely considered to sterilize the interiors of biological safety cabinets, operation theatres etc. Gamma radiation is considered for ionizing radiation sterilization and is commonly used for sterilization of disposable medical equipment, such as syringes, needles and food. It is emitted by radioisotopes.

On the other hand, the microorganisms are physically removed in the filtration process without the destruction of the cells. It is considered for sterilization of liquids and gases as it is capable of stopping the passage of both viable cells and non-viable particles. The nutritional quality of the solutions may be damaged or denatured at high temperatures or with chemical agents. The filtration mechanisms mostly involve adsorption, sieving, and trapping within the solid matrix of the filter. The filter pore sizes of the filter for the removal of bacteria, yeasts, and fungi are varied in the range of 0.22–0.45 μm.

Chemical sterilization is the destruction of all viable microorganisms and their spores using liquid or gaseous compounds such as alcohol, oxidizing agents (chlorine), salts, disinfectant agents etc. Sterilization may be done through either physical removal or destruction of the microorganism by heat. Selective destruction of the microorganism may adversely affect a specific biochemical process. Therefore medium sterilization is based on either removing or destroying all microorganisms or contaminants present in the medium. Thus sterilization of the medium is broadly classified as

- Filtration or physical removal of microorganisms and
- Destruction of microorganisms by heat

The heat treatment process is largely used for medium sterilization in biochemical processes due to the good heat conduction/convection properties of water. The choice of sterilization method is influenced by the following factors

- Effectiveness of the process, i.e. level of sterility
- Reliability
- Change in the nutritional quality of the medium
- Capital and operating cost

7.2.1 Filtration

The physical properties of the contaminants present in the liquid play an important role in designing a suitable procedure for the removal of microorganisms by filtration. Filamentous, mycelial organisms such as moulds, typically have dimension varying from 7 μm to several millimetres. As a consequence, a filtration process may be used for the removal of these organisms. On the other hand, the sizes of yeast, bacteria and phages vary from 0.5 to 7 μm. Usually, phage contamination does not take place in most of the fermentation processes. Therefore the filtration process is used for the removal of bacteria as the smallest organisms.

Physically the microorganisms are removed either via depth filter or membrane filter. The characteristics of the depth filter have been discussed previously. Usually depth filters are rarely used for medium sterilization. Membrane filters are considered to be absolute filters. 100% of microorganisms can be removed based on size exclusion. Different membranes may be considered for the membrane filter such as ultrafiltration, microporous or even macroporous membranes whose maximum pore size is less than the minimum size of the particles to be removed. Therefore the mechanism of membrane filtration is on the basis of absolute size exclusion. In general membrane filtration is more expensive than thermal methods due to high pressure drop. However, the main advantage of this filter is that it can be used to sterilize heat-liable materials. In the case of animal cell culture fetal bovine serum medium is used for the cultivation of the cell, which contains several heat labile nutrients. Membrane filtration techniques are found to be suitable for the sterilization of the medium. In addition, microporous and ultrafiltration techniques are considered for the preparation of pyrogen-free water.

7.2.2 Thermal Destruction

Microbial death follows in the first order kinetic as shown in Figure 7.7. Dead microbial cells have the inability to reproduce themselves in a given environment.

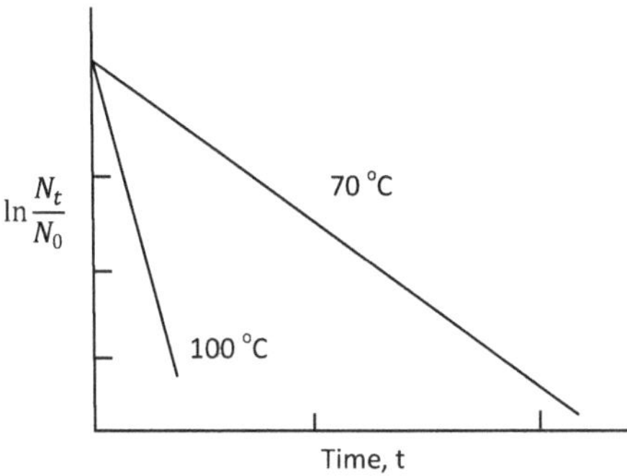

FIGURE 7.7 Plots of the the natural logarithm of the proportion of survivors of the microorganisms over a period of time.

In addition, in the case of cell death all enzymatic activity may not be destroyed. Enzymes may still have the potential to catalyse one or more reactions. The presence of active enzyme in a processed food may cause degradation after a long period of time. The kinetics of cell death may be described as first order as shown in Eq. 7.22.

$$\frac{dN}{dt} = -k\,N \tag{7.22}$$

where N = no. of viable cells, t = time, k = death rate constant.

Eq. 7.22 may be integrated with respect to time to reduce the number of viable organisms from N_0 to N_t. The resulting equation is shown in Eq. 7.13.

$$\ln \frac{N_t}{N_0} = -k\,t = \nabla_{total} \tag{7.23}$$

Where N_0 = initial number of viable cell, N_t = number of viable cells after the time t, k = thermal death rate constant, ∇_{total} = total cell death after some time interval.

This kinetic of cell death has two assumptions which contradict each other:

i) Sterile conditions are achieved after infinite time
ii) Less than one viable cell remaining after a certain time

The value of N_t is less than one organism. Therefore the probability of an organism surviving a treatment is to be considered for the design of the sterilizer; e.g. 1:1,000, 1:10,000, 1:1,000,000 etc. With a more stringent sterilization process the ratio will be less, e.g. one survival out of 1 million is more stringent than that of one in ten thousand.

Decimal reduction time (D-value) is defined as the time of exposure to heat to reduce the viable cell number to one-tenth value. Eq. 7.23 may be written as shown in Eq. 7.24.

$$\ln \frac{1}{10} = -k\,D \tag{7.24}$$

$$D = \frac{2.303}{k} \tag{7.25}$$

The value of k is different for vegetative bacterial cells and bacterial spores, e.g. k values for vegetative cells are much greater than those of the spores of the same microbial cell. The increased resistance power of the spore to heat is due to the presence of dipicolinic acid. Thermal death time (TDT) is defined as the time required to kill all microorganisms present in a sample at a specific temperature and under defined conditions.

The Arrhenius relationship deals with the effect of temperature on the thermal death rate constant (k) of the cells as shown in Eq. 7.26.

$$k = Ae^{-\frac{E_a}{RT}} \tag{7.26}$$

where E_a = activation energy, A = Arrhenius constant, R = universal gas constant, T = temperature in Kelvin.

The activation energy (E_a) for vitamins and amino acids varies from 84 to 92 $kJ\ mole^{-1}$ whereas that for microorganisms is from 250 to 290 $kJ\ mole^{-1}$. Therefore a relatively higher effect on cell death than on nutrient destruction will take place with a small increase in temperature. High temperature short-time (HTST) sterilization processes are used by the industry on the basis of this principle. Table 7.1 shows the effect of temperature on the thermal death rate constant and time of sterilization. The following assumptions are made for the determination of the microbial death rate contant:

- (i) Free of aggregation of the cells or spores suspension
- (ii) No lag time for heating or cooling
- (iii) Minimum inhibitory effect on the test microbial cell due to the composition of the medium and pH

TABLE 7.1
Typical Values of k for *Bacillus stearothermophilus* Spores

Temperature (°C)	k (*min⁻¹*)	Holding time* (*min*)
100	0.02	1730
110	0.21	164
120	2.0	17
130	17.5	2
140	136	0.25
150	956	0.04

* Holding time is nothing but the time of sterilization.

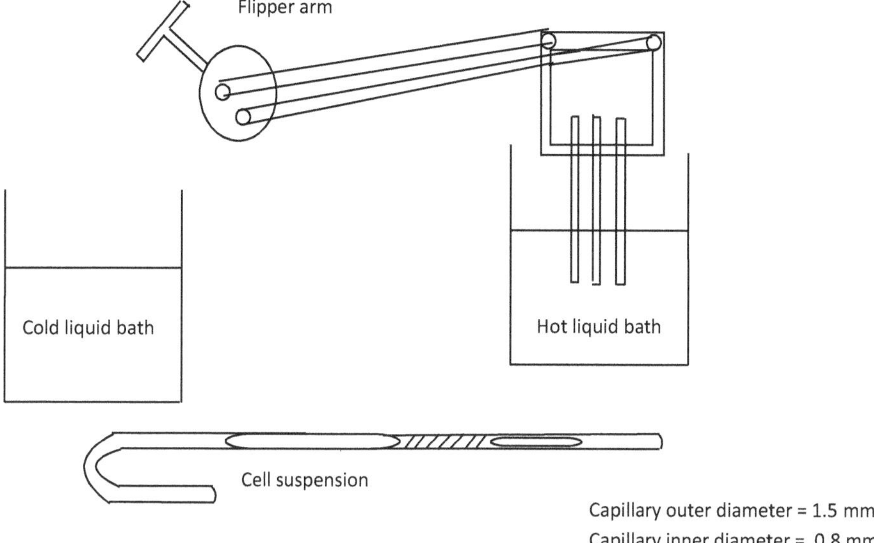

FIGURE 7.8 Schematic diagram for the determination of the thermal death rate constant of the microbial cell. (Source: Das D and Das D, Biochemical Engineering: An introductory Text Book, Jenny Stanford Pte. Ltd. Singapore, 2020.)

TABLE 7.2
The Values of the k and D of Spores of Different Bacterial Suspensions at 121 °C

Microbial Strains	k *(min^{-1})*	D *(min)*
Bacillus subtilis FS 5230	3.5–2.6	0.6–0.9
Bacillus stearothermophilus FS 617	2.9	0.8
Clostridium sporogenes PA 3679	1.8	1.3
Bacillus stearothermophilus FS 1518	0.07	3.0

The cell suspension is taken in a capillary and sealed. Figure 7.8 shows how the capillaries are hung over a cage on a ‘flipper’ arm. The purpose of the ‘flipper’ is to allow the capillaries to dip into either a hot or a cold bath for a given period of time. The number of viable organisms present in the capillaries before and after the exposure at an elevated temperature are determined by a plate count technique. The value of the viable cells at a particular temperature for a specific time depends on the type of microorganism used. Table 7.2 depicts the thermal death rate constant and the decimal reduction time (D) of spores of different bacteria (Aiba et al., 1973).

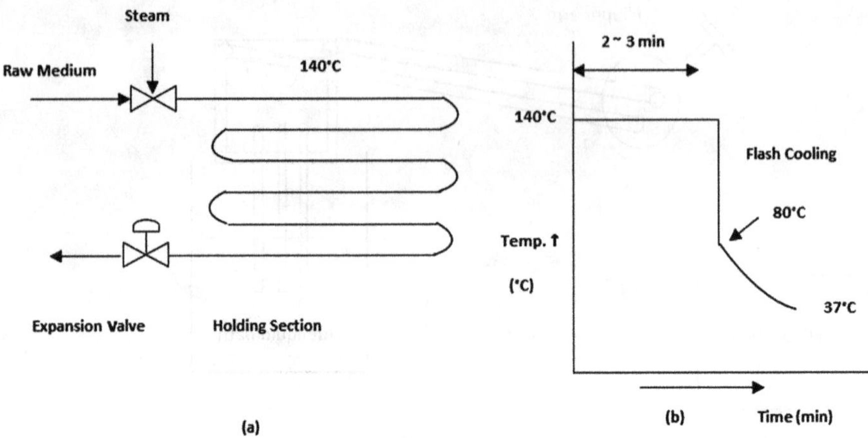

FIGURE 7.9 (a) Schematic diagram of CSI process for medium sterilization and (b) The temperature–time profile. (Source: Das D and Das D, Biochemical Engineering: An introductory Text Book, Jenny Stanford Pte. Ltd. Singapore, 2020.)

A batch sterilization process is mostly used for medium sterilization on a small scale. The major drawbacks of the batch sterilizer are listed below:

- Longer holding times are required to treat medium containing particles.
- Longer sterilization times are needed at the same holding temperature during scale up.
- The nutritional quality of the medium with respect to vitamins, proteins and sugars is reduced at elevated temperatures during heating and cooling.
- The batch process is tedious and energy intensive when a large amount of volume of medium is used.

Continuous sterilization processes are mostly considered for handling large volumes of medium. The continuous medium sterilization is based on the principle of high temperature short time (HTST). It has been observed that there is a relatively greater effect on thermal destruction of cells than on nutrients with the increase of sterilization temperature. In continuous sterilization steam consumption is usually 20 to 25% of that of a batch process. Five to six hours batch sterilization time is reduced to about two to three hours to sterilize media in case of continuous sterilization (Cooney, 2019; Block, 2001; Das and Bhattacharyya, 1989).

Two types of continuous sterilization process are available: continuous steam injection (CSI) and continuous plate heat exchangers (CPHE). In the case of CSI, the process is shown in Figure 7.9(a). Major features of this process are:

- ✓ rapid increase of temperature of the medium takes place in the absence of a heat exchanger
- ✓ foul heat exchanger due to the medium is avoided

✓ dilution of the medium causes due to condensed steam during the sterilization. Therefore it is difficult to control the pressure and temperature due to variation in medium viscosity.

The temperature-time profile is shown in Figure 7.9(b).

To avoid direct steam injection CPHE processes are used. Figure 7.10 shows the mode of operations and the temperature-time profile of CPHE. The features of this process are:

- raw medium entering the system is first pre-heated by sterile medium with high temperature in a heat exchanger, known as an economizer
- the economizer reduces the steam requirement for heating and also cooling the sterile medium
- effective with media containing suspended particles
- causes fouling of the surfaces of the heat exchanger and gaskets and seal leakage around the heat exchanger

The holding time of the medium will be required for a longer period of time for the batch sterilization in the fermenter as the volume increases from V_1 to V_2 at a particular temperature. Therefore, destruction of nutrients in the medium will be increased. Product titers concentration will be reduced during the scale-up if the nutritional quality of the medium plays an important role. Thus continuous sterilization processes (CPHE) are used for the medium sterilization for large scale fermentation processes because they drastically reduce the holding time as well as consumption of steam.

The type of fluid flow ensures adequate sterilization because the residence time of the medium plays a very important role. Perfect plug or piston flow is considered to be ideal flow, which is desired during the medium sterilization (Figure 7.11). In the real case, the medium flow will be between viscous and fully turbulent flow such that the u_{av} (average velocity) = 0.5 to 0.82 u_{max} (maximum velocity). A fully turbulent flow is desired in order for significant reduction of overheating of the medium. This is

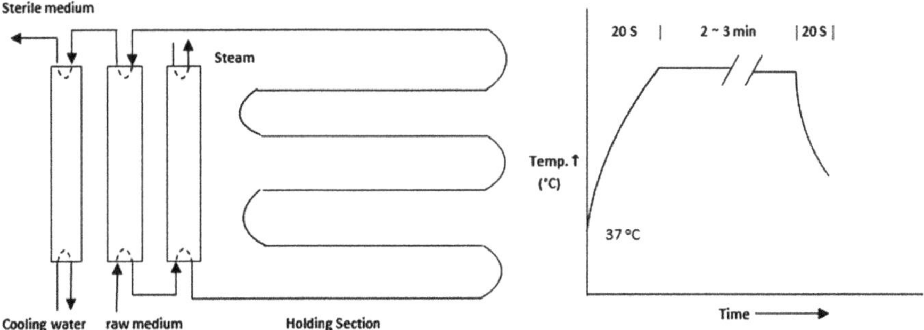

FIGURE 7.10 (a) CPHE for medium sterilization and (b) the temperature–time profile. (Source: Das D and Das D, Biochemical Engineering: An introductory Text Book, Jenny Stanford Pte. Ltd. Singapore, 2020.)

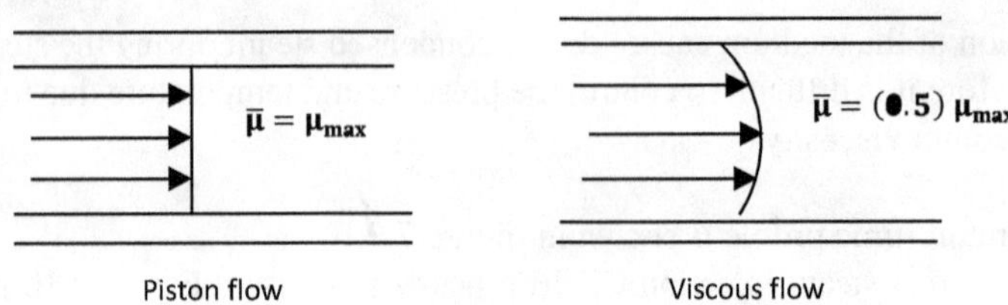

FIGURE 7.11 Flow pattern of the fluid. (Source: Das D and Das D, Biochemical Engineering: An introductory Text Book, Jenny Stanford Pte. Ltd. Singapore, 2020.)

possible when the Reynolds number (N_{Re}) is preferably more than 2×10^4. The extent of axial dispersion in the sterilizer will be minimized under this condition. The Peclet number includes the degree of axial dispersion which is shown in Eq. 7.27.

$$P_e = \frac{uL}{E_z} \tag{7.27}$$

Where u = average fluid velocity, L = length of the sterilizer, and E_z = axial dispersion coefficient. The Reynolds number of the fluid in the pipe line is

$$N_{Re} = \frac{Du\rho}{\mu} \tag{7.28}$$

Where D = diameter of the tube, ρ = specific density, and μ = medium viscosity.

Levenspiel (1958) has already reported the effect of axial dispersion on the residence time in a continuous sterilizer. The Peclet number is infinite in the case of plug flow where there is no change in outlet concentration. However, the usual Peclet numbers vary from 3 to 600. The cell concentration will be decreased with a normal residence time of 0.9. For the design of sterilizers, this deviation from plug flow is to be taken into account. Therefore the type of flow plays a very important role in designing the continuous sterilization of the medium. The Reynolds number should be greater than 2×10^4. A suitable diameter of the pipe is assumed for the calculation of liquid velocity. This is followed by calculation of the Peclet number (Levenspiel, 1958). The sterilization criterion $\frac{N_t}{N_0}$ is to be chosen. The desired tolerance for contamination is reflected in the value of N_t e.g. one batch out of one thousand. Typical value of $\frac{N_t}{N_0} = 10^{-12}$ may be considered for a batch fermentation. From Figure 7.12, the reaction number, N_r can be determined from the values of $\frac{N_t}{N_0}$ and the Peclet

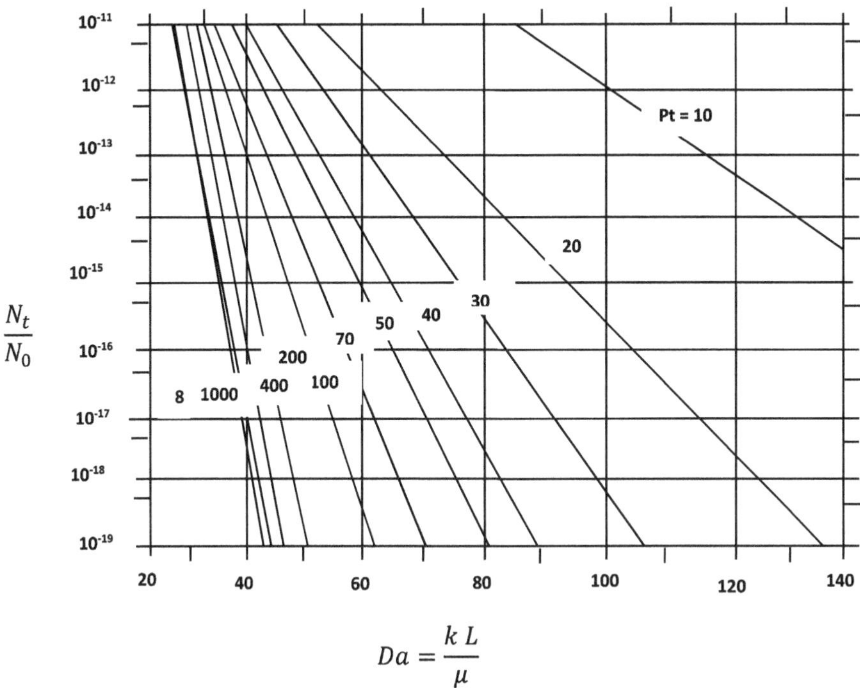

FIGURE 7.12 Correlation of Peclet number (Pe) and reaction number (N_r) during the destruction of the contaminants present in the medium by heat. (Source: S. Aiba, A.E. Humphrey and N.F. Millis, 1973, Biochemical Engineering, Academic Press, New York.)

number. N_r is expressed as shown in Eq. 7.29 (Aiba, 1973; Doran, 2012; Bailey, 2010; Das and Das, 2019).

$$N_r = \frac{kL}{u} \tag{7.29}$$

where k = thermal death constant.

The value of k at a $\frac{N_t}{N_0}$ holding temperature can be determined experimentally. Thus, from the reaction number, N_r, one can estimate the value of the length of the sterilizer (L).

7.3 ASEPTIC TRANSFER OF THE CULTURE

7.3.1 Transfer of Seed Culture to Inoculum Vessel

Industrial microbial strains are used in fermentation processes. These strains are produced in the medium in two different forms: unicellular and filamentous. Unicellular cells are easy to count. In a fermentation industry, there are at least two bioreactors: inoculum vessel (IV) and production fermenter (PF). The capacity of the PF is usually 10 to 20 times bigger as compared to IV. Therefore the inoculum for the PF is prepared in the IV in the fermentation plant. Seed culture is prepared in the R & D laboratory of the plant. The amount of cells is usually quantified with respect to number. In case of unicellular cells, one can count the number using a hemocytometer. In the case of filamentous cells such as *Aspergillus niger*, the spores are produced under stressed conditions (in the presence of less moisture). Spores can easily be counted. This optimum number of cells or spores is found through research work. The cells or spores suspension prepared in the laboratory is transferred in the seed can, under sterile conditions. The capacity of the seed can is usually 10 litres. This seed culture is aseptically transferred into the IV containing the cell growth medium as shown in Figure 7.13. The seed tank is fixed with the IV pipe line by tightening the nut and bolt as shown in Figure 7.13. Initially the steam supply line is on to sterilize the pipe line, which is followed by connecting with the sterilized air supply line to transfer the seed culture to IV.

7.3.2 Transfer of Culture from Inoculum Vessel to Production Fermenter

The inoculum from IV is transferred to the production fermenter by air pressure as shown in Figure 7.14. The sampling port is shown in the PF where live steam is injected, which is followed by drawing the sample. After sampling, the pipe line is to be sterilized by steam (Aiba, 1973).

PROBLEM 7.1: Air is sterilized by a glass wool filter. The initial values of the contaminants are to be reduced to 10^{-8} by using this filter. The following data is given:

$$d_f = 19\,\mu m, \text{filter diameter} = 1\,m, \alpha = 0.033,$$

$$d_p = 0.5\ \mu m,\ v_s = 10\ cm,\ \rho_{air} = 1.2 \times 10^{-3}\,g/cm^3$$

$$\text{Air viscosity} = 2 \times 10^{-4}\,g/cm\,s,$$

$$\rho_p = 1\ g/cm^3,\ \text{particulars diffusivity}\left(D_{BM}\right) = 2.78 \times 10^{-7}\,cm^2/s$$

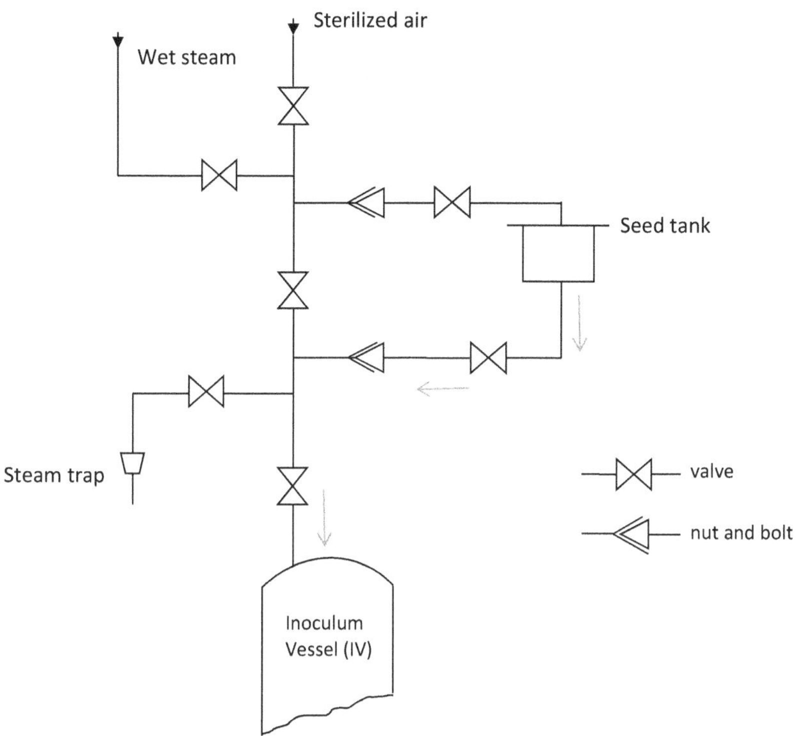

FIGURE 7.13 Aseptic transfer of seed culture prepared in the laboratory to inoculum vessel (IV).

SOLUTION

We know the Reynolds number (N_{Re}) $= \dfrac{D_f\, v_s\, \rho}{\mu(1-\alpha)}$

From the given data we can write

$$N_{Re} = \frac{(19\times10^{-4})(1.2\times10^{-3})(10)}{(2\times10^{-4})(1-0.033)}$$

$$= 2.357\times10^{-1}$$

$$N_{sc} = \frac{\mu}{\rho D_{BM}} = \frac{(2\times10^{-4})}{(1.2\times10^{-3})\times(2.73\times10^{-7})} = 6\times10^{5}$$

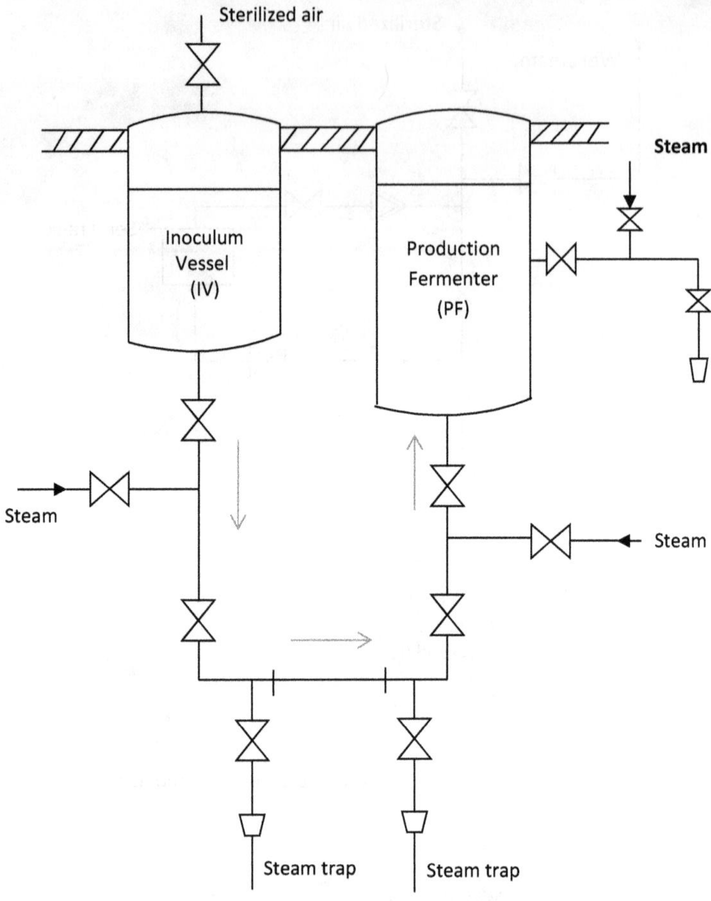

FIGURE 7.14 Transfer of the inoculum from IV to production fermenter (PF) and sampling port.

$$\text{Again } N_{pe} = N_{sc}N_{Re}$$

$$= \left(2.357\times10^{-1}\right)\times\left(6\times10^{5}\right) = 1.4\times10^{5}$$

$$\text{Therefore, } N_R N_{Pe}^{1/3} N_{Re}^{1/18} = \left(\frac{0.5\times10^{-4}}{19\times10^{-4}}\right)\times\left(6\times10^{5}\right)^{1/3}\times\left(0.2357\right)^{1/18} = 2.0474$$

From Figure 7.2. we get $\eta_o N_R N_{pe} = 2\times10$

$$\eta_o = \frac{2\times10}{\left(\frac{0.5\times10^{-4}}{19\times10^{-4}}\right)\times\left(6\times10^{5}\right)} = 0.0054$$

We know that, $\eta_{\infty} = \eta_{o}\ (1+4.5\ \alpha)$

Therefore, $\eta_{\infty} = 0.0054(1+4.5\times0.033)$
$= 0.0062$

$$\text{Therefore, } L = \frac{\pi\, d_f\ (1-\alpha)}{4\eta_{\infty}\alpha}$$

$$= \frac{3.14\times19\times10^{-4}\times(1-0.033)}{4\times0.0062\times0.033}\ln\frac{N_1}{N_2}.$$

$$= 7.049\ \text{cm}\times\ln 8\ =\ 14.65\ \text{cm}$$

Therefore, the length or depth of the glass-wool filter will be 14.65 cm.

PROBLEM 7.2: A production medium is to be sterilized in a continuous sterilizer. The flow rate of the medium in the steam heat exchanger is 3 m^3h^{-1}. The bacterial spores concentration in the liquid is 4 × 1013 m^{-3}. The activation energy and Arrhenius constant for thermal destruction of these contaminants are 283 $kj\,gmol^{-1}$ and 5.7 × 10^{39} h^{-1}, respectively. One organism surviving every 60 days' operation is considered as an acceptable contamination risk. The inner diameter and length of sterilizer pipe are 0.1 m and 36 m, respectively. The density and viscosity of the medium are 1000 $kg\,m^{-3}$ and 3.6 $kg\,m^{-1}h^{-1}$kg, respectively. Determine the sterilization temperature.

SOLUTION

Basis: one organism surviving every 60 days' operation.

Assumption: no cell death in the heating and cooling sections.

The total number of cells entering the holding section over 60 d is written as follows

$$N_0 = 3\,m^3h^{-1}\,(4\times10^{13}m^{-3}).\left(\frac{24\,h}{1\,d}\right).(60\,d) = 1.73\times10^{17}$$

According to the problem, N = 1, an acceptable number of cells remain after the sterilization.

$$\text{Therefore, } \frac{N}{N_0} = \frac{1}{(1.73\times10^{17})} = 5.8\times10^{-18}$$

The linear velocity (u) of the medium in the sterilizer

$$u = \frac{3\,h^{-1}}{\pi\left(\frac{0.1\,m}{2}\right)^2} = 382.16\,m\,h^{-1}$$

To calculate Pe, the value of D_z is determined by the correlation as shown in Figure 8.13.

$$\text{Again, Reynolds no. } (N_{Re}) = \frac{Dv\rho}{\mu} = \frac{(0.1\,m)(382.16\,m\,h^{-1})(1000\,kg\,m^{-3})}{3.6\,kg\,m^{-1}h^{-1}}$$
$$= 10.6 \times 10^3$$

For $Re = 10.6 \times 10^3$, D_z can be determined from Levenspiel (1958). The experimental curve is considered because this gives a larger value of D_z and a smaller value of Pe.

$$\text{Therefore, } \frac{D_z}{uD} = 0.63$$

$$D_z = (0.63)(382.16\,m\,h^{-1})(0.1\,m) = 24.1\,m^2\,h^{-1}$$

$$Pe = \frac{uL}{D_z} = \frac{(382.16\,m\,h^{-1})(24\,m)}{24.1\,m^2\,h^{-1}} = 381$$

Using Figure 7.6, we can find out the value of k for the desired level of cell destruction. At $\frac{N_2}{N_1} = 6.9 \times 10^{-17}$ and $Pe = 381$,

$$\frac{kL}{u} = 40$$

$$k = \frac{(382.16\,m\,h^{-1})(40)}{24m} = 637\,h^{-1}$$

From the Arrhenius equation

$$k = Ae^{-E_a/RT}$$

Rearranging the above equation

$$T = \frac{\left(\frac{-E_a}{R}\right)}{\ln\left(\frac{k}{A}\right)}$$

From the data given in the problem and above

$$E_a = 283\,\text{kJ g mol}^{-1} = 283\times10^3\ \text{J g mol}^{-1},\ A = 5.7\times10^{39}\ \text{h}^{-1}\ \text{and}$$
$$R = 8.314\ \text{J K}^{-1}\text{g mol}^{-1}, k = 637\,\text{h}^{-1}$$

$$T = \frac{\left(\dfrac{-283\times10^3\,\text{Jgmol}^{-1}}{8.314\,\text{JK}^{-1}\text{gmol}^{-1}}\right)}{\ln\left(\dfrac{637\,\text{h}^{-1}}{5.7\times10^{39}\ \text{h}^{-1}}\right)} = 400\ \text{K}$$

Therefore the temperature of the sterilization is 127 °C.

7.4 CONCLUSIONS

The special feature of most production fermentation processes is to maintain sterile conditions. Air, medium and bioreactor are the major sources of contamination. Air may be available in two forms: stagnant and moving. Stagnant air is required in the culture preparation room and moving air is passed through the fermentation medium to increase the DO concentration because the aerobic microorganisms can utilize DO for their growth and metabolism. Stagnant air is usually sterilized by either UV-rays or sparging germicides like chlorine, iodine etc. Physical separation techniques are found to be suitable for the sterilization of air used in the fermenter. Heat is found to be an effective medium for the sterilization of the medium. Continuous sterilization processes are found to be suitable as compared to batch sterilization processes with respect to steam consumption and the time of sterilization. HTST techniques are found to be effective for medium sterilization.

REFERENCES

Aiba S, Humphrey AE and Millis NF, Biochemical Engineering, University of Tokyo Press, Tokyo, 1973.

Bailey JE and Ollis DF, Biochemical Engineering Fundamentals, McGraw-Hill, New York, 2010.

Block SS, Disinfection, Sterilization and Preservation (5th Ed.). Lippincott Williams & Wilkins, 2001.

Cooney CL, Media Sterilization, In: Comprehensive Biotechnology (Volume-2) (Editor-in-Chief Murray Moo-Young), Pergamon Press, 2019.

Das D, and Bhattacharya BC, Short term course on "Analysis and Design of Novel Bioreactors", Indian Institute of Technology Kharagpur, India, 1989.

Das D and Das D, Biochemical Engineering: An Introductory Text Book, Jenny Stanford, Singapore, 2019.

Doran PM, Bioprocess Engineering Principles, Second Edition, Academic Press, Waltham, 2012.

Levenspiel O, Longitudinal mixing of fluids flowing in circular pipes, *Industrial & Engineering Chemistry,* 50, 343, 1958.

Richards JW, Air sterilization with fibres filters, *Process Biochemistry*, 2(9), 21–25, 1967.

Whittet TD, Hugo WB, and Wilkinson GR, Sterilization and Disinfection, William Heinemann Medical Books, London, 1965.

8 Downstream Processing

8.1 ISOLATION OF FERMENTATION PRODUCT

Not only expenditure in recovery systems is high but also extraction constitutes a large cost (up to 60 %) of the resultant product. For one antibiotic project, recovery machinery costs four times more than the fermenter. There is a need for an organized and stable recovery method and an effective production system. The primary issue in the separation of fermentation products from 'beer' or broth fermentation is that the necessary product generally forms a small percentage of the complex heterogeneous combination of cell debris, other metabolic products, and unused sections of the medium. In determining the process of extraction to be used, the following factors are taken into consideration:

1. Amount of final product.
2. The final product purity level should be high enough to also consider yielding capacity.
3. The chemical and physical characteristics.
4. Position of the product in the mixture i.e. either it is cell attached with cell or free in a medium.
5. The impurity location and properties.
6. The cost-benefit.

The different aspects applying in the extraction of fermentation products along with the estimated degree of purification achieved at each step are shown in Figure 8.1. The operation followed at each stage depends on the material being extracted and mentioned below. The substance sought may be the cells themselves (yeast), trapped in cells (e.g. streptomycin, enzymes) or free of penicillin in the system (Belter et al., 1987).

8.1.1 Recovery and Purification of the Product

The recovery and purification of the fermentation product are important for any industrial method. As the biochemical structure of the fermentation medium is difficult and an exceptionally high degree of quality is needed for recovery and purification, this

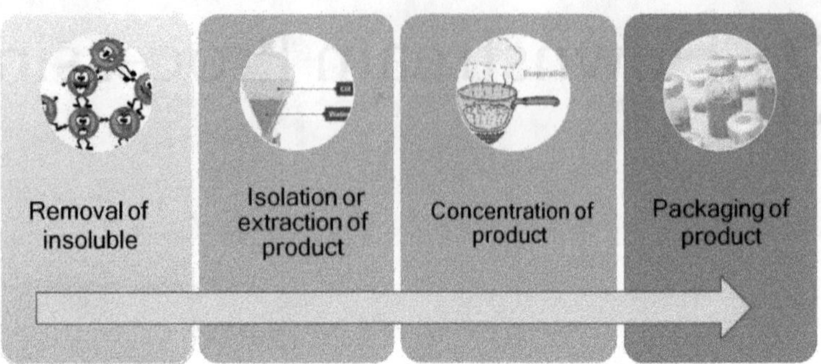

FIGURE 8.1 Steps involved in DSP.

also requires multiple stages and sometimes the cost of production is greater for the purification process. Isolation techniques differ with the size and type of the material and various procedures that need to be used for fermentation of broth comprising soluble compounds of various molecular sizes. Figure 8.1 describes the main steps included in the isolation and purification of the enzyme from the fermentation medium, and this procedure is typically acceptable to several protein components (Weuster-Botz et al., 2007). This can be extended to include four main operations: (1) isolation of insoluble substances as solids; (2) initial isolation or accumulation of the substance and elimination of much of the water; (3) elimination of polluting chemical compounds; and (4) processing of the product. Water should be drained at the beginning of the system as that helps to reduce the size of the machine. Since steps designed mainly to concentrate the substance and those designed to extract polluting chemicals also include solvents, there is not much difference between steps 2 and 3. It is necessary to handle these steps by defining in the same section operations to recover product. Firstly, it includes certain steps that eliminate solutes like cells, unutilized substrates, metabolites etc. (Weuster-Botz et al., 2007).

Solids (insoluble) exclusion: The first phase generally divides solids from the fluid portion thus encouraging the next extraction process. If the necessary product is solid or is embedded in the insoluble component; liquid removal concentrates the solid's components; sometimes a product cannot separate e.g. acetone-butanol fermentation, in which the whole beer is utilized. The common separation techniques used are filtration, centrifugation, decantation, and foam fractionation. If the appropriate proportion is the cells, then most of the contaminants are extracted with the filtrate. The different techniques employed in solids elimination are mentioned below.

8.2 FILTRATION PROCESS

Solids separation involves a filter, in which solid and liquid are classified in many different ways. The technique is called filtration. For present purposes, a division into those in which cakes are formed and those in which the particles are captured in the depth of the medium is adequate. Cake filters can be further divided into pressure, vacuum, centrifugal and gravity operations. They may be defined as a process of separation of solids from a fluid by passing it through a porous medium that retains the solids but allows the fluid to pass through. When solids are present in very low concentration i.e. not exceeding 1.0%w/v, the process of their separation from liquid is called clarification.

8.2.1 Filtration Mechanism

The process of filtration involves four steps: i. Filter medium (filter paper) is placed on a support (mesh); ii. Slurry is placed over the filter medium; iii. Due to pressure differences across the filter, fluid flows through the filter medium (The pores of the filter medium are smaller than the size of the particles to be separated; gravity is acting over the liquid medium); iv. Solids are trapped on the surface of the filter medium. There are four different mechanism involved in filtration.

1. Straining – Similar to sieving i.e. particles of larger size cannot pass through smaller pore size of filter medium. 2. Impingement – In this case, the solids having momentum move along the path of streaming flow and strike (impinge) the filter medium. Thus the solids are retained on the filter medium. 3. Entanglement – In this case, the particles become entwined (entangled) in the masses of fibres (of cloths with fine hairy surface or porous felt) due to smaller size of particles than the pore size. Thus solids are retained within the filter medium. 4. Attractive forces – In this case, the solids are retained on the filter medium as a result of attractive force between the particles and the filter medium, as in the case of electrostatic filtration.

8.2.2 Theory of Filtration

The flow of liquid through a filter follows the basic rules that govern the flow of any liquid through a medium offering resistance. The rate of flow may be expressed as

$$\text{Rate} = \frac{\text{driving force}}{\text{resistance}}$$

The rate of filtration may be expressed as volume (L) per unit time (dv/dt).

The rate of flow will be greatest at the beginning of the filtration process, since the resistance is minimal. After the forming of a filter cake, its surface acts as filter medium and solids continuously deposit adding to the thickness of the cake. The resistance is not constant. It increases with the deposition of solids on the filter medium. Therefore, filtration is not a steady state. The pressure drop can be achieved in a number of ways: the simplest method being to pump the slurry into the filter under pressure. The gravitational force could be replaced by centrifugal force in particle separation.

Based on the application of external forces, the filtration equipment can be classified broadly into three categories: Pressure filters (plate and frame filter press); centrifuge filters and vacuum filters (rotary vacuum filter).

The mechanical device which is specially used in solid/liquid separation using the principle of pressure driven which is provided by a slurry pump is called a plate and frame filter press. The frame contains an open space inside which the slurry reservoir is maintained for filtration and an inlet to receive the slurry. The plate has a studded or grooved surface to support the filter cloth and an outlet. The system is usually made of aluminium alloy. The mechanism of the plate and frame filter press is based on surface filtration. The slurry enters the frame by pressure and flows through the filter medium. The filtrate is collected on the plates and sent to the outlets. A number of frames and plates are used so that the surface area increases and consequently large volumes of slurry can be processed simultaneously with or without washing. Filter presses are used in a huge variety of different applications including the food industry, the mining industry, the pharmaceutical industry, the chemical industry, wastewater treatment etc.

In industry, a rotary vacuum filter is largely used. The filter comprises a hollow rotating cylinder split into 4 sections and filled with a metal or fabric gauze as described in Figure 8.2. In the cylinder, the vacuum is applied and as the rotary cylinder rotates, liquid materials are extracted from the shallow trough. Inside the shallow trough a revolving cylinder is situated. For thick slurries that are challenging to sieve (e.g. aminoglycoside broth) a thin surface of filter (e.g. kieselguhr) is first permitted to be consumed into the cylinder (Kalyanpur, 2002). Next, the filter cylinder with its fine layer of filter material is permitted to move in the passage where the broth entered. The moving cylinders, still under vacuum, are sprayed with water; a knife with an edge just below the filter aid layer cuts the solids extracted from the broth away. No filter aid is used for easy-filtering liquids such as penicillin broth.

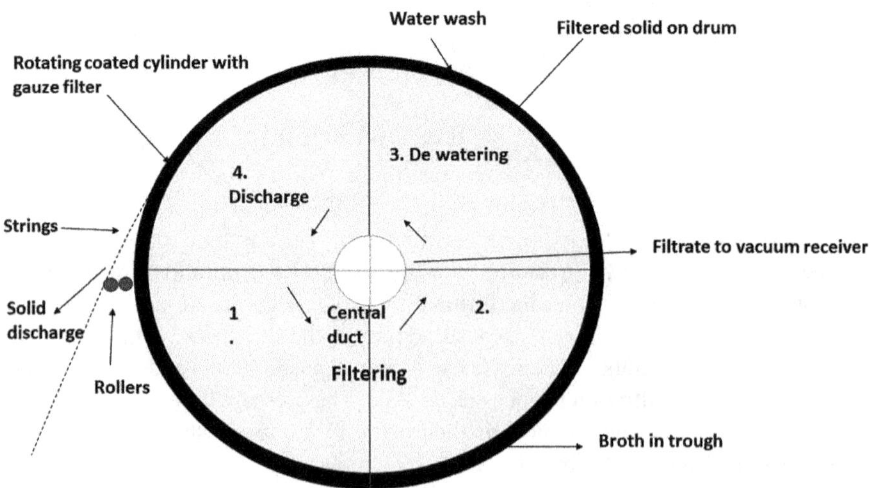

FIGURE 8.2 A schematic representation of a rotary vacuum filter.

Filters with a ring and wire type contain a diatomaceous earth covering on a wire mesh, which is held up on a metal rod frame. The fluid that needs to be filtered is applied at a pressure of 75 p.s.i. This method is mostly used when the weight of substances is less, for example for polishing beer or fruit juices. They may be washed with water by flushing the back.

8.3 CENTRIFUGATION

Centrifugation is done using centrifugal forces to isolate objects ranging approximately 100 and 0.1 mm in size from air. Centrifugation is not commonly utilized for primary isolation of solids from the broth in fermentation beer (Jungbauer, 2013). However, centrifugation is favoured in the enzyme-isolation industry, possibly because this process eliminates unnecessary cell debris very rapidly. During fermentation, centrifugation can be applied as an initial step. Particulate settlement in a suspension with high particle density is known as impeded settlement, which resembles solid–liquid. Gravitational force (F_G), drag force (F_D), and buoyant force (F_B) are the main forces operating on a rigid body resting in a liquid through gravitational forces. Once the particles hit a terminal settling velocity they cause each other to behave on a particle equilibrium, resulting in a net cause of zero.

$$F_G = F_D + F_B \tag{8.1}$$

Where

$$F_G = \frac{\Pi}{6} D_p^3 \rho_p \frac{g}{g_c} \tag{8.2}$$

$$F_B = \frac{\Pi}{6} D_p^3 c \frac{g}{g_c} \tag{8.3}$$

$$F_D = \frac{C_D}{2g_c} \rho_f U_0^2 A \tag{8.4}$$

F_D is the drag force exerted by the fluid on solid particles, C_D is the drag coefficient, ρ_f is the fluid density, U_0 is the relative velocity between the fluid and particle or the terminal velocity of the particle, and A is the cross-sectional area of the particles perpendicular to the direction of fluid flow; for a sphere, $A = (\pi/4)D_p^2$.

8.4 COAGULATION AND FLOCCULATION

Coagulation is the cohesion of dispersed colloids into small flocs, which are aggregated to create bigger masses during flocculation. The coagulation is caused by electrolytes, and flocculation by polyelectrolytes, higher molecular weight, water-soluble substances which can be produced in ionic, anionic, or cationic types. Bacteria and proteins that are negatively charged by colloids are quickly flocculated by electrolytes or polyelectrolytes (Belter et al., 1987). Clay or activated charcoal may also be included. The result of flocculation is that the disposal of colloids encourages the filtration process. Since coagulation relies on the features of the cell wall, the operators must comply with the preceding criteria, particularly if the necessary ingredients are the cells and not the liquid. The flocculants must have these important attributes:

1. They should respond quickly to the cells.
2. They should be non-hazardous.
3. They do not change the chemical components of the cell.
4. They must have the least coherent power to enable successful filtration to eliminate water.
5. Their incorporation does not lead in either a high acidity or high alkalinity.
6. They must be available in large quantities and cost-effective.
7. They must ideally be washable for use again.

8.5 FOAM FRACTIONATION

The concept of foam fractionation says that in the liquid foam process, the chemical properties of a liquid mass (bulk) generally differ from the chemical properties of some components in foam. Foam is developed by sparging a bulk liquid that usually contains the material to break down with an inert gas (Weuster-Botz et al., 2007). The gas is supplied at the bottom (Figure 8.3) of the tower and the foam produced overloads at the top. Additionally, they carry the solids to break down. Active compounds that decrease surface tension, e.g. Teepol, may be introduced in liquids which do not produce foam. This technique was used to accumulate a broad variety of microorganisms.

8.6 WHOLE BROTH METHOD

As previously stated, in certain fermentations like the fermentation of acetone-butanol unseparated broth is strip off the composition of the needed material. An equivalent condition has been obtained in the antibiotic industry before it was extracted directly by the antibiotic streptomycin. Resins isolate the antibiotics after crystallization occurs. This method avoids the investment and maintenance costs of the actual isolation of solid particles through the broth.

8.7 PRIMARY INTRACELLULAR PRODUCT EXTRACTION

After the division of the broth into soluble and insoluble parts, the next step relies on the position of the required material as shown in the yeast cells, in which the yeast themselves can be a desired item; these are dried or refrigerated as well as

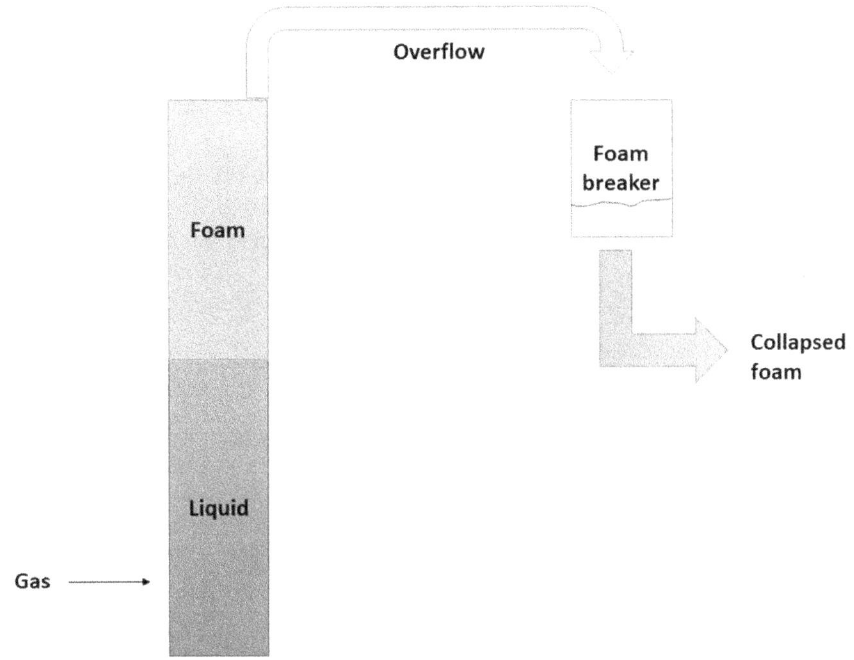

FIGURE 8.3 Diagrammatically representing the foam fractionation method.

the fluid removed. The necessary compound can be attached to mycelia or bacterial cells, as with fused enzymes or antibiotics. After that cells are disturbed by the following methods– high temperature, physical methods for interruption of cells etc. Cell residue is now extracted by centrifugation, filtration, etc. Where the content is extracellularly accessible or where it can be collected by leaching with or without cell interruption, it is handled with the following processes: fluid extraction, dissociation, sorption or accumulation (Belter et al., 1987).

8.7.1 Mechanical Technique for Cell Disintegration

A great number of biomolecules are within the cell and need to be extracted from it. This is done by cell interruption (lysis). Cell interruption is a critical method because it does not allow high osmotic pressure. Moreover, problems occur from unregulated cell disruption resulting from the discharge of undesirable intracellular products (nucleic acid proteins, cell debris) and criteria for cell disruption without the necessary component denaturation (Zydney, 2016). If the main component is intracellular, the microbes are primarily affected by the physical disturbance of the cells. The cell interruption method consists of:

Homogenizers: It also pumps slurry via constricted orifices or valves at very high pressure (up to 1500 bar) accompanied by immediate enhancement via a particular outlet nozzle. The rapid fall of pressure on removal results in the explosion of cells. The technique is primarily used for the removal of intracellular chemical compounds.

Ball mills: In a ball mill, the cells are disturbed in a tiny abrasive crystal. The cells divided due to shearing forces, friction with the beads, etc. Beads interrupt the cells for the extraction of biomolecules.

Ultrasonic interruption: This process of cell lysis is accomplished with a high frequency of sound generated by electronic methods and transmitted via a metallic tip to a properly focused cell suspension. It is costly and is primarily required in research laboratories.

8.7.2 Non-Mechanical Methods

Techniques widely employed include osmotic pressure and ice crystal breakup. The cell wall and membrane may be destroyed by gradually melting, and eventually thawing a cell paste, releasing enzymes into the environment. Changes in the medium's osmotic pressure may result in the release of certain enzymes in Gram negative cells, especially periplasmic proteins (Soetaert and Vandamme, 2010). Heat shocking, freeze-drying after treatment with acetone, butanol, and buffers result in cell-wall disruption. Enzymes such as lysozyme (a carbohydrase) can be used to lyse cell walls of bacteria. Gram-positive bacteria are far more susceptible to enzymatic lysis than Gram-negative bacteria. The cells might be treated with EDTA or by freezing and thawing before treatment with lysozyme. Ethylenediaminetetraacetic acid (EDTA) is a chelating agent; high levels of EDTA will extract divalent ions that are part of the cell envelope (Kalyanpur, 2002). Enzymatic hydrolysis is an expensive method and is not very widely used in industry. Actively growing cells can be treated with an antibiotic, such as penicillin or cy-closerine, which interferes with cell-wall synthesis and, coupled with the correct osmotic conditions, can lead to cell disruption. After cells have been lysed and the products are released into the medium, cell debris can be separated by ultracentrifugation or ultrafiltration, and soluble products are recovered using the following methods.

Cell Interruption

Chemical permeation: Several chemical techniques are used to isolate intracellular constituents from microbes by allowing chemicals to enter the outer cell wall membranes. Natural solvents such as toluene, ether, phenylethyl alcohol DMSO, benzene, methanol, and chloroform can be attained by the formation of canals across the cell membrane. Chemical permeability may also be attained with antibiotics, thionins, surfactants (Triton, Brij, Duponal), chaotropic factors, and chelates. EDTA (chelating agent) is a very powerful compound and is commonly utilized for the distribution of Gram-negative bacteria (Belter et al., 1987). Permeability of Gram-negative microorganisms; this efficacy is due to this capacity to bind the divalent cations of Ca^{++}, Mg^{++}. Such cations strengthen the framework of the cell membrane by connecting the lipopolysaccharides. To eliminate these cations, EDTA enhances the permeability of the cell wall.

Mechanical permeability: Another type of osmotic shock. Slow changes in osmotic pressure cannot disrupt cell walls but with high pressure can disrupt them easily. Such a method is usually performed by permitting the cells to balance internal and external osmotic stress in a high-sucrose solution, as well as quickly diluting the sucrose (Jungbauer, 2013). The resultant rapid overpressure of the cytosol is expected to harm the cell wall. Enzymes produced by such a technique are assumed to be periplasmic or situated close to the cell membrane.

Enzymatic permeability: Enzymes may also be used to permeate cells; however, this process is mostly restricted to the release of periplasmic or surface enzymes. In such methods, EDTA is also required to disrupt the cellular membranes (Belter et al., 1987). Enzymes such as beta (1-6) and beta (1-3) glycanases, proteases, and mannase are used for permeability.

8.8 LIQUID–LIQUID EXTRACTION

This is recognized as solvent extraction or liquid–liquid extraction; this method is commonly utilized by various companies. It requires moving a solute from one solvent into another solvent to make it soluble. It may be feasible to segregate soluble solutes from a mixture of unsolvable materials by liquid methods (Kalyanpur, 2002). The law of liquid–liquid extraction states that whenever the organic solid is revealed to a two-phase immiscible liquid framework, the proportion of the solute concentration in the two phases is stable at a specific temperature. In these K, the distribution coefficient, may be written as

$$K = \frac{C_1}{C_2} \tag{8.5}$$

C_1 = Concentration of phase 1
C_2 = Concentration of phase 2

However, the equation is useful for dilute solutions. The selection of solvents depends on several aspects such as characteristics of the solute, volatility, recovery rate, and expense. In this mechanism, the liquid to be collected is agitated with a hydrophobic solvent (i.e. insoluble in water), permitted to sink and the solvent that consists of a high proportion of components is isolated and then taken out.

Light and heavy process mass circulate are maintained to ensure that $L_0 = L_1 = L$ and $H_0 = H_1 = H$, the mass maintained at the extracted solute output where E = LKD/ H is the extraction variable, and where Y_L and X_H are concentrations of the solute in light and heavy phases, respectively.

$$K_D = \frac{Y_L}{X_H} \tag{8.6}$$

Assuming that K_D is constant and the solvents are totally immiscible (i.e. the mass flows of the light and heavy phases are conserved so that $L_0 = L_1 = L$ and $H_0 = H_1 = H$), a mass balance on the extracted solute yields.

$$H(x_1 - x_0) = LY_1 \tag{8.7}$$

$$x_1 = x_0 - \frac{L}{H} y_1 \tag{8.8}$$

Since $kD = \frac{y_1}{x_1}$ Eq. 8.8 may be written as

$$x_1 = x_0 - \frac{LK_D}{H} x_1 \tag{8.9}$$

$$\frac{x_1}{x_0} = \frac{1}{1+E} \tag{8.10}$$

where $E = LK_D/H$ is the extraction factor.

For a counter-current multi-stage activity, as content inequal proportion for production of solute.

$$R = \frac{E(E^n - 1)}{E - 1} \tag{8.11}$$

The connection among E, x_n/x_0, Here R is the factor of rejection, that is the proportion of the mass of the solute in the light stage to the heavy stage, and n is the total number of levels of balance. The percentage of the solute produced is then

$$\% \text{ of extraction} = 1 - 1/R + 1 \tag{8.12}$$

As the solute reaches the cycle during the light process, the amount of rejection is

$$R = \frac{E^n(E - 1)}{E^n - 1} \tag{8.13}$$

Many other antibiotics are obtained from the fermentation process utilizing reagents like amyl acetate or isoamyl acetate. The continuous centrifugal extractors of Podbileniak are generally used during the recovery of antibiotics. Penicillin is soluble in the organic stage at low pH (pH = 2 to 3) and is extremely stable in the aqueous phase at higher pH (pH = 8 to 9). Penicillin is collected in multiple ways among organic and aqueous processes by changing the pH to increase the quality of the substance. The variance of the distribution ratio (solvent – aqueous) with pH in the amyl acetate production of penicillin.

Aqueous Two-Phase Extraction

Aqueous bi-phase separation is a method under effective progress for the isolation of dissolved proteins, including such enzymes through two aqueous phases comprising incompatible polymers, such as polyethylene glycol (PEG) and dextran (Figure 8.4).

The phases including PEG and dextran are greater than 75% water and are insoluble. Normal aqueous phases required for this reason are PEG – water/dextran – water and PEG – water/ K – phosphate – water. PEG/dextran and PEG/ K phosphate are very insoluble (Kalyanpur, 2002). The fraction factor, Kp, differs from the molecular weight (MW) of the total solids as an exponential factor.

$$K_p = e^{AM/T} \tag{8.14}$$

While M stands for MW of protein, T stands for absolute temperature, and A is the constant.

The partition coefficient for Kp, of several enzymes between the 2 stages (CPEG-CDEX) differs between 1 and 3.7, leading to weak differentiation at one level. The distribution of factors can be strengthened by using ion-exchange resins or other salts, such as $(NH_4)_2SO_4$ and KH_2PO_4, in one step. Rather than tenfold improvement in Kp levels can be enhanced by increasing the intensity of KH_2PO_4 from 0.1 to 0.3 M. The quality of a partition factor can be increased with ion-exchange resins. The cation or anion interaction features of the PEG can be derived and employed in a two-step K method (Ratledge and Kristiansen, 2001). Interaction ligands will be utilized to raise Kp; PEG – NADH and PEG – cibacron blue systems. This process is called the extraction of two-water phase affinities. Partial hydrolysis of dextran and PEG also lead to a rise in the Kp value, as lower-MW materials could communicate more successfully with proteins. Through combining two types of PEGs (PEG400

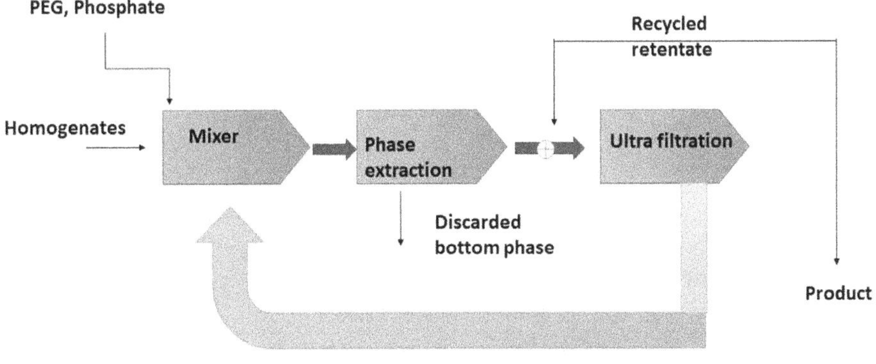

FIGURE 8.4 Two-phase extraction process with PEG recovery.

and PEG4000), the Kp quality for fumarase partitioning may be improved by a ratio of 6.It is also applicable in the retrieval of cell residues, polysaccharides, and nucleic acids. The concentration gradient for whole cells and DNA is around 100 and 0.01; for proteins it is around 10 and 0.1, and small ions it is between1. The next stage of isolation can be collected by centrifugation or decantation, and the PEG can be retrieved by ultrafiltration. This segregation process is quick and can be performed at mild temperature, pressure, and pH environments. Dextran and PEG are retrieved and recycled.

Dissociation Extraction

Products of fermentation are weak bases or acids (Kalyanpur, 2002). During solvent extraction, the pH is used to unionize the extracted product. Since the ionized type is dissolved in the aqueous state as well as the non-ionized type is soluble in the solvent state. Thus, weak bases are collected at a strong pH and weak acids at a low pH. As a result, the extraction of solute is quick.

Ion-Exchange Adsorption

Fermented broth can be filtered and concentrated using column-packed ion exchange resins. Further synthetic ion substitute resins are typically formed as porous crystals with a large outer and porous surface where ions can hold (Labrou and Clonis, 1994). Adsorption performs a function any time there is a wide surface area. When an ion exchange resin is bound to a particle, no ion is released. Monitoring the ions in the industrial effluent differentiates results between adsorbents and ion exchange. Common ion-exchange polymers (artificial) are

- -COOH poorly ionized to $-COO^-$
- $-SO_3H$ with a heavy ionization to $-SO_3^-$
- $-NH_2$ which attracts H^+ to form NH_3^-
- -The NR_3^+

It has along-lasting ionic charge(R is for some organic group). Such entities are enough to choose a resin irrespective of the strength of positive or negative charge. These polymers are normally branched ones (Labrou and Clonis, 1994). The polymers lose the unstable ions and then bind with suitable ions in the fluid that percolates throughout the column. Even so, resins are typically included in neutral types: Na- formed cation exchangers and Cl- formed anion exchangers. The trade performance relies on the following factors:

A. The resin's ability to adsorbions generally demonstrated in milliequivalents.
B. The resin spheres' size: the smaller, the more they are replaceable.
C. The rate of flow; the lower, the higher the adsorption.
D. The greater the temperature, the quicker the transfer.

The selection of resin relies on chemical and physical characteristics, and spoilage substances. For example, sometimes $CaCO_3$ is excluded from streptomycin fermentation medium since Ca^{++} ions can replace streptomycin cations (Labrou and Clonis, 1994). Streptomycin is isolated utilizing a resin (-COOH resin). The broth is progressively moved via two resin columns and washed with salt to create Na stage. The resin extracts a significant quantity of streptomycin which is washed by HCI transferring the streptomycin towards chloride and also the polymer to the H^+ type. Thus, streptomycin is obtained in cleared as well as condensed form.

The insolubility of certain salts is non-soluble; this nature can be used for the extraction of many industrial goods and is especially helpful in the removal of protein contaminants. Precipitation is done by various processes: by incorporating inorganic salts or by decreasing sorption with the incorporation of organic solvents such as alcohol (enzymes) and in purification of antibiotics the lactate and oxalate salts of erythromycin was extracted, and the calcium salts of citric acid were extracted (Kalyanpur, 2002). Precipitation is an important stage in the clearance of intracellular proteins following a cell disturbance. Proteins in fermentation medium (earlier or after cell lysis) can be isolated from several components by accumulation by utilizing some salts. Observations involve streptomycin sulfate and ammonium sulfate.

The two main protein precipitation processes:

1. Salting removed by adding inorganic salts $(NH_4)_2SO_4$ at high ion power.
2. Reduced solubility at cold temperatures by introducing organic solvents (T < –5 °). Isolating.

Salting with proteins is accomplished by raising the ionic intensity of a protein including solvent by inserting salts $(NH_4)_2SO_4$ or Na_2SO_4. The ionic intensity of a protein-including medium is improved by introducing such ions as $(NH_4)_2SO_4$ or Na_2SO_4. The ions introduced associate quite powerfully to water and accumulate the protein complexes. Protein permeability in a liquid relies on the ionic intensity. Whereas S is a solvent-soluble protein (g/L), S_0 is protein-soluble when I = 0, I is liquid ionic intensity, and K-K is the temperature and pH dependent salting constant. The ionic quality of a solution is known as

$$I = \frac{1}{2}\sum C_i z_i^2 \tag{8.15}$$

As *Ci* is ion molar intensity (mol/l); *Zi* is ion charge (valence).

The absorption of the protein declines logarithmically with high ionic concentration.

The solubility of protein as a function of the dielectric constant of a solution is given by

$$\log\frac{s}{s_0} = k'/D_s^2 \tag{8.16}$$

So D_S is the dielectric value of the water-solvent. Reducing the dielectric constant of the fluid helps in greater electrostatic interaction among protein complexes and promotes the accumulation of proteins. The introduction of solvents even reduces protein–water interaction and thus reduces protein solubilization. Solvents can induce coagulation of proteins. Isoelectric accumulation is the deposition of proteins at their isoelectric level.

This is the pH where the proteins do not have a total charge. The iso-electric value of the protein is described as pI = ($pK_1 + pK_2$). When pH = pI, the protein is free of charges and precipitates. This approach is appropriate for large specific hydrophobic proteins (i.e. non-polar). Ionic polyelectrolytes such as ionic polysaccharides and polyphosphate may be used to modify the ionic activity of the medium and induce protein accumulation. However, polyelectrolytes can trigger protein coagulation and harm to cells. The usage of non-ionic polymers including dextrans and polyethylene decreases the amount of water hence losses of interaction with proteins lead to denaturation (Belter et al., 1987).

8.9 DIALYSIS

Dialysis is a surface isolation process done by using a membrane that is utilized to extract low-MW solids such as organic acids (100 < MW < 500) and inorganic ions (10 < MW < 100) from the system (Figure 8.5). The utilization of dialysis filters is to extract urea (MW = 60) from urine in synthetic devices. In the biotechnology industry, dialysis could even be utilized to eliminate salts from protein. The dialysis layer divides two stages of low-MW as well as high-MW solutions.

$$\mu_\alpha^1 = \mu_\beta^1 \tag{8.17}$$

Here μ is the chemical strength of its diffusing substance.

$$RT\ ln\ C_1^\alpha \gamma_1^\alpha = RT\ ln\ C_1^\beta \gamma_1^\beta \tag{8.18}$$

$$C_1^\alpha \gamma_1^\alpha = C_1^\beta \gamma_1^\beta \tag{8.19}$$

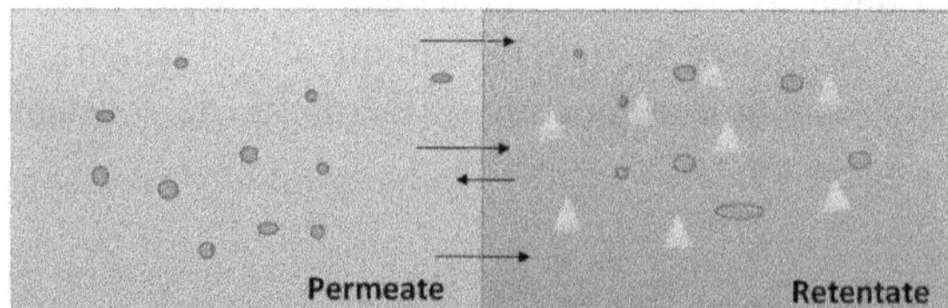

FIGURE 8.5 Schematic representation of dialysis.

Whereas C_l is the density and γ_1 is the diffusing factor activity coefficient 1. For optimal (very dilute) solutions, γ_1 and

$$C_1^{\alpha} = C_1^{\beta} \tag{8.20}$$

Concentrations are expressed in Eqs. 8.17 and 8.20 are amounts of the soluble (unbound) compound 1. This dialysis balance is focused on the expectation of the soluble (unbound) compounds. When there are macromolecules/polyelectrolytes, like proteins and nucleic acids, so the composition of the macromolecules must be known. That is the equilibrium defined as Donnan's equilibrium (Weuster-Botz et al., 2007).

8.10 REVERSE OSMOSIS

In fermentation broths, osmosis is the transfer of water ions from either a higher to a lower concentration area (i.e. from an aqueous phase to a salt-including aqueous medium) when such two parts are divided by a selectively permeable boundary. Water moves through the membrane quickly, although salt does not pass through it. At balance, the chemical ability of water should be similar to either section of the membrane (Zydney, 2016). When the water transfers into the salt form, the pressure rises. This osmotic pressure can be described as

$$\pi = CRT(1 + B_2C + B_3C^2 + \cdots) \tag{8.21}$$

Where C is the amount of the solute, T is the temperature, R is the gas constant, and B_2, B_3 is the virial factor for the solute. For dilute ideal solutions, $B_2 = B_3 = 0$.

$$\pi = CRT \tag{8.22}$$

In reverse osmosis (RO), a pressure is introduced onto a salt layer, that leads fluid compounds from a low- to the high-concentration part and leads to the concentration of solute (salt) compounds on one part of the membrane. The pressure needed to transfer solvent from a low- to high-concentration process is equivalent or a little greater compared to osmotic pressure. As Dp > p, the solvent flow usually occurs opposite to the concentration gradient.

$$\text{Solvent: } N_1 = K_p\left(\Delta P - \pi\right)^{\infty} \tag{8.23}$$

$$\text{Solute } = N_2 = C_i\left(1-\sigma\right)N_1 + K'_{pi}\Delta C_i;\ i \geq 2 \tag{8.24}$$

where K_p and $K_{p'}$ are permeability coefficients for solvent and solute, respectively, C is the average solute concentration in solution, and ΔC is the solute concentration difference across the membrane.

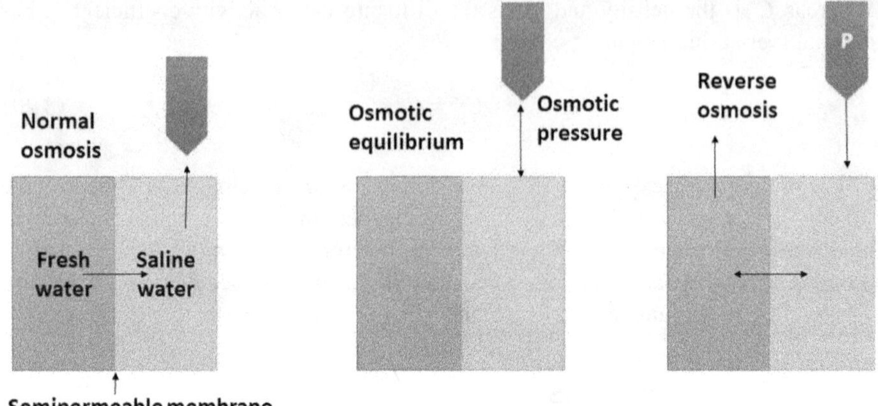

FIGURE 8.6 Diagrammatically explaining normal osmosis and reverse osmosis.

In certain reverse osmosis systems, membranes can permit the entry of solute components with solvents. In a solute transfer by diffusion or evaporation, the coefficient(s) of reflection for every solute could be described as the proportion of the solute compounds held on one part of the membrane in the existence of a solvent flux (Figure 8.6). Thus, for s = 0, the full flow of the solute is achieved, and also for s = 1 no flow of the solute is obtained (i.e. perfect reflection). Where Kp and Kp′ are permeability constants for solvent and solute, C is the total amount of solute in a sample, and DC is the variation in the concentration of solute passing out from the membrane. The amount of pressure needed differs from the amount of sample. Pressures of 30 to 40 atm are necessary for a 0.6 M salt solution. Salt rates in fermentation liquids are much greater compared to 0.6 M, needing high pressure for RO differentiation. The functions of RO in bioseparations are minimal because the process needs large pressure and is focused on the extraction of solvents (Kalyanpur, 2002).

RO membranes are typically utilized for dewatering and concentration reasons, yet not for protein extraction. Another issue with reverse osmosis is the accumulation of solid particles on membrane surfaces, resulting in a barrier to solvent movement. This effect, recognized as concentration polarization, could be resolved by the level of turbulence on the membrane surface. The osmotic pressure for multi-element processes is equivalent to the total of the specific osmotic pressures:

$$\pi = \sum \pi_i = \sum C_i RT(1 + B_{2i} Ci + B_{3i} Ci^2 + \cdots) \tag{8.25}$$

$$\text{Solvent: } \mathrm{N_1 = K_{pi}\left(\Delta P - \pi\right)} \tag{8.26}$$

$$\text{Solute: } \mathrm{N_1 = C_i\left(1 - \sigma_i\right)N_1 + K'_{pi}\Delta C_i;\ i \geq 2} \tag{8.27}$$

8.11 PURIFICATION AND CONCENTRATION

These methods are better for removing contaminants, making the product purified.

8.11.1 Chromatography

Chromatography separates mixtures into components by passing through a bed of adsorbent material over a fluid mixture. In elution chromatography, a column is packed with adsorbent particles that can be solid, porous, gel, or liquid phase that is immobilized in or on a solid. A mobile process or fluid layer is filled with a combination of solutes. A liquid or eluent follows on from this wave. The pulse reaches a small condensed point but the extra solvent disperses and dilutes the exits. Specific solutes in the mixture interact differently with the stationary phase adsorbent material; some interact weakly with each other and some interact intensely. During chromatography, two or more solutes move to different levels of solids because of the diverse solubility of the solutes in a particular solvent. A mixture of solutes and a solvent are applied from one solid phase stage. Displacement chromatography can be attractive when processing large amounts of material. In this process the column undergoes sequential phase adjustments in inlet conditions (e.g. fluid nature). In this process is added the feed combination, accompanied by a continuous displacer fluid infusion. For the stationary phase the displacer has to have a higher affinity than any compound in the feed solution. The ‘drive’ displacer eliminates the stationary process and back into the active zone. Fermentation substances are segregated as follows by either of the chromatographic techniques below, in which isolation of the solutes occurs for the purposes mentioned. When conditions are correctly chosen, the feed components are pushed into neighbouring square wave-like zones of condensed, pure solutes. Such areas then split through the end of the column with the region having first left the solute with the lowest propensity for the stationary process. The primary benefit of displacement chromatography over elution chromatography is the capacity for better efficiency, but the process is more complex, and in certain cases high precision (separation of solutes) may be hard to achieve (Freitag, 2014).

Chromatographic adsorption: Differences occur in the low forces such as Van der Wall forces that connect solutes to its solid phase (e.g. paper chromatography).

Chromatographic partition: A mobile solvent moves via a column holding a steady liquid phase; solvent and liquid phases are not mixed (Freitag, 2014). Segregation takes place among mobile phases and liquid phases through separate distribution or partition coefficients of solutes.

Chromatography of ion exchange: This approach is based on the variation in chemical force among the different solids as well as the resin polymer.

Filtration gel: This relies on the capacity of particles of varying dimensions and types in the ideal solvent to encompass in a matrix. The gel could be regarded as having two forms of solvent: the inside and outside of the gel matrix (Freitag, 2014). Big particles that cannot enter the gel occur in the column effluent. In the effluent after an amount equivalent to the overall amount of fluid outside the gel, the matrix has appeared. Small particles that penetrate the gel matrix occur in the effluent after an amount equivalent to the overall amount of fluid inside the matrix has appeared.

8.11.2 Decoloration of Carbon

A few solids are capable of adsorption and condensing some compounds on their coatings when they connect a liquid medium (or a gas mixture that includes activated carbon, aluminium, and titanium oxides and different forms of absorbent clay (Belter et al., 1987). Absorbents are applicable to adsorb antibiotic products from broths, coloured contaminants, etc. Activated Charcoal is commonly used due to its big pores giving it a high surface area. However, the pores are too big for the fluid to flow through it.

Activated carbon is mostly applicable to the elimination of colour. Thereby, the penicillin solvent is generally handled with activated carbon before the crystallization of the amino salt. While adding an adsorbent, depth of the absorption area must be known or should be performed experimentally.

8.11.3 Crystallization

This process applies to components that can withstand high temperatures. The solvent is concentrated by heat treatment and evaporates at atmospheric pressure to generate a saturated liquid (Zydney, 2016). Even so, several fermentation substances do not tolerate high temperature and the initial extraction of water is achieved by heating at a lower pressure or by decreasing the temperature to create crystals that can be centrifuged by retaining a condensed liquor (Kalyanpur, 2002). This produces substances that are extremely active, relatively durable and no coloured impurities. To acquire crystals, a saturated solution is first generated; secondly, a minute nucleus or seeds are established, and thirdly, the compounds of the solute are developed on the nucleus. Crystals from the prior process may be intentionally added to produce the nucleus. In the manufacturing of procaine penicillin, tiny crystals are utilized to stimulate crystallization, while in dehydro streptomycin sulfate the introduction of methanol causes crystallization.

8.11.4 Drying

The last separation of the product is carried out by either of two processes:

A. Treatment of crystalline products.
B. Drying out products from solution

Crystalline Processing: Crystalline materials are incompressible; there is no need to filter. Thus, they are filtered on dense beds at high pressure (Jungbauer, 2013). That is normally achieved on a centrifugal system able to produce a very strong (about 1,000-fold) gravitational force. The crystals are cleaned to extract the sticky liquor. Further, they are drained by spinning for the drying processor extraction of liquid.

Drying comprises the removal of liquids (either organic solvent or water) from wet crystals as mentioned previously from a fluid, or from solutes or cells which have been extracted since the earliest action. There are many drying solutions available but it depends on factors including the physical nature of the final result, its temperature tolerance, the type appropriate to the customer, and the quality and cost by using a different product (Kalyanpur, 2002). Drying may be listed under two titles: (i) moist elimination in the liquid phase and (ii) moist elimination in the solid phase.

Liquid-phase moisture elimination requires a dry heat process. Once drying is achieved at high temperatures, the mechanisms can be subdivided to provide heat to the component and remove the associated vapours (Kalyanpur, 2002). The easiest approach is active heating in which natural air is heated; both the substance and vapour. For other situations, heating is performed at a lowered pressure to allow the measurement of water vapour. In these situations, external heating by a hot layer, radiation (e.g. infrared), and either method is employed to enhance the heat generated by lowered vapour pressure. Other heating methods are as follows.

Tray dryer: The most frequently used method. It is flexible and comprises heated racks in a large unit that can be flooded by vacuum. In certain cases, the trays could have the availability of vibration to promote evaporation. Also, they can be heated at a relatively low temperature feasible for heat-resistant components (Weuster-Botz et al., 2007).

Drum dryer: In this process, a broth or slurry is added to the edge of a rotating hot drum. Maximum heat is transferred to the component to be treated for a short period as damage can happen (*Handbook of Microalgal Culture*, 2003). The design of the drum dryers is shown in Figure 8.7.

Spray drying: This technique is widely utilized in food and fermentation companies. These techniques are used for drying heat-sensitive products such as medications, plasma, and dairy industry products. The standard spray comprises a mechanism for the application of a spray nozzle of the liquid to be dried toward a counter-current of heated air (Kalyanpur, 2002). Because the material has been subjected to high temperatures at a short period, a sometimes small amount of harm typically arises. Often the substance is added at the same time as air (Figure 8.8).

8.11.5 Electrophoresis

Electrophoresis is used in an electric field to isolate the charged biomolecules according to their size and energy. Within an electric field, electrostatic forces equal the drag force on a charged particle as the object is travelling at a steady terminal velocity (Belter et al., 1987). A force balance on a charged particle moving with a terminal velocity in an electric field yields the following equation:

$$qE = 3\prod \mu D_p V_t \tag{8.29}$$

where q is the charge on the particle, E is electric field intensity, D_P is particle diameter, m is the viscosity of the fluid, and Vt is the terminal velocity of the particle. Depending on the pH of the medium, electrostatic charges on protein molecules will

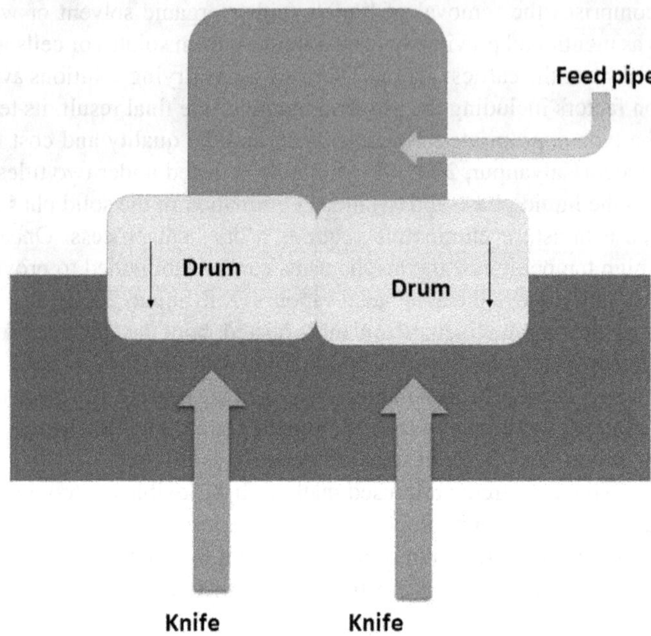

FIGURE 8.7 Diagrammatic representation of a drum dryer.

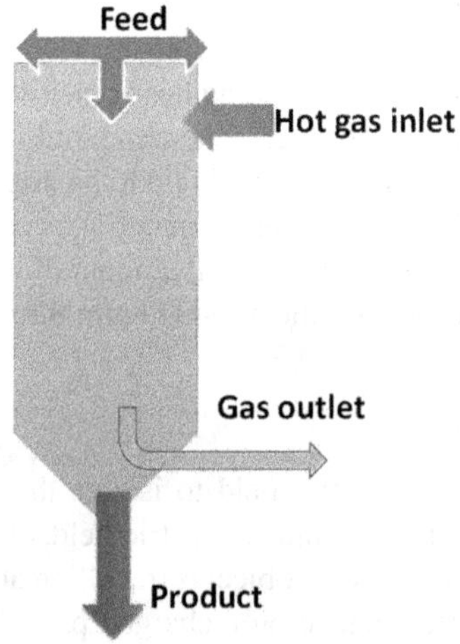

FIGURE 8.8 Schematically representing the spray drying method.

be different. When pH > pI, the protein will be negatively charged; and when pH < pI, the charge on the protein will be positive. The net charge on the protein will determine the velocity of the protein. When a protein molecule is placed in a pH gradient, the electrophoretic velocity becomes zero when pH = pI, since the net charge on the protein is zero at pH = pI. Precipitation of proteins in a pH gradient at their isoelectric point is known as isoelectric focusing. Proteins will be separated from each other in an electric field. Certain gels, such as agar or polyacrylamide, are used for protein separations by gel electrophoresis. Gel electrophoresis is an important analytical separation technique. Scale-up is problematic due to thermal convection resulting from electrical heating. One analytical and micropreparative version of electrophoresis that has an excellent resolution is called two-dimensional protein electrophoresis (2DE). The 2DE procedure is actually the series combination of two electrophoretic separations which resolve protein mixtures based on two independent characteristics – charge and size (Kalyanpur, 2002). Proteins are first separated in a polyacrylamide gel matrix using isoelectric focusing, an equilibrium separation technique that resolves proteins based on their respective isoelectric points. Proteins are subsequently coated with an anionic surfactant and separated by size in another polyacrylamide gel. Finally, proteins are detected using a chemical stain, by autoradiography or by other methods. The result of this two-dimensional separation is a high-resolution fingerprint of protein expression that is characteristic of a particular biological system. This technique has at least two orders of magnitude better resolution than any other analytical tool for protein analysis. Separated protein spots can be cut out of the polyacrylamide gels and subjected to further microchemical analysis to determine the amino acid sequence (Zydney, 2016).

8.12 CONCLUSIONS

The global market for green packaging and sustainable growth microbial growth has gained more and more interest today, but its production is constrained by high costs. Downstream processing costs add a great deal to the total cost of the product output. Effective separation methods are needed for performance in the DSP tasks. These include outlining the concepts, the procedure configurations utilized, the main process parameters and relevant implementations. They address benefits and drawbacks of various approaches as demonstrated in realistic implementation and examine process variations for assembling full DSP sections. Thereby foundations are given for choosing unit activities for product separation. Requirements for products and processes (e.g. quality, quantity of production) and economic factors (e.g. expenditure and cost of operation) will be addressed.

REFERENCES

Belter PA, Cussler EL, Hu W, Bioseparations: Downstream processing for biotechnology, 1987.

Freitag R, Chromatographic Techniques in the Downstream Processing of Proteins in Biotechnology, in Pörtner R (Ed.), Animal Cell Biotechnology: Methods and Protocols, Methods in Molecular Biology. Humana Press, Totowa, NJ, pp. 419–458, 2014.

Handbook of Microalgal Culture, 1st ed, John Wiley, 2003.

Jungbauer A, Continuous downstream processing of biopharmaceuticals. Trends Biotechnol. 31, 479–492, 2013.

Kalyanpur M, Downstream processing in the biotechnology industry. Mol. Biotechnol. 22, 87–98, 2002.

Labrou N, Clonis YD, The affinity technology in downstream processing, J. Biotechnol. 36 (2), 95–119, 1994.

Ratledge N, Kristiansen B, Basic Biotechnology. Cambridge University Press, 2001.

Soetaert W, Vandamme EJ, Industrial Biotechnology: Sustainable Growth and Economic Success. John Wiley, 2010.

Weuster-Botz D, Hekmat D, Puskeiler R, Franco-Lara E, Enabling Technologies: Fermentation and Downstream Processing, In: Ulber R, Sell D (Eds.), White Biotechnology, Advances in Biochemical Engineering/Biotechnology. Springer, Berlin, Heidelberg, pp. 205–247, 2007.

Zydney AL, Continuous downstream processing for high value biological products: A Review. Biotechnol. Bioeng. 113, 465–475, 2016.

9 Ethanol Production and Alcoholic Beverages

9.1 INTRODUCTION

Ethanol fermentation is known as alcoholic fermentation which converts sugars such as glucose, fructose, and sucrose into ethanol and carbon dioxide as the main products. Because yeasts perform this conversion in the absence of oxygen, alcoholic fermentation is considered an anaerobic process. The alcohol can be classified as taxable and non-taxable alcohol. Taxable alcohol is used in its pure form, mainly for human consumption as different alcoholic beverages. In addition, it is used in the pharmaceutical industry, the perfume industry etc. Non-taxable alcohol is utilized as chemical feedstock e.g. acetic acid, the polythene industry, the rubber industry etc. These are unfit for human consumption (due to addition of certain chemicals). This type of denatured ethanol contains additives to make it poisonous, foul smelling or nauseating.

Alcohol is named to indicate the source of raw material from which it is manufactured or to indicate the general purpose for which it is to be used. For example, grain alcohol is made via fermentation from grains such as corn, wheat, or rice, while molasses alcohol is produced from sugar cane molasses. The name 'industrial alcohol' indicates ethyl alcohol used for industrial purposes. Power or fuel alcohol is used in combination with gasoline (known as gasohol) or other motor fuel. This type of bioethanol is considered to be a very efficient alternative for liquid fuel. Nowadays, the main source of fuel to meet demand is petrochemicals and due to fossil fuel depletion an alternative source for fuel production is currently needed. Bioethanol turns out to be an excellent source for this.

Any sugar fermentable by yeast can act as raw material for alcohol fermentation. The process of manufacture of ethanol depends mainly on the raw materials. There are basically three types of raw material from which alcohol is produced. Saccharine materials such as sugar cane, sugar beets, molasses and fruit juices etc. need little or no pre-treatment other than dilution. However, starchy materials and cellulosic materials do require pretreatment.

Bioethanol is mainly produced from biomass which contains a large amount of sugar. Overtime there arose three generations of bioethanol production. Firstly, it was produced from starch followed by the second generation which uses lignocellulosic biomass for bioethanol production. The production of bioethanol from starch has three steps; these are hydrolysis of the starch and fermentation of the glucose present followed by purification, whereas production of bioethanol from lignocellulose involves four steps; these are pretreatment of lignocellulose, hydrolysis of cellulose, fermentation of the sugar present and finally purification.

9.2 BIOETHANOL PRODUCTION

Bioethanol is used as a gasoline equivalent and can be used as fuel for transport purposes. Overtime, there has been an increase in the production of bioethanol globally. Enzymatic hydrolysis of cellulose has improved production efficiency. The production of bioethanol from biomass is environmentally friendly which makes it sustainable. Residues from agriculture have increased the production level of bioethanol sixteen times when used as a raw material for its production (Alvira et al., 2010).

9.2.1 Production of Bioethanol from Starch

Starch acts as storage of energy in plants and hence is a very high-yielding feedstock in terms of ethanol production.

The three steps in bioethanol from starch include

a. Hydrolysis.
b. Fermentation.
c. Purification/separation of the product.

In the first step, hydrolysis of polysaccharide to monosaccharides is achieved. This is a chemical reaction in which the molecule is divided into two sections with an addition of a water molecule. This disintegrates the polymer into monomeric sugars (Figure 9.1).

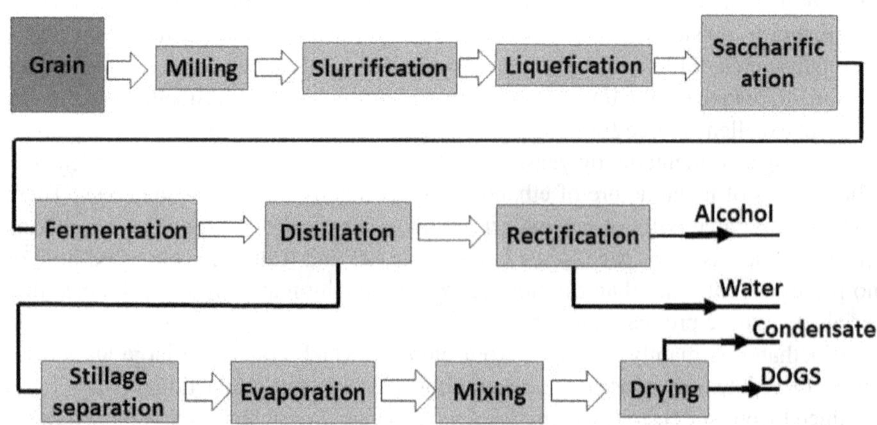

FIGURE 9.1 Ethanol production from starch grain.

The hydrolysis of starch can be achieved by chemical reactions using acids (mainly sulfuric acid) or by using enzymes such as cellulase and amylase. Hydrolysis when carried out enzymatically can be done by using either soluble enzymes or immobilized enzymes (Aditiya et al., 2016). Biotechnology has played a role in increasing the efficiency of bioethanol production by genetically modifying plants such as corn which has led to efficient and cheaper production of feedstock (Wei et al., 2014).

In the second step, which is fermentation, sugars are converted into mainly ethanol and carbon dioxide gas. Microorganisms are used for this process which is carried out under anaerobic conditions. For the production of ethanol from starch, strains of *Saccharomyces cerevisiae, Schizosaccharomyces pombe, Kluveromyces lactic* and *Saccharomyces amucae* are extensively used, which results in the fermentation of sugars into ethanol and organic acids. To increase efficiency in the production of ethanol through the process of fermentation, yeast is added and this is also achieved by using genetically engineered microorganisms (Aditiya et al., 2016). With the onset of genetic modifications in yeast and bacterial strains, these became capable of fermenting hexose and pentose sugars into ethanol to get higher yields. The last step involves the purification of the final product ethanol by distillation methods.

9.2.2 Production of Bioethanol from Lignocellulose

Conversion of lignocellulose into bioethanol is comparatively more difficult than other sugars. This is the second generation of bioethanol production. There are four steps involved in this process, which are

a. Pretreatment.
b. Hydrolysis.
c. Fermentation.
d. Purification/separation.

Only one additional step is added in this process, which is pretreatment; otherwise all subsequent steps are similar to bioethanol production from starch. Pretreatment is done in order to destroy and damage the structure of the lignocellulosic waste. There are some specific requirements for this step, which include an enhanced recovery of fermentable sugars, avoiding the degradation of carbohydrates, cost-effective process and usage of products which inhibit the upcoming processes (hydrolysis, fermentation) (Tomás-Pejó et al., 2008). The pretreatment of lignocellulose hydrolysis is carried out so that lignin and hemicellulose are removed from the lignocellulosic biomass which leads to reduction of cellulose crystallinity and an increased porosity.

There are two types of pretreatment; chemical and biological.

Pretreatment by chemical methods:

This form of pretreatment can be acidic or alkaline. This includes ozonolysis, alkaline hydrolysis, acid hydrolysis, oxidative delignification and the organosolv process. Acid causes chemical hydrolysis, which mainly works on hemicelluloses and lignin. These are dissolved in the plant's cell wall which favours enzyme activity in the cellulose (Wei et al., 2014). One limitation observed during this process is the generation

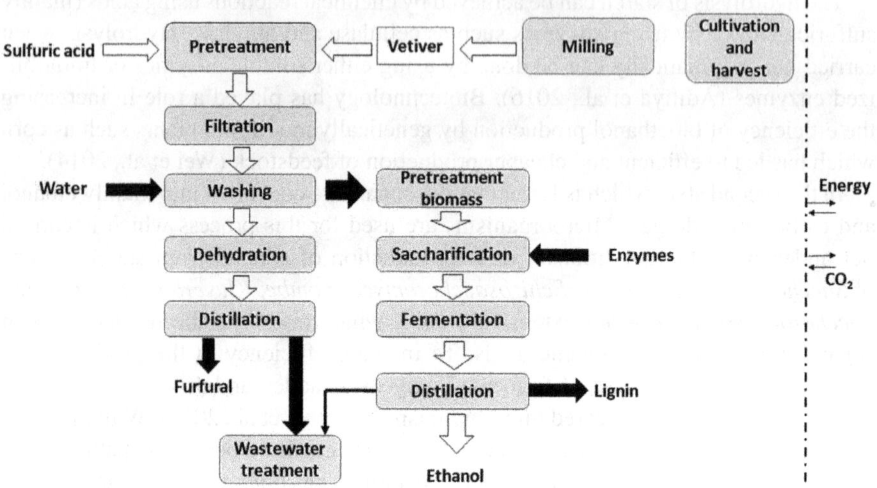

FIGURE 9.2 Ethanol production from lignocellulosic source.

of a number of degraded products such as furfural, which inhibits the successive steps in the production of bioethanol (Figure 9.2).

Acid hydrolysis is carried out in order to convert lignocellulosic products into fermented sugars. The two types are diluted and concentrated acid hydrolysis. Both these processes degrade sugars and hence are done at higher temperatures (Yılmaz and Selim, 2013). Concentrated acids are used on the lignocellulosic biomasses where the acid goes inside the hemicellulosic part and leaves the cellulose and lignin part intact. The other limitations of this process are its corrosivity and toxicity.

Cellulose can be degraded to sugar by concentrated HCl

(a) $\text{Cellulose} \xrightarrow[\text{Conc. HCl}]{30\text{–}35^\circ\text{C}} \text{Sugars (glucose) and browning reaction}$

(b) $\text{Cellulose} \xrightarrow[\text{dil. } H_2SO_4]{150\text{–}160^\circ\text{C}} \text{Sugars (glucose) and browning reaction}$

To tackle these, diluted acids are used for the pretreatment of the lignocellulosic materials and these are of two types:

a. Continuous flow which is done at higher temperature.
b. Batch process which is done at lower temperature.

Pretreatment with dilute sulfuric acid leads to hydrolysis of the hemicellulose to make cellulose available and this process is widely accepted for pretreatment methods. Pentose sugar is degraded more quickly than hexose and to decrease the sugar degradation level a two stage process is implemented where in the first stage, mild conditions are maintained to retrieve pentose which is followed by the second stage where the conditions are made harsh in order to retrieve hexose. The only limitation this process has is its higher cost efficiency (Khoo et al., 2013).

In alkaline pretreatment, catalysts such as calcium oxide and sodium hydroxide are used to target specifically the hemicellulosic acetyl groups. Lignin-carbohydrate ester links are used in the alkaline pretreatment. In this process, saponification is done of the ester bonds which are present inter-molecularly and have crosslinked xylan hemicelluloses and lignin. Using diluted NaOH has many limitations and therefore calcium hydroxide is used, which is a better solution. Lignin and hemicellulosic elements are removed by lime pretreatment, which increases the crystallinity of the substance; this in turn favours the hydrolysis rate (Khoo et al., 2013).

To pretreat the lignocellulosic biomass, biological methods are adopted, which includes microbial enzymes. Microbes secrete extracellular enzymes including laccase and lignin peroxidases which dissolve the hemicellulose and lignin. There are many advantages offered by this type of pretreatment, such as lower requirement of energy and requirement of mild environments to function. Limitations include a lower rate of hydrolysis and higher residence time. After pretreatment, the subsequent steps are similar to the production of bioethanol from starch where hydrolysis is done, and that is followed by fermentation and the end product is filtered in order to get the final product.

Ethanol is produced mainly by sugar beet, cassava, potatoes and sugarcane. Microbial fermentation is performed in order to produce ethanol. Yeasts have been utilized for industrial ethanol production where the strains mainly used are *Saccharomyces cerevisiae, S. uvarum, S. carlsbergensis, K. lactis, Candida brassicae, C. utilis* and *Kluyveromyces fragilis* and also certain bacterial strains such as *Zymomonas mobilis*. Efforts are being made to produce ethanol using Schizosaccharomyces; this is giving satisfactory results. Strains of *K. fragilis* are mainly used when whey is used as substrate (Demirbaş, 2001). Heat generated during anaerobic respiration is retained in the fermentation process where the heat released is 1.371 MJ mol^{-1}, whereas the total heat of combustion of both sucrose and glucose is 8.463 MJ mol^{-1}. According to theory, 97% of the sugar gets transformed into ethanol but in practice only 48.4 grams of ethanol is yielded from 46% of pure glucose (Demirbaş, 2001).

$$C_6H_{12}O_6 \dashrightarrow 2\ CH_3CH_2OH + 2\ CO_2$$

Sugar (Glucose) ====> Alcohol (Ethyl alcohol) + Carbon dioxide + Cell mass

$$100\ \text{kg} \longrightarrow 51.1\ \text{Kg} + 48.9\ \text{Kg}$$

Therefore the theoretical ethanol yield: 51.1%g/g

Two theories have been established for alcohol fermentation: i. Neuberg theory and ii. Embden–Meyerhof–Parnas (EMP) pathway. While the Neuberg theory is mainly on a theoretical basis, EMP pathways are mainly based on experimental facts. Several factors influence alcohol fermentation: efficiency of pretreatment; optimum concentration of sugar; optimum pH and temperature; addition of nutrients; use of a vigorous strain of yeast; maintenance of anaerobic conditions during the fermentation are among them.

There are three steps in this process which include

a. Medium preparation
b. Fermentation
c. Recovery

For medium preparation, there are three types of substrate that can be used for the production of ethanol. These are (i) substrate containing starch, (ii) juice which is recovered from sugarcane or sugar beet and (iii) waste products which are recovered from wood (Tomás-Pejó et al., 2008).

Continuous fermentation is used for the production of ethanol and, therefore, large fermenters are preferred for the continuous production of ethanol. The conditions required for fermentation are a pH of 5 and temperature of 35 °C. Fermentation is started within 12 h of production starting and is carried on for several days. After the process of fermentation is finished, the cells are separated to recover the biomass of yeast cells; these are then used further as single cell protein (SCP) which then acts as a nutrition product for animals. The recovered medium is further processed for ethanol recovery.

9.3 DOWNSTREAM PROCESSING

The final step is recovery, where up to 95% of ethanol can be recovered by successive distillations. For 100% recovery, an 'azeotropic mixture' needs to be formed which contains 5% water. This 5% water is then removed from the azeotropic mixture of water, benzene, and ethanol. This is achieved after distillation. After this the mixture is removed to get absolute alcohol (Alvira et al., 2010).

Problem 9.1 Production is planned to output2000 m^3 of rectified spirit per day in a chemostat. Cane molasses contains about 45%w/w sucrose. The rectified spirit contains 85%v/v of ethanol. *Saccharomyces cerevisiae* is used for the production of ethanol. The temperature and initial pH of the fermentation medium are 35 °C and 6.5, respectively. Answer the following:

a) Volume of the bioreactor,
b) Total amount of cane molasses required per day,
c) Amount of cell mass produced per day,
d) Amount of carbon dioxide produced per day,

The following data is given:

1. $\mu_{max} = 0.15\ h^{-1}$, $K_S = 2\ g\ L^{-1}$, $\mu_{death} = 0.03\ h^{-1}$
2. $Y_{X/S} = 0.1$, $Y_{P/S} = 0.5$
3. Maximum ethanol concentration = 8%v/v
4. Density of absolute alcohol = 0.78 g mL^{-1}

Assume suitable data as required.

SOLUTION

Given data,

$$\mu_{max} = 0.15\ \mathbf{h^{-1}},\ \mu_{death} = 0.03\ \mathbf{h^{-1}}, \mathbf{K_s} = 2\ \mathbf{gL^{-1}}, \mathbf{Y_{x/s}} = 0.1, \mathbf{Y_{P/S}} = 0.5$$

Maximum ethanol concentration = 8%v/v= 80 mL/L
Density of ethanol = 0.78 g/mL
Substrate conversion = 95% (assumed)

$$\text{Initial substrate concentration } (S_0) = \frac{80\ {}^{mL}\!/\!_{L} \times 0.78\ {}^{g}\!/\!_{mL}}{0.95} = 65.7\ \mathbf{{}^{g}\!/\!_{L}}$$

The present process is a chemostat.
In case of sterile media, inlet biomass concentration, $X_0 = 0$
Steady-state biomass concentration, $X = X_0 + Y_{X/S}(S_0 - S)$
It is considered that the reactor is operated at the dilution rate of D_{max}.

$$\mathbf{D_{max}} = \mu_{max}\left(1 - \sqrt{\frac{\mathbf{K_S}}{(\mathbf{K_S} + \mathbf{S_0})}}\right)$$

$$= 0.15\left(1 - \sqrt{\frac{2}{(2 + 57.5)}}\right) = 0.12\ \mathbf{h^{-1}}$$

Also, steady-state substrate concentration

$$\mathbf{S} = \frac{\mathbf{K_S}(\mathbf{D_{max}})}{\mu_{max} - (\mathbf{D_{max}})} = \frac{2 \times (0.12)}{0.15 - 0.12} = 8\ \mathbf{g/L}$$

Cell death is negligible compared to cell growth

$$\mathbf{X} = 0.1(57.5 - 8) = 4.95\ \mathbf{g/L}$$

(a)

Basis: 2000 m^3 spirit/day ≡ 2000 m^3 × 0.85 ethanol/d = 1700 m^3 ethanol/d (since spirit content 85%v/v ethanol) = 17,00,000 L ethanol per day

Density of ethanol = 780 g/L

Amount of ethanol = 780 × 17,00,000 g =132.6 × 10^7 g

= 32.6 × 10^4 kg ethanol per day

$$\text{Substrate required} = \frac{\mathbf{132.6 \times 10^4\ kg/d}}{\mathbf{Y_{P/S}}} = \frac{\mathbf{132.6 \times 10^4\ kg/d}}{\mathbf{0.5}} = \mathbf{265.2 \times 10^4\ kg/d}$$

$$\text{Actual substrate required} = \frac{\mathbf{265.2 \times 10^4\ kg/d}}{\mathbf{0.95}} = \mathbf{279.16 \times 10^4\ kg/d}$$

Volumetric feed flow rate =

$$F = \frac{\textbf{substrate required}}{\textbf{initial substrate conc}} = \frac{\mathbf{279.16 \times 10^4 kg/d}}{\mathbf{57.5\ g/L}} = \frac{\mathbf{265.2 \times 10^4\ kg/d}}{\mathbf{57.5\ kg/m^3}}$$

$$= \mathbf{4859.2 \frac{m^3}{d} = 2023 \frac{m^3}{h}}$$

$$\text{Now, } \tau_{CSTR} = \frac{S_0 - S}{(-r_s)}$$

$$\left(-r_s\right) = \frac{1}{Y_{X/S}} \mu X$$

Assuming ethanol production as a growth-associated product.
We substitute $\mathbf{D_{max}}$ for μ

$$\left(-r_s\right) = \frac{1}{Y_{\frac{X}{S}}} D_{max} X = \frac{1}{0.1} \times 0.12 \times 4.95 \frac{g}{L.h} = 5.94 \frac{g}{L.h}$$

$$\tau_{CSTR} = \frac{S_0 - S}{\left(-r_s\right)} = \frac{V}{F}$$

$$\frac{V}{F} = \frac{(57.5 - 8)}{5.94} h = 8.33\ h$$

$$V = 2023 \frac{m^3}{h} \times 8.33\ h = 16851\ m^3$$

(b)

Sugar required per day = $\mathbf{279.16 \times 10^4\ \ kg/d}$

Cane molasses content 50%w/w sugar

Therefore, cane molasses required = $\mathbf{558.32 \times 10^4\ \ kg/d}$

(c)

$$\text{Cell mass productivity} = \mathbf{D_{max}X = 0.12 \times 4.95}\ \mathbf{g/_{L.h}} = \mathbf{0.594}\ \mathbf{g/_{L.h}}$$

$$\text{Cell mass produced per day} = \mathbf{0.594}\ \mathbf{kg/_{m^3.h}} \times \mathbf{24\ h \times 16851\ m^3 = 240\ MT/d}$$

(d)

2000 **m**^{3} spirit = 132.6 × 10^4 kg ethanol

$$C_{12}H_{22}O_{11} \longrightarrow 2\ C_6H_{12}O_6 \dashrightarrow 4\ CH_3CH_2OH + 4CO_2$$

Sucrose ====> Glucose ====> Alcohol + Carbon dioxide
(Ethyl alcohol)

342 kg 360 kg 184 kg 176 kg

To produce 132.6 × 10^4 kg ethanol, (176/184) × 132.6 × 10^4 kg = 1268 MT CO_2 is produced per day.

9.4 WINE PRODUCTION

Alcohol beverages comes under the category of taxable alcohol. Alcohol beverages can be classified into two categories: distilled alcohol and non-distilled alcohol. Wines and beers and ciders are among the most common non-distilled alcohol. Examples of distilled alcohol include whisky, rum, vodka, brandy, gin etc.

Wine is characterized as an outcome of a 'normal alcoholic fermentation of grape juice'. Any fruit with high sugar content can be utilized for the manufacture of wine. Therefore it is possible to use citrus, bananas, apples, pineapples, strawberries etc. Such types of wines are termed fruit wines. Wine production is easier compared with beer production because there is no requirement for the malting process since sugars are available in the fruit juice. However, this exposes winemaking to higher risks of spoilage (Wissemann and Lee, 1980).

Wine is developed mainly in mild-winter regions with cool summers and with even rainfall distribution in the overall year. The United States is the leading wine producer; in North America this is mostly in the State of California with some production in New York. In Europe, the main suppliers are Italy, Spain, and France. Other manufacturing regions are Turkey, Syria, Iran, and Australia.

9.4.1 Methods of Wine Production

Wine production includes several distinct steps (Figure 9.3).

Initially, stalks should be eliminated, as they contain tannins which provide a harsh taste to wine. The skins of black grapes are included for production of red wines, for colour appearance. Glucose and fructose are the main sugars in grapes which have an approximately similar ratio in ripe fruits (Fleet, 2008).

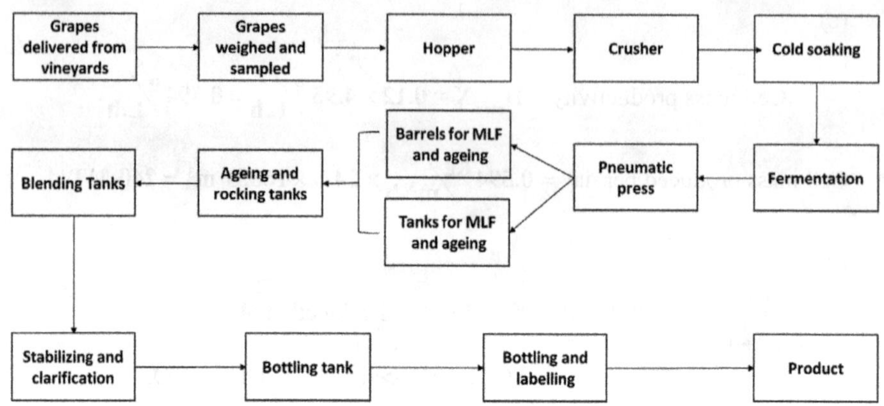

FIGURE 9.3 Block flow diagram of grape wine production.

Grapes contain malic and tartaric acids with a small citric acid concentration due to which the acidity range is about 0.60–0.65 and the pH range is 3–4. The ripening process increases both levulose concentration and tartaric acid concentration. Additionally, grape juice contains nitrogen concentrations in different forms such as amino acids, peptides, purines, and a small number of ammonium components and nitrates (Jones and Davis, 2000).

9.4.2 Fermentation

Grapes contain a natural microorganism which induced the fermentation process in previous times and helps in the production of different types of wines. Nowadays, sulfur dioxide, bisulfate and metabisulfite are used for partially sterilizing the wine (killing of several microbes) except for wine yeasts which are then inoculated. The applicable yeast for inoculation is *Saccharomyces ellipsoideus.* Special wine-making yeasts such as *S. fermentati, S. oyiformis and S. bayanus* are used (Fleet, 2008).

The characteristics of wine yeast:

(a) growth at low pH of grape juice
(b) high alcohol tolerance (more than 10%)
(c) sulfite tolerance

During the fermentation process, heat is liberated which is estimated to be 24 cals per 180 g of sugar. However, 'must' contains 22% of sugar which may increase the temperature by 52°F ('must'is from the Latin vinum mustum, 'young wine' which is freshly crushed fruit juice – usually grape juice – that contains the skins, seeds, and stems of the fruit). If the starting temperature is 60 °F (16 °C), at this stage fermentation will be halted at a temperature of 100 °F (38 °C), so just 5% of the alcohol will remain. For cooling purposes, the temperature is kept at about 24 °C via the cooling system within the fermenter (Sablayrolles, 2009).

Generally, yeast ferments glucose but special yeast like *Saccharomyces elegans* favours fructose. Glucose-fermenting wine yeasts are used to generate sweet wine. Many other macro and micro-nutrients are present in the 'must'; however, nitrogen is limited. Thus small volumes of $(NH_4)_2\ SO_4$ or $(NH_4)_2\ HPO_4$ are added (Sablayrolles, 2009).

Oxygen is needed for yeast growth; this is an initial step of the fermentation process. In the second stage, the growth condition is shifted to anaerobic which leads to the utilization of intermediate compounds such as acetaldehydes as hydrogen-acceptors by yeast, thus stimulating alcohol production.

9.4.3 Ageing and Storage

During fermentation, skins are not permitted. Hence, at the final step, the wine is allowed to flow via the perforated filter where pomace is removed. Then the next step is 'racking' where the wine is permitted to stay still until a significant proportion of the yeast cells or other materials have accumulated as sediment or 'lees' at the bottom of the flask. After this, the wine is pumped or diverted off without destroying the precipitate. This wine is then transferred to wooden barrels (100–1,000 gallons) or tanks (a thousand gallons). The wood allows the wine to have minimal access to oxygen. However, water and ethanol slowly evaporate and create air pockets that are suitable for production of aerobic wine toxins; for example, acetic acid bacteria and yeast (Sablayrolles, 2009).

Casks (barrel-shaped containers) are frequently used to avoid air pockets. In recent technology, stainless-steel containers are being used to solve such air pocket formation by loading the space with an inert gas such as carbon dioxide or nitrogen. During the ageing process, beneficial improvements occur in the wine. Such improvements occur because of the following factors:

a) Oxygen will gradually diffuse; also, small quantities of oxygen penetrate during the filling process. For formation of esters, alcohols react with acids as well as tannins being oxidized.
(b) The wood extract also helps in the ageing process by changing the flavour.
(c) Microbial malolactic fermentation exists in some wines. Within this fermentation, malic acid is first transferred into pyruvate after conversion into lactic acid (Aleixandre-Tudo and du Toit, 2018). The process is essential for the richer taste that forms through the ageing of certain wines, e.g. Bordeaux. The colonies involved in this process are *Lactobacillus sp and Leuconostoc sp.*

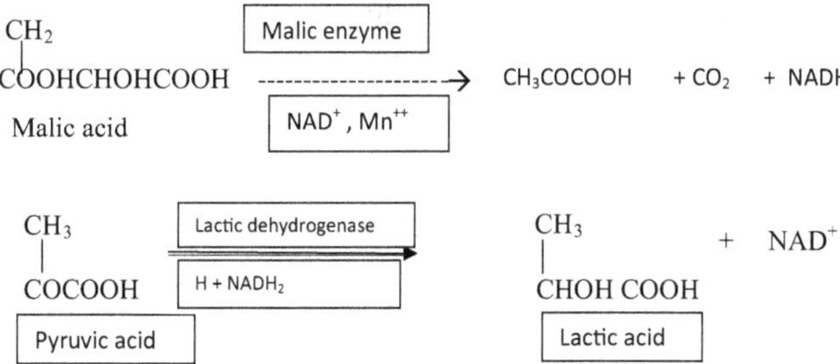

For ageing wines, a temperature of 11 to 16 °C is ideal, and higher temperatures probably increase the rate of the oxidation process.

9.4.4 Clarification

Based on the type of wine, wine is permitted to age from two to five years. Sometimes an artificial clarification process is required. Further fining agents are also used in the clarification process. Fining agents react with tannins, acids, proteins or with some extra substrates to produce heavy quick-setting coagulum (Fleet, 2008). Different suspended components are adsorbed during the process of settling. Traditional fining compounds for wine are gelatin, casein, tannin, isinglass, egg albumin and bentonite. In some cases the elimination of metal ions is carried out with potassium ferrocyanide known as 'blue fining'; this eliminates extra ions of copper, iron, manganese, and zinc from the wines.

9.4.5 Packaging of Products

Initially, wine is mixed then pasteurized, after which the packaging is done. In certain wine industries, sterilization is conducted through the filtration process. Wine is packed and sold in casks in several countries.

9.4.6 Defects in Wine

The main reason for the contamination of wine is microbial load along with other defects such as acidity and cloudiness. The following aspects contaminate the wine:

(1) The contents of the wine i.e. the concentration of sugar, alcohol, and sulfur dioxide
(2) Storage temperature e.g. temperature and air space in the container
(3) The level of actual microorganism penetration during bottling

Bacterial spoilage is uncommon when good hygiene is maintained. If spoilage occurred due to acetic acid bacteria then it gives a sour taste to the wine. Lactic acid bacteria, mainly *Leuconostoc*, and often *Lactobacillus,* contaminate wines. Different yeasts sometimes contaminate wine (Fleet, 2008). *Brettanomyces* are abundant, slow-growing yeasts that develop in wine and cause turbidity and bad flavour. Additionally, *Saccharomyces bayanus* can result in turbidity and sedimentation of some residual sugar in sweet wines (Aleixandre-Tudo and du Toit, 2018). Also, aerobic yeasts such as *Pichiamembanae facie* grow with sufficient oxygen, particularly in young wines.

9.4.7 The Preservation of Wines

Wine is preserved by utilizing chemicals involving bisulfides, diethylpyrocarbonate, and sorbic acid, or by several physical processes such as pasteurization and sterile filtration. However, pasteurization is prevented due to its negative effect on the taste of wine.

9.4.8 Wine Categories

In various ways, grape wines can be graded involving several parameters such as country of production, colour, amount of alcohol and sugars. The method introduced here is predominantly used in the U.S. This system divides wine into two categories: natural wines and fortified wines.

Natural wines are the outcome of fully natural fermentation. Additional fermentation is avoided as the sugar is largely utilized. Spoilage microbes, e.g. acetic acid bacteria, are not allowed to grow when the air is removed. Due to the sugar composition of the grapes, the alcohol level may not increase to more than 12%.

Natural wines are classified into still wine (without CO_2) and sparkling wine (with CO_2). Pink or red colour wines are developed from the skin colour of the grapes; with white wine the grape skin is light green, but the juice is clear (transparent) because grape skins can be excised instantly after the pressing method and before fermentation. Additionally, black grapes are permitted to ferment for red wine production. The skins are left for a prolonged time to isolate the colour.

Natural wines are normally used at the time when they are opened. They are therefore known as 'table wines' and are planned to be included in a meal. Table wine is normally offered in good quantity, as they carry less alcohol content than dessert wines and appetizers; also table wine has a much shorter shelf life, once opened, in comparison to appetizers and dessert wines.

Dessert and appetizer wines are the second largest category of wines. These are offered at the start of meals (appetizing) or the finish (dessert), as indicated by their descriptions. These are produced in the presence of alcohol from distilled wines, partially to enhance the production, as well as prevention from yeast spoiling. They are classified in three ways:

(A) Sweet, e.g. port
(B) Sherries are sweet or dry, originating from Portugal, with flavours introduced by different oxidation levels.
(C) Flavoured wine, e.g. vermouth, with herbs and special ingredients which are a secret of this wine, sparkling wines (especially champagne), sherry, and flavoured wines.

Sparkling wines are produced by compressing CO_2 under pressure. These are known as sparkling because after the bottle has been opened a slight release of carbon dioxide from the wine shows sparkle (Jones et al., 2014).

Champagne is a transparent, sparkling wine produced from white wine. Other pink or red champagnes are produced from wines of a similar colour. Champagne is made by second fermentation of what is already fine wine in the bottle. To manufacture champagne, it takes a prolonged period of time to complete, but it also involves a complex process; therefore it is easier to handle the process in a manual way rather than using an automated system. Because of this, the drink is costly (Alexandre and Guilloux-Benatier, 2006).

Champagne production parlance is predictably French. The technique to be explained is Champenoise which is utilized to make better sparkling wines. The wines to be considered for champagne – after the 'must' is completely fermented – are racked, clarified, and fined. A mixture of several different wines is used to provide the ideal flavour. The mixture is called cuvée and will have a 9.5 to 11.5% alcohol concentration, be adequately acidic (0.7 to 0.9% titratable as tartaric acid), and will be a light straw or light-yellow coloured wine. The concentration of SO_2 is small because of the scent of SO_2 being detectable as the wine is poured; the yeasts may turn SO_2 into types of hydrogen sulfide which will give the wine a rotten scent of egg (Alexandre and Guilloux-Benatier, 2006). The cuvée is put in thick-walled bottles which can tolerate the high CO_2 pressure that will eventually develop in the bottle. Some yeast, additional fermentable sugar – typically sucrose – and small amounts (0.05 to 0.1%) of ammonium phosphate are introduced into the secondary fermentation where *Saccharomyces bayanus* is commonly used as yeast. This must develop at a relatively high alcohol concentration (10–12%), at high pressures and low temperatures. The concentration of sugar varies depending on the CO_2 pressure required for the sparkling wines. As a rough guide, 4 g of sugar per litre will generate one CO_2 pressure atmosphere. Hence the addition of sugar is around 24 g/L. All the sugar found inside the wine is taken into consideration. Although the bottle is thick, when the fermentation is too fast it can break at high temperatures or high sugar content (Torresi et al., 2011).

The bottle is positioned horizontally and with a combination of wine, sugar and yeast the fermentation process is carried out at about 15–16°C. Two to three months are usually required for the secondary fermentation, a product of secondary fermentation is defined as tirage. It is preserved at 10 °C in a horizontal position for about 1–5 years. Most of the very well-aged champagnes release aroma which appears to come from the substances produced by the yeasts. The next step is to bring the sediment from the side to the bottleneck. This is done by putting the bottleneck downward at an angle in a rack using various jolting rates. The bottles are turned further in a clockwise and anticlockwise direction on alternating days throughout which the bottle is slowly straightened to the perpendicular. The duration of the cycle varies from two to six weeks after which the sediment makes its way to the bottle neck (Torresi et al., 2011). The neck of the container must be frozen at between 1°C to 15°C to eliminate the sediment in the yeast cells. The bottle is rotated and opened at an angle of around 45 °. The bottle pressure pushes the ice plug out. The lost wine is substituted from another bottle and the dosage is introduced. The dose consists of approximately 60 g of sugar per 100 mL of wine. All champagnes involve dosage; instead of that they would turn sour. Up to 10% sugar is found in sweet champagne.

Sparkling wines may be manufactured in bulk in a big container instead of individual bottles. The output of bulk is identified as the Charmat process in which the tank is lined with an inert substance such as glass, holding between 500 and 25,000 gallons and is designed to handle 10–20 atmospheres of pressure as a security precaution. Valves regulate the pressure and cooling coils control the temperature. As the containers are aerated, a quick profit margin is feasible, and 6–12 fermentations are produced per year. A further reason for such high activity is that there is a massive yeast growth which could contribute to the development of flavours, particularly of H_2S.

Tank-fermented sparkling wine is generally produced in a cold-stabilizing process to eliminate extra tartrates. It is filtered in cold temperatures and under pressure to eliminate yeast cells. Then the wine is loaded with the correct dose in the bottles. It is normal to add sulfur dioxide to the sparkling wine system if all the yeast has not been eliminated by filtration. Sulfur dioxide even allows avoidance of dark oxidation, as the wine consumes oxygen as it is transferred. When Charmat-prepared sparkling wine bottles are opened, a sulfur dioxide odour is typically notable. They avoid the fragrance provided by autolysis of yeast on well-aged bottle labels. Bulk sparkling wines are ideal for manufacturing as they are comparatively less complex to ferment; however, due to their stronger tannin material; higher fermentation temperatures may also be needed (Torresi et al., 2011).

The warmer southern part of the Iberian (Spain–Portugal) peninsula in Europe is the origin of the names of various fortified wines. They are classified into three types: sherry (Jerez de la Frontera region in Spain); port (Douro Valley in Portugal); madeira (Island of Madeira). Fortified wine regulates the oxidation of wines, accomplished by continuous ageing under air pressure by increasing the population of aerobic yeast or by increasing temperature (Sablayrolles, 2009). The outcome of this oxidation is a liquid with a deep, reddish-brown colour with a distinctive flavour; the original wine is white or red. With sherry, white wine is chosen. However, for port or madeira, red or white wine maybe used. These three fortified wines contain high alcohol concentration of about 15–20%(v/v). They are typically categorized into two parts: (a) vermouth (b) other flavour wines (special natural wines). Vermouth maybe of Italian (sweet) or French (dry) kinds. Wormwood is an ingredient of vermouth, from the German 'wermut' (Artemesia absinthium, a common herb). Other flavoured fortified wines, such as Campari, Dubonnet, Byrrh, as with vermouth, possess 15–21% alcohol.

Fruit wine: Cider and Perry

Frequently, fruits containing sufficient sugar concentrations can be used to produce a strong drink. In these situations, additional sugar is included to facilitate fermentation by utilizing sucrose. Fruit wines have become famous in many places where grapes cannot survive. Apple, malus pumila, or pear or a combination of pears and apples are extracted from cider. Fruit wine contains low alcohol concentration (4–5%v/v) with up to 7–8%v/v as sugar inappropriate concentration. They are also different from other fruit wines. The basic steps of pressing the juice, fermentation, maturation, and bottling are identical to those of grape wines. Fruit wines can also be produced using cashews and pineapples (Aleixandre-Tudo and du Toit, 2018).

Palm wine is a common term in alcoholic drinks which typically originate from the sap of palm trees. It is different from grape wines because it is hazy in appearance. It is consumed in Africa, Asia, and South America. Palm wine is traditionally a whitish and energetic drink. The palm sap is harvested from several locations: the stem of the growing tree, the tip or trunk of the dead tree and the bottom of the immature male inflorescence. The preferred approach is based on the country, but most studies have focused on inflorescence wine. The sap generated by this process in *Elaeisguiniensis* comprises about 12% sucrose, about 1% each of fructose, glucose, and raffinose, and small amounts of protein and some vitamins. Itis a clear, sweet, syrupy liquid. To generate palm wine a sequence of microorganisms is required; mainly Gram-negative bacteria, lactic acid bacteria and yeasts and finally acetic acid bacteria. Yeasts in palm wine are described as originating from specific genera. Microbes are not added but they make their way into the wine from a variety of sources, such as air and tapping utensils, including the last few brews. The wine contains around 3%v/v alcohol and allows easier growth of bacteria and yeasts because it is a source of protein (single cell) and various vitamins (Lea and Piggott, 2012). The main drawback with palm wine is that it has a relatively limited shelf life. It is better to use within 48 h, but not more than five days after tapping. For this purpose, different strategies have been invented to preserve palm wines. Pasteurization has been effective, but methods that minimize the bacterial growth of wine by centrifugation or filtration method can also be used. However, they require the application of milder pasteurization temperatures and lower amounts of chemicals.

Distilled alcoholic (or spirit) beverages are certain drinks for which the alcohol concentration is raised using the distillation process. With the method of distillation, volatile materials emerge from the fermented substratum directly or by a yeast metabolism which is used to incorporate materials that have a huge impact on the quality of the drink. Congeners are believed to be the ingredients of spirit drinks that impart certain flavours (Lea and Piggott, 2012).

Evaluating the Alcoholic Content of Distilled Drinks

The alcoholic concentration of spirit drinks (and non-potable alcohol) in the United States, Canada, the United Kingdom and Australia is known as 'proof'. The purpose of the word is historic; for alcohol concentration measurement, spirit drinks were mixed with gunpowder before the invention of the hydrometer occurred. If the gunpowder caught fire, then it is acceptable as it has less than 50%v/v alcohol. It was 'underproof' if it didn't ignite because it contained less than 50% water. In the United States, proof spirit present one half its volume of alcohol of specific gravity of 0.7939 at 15.6 °C. Alternatively, the proof is twice the alcoholic concentration by volume (Plutowska and Wardencki, 2008). As a result, a 100 proof spirit includes 50% alcohol. In the British system, proof spirit includes 57.1% by density and 49.28% by weight of alcohol and a conversion factor of 1.142 is implemented to transform the United States proof to British proof.

In Germany, weight percentage is applied for calculation of ethanol concentration of a drink. The hydrometers graduate at 15 °C, and the comparison is with the density table (Mendeleev's).

Various countries use this volume method, particularly in Europe. The hydrometer is mostly 15 °C calibrated. France, Belgium, Norway, Finland, Switzerland (which also uses percentage of weight), Brazil and Egypt, and Russia use 20 °C, Denmark, and Italy 15.6 °C, while other countries in South America use 12.5 °C.

9.5 GENERAL PRINCIPLES IN SPIRIT BEVERAGE MANUFACTURING

The basic methods are usually applied in preparing the drinks listed above. The information depends on the drink.

Medium preparation: Grain starch is converted by using microbial enzymes or barley malts enzymes (hydrolysed) into sugars. Hydrolysis is not needed due to the grape sugars which are used in brandy while cane sugar is used in rum (Lea and Piggott, 2012).

Propagation of yeast inoculum: There are large distilleries yielding hundreds of litres of spirits per day. Hence fermentation broths are required in much greater quantities. Such broths are inoculated with a dense yeast broth up to 5%v/v. Even if yeast is re-used, regular inoculant is still needed. Inoculum should be overall alcohol tolerant. Normally *S. cerevisiae* is grown aerobically with agitation on molasses.

Fermentation: In case of inadequate nitrogen content in the medium, an ammonium salt can generally be added. The heat produced during the process must be decreased by cooling, and normally temperatures should not exceed 35–37 °C with an optimum pH ranging from 4.5 to 4.7, if the medium buffering capacity is increased.

Larger values of pH tend to increase the production of glycerol. When the buffering power is smaller, the current pH is 5.5, but it generally starts falling to about 3.5. Contaminations can have drastic effects on the system. Sugar is used and produces less yield, yet the taste of the result can be altered by metabolic products. Lactic acid impacts the taste of the resultant product; it is the most significant contaminant in distilling factories (Sablayrolles, 2009).

Distillation: Distillation is a system utilizing a vaporization and condensation process which separates more volatile substances from less volatile substances. There will be a discussion of three methods for spirit distillation as follows.

Rectifying still:

When the condensate is repetitively distilled, the succeeding distillates will involve more volatile elements. Repeated distillation processes are referred to as rectification. The combination of alcohol and water flows down, and alcohol is discharged from the downside by steam as it flows upside. The top of the column is an alcohol-rich distillate. Fuel oils contain higher alcohol concentration isolate at the upper level to the entry point of the mixture and are driven to some other line (Lei et al., 2003). Volatile fractions are made up of esters and aldehydes. Whisky and brandy may still be effectively distilled in two columns, but for high potency distillates, at least three to five columns are needed.

Pot still: These are typically copper stills. At the lower part of the cooling coil, they are spherical in shape. This is a batch process. The initial part or 'heads' and the final part, 'tails' are usually removed from 'low wines' only the centre portion is collected (Lei et al., 2003). Malt whisky, rum, and brandy are manufactured in pot stills (Huang et al., 2008). The advantage is that a lot of the lower fragrance substances are produced in the beverage, thus giving it a unique aroma.

Coffey still: In 1830 the Coffey was patented, and since then various changes have been introduced to the initial model (Lei et al., 2003). Its key feature is that next to the wash column it has a rectifying column where the beer is first distilled.

Maturation: A few distilled alcoholic drinks are matured for a certain number of years, frequently determined by legislation.

Mixing: Samples from several cartons of various types of drinks are mixed to produce a specific fragrance then packaging is done.

Spirit Beverages

Whisky, brandy, rum, vodka, kai-kai (or akpeteshi), schnapps, and cordials are the beverages in question.

Whisky/whiskey: This is an alcoholic drink obtained from the distillation of fermented grain. Different forms of whisky are manufactured. They vary mainly with which type of grain is utilized. The nations connected with whisky are, Scotland, Ireland, the United States and Canada, and now even Japan and Australia. In all whisky-manufacturing regions the alcoholic concentration, components, and formulation technique is regulated by governmental rules. Rye and bourbon whiskey are the major types in the United States (Fleet, 2008).

Rye whiskey is made from rye and rye malt, barley malt and barley. A typical mash containing 51% of corn; the composition is in the form of 70% maize, 15% rye, and 15% barley malt. Unmalted rye or maize is cooked to gelatinize it, thus facilitating the saccharification (or transformation to sugar) of malt enzymes. The solids are not separated from the mash, and often *Lactobacillus* is used as the inoculating yeast; these lactic acid bacteria enhance the taste of the whiskey. Fermentation generally takes place in a two-column Coffey-type still. All of the whiskeys are aged in wooden barrels, for three or more years. Before packaging, they may be blended with different types (Lea and Piggott, 2012).

Brandy: Brandy is a distillate of fermented fruit juice. Therefore, it is possible to make brandy from certain fruit; strawberries, pawpaw, or cashew. However, technically, the term brandy relates to the distillate of fermented grape juice. At least two years of maturation in oak barrels is needed (Lea and Piggott, 2012).

Rum: Rum is generated from the by-products of cane or sugar, particularly molasses. In the Caribbean, Jamaica, Cuba and Puerto Rico are known for the development of rum. It is also made in the east of the United States. Rum is made from molasses with a heavy body, whereas light rum is made from cane syrup by constant distillation. The molasses are filtered during fermentation. Furthermore, addition of sulfuric acid eliminates colloidal material which may obstruct further processes. The pH level is maintained at approximately 5.5 and ammonium sulfate or urea, a nitrogen source, can be introduced. For heavy rum, *Schizosaccharomyces pombe* is chosen and for lighter types *Saccharomyces cerevisiae*. Rum is preserved in oak barrels for ageing for 2–25 years (Lea and Piggott, 2012).

Gin, Vodka, and Schnapps: A) Wooden casks (in which beverages have been stored) extractives give a pale-yellow to deep brown colour to brandy, rum, and whisky. Caramel is also used to achieve a uniform colour. Gin, vodka, and schnapps are transparent like water.

B) Cereal is generally the raw material for their manufacture, but sometimes potatoes or molasses may be utilized. Maize is utilized for gin, while rye is used for vodka. The gelatinized cereals are cooked and mashed with malted barley. Recently amylases have been manufactured by fungi or bacilli because the malt taste is not needed in the drinks. Russian vodka is manufactured from rye spirit. It is carried on particularly activated wood carbon. *Schnapps* is a gin flavoured with blended herbs (Lea and Piggott, 2012).

9.6 BEER PRODUCTION

The method of production of beer is termed brewing. The word beer originates from the Latin word 'bibere'. 4,000 years ago, ancient Egyptians produced beer from barley. However, some studies suggest that Egyptians learned the method from the Tigris and Euphrates communities where it is said that human civilization started. Nevertheless, the addition of hops in beer is a new concept. Barley beers are categorized into two groups: 1) Top-fermented beers; 2) Bottom-fermented beers (Parker, 2012).

9.6.1 Types of Beer

Bottom-fermented beers: In this method, after fermentation the beer is stored in cold chambers for clarification and maturation; such beers are named lager beers. *Saccharomyces uvarum* is mostly used for bottom-fermented beers (previously *Saccharomyces carlsbergensis*). Many forms of lager beer are recognized; for example Pilsener, and those originating from Dortmund and Munich. The Pilsener category of lager beer is the most consumed in the world.

Bottom-fermentation's history began in Germany's Bavarian region, whose capital is Munich. It is ironic that in 1842 monks transferred the methods and the yeasts (inoculum) to Pilsen. After three years the knowledge reached Copenhagen, Denmark. Relatively soon thereafter, German immigrants took the bottom-fermentation brewing technique to the US (Unger, 2011).

Pilsen Beer contains an alcohol concentration of about 3.0–3.8 by weight. It is a pale beer with medium hop flavour. In modern breweries, lagering time is less as compared to the traditional method. The beer is lagered for two to three months. The *Pilsen* brew requires soft water with relatively little concentration of calcium and magnesium ions (Unger, 2011).

Dortmund beer is also a type of pale beer although it has less hop concentration compared with Pilsen. It is thicker and has flavour. It is traditionally lagered for a little longer than usual; 3–4 months. The brewing utilizes hard water made up of a higher concentration of carbonates, sulfates, and chlorides (Parker, 2012).

In Germany, *Weiss beer* is made from wheat. Steam beers (California and USA) are bottom-fermented beers. They are noticeable because they are highly effervescent (Parker, 2012).

Top-fermented beers: Saccharomyces cerevisiae strains are used for brewing top-fermented beers. Ale originates from England. After the UK joined the European Economic Community lager gained popularity in the UK. English ale is a pale, high concentration of hopped beer also with 4.0 to 5.0%w/v alcohol concentration. Occasionally the concentration of alcohol may increase to 8%. Hops are added during or after fermentation to give a bitter taste; it also has an acidic taste, with a presence of fragrance of wine due to the high concentration of esters (Briggs et al., 2004). Mild ale is sweeter, as it has a lower content of hops compared with standard pale ales which began in Burton-on-Trent, where the water contains a high concentration of gypsum (calcium sulfate).

Porter is a heavy-bodied, dark brown, strongly foaming beer obtained from dark malts. It contains less hop concentration than ale and therefore it is sweeter (Briggs et al., 2004). The alcohol level is 5.0%.

Stout is a heavy-bodied, high concentration of hops; a very dark beer with a strong malt odour. Stout is made from dark or caramelized malt; sometimes caramel is included. It has a considerably high alcohol concentration at 5.0–6.5%w/v. Traditionally it is preserved for approximately six months, and often is fermented in the bottle. A few stouts are sweet, and hopped much less than usual (Wilson et al., 1998).

9.6.2 Raw Materials Used for Brewing

Raw materials such as barley, malt, adjuncts, hops, yeasts, and water are used for the brewing process (Unger, 2011).

Barley when used for brewing has significant benefits as follows:

Its husks are dense, hard to crush, and stick to the kernel. After mashing, barley makes this process easier for the malting and filtration process. A further benefit is that the thick texture of the husks during storage provides an antifungal environment. Generally, the gelatinization temperature is 52–59 °C; at this temperature, starch is transformed into a water-soluble gel but this is less than the ideal (optimum) temperature of alpha-amylase (70 °C) and also of beta-amylase (65°C). Finally, barley grain includes large concentrations of beta-amylase (alpha-amylase is only produced in germinated seeds). There are two different forms of barley known: six fertile kernel rows (*Hordeum vulgare*) and two fertile kernel rows (*Hordeum distichon*). They vary in characteristics and exhibit thousands of variants. The six-row variety is utilized largely in the United States, while the two-row variety is utilized in Europe also in some parts of the U.S. The six-row varieties contain a greater amount of protein and enzyme concentration than the two-row varieties. This contains high enzyme concentrations therefore adjuncts are widely used in the U.S. Adjuncts dilute the proteins (raise the proportion of carbohydrates/proteins). When an all-malt beer is brewed by using malt it has properties such as the higher protein content belonging to six-row varieties. This protein will make its way into the beer and cause hazes. This malting method helps to germinate seedlings that produce enzymes (amylases and proteases) (Unger, 2011).

Six-row barley varieties cultivated in the USA generate malt that contains a large amount of amylase. Therefore, it necessary to hydrolyse the starch present in malt. After that, other components such as malt adjuncts are hydrolysed by amylase. Starchy adjuncts generally containing a small quantity of protein relate to fermentable sugars after hydrolysis, which in turn improves the alcohol content. Thus, adjuncts help reduce brewing costs because they are significantly cheaper than malt. The aroma, colour, or taste are not contributed to by adjuncts. Based on the cost, starch components such as sorghum, maize, rice, unmalted barley, cassava, and potatoes can be used. In the United States, the most frequent use is of corn grit (defatted and ground), maize syrup, and rice, when maize is used as an adjunct. Maize is heavily milled then the germ and husk containing the most maize oil, which can form 7% of maize grain, are removed as much as possible. The oil in beer can become rancid and thus negatively impact on taste unless it is excluded. In the brewing process corn syrup is produced from enzymatic action or acid hydrolysis. This defatted maize is also known as maize granules. As there is little nitrogen in the additives, the malt must meet all the requirements to grow the yeast. The ratio of malt and adjuncts does not exceed 60/40. Sometimes soybean powder added to the brews helps to feed the yeast (Briggs et al., 2004).

Hops are cultivated in the northern regions of Europe, Asia, and North America. These originate from *Humulus lupulus* (synonyms: *H. americanus, H. heomexicams, H. cordifolius*). *Humulus lupulus* is a dried cone-shaped female flower. It is connected botanically to the genus *Cannabis*; another member is *Cannabis sativa* (Indian hemp, marijuana, or hashish). At present, hop extract is used instead of dried hops. Hop resins produce bitterness in beer and the essential (volatile) oils give fragrance to hops. Resins also contain essential oils embedded in the flower in the lupulin glands (Unger, 2011).

The quantity of hops added to the original Pilsen beer is about 4 g/L. The many impacts of hops on addition to beers are as follows:

1. Hops replace the flat flavour of beer with a distinct bitter taste and also give the sweet fragrance to beer.
2. They possess antibacterial properties, particularly towards beer sarcina (*Pediococus damnosus)* and also towards contaminating bacteria.
3. The tannins present in hops assist in precipitation of proteins during the process of boiling of the wort. If proteins are included in beer this causes haze at low temperature.

Water quality: The mineral quality of the water and its pH have a significant impact on beer manufacture. Some ions are unacceptable in brewing water. a) Nitrates slow down the rate of fermentation; b) iron destabilizes the beer colloidal stability; c) calcium ions generally give a better taste as compared to magnesium and sodium ions. The pH of the water, and malt extraction, help to regulate

A) different enzymes
B) the degree of malt extract of soluble components
C) the solution of tannins
D) other colouring materials

Calcium and bicarbonate ions show an impact on pH. The natural water present in the world's major brewing centres has given special tastes to beers (Briggs et al., 2004). Water with a high content of calcium and bicarbonate ions is used in Munich, Copenhagen, Dublin, and Burton-on-Trent to produce darker, sweeter beers (Table 9.1). The purpose of this is unclear but carbonates in particular increase the pH, and also help in significantly raising the extraction of dark-coloured malt. Pilsen beer is made by using water which contains a low level of minerals. It is popular because of its pale, light-coloured appearance (Lea and Piggott, 2012). Proper composition of water is not available naturally, and therefore water treatment is important. If the concentration of carbonates is high in the water and the aim is the production of Pilsen beer with less concentration of hops then treatment of the water is required as follows:

TABLE 9.1
Mineral Concentration (ppm) of Water in Some Cities and Breweries

Location	Total solids	Ca^{2+}	Mg^{2+}	SO_4^{2-}	NO_3^-	Cl^-	HCO_3
Milwaukee	148	34	11	20	0.8	6.6	
New York	28	6	1	8	0.5	0.5	11
St. Louis	201	22	12	77	4	10	65
Pilsen	63	9	3	3		5	37
Munich	270	71	19	18		2	283
Dublin	3	100	4	45		16	266
Copenhagen	480	114	16	62		60	347
Burton-on-Trent	1206	268	62	638	31	36	287

1. Adding calcium sulfate (gypsum) can 'burtonize' the water. The introduction of gypsum in an equation neutralizes the alkalinity of the carbonates which works as follows:

$$2Ca\left(HCO_3\right)_2 + 2CaSO_4 \rightarrow 2Ca^{++} + 2H_2SO_4 + 4CO_2$$

2. Lactic acid, phosphoric, sulfuric, or hydrochloric acid may be introduced. CO_2 is released, but there is also a possibility that salt could be left in the system. Gas stripping excises the CO_2.
3. Sometimes the water is decarbonated by heating and also by the addition of lime calcium hydroxide.
4. The ion exchange process also helps to increase the quality of water.

Brewer's yeasts: Yeasts typically manufacture alcohol from sugars although not all yeasts are ideal for anaerobic brewing. Besides producing beer they also produces from wort sugars and also via proteins a proper concentration of esters, acids, higher alcohols, and ketones which provide flavour to the beer (Briggs et al., 2004). There are different ways to differentiate the two types of yeast.

A. Under the microscope *S. uvarum* (*S. carlsbergensis*) is observed singly or in pair form. *S. cerevisiae* normally form chains and often a crossed chains shape. These features, as well as other diagnostic tests used, are as follows (especially biochemical tests).
B. *S. cerevisiae* forms spore quickly compared to *S. uvarum.*
C. The critical separation of *S. uvarum* and *S. cerevisiae* is as follows:

   S. uvarum can ferment the trisaccharide raffinose, which is composed of galactose, glucose, and fructose. *S. cerevisiae* can ferment just fructose because *S. cerevisiae* does not contain the enzyme system necessary for fermenting melibiose formed from galactose and glucose.

D. *S. cerevisiae* strains contain a stronger respiratory system than *S. uvarum* and this is displayed in the various cytochrome spectra of the two categories.
E. Bottom-fermenters will flocculate and sink within the brew. Bottom fermenters are categorized as quick-settling or slowly settling (powdery); the settling qualities impact on the amount of yield. Yeasts are recycled several times after fermentation, depending on the brewery's procedure. During this activity, mutation and contamination are risks but unavoidable in every inoculum.

9.6.3 Brewery Processes

The different steps included for the conversion of barley malt into beer are as follows:

A. Malting
B. Cleaning and milling of the malt
C. Mashing
D. Mash treatment
E. Wort boiling treatment
F. Fermentation process
G. Storage or lagering
H. Packaging

Breweries buy malt from highly specialized malt manufacturers (Unger, 2011). The target of malting is to produce amylases and proteases in the barley. The germinated barley produces these enzymes to degrade the carbohydrates and proteins, and immediately nourish the germinated seedling before its photosynthetic process turns on. Nonetheless, as soon as the enzymes are produced and before the young seedlings have established, any noticeable change in the seedling growth is stopped by drying at low temperature, but it does not fully inactivate the enzymes. While mashing, such enzymes are activated again and hydrolyse starch and proteins, which also provide the nutrition to yeast. Barley grains are washed during malting; broken barley grains, foreign seeds, sand etc. are eliminated. Instead, the grains are soaked at 10–15 °C. The grain consumes water and eventually increases by about 4% in length. Embryo respiration starts after the water is consumed. Microbes are produced in the steep, and the steep water is replaced approximately every 12 h to prevent grain degradation until the grain is around 45% moisture. It takes two or three days for the steep process.

The grains are then dried (remove moisture) in order to germinate and may be transported to a malting floor or a rotating drum. The sprouts produce heat thereby accelerating germination. Frequently moist warm air is passed up about 30 cm deep through the beds of germinating seedlings. Water is sprayed onto them. To shorten the germination time, the plant hormone gibberellic acid is occasionally included in the grains. The grain itself also utilizes gibberellic acid. It inhibits the production of different hydrolytic enzymes via the aleurone layer located on the grain's peripheral

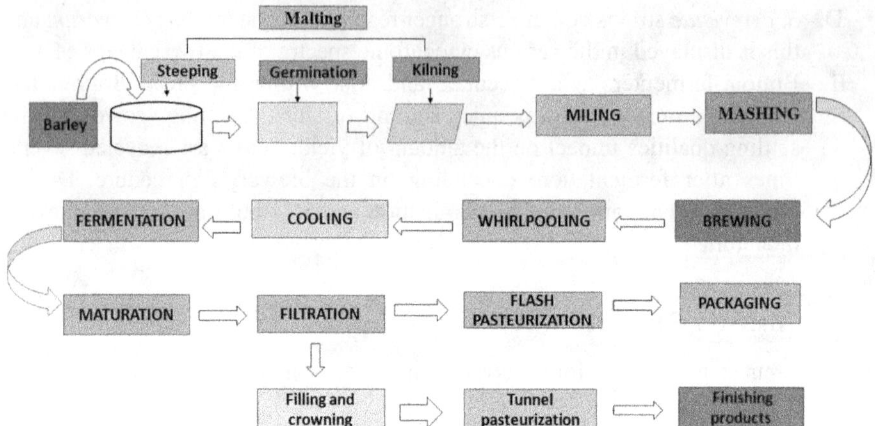

FIGURE 9.4 Steps involved in beer production.

side. Amylase and protease enzymes are permeable at the centre of the grain, where the endosperm is situated.

Starch granules are located inside the cell of the endosperm. Grain cell walls composed of hemicellulose are degraded by hemicellulase enzymes before the starch can be attacked by amylases. Alpha-amylase is also produced by the grain (see discussion on mashing). Beta-amylase already exists, also attached to protein, and is removed by proteolytic enzymes.

Malting: Malting is the first step in beer production (Figure 9.4). 'Modification' or enzymatic activities are completed within four to five days of seedling growth. The effect is tested approximately via the sweetness produced in the grain as well as by measurement of the young plumule. The different enzymes help in disruption of a particular substrate, but the important disruption process occurs at the time of the mashing process. Kilning stops further processes. Kilning is done by placing the 'green' malt in an oven, set initially at mild temperature to reduce moisture concentration ranges from 40% to about 6%. Then the heating temperature increases; this relies on the nature of the beer being generated (Unger, 2011).

The malt is pale for Pilsen types of beer and it contains no fragrance; it requires 20–24 h at 80–90 ° C for the kilning process. The darker Munich beers take 48 h at 100–110 °C for the kilning process. Sometimes the kilning temperature can be as high as 120 °C. Dark beers also caramelize the malts which are used for stout. There is a small amount of enzyme activity for such malts. A few alterations in the gross composition of the barley grain appear at the level of malting as seen in Table 9.2. The small roots are excised from malt and used for cattle feed. The weight of malt decreases because at each malt step a small quantity of malt is lost, and the cumulative loss can be as high as 15%. Barley malt is like swollen grains (filled with enzymes) of unthreshed rice (Figure 9.5); it can also be preserved for some time (Briggs et al., 2004).

The barley grain processing is carried out at the upper level of the barley plant. The consequent process occurs on the lower floors. On the ground floor, lagering

TABLE 9.2
The Concentration of the Grain of Barley before and after the Malting Process

Fraction	Barley	Malt
Starch	63–65	58–60
Sucrose	1–2	3–5
Reducing sugars	0.1–0.2	3–4
Other sugars	1	2
Soluble sugars	1–1.5	2–4
Hemicellulose	8–10	6–8
Cellulose	4–5	5
Lipids	2–3	2–3
Crude proteins	8–11	8–11
Albumin	0.5	5
Globulin	3	-
Hordein-protein	3–4	2
Glutelin-protein	3–4	3–4
Amino acids and peptides	0.5	1–2
Nucleic acids	0.2–0.3	0.2–0.3
Minerals	2	3
Others	5–6	6–7

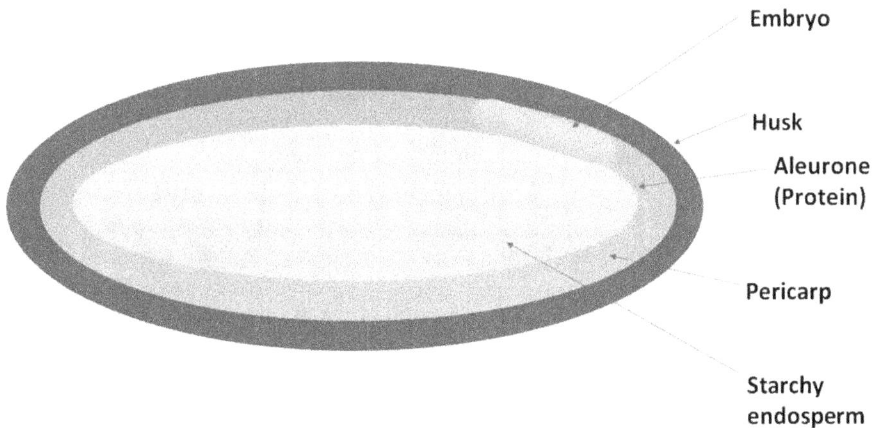

FIGURE 9.5 Structure of barley grain.

and bottling occur. In this manner, the materials are transported with gravity and the pumping costs are reduced. The barley malt is washed at the top of the brewing tower and passed over a magnet to eliminate metal pieces, especially iron. After that it is ready for milling (Figure 9.5). Malt particles are broken down by hydrolytic reaction by malt enzymes that occur in the mashing method (Lea and Piggott, 2012).

Therefore, the smaller the particle size the bigger the malt extract. However, very fine particles hinder filtration and excessively extend it. Therefore, the brewer must know the ideal particle size that will give optimum extraction results, and also allow a high filtration rate.

Mashing is the cornerstone of brewing. The quality of the wort depends on the mashing process. The objective of the mashing process is to absorb the soluble part of malt as much as possible and to breakdown the insoluble parts of malt and adjuncts enzymatically. Mashing is essentially a mixture of ground malt and adjuncts at optimum temperature for amylase and protease enzyme production. The resulting aqueous substances of mashing are recognized as wort (Briggs et al., 2004).

The main elements are starch (55%) and protein (10–12%) in dry grain. The regulated deterioration of these two factors has an enormous impact on the nature of the beer. Starch is about 55% of the dry weight of the malt. In overall starch specifically, amylose occupies 20–25% of the malt starch. The alpha and beta-amylases are important enzymes in breaking down malt starch (Wilson et al., 1998). Basically, mashing is an extension of the malting process.

During malting, the deterioration of malt proteins, albumins, globulins etc. begins while malting is a method and process before the mashing process. Proteases which convert proteins into peptones then into polypeptides and polypeptides convert to amino acids (Briggs et al., 2004).

There is no perfect temperature for the disintegration of proteins, yet they are placed at 60 °C unifor Copenhagen, Dublin, and Burton-on-Trent to produce mly. If the temperature rises to more than 60 °C it inhibits the activity of the protease and polypeptidase activity. However, the proteolytic activity in wort relies on the pH, and thus the pH of the wort is maintained at 5.2 to 5.5 with lactic acid, minerals, and calcium sulfate.

The higher pH for amylases and proteolytic enzymes has previously been studied. The optimum pH for beta-amylase activity is similar to proteolysis, a happy coincidence for optimum maltose manufacturing and protein breakdown. Mash concentration is significant. Thinner the mash, more will be the extract (i.e. the dissolved materials from the malt).

Mashing: There are three methods:

1. Method of decoction in which a portion of the mash is moved from the mash container into the mash kettle for the boiling process.
2. Methods of infusion; in this process the mash is not supposed to boil. The temperature is just kept increasing.
3. Method of double mash in which starchy adjuncts are boiled and introduced to malt.

Decoction: This method mostly occurs in continental Europe. With this technique, mash is heated initially at a temperature of 35–37 °C; after that the temperature will be increased to about 75 °C. Approximately one-third of the initial mash is removed, and transported to the mash kettle until it boils slowly and is then brought back to the mash containers, raising the mash temperature. During heating, enzymes are damaged but grains are cooked and gelatinized (Lea and Piggott, 2012).

The remaining portion can be excluded, cooked, and brought back. In the three-mash process, for proteolysis (Table 9.2), a temperature of 35–40 °C is favourable; for full proteolysis 50°C for half an hour is required; also saccharification and maltose degradation requires 60–65 °C. Furthermore, for dextrin formation 70–75°C and a duration of two or three hours is required (Briggs et al., 2004).

Infusion: In Britain, the technique of infusion is widely used; especially for making top-fermenting beers. This technique is performed in a mash tun similar to a lauter lager beer tub but is deeper in configuration. The technique includes crushing malt and a smaller quantity of unmalted cereal, which is often precooked. The ground material, or grist, is extensively combined with hot water (2:1 by weight) to generate a thick porridge-like mash and the heating temperature is accurately elevated to around 65 °C. At that temperature it then remains in the system for from 30 minutes to several hours. Additional water boiled at 75–78°C is sprinkled onto the mash to get extract and then to stop the activity of the enzymes. Double mash or decoction methods are more effective compared with the infusion method. Nevertheless, this is easier to automate, but malt with already degraded proteins is used because the high heat of mashing easily damages the proteolytic enzymes (Lea and Piggott, 2012).

Double-mashing is known as the cooker process. Due to the use of adjuncts this method has been developed in the US. Traditionally the US have used a double mash technique, mashing the ground malt and heating at 35 °C with water (Lea and Piggott, 2012).

Next, this process continues for one hour for proteolysis. Adjuncts are then heated for 60–90 min in an adjunct cooker. At the time of cooking, approximately 10% of malt is introduced. Heat-cooked adjunct is then added to ground malted mash to increase the heating temperature to 65–68 °C for the process of starch hydrolysis and maintained for about half an hour. An iodine test is used to check starch hydrolysis. The temperature is raised to 75–80 ° C after that process is finished. Husks and other insoluble components are excluded from the wort in two processes at the end of mashing. Firstly, this differentiates the wort from the solid materials. Secondly, the solid components themselves are released by washing or sparging with hot water. The typical method of partition of the husks and some solids from the mash is to spread the mash in a lauter (German for 'clarification') tub, which is a container with a perforated pseudo bottom about 10 mm situated above the real floor upon which the husks create a bed (Wilson et al., 1998). In modern years, the Nooter strain master has been used in large breweries, particularly in the United States.

It is similar to the Lauter tub; filtration is via a husk bed, but rather than a pseudo bed, straining is by a pattern of perforated triangular pipes situated at diverse bed distances. This tub (master) is rectangular in shape with a conical base while the Lauter tub is cylindrical in shape. An additional benefit is that it can manage greater amounts than the Lauter tub. Sometimes cloth filters situated in plate filters and screening centrifuges are applicable. The mash solids are sparged (or washed with hot water) at approximately 80 °C with water and are processed until the extraction is complete. The substance remaining after sparking is known as spent grain and is utilized as feed for animals. Occasionally through centrifugation liquid is absorbed from the spent grain, and the extract is cooked for adjuncts (Wilson et al., 1998).

The wort is boiled in a brew kettle (or copper) for 1.5 h which were previously made of copper but are now made of stainless steel in various advanced breweries (Parker, 2012). When corn syrup or sucrose is utilized as an adjunct, it is introduced at the start of the boiling process (Lea and Piggott, 2012). Hops are also included in boiling; they are added before the process and some at the end of process. The boiling objective is as listed below:

A) Concentrating the wort; boiling decreases it by 5–8% of its volume via the evaporation process.
(B) Sterilizing the wort to decrease its microbial population before it is put into the fermenter.
(C) The content of the wort does not change by inactivating any enzymes.
(D) Extraction of soluble materials from hops helps in the elimination of proteins as well as for the production of a bitter taste.
(E) Precipitation of proteins producing large flocs due to the degradation and tannins obtained from the hop and malt husk. Residual proteins make hazes in the beer; a lower quantity of protein results in lower generation of haze (Lea and Piggott, 2012).
(F) In some beers, colour arises from the malting process. However, at the time of wort boiling the bulk grows. Colour is made via various chemical processes such as caramelization of sugars, oxidation of phenolic components, and amino acid interactions and sugar reduction (Parker, 2012).
(G) Exclusion of volatile components: removal of volatile components such as fatty acids which may cause rancidity in the beer.

Agitation and circulation of the wort assist in raising the proportion of precipitation and floc generation during boiling. The hot wort does not instantly pass to the fermentation container. Proteins and tannins are precipitated during boiling. A few more precipitations occur when it is chilled to around 50 °C. The warm precipitate, known as 'trub', also comprises 50–60% protein, 16–20% hop resins, 20–30% polyphenols, and about 3% ash. Centrifuge or a whirlpool separator remove trub. Within this system, the wort that includes a flat centrifuge is removed from the wall of the system and drained from the outside via the outlet. The heavy material (the trub) is placed in the centre and removed through a central outlet (Briggs et al., 2004). The resultant wort is chilled in a heat exchanger device. If the temperature drops to around 50 °C, sludge called a 'cold break' starts to settle but cannot be segregated in a centrifuge because it is too small. The wort is purified at this step with kieselguhr, white diatomaceous earth. The refrigerated wort is prepared for fermentation. It does not incorporate enzymes but, however, is a rich source for the fermentation process. This should be protected from spoilage. The wort needed oxygen at approx. 8 mg/L during transmission to the fermenter. The oxygenated environment is important for the initial growth of yeast (Lea and Piggott, 2012).

Fermentation: The cooled wort is either pumped or permitted to pass through gravity to fermentation containers, while the yeast is introduced in $7–15 \times 10^6$ yeast cells/mL, typically obtained from the original brew system.

Top fermentation processes are applied in the United Kingdom for the manufacture of stout and ale by *Saccharomyces cerevisiae* strains. Conventionally, an open fermenter is used. A spray or fishtail is incorporated so that a medium of aerated 5–10 mL/L of oxygen is maintained for the production of the yeast. At a temperature of 15–16 °C, yeast is processed at a level of 0.15 to 0.30 kg/h L. For a duration of about three days, the temperature can gradually increase to 20 °C. At this point it is chilled to a steady temperature point. All primary fermentation processes requires six days. Yeasts float on the top of the surface then are taken out and used for the next inoculum (Wilson et al., 1998).

After the final three days, the yeasts have converted into a hard layer which is removed. In certain cases, after the first 24–36 h, the wort is shifted to another vessel called a drop system. The transmission allows aeration of the medium; aeration is also often obtained by the circulation of paddles and by pumps. In recent systems, cylindrical vertical closed containers are replacing typical open containers (Hoalst-Pullen et al., 2014).

In bottom fermentation, wort is added at about 7–15 × 10^6 yeast cells per mL of wort. In three to four days of operation, the yeasts then increase in growth four to five times (Briggs et al., 2004). Yeast is set at 6–10 °C and raised to 10–12 °C, then chilled at the final fermentation phase for about three to four days. CO_2 is emitted and generates krausen, and after four to five days the yeast begins to die. The average period of fermentation is 7–12 d.

An anaerobic environment predominates during wort fermentation for both upper and lower fermentation. Initially, oxygen is needed just for cell development. Fermentable sugars are transferred into alcohol, CO_2, and heat which is lost by cooling. Higher alcohols (sometimes referred to as fuel oils) such as propanol and isobutanol are produced from amino acids. Organic acids like acetic, lactic, pyruvic, citric, and malic are also obtained from carbohydrates through the tricarboxylic acid cycle (Hoalst-Pullen et al., 2014).

The advancement of fermentation is preceded by wort-specific gravity. Wort gravity starts to decline slowly during the fermentation process due to the extract being used by the yeast strains. Meanwhile, alcohol is produced in the system. Since alcohol has lower gravity than wort, the results of a saccharometer (special hydrometer) are therefore also lower. The reading of the saccharometer represents not the actual extract but apparent extract. Apparent extract is smaller compared to the actual extract due to the involvement of alcohol. In the United Kingdom and a few specific regions, the extract is evaluated as a direct specific gravity of 60 ° F (15.5 ° C) × 1000. These methods calculate the percentage of sucrose needed to achieve the same specific gravity (Briggs et al., 2004).

Von Balling constructed the original tables. Better and more precise changes were produced initially by Brix and then by Plato in Von Balling's tables, but the results were not significantly modified. Therefore, except for fifth and sixth decimal places Balling, Brix and Plate are similar. The sugar companies use Brix, while the brewing industry uses Balling (US) and Plato (continental Europe). The level of attenuation is the quantity of fermented extract, expressed as a percentage of the initial or complete extract, and therefore also an apparent and a real extent of attenuation occur. Lagering (bottom-fermented beers) and treatment (top-fermented beers):

Lagering: This beer, also called 'green' beer, is rough and bitter at the end of the primary fermentation. It has a yeasty flavour that presumably originates from higher alcohols and aldehydes. Green beer is kept in closed containers at low-temperature levels (around 0 °C) and stored for six months, to mature and prepare for consumption. The secondary fermentation saturates the beer with CO_2, and depends on a particular rate of CO_2 release from a safety valve. Often actively fermenting the wort or krausen is included. In many cases, CO_2 is artificially mixed into the lagering beer. During secondary fermentation, materials that can unpleasantly impact the taste and are present in green beer such as diacetyl, hydrogen sulfide, mercaptans, and acetaldehyde are reduced by evaporation. An increase is observed in the desirable elements of beer including esters. Throughout the lagering phase, any tannins, proteins, and hop resins remaining are precipitated out (Lea and Piggott, 2012).

The lagering process requires approximately nine months of duration. Nowadays the duration is less, requiring three weeks for the turnover from brewing, lagering, and tasting. This reduction has been accomplished because of increased knowledge of the lagering methods, and it is done by synthetic carbonation and manipulation of conditions. Thus beer is preserved at high temperatures (14 °C) to drive off volatile compounds (Parker, 2012). After that the beer is cooled at 2 °C to eliminate contaminants such as chill haze components, and then carbonated. In this way, it decreases the duration of the lagering period from 2 months to 10 d. Lagering provides beer with beneficial organoleptic properties, although the complexes of protein tannin and yeast cells make it foggy. To remove haze, the beer is filtered via kieselguhr or membrane filters.

Top-fermented beers do not perform advanced bottom-fermented beer lageration. They are handled in multiple ways such as via barrels or bottles. In other methods, the beer is moved to casks with a weight of 0.2–4.0 million yeast cells/mL at the end of fermentation. It is 'primed' by adding a small quantity of sugar combined with caramel, to enhance its flavour and presence. The yeasts multiply in the sugar and in the carbonated beer. At this point, hops are sometimes added. The beer is held at approximately 15 °C for seven days or less.

The beer is 'fined' by supplementation with isinglass. Isinglass, a gelatinous content from the fish swim bladder, precipitates complexes of yeast cells, tannins, and protein-tannins. After that, the beer is pasteurized and dispersed (Lea and Piggott, 2012). The bottles take approximately half an hour to hit pasteurized temperature, stay for half an hour in the pasteurizer, and cool for the next half an hour. This pasteurization process also creates risks, and several of the major breweries now perform bulk pasteurization, as well as aseptically filling tanks. Wild or unwanted yeasts found in beer spoilage are scattered across several genera namely *Kloeckera, Hansenula, and Brettanomyces,* but *Saccharomyces sp.* appears to be the most frequent, specifically in top-fermented beers (Briggs et al., 2004). These encompass *S. cerevisiae, V. turblidans, and S. diastaticus. Lactobacillus, Streptococcus, Acetobacter,* and lactic acid bacteria are among the most significant of all. This last is significant due to its general capacity to develop on dextrins in beer, causing haze and bad odour. The above is acidophilic, hop antiseptic, micro-aerophilic etc. Acetobacter generates acetic acid from alcohol thus leading to a sour taste in beer. In top-fermented beers, *Lactobacillus pastorianus* is a contaminant that generates a sour taste and a silk thread form of turbidness. *Streptococcus damnosus* (*Pediococcus damnosus,*

Pediococcus cerevisiae) is defined as 'beer sarcina' causing 'sarcina sickness' or beer smelling like honey.

Packaging: The beer is transmitted to vessels under pressure, where the beer is packed in bottles, cans, or other containers. Oxygen cannot interact with beer throughout this operation; CO_2 or spoilage by microbes are also inhibited. To reach such goals, the beer is placed in containers with a CO_2-counter pressure, and the entire system is washed and disinfected periodically. The filled and crowned bottles are carried to pasteurization, for half-hour heating of the bottles at 60 °C. The bottles require half an hour to reach the ideal pasteurizing temperature, and continue to stay in that process for half an hour. They require another half an hour to achieve normal temperature (Briggs et al., 2004).

Defects in beer:

Physico-chemical turbidities

Non-biological hazes form in beer for the following reasons:

(I) Metal-generated hazes.
(ii) Hazes with protein-tannin.
(iii) Sedimentation of polysaccharides.
(iv) The hazes and sediments of oxalates.

Aluminium, iron, and copper are responsible for forming haze in beer. A small 0.1 ppm of tin is able to create haze in beer instantly. It does not operate as an oxidation catalyst unlike other metals but immediately precipitates haze cofactors. Copper and iron perform a significant part in the degradation of the polyphenolic moiety of beer protein-haze activators. These even seem to be produced by the brewery via malts and hops (copper insecticides). It has been suggested to use EDTA (ethylene diamine tetraacetic acid) to form chelates with copper and iron, in order to help avoid their dangerous behavior (Lea and Piggott, 2012).

Sometimes beer polyphenols have been related exclusively and inaccurately to tannins. Tannins are used to convert leather skins, but alcohol polyphenols cannot be used as well. Polyphenols are abundant in plants. Hops and barley husks are the results of beer tannings or polyphenols. They respond with proteins to build complex, insoluble, haze-shaped molecules (Hoalst-Pullen et al., 2014). Hazes comprise polypeptides, polyphenols, carbohydrates, and a small number of minerals. Beer hazes are split into two: chill hazes (particles with a diameter of 0.1–2 nm) form at 0 °C and dissolve again at 20 °C. Stable hazes (1.0–10 nm) persist above 20 °C.

Protein-tannin hazes may be excluded by:

(a) the addition of papain that hydrolyses polypeptides to low molecular weight; elements that cannot make hazes.
(b) the absorption of polypeptides by silica gel and bentonite.
(c) precipitation of polypeptides by tannic acid.
(d) polyphenol adsorption by polyamide resins e.g. Nylon 66.

Freezing and thawing of beer may trigger an unexpected haze that may occur in the form of flakes. This haze is different from cold haze, which is uniquely made up of carbohydrates. It has been reported in lager cooled to –10 °C and is composed primarily of beta-glucans derived from malt (Wilson et al., 1998).

An off-flavoured beer is formed when it is exposed to sunlight; this is because of the development of mercaptans by a photocatalytic process in the blue-green range of visible light, i.e. 420–520 nm (Briggs et al., 2004).

9.6.4 Advancement in Brewing

The above explanation is a standard brewing of beer. There have been some advances in both the production of beer and beer categories: some of these will be discussed here briefly.

Even though continuous brewing is not yet commonly used, nowadays in several countries it is gradually gaining recognition. In conventional continuous brewing methods this is a continuous type of fermentation. Secondary fermentation and lagering are a batch type of fermentation. Two continuous fermentation processes are identified: the open and partially closed.

In the continuous fermentation through open systems, the wort is consistently added inside the fermenter, and also the beer passes from the outlet at the same speed. The yeast is permitted to reach its normal level. Furthermore, the wort is obtained prudently from the brewing house and can be preserved at 2°C for 14 days before use. In a heat exchanger the wort is sterilized, prior to oxygenation. The wort is then relocated through the bottom to the first tank, which is continuously stirred and also produces aerobic development. It is then moved to a second container offering an anaerobic environment, as well as alcohol and CO_2.

The beer streams into a third tank for sedimentation including its suspended yeasts. The final beer is eliminated from the upper level and yeast cells from the bottom level. For secondary fermentation, the quantity of yeast is important (Lea and Piggott, 2012). CO_2 is obtained from the upper level. The yeast used is a special type that gives proper taste, and active fermentation in an anaerobic condition. Also, it is ready to flocculate quickly once in a chilled sedimentation tank. One or two tanks may be used for the sedimentation process. Three tanks are used in some systems, but two provide versatility in layout and applications (Figure 9.6).

In partially closed continuous fermentation processes, only a certain amount of yeast is stored. The open system on its own may revise the use of yeast to obtain higher yeast content from the sedimentation container into the first container (Lea and Piggott, 2012). The drawback of the alteration is the chance of spoilage. Lastly, the returned yeast is in a separate physical state of growth from the yeast effectively associated with fermentation, so the wort and the efficiency of the beer may be affected. In the closed method, characterized by the tower fermenter (Figure 9.7), sterilized wort is pumped with oxygenation into the bottom of the cylindrical device, and the beer is drained off at the top (Briggs et al., 2004). Yeasts reach a very high density, approaching 350 g/L, and the wort is nearly prepared beer. The upper areas contain less yeast concentration and partially work as a final fermentation method, to

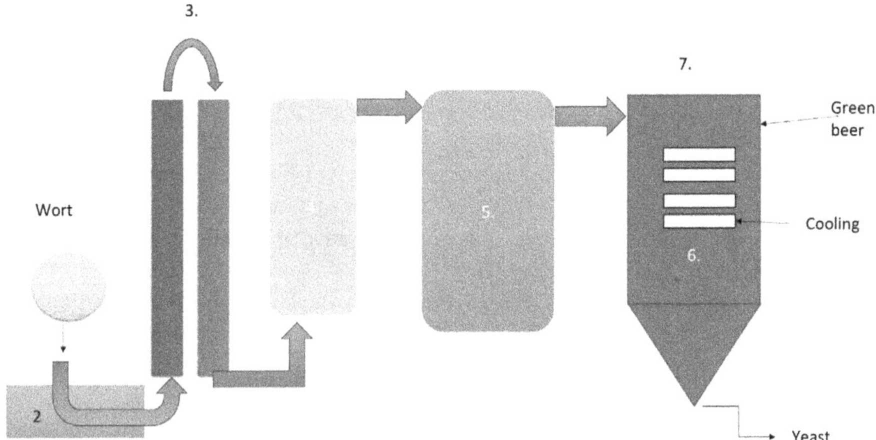

FIGURE 9.6 The open system of continuous brewing. (1 = Pump; 2 = Flow regulator; 3 = Sterilization; 4 = Perforated plates; 5 = Control of temperature; 6 = Yeast separator)

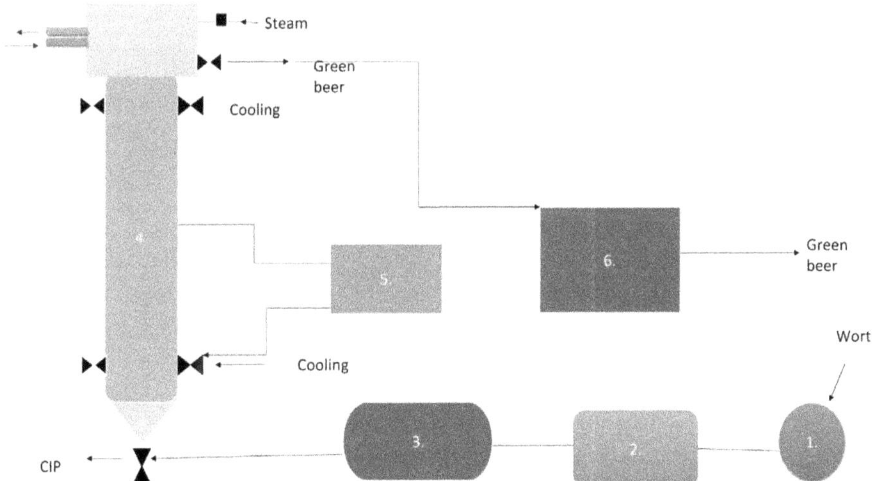

FIGURE 9.7 The tower fermenter (closed) of continuous brewing. (1 = Pump; 2 = Flow regulator; 3 = Sterilization; 4 = Perforated plates; 5 = Control of temperature; 6 = Yeast separator)

distinguish yeasts. The baffles cause the growing CO_2 and beer to be diverted from the beer outlet. The main advantage is that optimal level yeast concentration of 350–400 g/L can be formed within four hours. When enough yeast is utilized, special yeasts are capable of maintaining high mass at ground level and are still capable of passing outside of the fermenter (Lea and Piggott, 2012). While continuous brewing is not traditionally used, brewers are forced to make batch brewing more effective.

Several companies are involved in selling enzymes extracted from bacteria and fungi that could fulfil the action of malt. The benefit of using such enzymes is to decrease costs, as well as removing the malting step. Instead of higher potential, brewers are still not ready to approve this because it impacts on the need to reduce the barley farming and malting industry. Therefore it makes it difficult to accept the method. When enzymes are normally used, this must ensure that the enzyme mixture includes the main enzymes such as amylases and proteases and also other enzymes including beta-glucanases that hydrolyse barley gums. It should also remove toxic microbial products (Sablayrolles, 2009).

Kaffir beer and other traditional sorghum beer: Barley cultivated in a temperate region. For centuries in many areas of tropical Africa, beer has been brewed with local cereals. The most widely used cereal is Sorghum bicolor (= Sorghum vulgare) called milo in the United States, known as kaffir corn in South Africa, and as Guinea corn in a few regions of West Africa. The African cereal is particularly resistant to drought conditions. Sorghum is sometimes combined with maize (*Zea mays)* or millet *(Pennisetum sp.).* Across Africa sorghum is not commonly applicable for brewing. In the United States it is used as an adjunct on demand (Nanadoum and Pourquie, 2009).

The processes by which the African continent produces these sorghum beers are nearly identical. They are as follows:

(i) pinkish in colour, sour in taste; and of relatively heavy due to the presence of starch particles.
(ii) intake with the presence of microbes.
(iii) are not aged or explained.
(iv) contain lactic fermentation.

Tropical beers are identified by various labels in several places around the world: 'burukutu,' 'otika', and 'pito' in Nigeria, 'maujek' among the Nandi in Kenya, 'mowa' in Malawi, 'kaffir beer' in South Africa, 'merisa' in Sudan, 'bouza' in Ethiopia and 'pombe' in various regions of East Africa.

Manufacturing in large breweries has been conducted only in South Africa. Kaffir beer is manufactured and consumed more in South Africa compared to barley beers. Beer production methods include malting, mashing, and fermentation.

For malting, periods ranging from 16–46 h are required for sorghum grains to soak in water. After that grains are soaked and germinated for 5 to 7 d. Water is sprinkled on the spread grains. The grains are generally dried by sun or driers at 50–60 °C (South Africa). Kilning is not used. Sorghum malt is higher in alpha-amylases, while the non-germinated grains do not contain beta-amylase as in barley. In certain areas, dried malt can be utilized for several months. As a brewing ingredient, sorghum has not gained much consideration, even in the United States where it is occasionally used as adjunct. Saccharification of sorghum starch was done by fungi as well as by sprouting. These fungi include *Rhizopus oryzae, Aspergillus flavus, Penicillium funiculosum, and P. citrinum* (Nanadoum and Pourquie, 2009).

In mashing, the malt is ground roughly and mixed with water to maintain a proportion of 6:1 (v/v) and heated for a duration of 2 h. At the time of the boiling process, starchy adjuncts in the form of dried powder of cassava ('gari') or unmalted cereal are included in order to maintain a 1:2:6 ratio of malt with water. This is then filtered and prepared for the fermentation process.

There are two fermentations during the processing of sorghum: alcohol and lactic acid fermentation. In classical fermentation, the dregs of the previous fermentation are inoculated into the distilled, washed, and cooled wort. This inoculum consists of a combination of yeast, lactic acid bacteria and acetic acid bacteria. The first stage of fermentation includes lactic bacteria, primarily *Lactobacillus mesenteroides* and *Lactobacillus plantarum*. In sorghum beer breweries in South Africa, mash temperatures are initially held at 48–50 °C to promote the development of naturally occurring thermophilic lactic acid bacteria between 16–24 h (Nanadoum and Pourquie, 2009).

Then the pH falls to roughly 3–4. The sour malt is introduced in heated adjuncts of unmalted sorghum or maize, and additional malt is also sometimes used. The malt is then cooled to 38 ° C and high fermentation yeast pitched in. The conventional approach involves yeasts and lactic acid bacteria (dregs). The yeasts that have been recognized in the fermentation of Nigerian sorghum beer are *Candida sp., Saccharomyces cervisiae, and S. chevalieri.* Fermentation requires approximately 48 h where lactic acid bacteria start their growth. Thereafter they distribute and function. No secondary fermentation is performed in lager beer, lagering, or clarification. The yeasts and the lactic acid bacteria are grown in a similar way as in palm wine (Nanadoum and Pourquie, 2009). In certain regions, fermentation should last for some period and there is a slight taste of vinegar because of the discharge of acetic acid by acetic acid bacteria. Sorghum beers typically comprise a high amount of solids, primarily starch. Therefore, some researchers consider them to be food, as much as they are alcoholic beverages.

Market trends of bioethanol: The global market size for bioethanol is expected to rise from USD 33.7 billion in 2020 to USD 64.8 billion by 2025, at a compound annual growth rate of 14 % from 2020 to 2025. Bioethanol demand can be due to compulsory use of bioethanol fuel mixtures in many countries, reducing greenhouse gas (GHG) emissions and increasing vehicle fuel efficiency. The North American region is projected to account for the highest share in the global bioethanol market during the forecast period. Key players include Archer Daniels Midland (US), POET, LLC (US), Green Plains (US), Valero Energy Corporation (US), Tereos (France), and Raizen (Brazil).For diesel or petroleum powered vehicles, bioethanol is a cheaper and environmentally friendlier option. To prepare bioethanol blends, a small percentage of bioethanol is blended with pure gasoline, which burns more effectively and emits zero carbon emissions. As a consequence, in many countries around the world, the use of bioethanol fuel blends is compulsory.

The worldwide intake of beer in 2018 was 188.79 million kilolitres (about 298.2 billion 633 ml bottles equivalent), with a rise of about 1,540,000 kilolitres, equal to approximately 2.4 billion 633 ml bottles (Figure 9.8). There was an average growth of 0.8% compared with 2017 (Hoalst-Pullen et al., 2014).

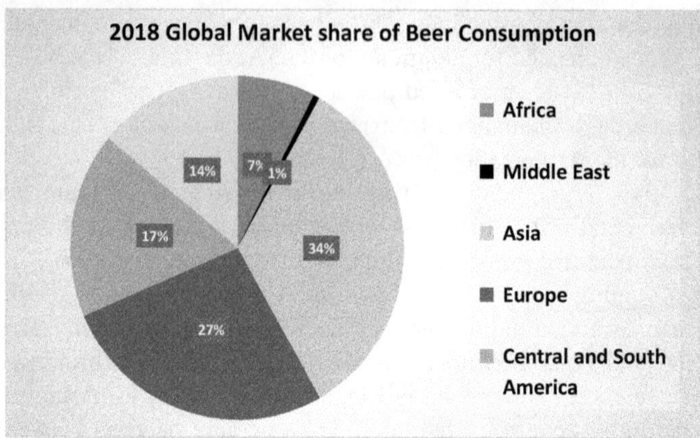

FIGURE 9.8 Graphical representation of 2018 global market share of beer consumption.

With a CAGR of 5.8%, the global wine market is expected to rise during the forecast time frame (2020–2025). Owing to its nutritional advantages and premiumization of wine products, along with creativity in quality and increasingly modern distribution networks, the wine sector is powered by rising demand for wine globally. The growth of the wine industry is expected to be fuelled by shifts in taste and emerging customer tastes, along with growing demand for new and exotic flavours, such as Riesling wine and other tropical fruit wines. During the projected years Asia-Pacific is expected to experience fast growth. This is attributed to a growth in the number of wine production firms in China, India, and Japan. Rising demand for luxury wines and increasing westernization both serve as factors driving the development of the region's wine industry.

9.7 CONCLUSIONS

In our society the use of alcohol has become entrenched and is a feature of daily life. The present chapter described the production of bioethanol, wine and beers. It discussed different types of alcohol beverages and their processing, and explained why the business of alcohol products has sought to increase quality and deliver a quality product. Such techniques have been adopted and developed by the alcohol consumption sector, centred in conventional or artisanal approaches. The method was not just the testing strategy but also the usage of microorganisms that can be re-used. The pattern in future years may be the use of nanoparticles to recover yeast. It is well established that the raw material is essential according to the type of drink. The selection of the right microorganisms is important for research that seeks to enhance this method. Nevertheless, techniques and products have to coincide if they are going to be marketed or manufactured with the correct legislation. One thing to bear in mind is the need to remind the customer of the nutrient details on labels. All these factors have driven us to understand the business and there are tremendous incentives for researchers to produce new developments.

REFERENCES

Aditiya HB, Mahlia TMI, Chong WT, Nur H, Sebayang AH, Second generation bioethanol production: A critical review. Renew. Sustain. Energy Rev. 66, 631–653, 2016.

Aleixandre-Tudo JL, du Toit W, Cold maceration application in red wine production and its effects on phenolic compounds: A review. LWT 95, 200–208, 2018.

AlexandreH, Guilloux-BenatierM, Yeast autolysis in sparkling wine – a review. Aust. J. Grape Wine Res. 12, 119–127, 2006.

Alvira P, Tomás-Pejó E, Ballesteros M, Negro MJ, Pretreatment technologies for an efficient bioethanol production process based on enzymatic hydrolysis: A review. Bioresour. Technol., Special Issue on Lignocellulosic Bioethanol: Current Status and Perspectives 101, 4851–4861, 2010.

Briggs DE, Brookes PA, Stevens R, Boulton CA, Brewing: Science and Practice. Elsevier, 2004.

Demirbaş A, Biomass resource facilities and biomass conversion processing for fuels and chemicals. Energy Convers. Manag. 42, 1357–1378, 2001.

Fleet GH, Wine yeasts for the future. FEMS Yeast Res. 8, 979–995, 2008.

Hoalst-Pullen N, Patterson MW, Mattord RA, Vest MD, Sustainability Trends in the Regional Craft Beer Industry, in:Patterson, M., Hoalst-Pullen, N. (Eds.), The Geography of Beer: Regions, Environment, and Societies. Springer Netherlands, Dordrecht, pp. 109–116, 2014.

HuangH-J, RamaswamyS, TschirnerUW, Ramarao BV, A review of separation technologies in current and future biorefineries. Sep. Purif. Technol. 62, 1–21, 2008.

Jones GV, Davis RE, Climate Influences on Grapevine Phenology, Grape Composition, and Wine Production and Quality for Bordeaux, France. Am. J. Enol. Vitic. 51, 249–261, 2000.

Jones JE, Kerslake FL, Close DC, Dambergs RG, Viticulture for Sparkling Wine Production: A Review. Am. J. Enol. Vitic. 65, 407–416, 2014.

Khoo HH, Koh CY, Shaik MS, Sharratt PN, Bioenergy co-products derived from microalgae biomass via thermochemical conversion–life cycle energy balances and CO2 emissions. Bioresour. Technol. 143, 298–307, 2013.

Lea AGH, Piggott JR, Fermented Beverage Production. Springer Science & Business Media, 2012.

Lei Z, Li C, Chen B, Extractive Distillation: A Review. Sep. Purif. Rev. 32, 121–213, 2003.

Nanadoum M, Pourquie J, Sorghum Beer: Production, Nutritional Value and Impact upon Human Health, in: Preedy VR (Ed.), Beer in Health and Disease Prevention. Academic Press, San Diego, pp. 53–60, 2009.

Parker D, Beer: production, sensory characteristics and sensory analysis, in: Piggott, J. (Ed.), Alcoholic Beverages, Woodhead Publishing Series in Food Science, 2012, Technology and Nutrition. Woodhead Publishing, pp. 133–158.

Plutowska B, Wardencki W, Application of gas chromatography–olfactometry (GC–O) in analysis and quality assessment of alcoholic beverages – A review. Food Chem. 107, 449–463, 2008.

Sablayrolles JM, Control of alcoholic fermentation in winemaking: Current situation and prospect. Food Res. Int. Bioprocess Food Industries 42, 418–424, 2009.

Tomás-Pejó E, Oliva JM, Ballesteros M, Realistic approach for full-scale bioethanol production from lignocellulose: a review. JSIR 6711, 2008.

Torresi S, Frangipane MT, Anelli G, Biotechnologies in sparkling wine production. Interesting approaches for quality improvement: A review. Food Chem. 129, 1232–1241, 2011.

Unger RW, Beer production, profits, and public authorities in the renaissance. Econ. Beer 29–50, 2011.

Wei P, Cheng L-H, Zhang L, Xu X-H, Chen H, Gao C, A review of membrane technology for bioethanol production. Renew. Sustain. Energy Rev. 30, 388–400, 2014.

Wilson RG, Gourvish T, Gourvish TR, The Dynamics of the International Brewing Industry Since 1800. Psychology Press, 1998.

Wissemann KW, Lee CY, Polyphenoloxidase Activity during Grape Maturation and Wine Production. Am. J. Enol. Vitic. 31, 206–211, 1980.

Yılmaz S, Selim H, A review on the methods for biomass to energy conversion systems design. Renew. Sustain. Energy Rev. 25, 420–430, 2013.

10 Citric Acid, Lactic Acid, and Acetic Acid Production

10.1 CITRIC ACID

Citurgia Biochemicals Ltd., Surat, India was established in collaboration with John & E Sturdia Ltd., UK for manufacturing citric acid on a commercial scale in India. Citric acid is the weak organic tricarboxylic acid having the empirical formula, $C_6H_8O_7$. It is a tricarboxylic acid: $CH_2COOH - C(OH)COOH - CH_2COOH$. 75% citric acid is used in food confectionery and beverages. It is used as a food acidulant in jam, jelly, orange squash, soft drinks etc. It has a pleasant taste and enhances flavour. 10% citric acid is used in pharmaceutical industries e.g. ferric ammonium citrate tablet for anaemia patients. The remaining 15% is used by other industries, e.g. for cleaning power station boilers to remove scale formation in the boiler tubes. It is available in two forms: citric acid anhydrous (CAA) and citric acid monohydrate (CAM). CAA is mostly used in the pharmaceutical industry and CAM is used for other purposes. The properties of citric acid are shown in Table 10.1 (Moo-Young, 2018, Prescott and Dunn, 1959).

Citric acid is used for cleaning boiler tubes in power stations to remove the deposition of solid material. Mineral acids such as HCl or H_2SO_4 affect the material of construction. Therefore, citric acid is recommended for cleaning scale formation in the boiler tubes.

Stoichiometry of conversion of sugar to citric acid in the fermentation process is shown below.

$$\begin{array}{llll} C_{12}H_{22}O_{11} + H_2O = & 2C_6H_{12}O_6 = & C_6H_8O_7.H_2O \\ 342 & 2\times180 & 2\times210 \end{array} \tag{10.1}$$

$$\text{CAM yield} = \frac{2\times210}{342} = 123\% \tag{10.2}$$

$$\text{CAA yield} = \frac{2\times192}{342} = 112\% \tag{10.3}$$

TABLE 10.1
Characteristics of Citric Acid

Properties	CAA	CAM
Formula	$C_6H_8O_7$	$C_6H_8O_7 . H_2O$
Molecular weight	192	210
Specific gravity (at 20 °C)	1.665	1.542
Melting point	150 °C	Loses water at 70–75 °C
Transition temperature	36.6 °C	

In 1784, Scheele first isolated citric acid as solid crystal from lemon juice. In 1893, Wehmer produced citric acid by mould fermentation. In 1917, Currie reported that *Aspergillus niger* can produce citric acid. Pfizer, a pharmaceutical company, produced citric acid at industrial level using this fermentation process. John & E. Sturge Ltd., UK took the technology from Pfizer and marketed it throughout the world. In 2007, it has been reported that approximately 1,600,000 tons of citric acid are produced annually worldwide (Prescott and Dunn, 1959).

10.1.1 Citric Acid Fermentation Pathways

Citric acid fermentation is carried out under aerobic conditions. Glucose is converted to pyruvic acid through glycolysis or Embden–Meyerhof pathway (EMP). Citrate is the intermediate product of the Krebs or tricarboxylic acid (TCA) cycle, a central metabolic pathway of animal, plants and bacteria – particularly aerobic organisms. The pyruvic acid produces acetyl CoA which, in combination with oxaloacetate, condenses to give citric acid. In the TCA cycle, citric acid produces isocitrate, α-ketoglutarate, succinyl CoA, succinate, fumarate, malate and oxaloacetate. The subsequent two steps involved after the citric acid formation are given below (Patel, 2012; Doran, 2012).

$$Glucose \xrightarrow{\text{EMP pathways}} 2\,Pyruvic\,acid \xrightarrow[\text{2NADH}]{\text{Pyruvate dehydrogenase}} Acetyl-CoA \xrightarrow{\text{Citrate synthase}}$$

$$> Citric\,acid \xrightarrow{\text{Aconitase}} Isocitric\,acid \xrightarrow{\text{Isocitrate dehydrogenase}} \alpha-keto\,glutaric\,acid$$

Citric acid inhibits enzymes such as aconitase and isocitrate dehydrogenase, which are responsible for the formation of isocitrate and α-ketoglutarate, respectively. In the first 24 h of fermentation, there is no accumulation of citric acid. However, after about 46 h to 48 h of fermentation there will be accumulation of citric acid which inhibits isocitrate dehydrogenase. After 96 h of fermentation the total activity of aconitase and isocitrate dehydrogenase will be inhibited, which is responsible for the accumulation of citric acid in the fermentation broth to a great extent. The normal time for

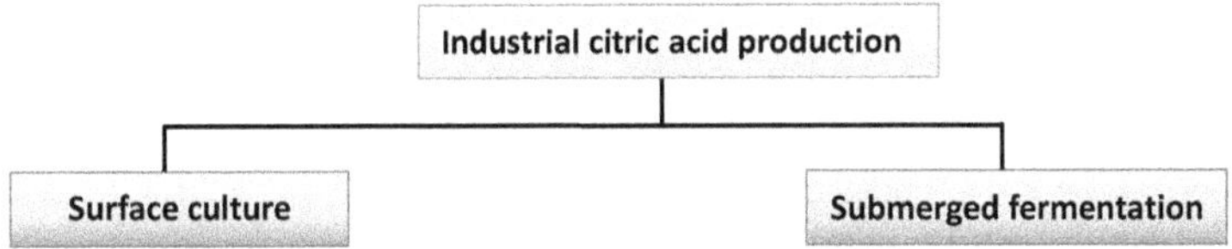

FIGURE 10.1 Modes of citric acid fermentation processes.

citric acid fermentation is 124 to 144 h, which is approximately 6 d (Prescott and Dunn, 1959).

10.1.2 Citric Acid Fermentation Processes

Citric acid fermentation process is carried out two ways: surface culture and submerged culture (Figure 10.1). In the surface culture, the diffusion of oxygen depends on the surface area of the culture exposed to the air because organisms are grown on the surface of the medium. This is done with either solid or liquid medium. Increase of surface area causes contamination problems to a great extent. This is a slow fermentation process as compared to the submerged process mainly due to lower dissolved oxygen (DO) concentration. The media is placed in simple thermostatic boxes on baking-tray like plates to which the microorganisms are inoculated. Therefore, citric acid production by surface culture is much less used in industry. On the other hand, submerged culture is mostly used by industries where the culture can grow throughout the medium mainly due to sufficient availability of water, nutrients, DO, and uniform suspension of the culture in the medium. DO concentration of the medium can be increased by sparging air. Batch-stirred tank reactor (BSTR) and air-lift reactor (AR) are used for submerged fermentation. In BSTR, a mechanical stirrer is used for maintaining homogeneity of the medium so that microorganisms can freely interact with nutrients for the and metabolism. Vegetative cells of *Aspergillus niger* are filamentous in nature. Therefore, it was observed that mechanical shear force is detrimental to the growth of the cell. Hence initially the mechanical stirrer is kept in switch-off mode for some time both for the Inoculum Vessel (IV) and Production Fermenter (PF). The major steps involved in citric acid production are shown in Figure 10.2.

10.1.3 Preparation of Seed Culture

Different organisms can be used to produce citric acid, such as *Aspergillus niger*, *Aspergillus clavatus*, *Penicillum luteum. Aspergillus niger* is mostly used for the industrial production of citric acid. This organism is available in rotten citrus fruit with black spores (Figure 10.3).

Industrial strains of *Aspergillus niger* are used for the commercial production of citric acid. The productivity of the strain is very high, and it produces a high concentration of citric acid. It can make use of higher concentration of sugar for the production of the citric acid. Uniform biochemical characteristics, easy cultivation and

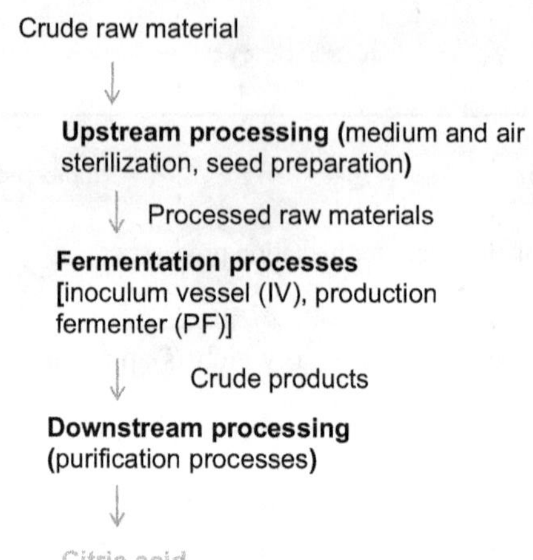

FIGURE 10.2 Major steps involved in the citric acid fermentation process.

FIGURE 10.3 Spores of *Aspergillus niger.*

production of negligible quantity of undesirable by products are the other advantages. Seed culture preparation for the inoculum preparation (volume is usually 5–10% of PF) for the production fermenter is done in the R & D laboratory of the factory. The quantification of organisms is essential for seed preparation, which is determined by intensive R & D activities. *A. niger* has a filamentous structure. Therefore, vegetative cells are not used for the seed culture preparation and thus the sporulation of *A. niger* is carried out under moisture-stressed conditions. The procedure for the preparation of spores of the *A. niger* is different. Spores can be produced on the surface of sliced bread in a wide-mouth test tube. One small piece of bread is taken in a sterile test tube under the aseptic condition. This is followed by the inoculation of the cell suspension on the surface of the bread. After 1–2 d of incubation, the bread will be totally covered with spores. These spores are transferred into the 0.85%w/v sterile saline water in a seed tank of 10 L capacity. Counting of the spores is done with a haemocytometer. The spore suspension present in the seeds is transferred to the IV located in the fermentation plant under aseptic conditions in liquid medium. After 30–33 h fermentation, the spores will be converted to vegetative cells. These cells are used as inoculum for the PF. This is transferred under aseptic conditions. Vegetative cells of *Aspergillus niger* are filamentous in nature. It has been observed that mechanical shear force is detrimental to the growth of cells. Therefore initially the mechanical stirrer is kept in switch-off mode for some time initially, both for the Inoculum Vessel (IV) and Production Fermenter (PF) to allow the cells to grow. The idle time for the IV is around 10 h and that of PF is 30 h. Temperature of fermentation varies from 30–33 °C. Since the time for the production fermenter is 124–144 h, morphological observation of the cells can give information about the fate of the fermentation as listed below.

- The success and yield of citric acid production mainly depend on the structure of mycelium.
- Mycelium with forked and bulbous hyphae which aggregate into pellets is ideal for citric acid formation.
- If the mycelium is loose and filamentous with limited branches, no citric acid production occurs.
- The hyphae of organisms with lot of vacuoles indicate that the fermentation process is expected to give a good amount of citric acid production.

10.1.4 Nutrient Requirement

Citric acid yield in the fermentation broth depends on the concentration of soluble carbohydrates such as sucrose, glucose, maltose etc. Cane molasses and beet molasses containing about 50%w/w sugar are commonly used as major raw materials for citric acid production. Minerals present in the medium play an important role in the citric acid fermentation process. Certain minerals (Fe, Cu, Zn, Mn, Mg, Co) are essential for the growth of *A. niger*. Manganese ions promote glycolysis, and iron is a cofactor for aconitase (TCA cycle). Fe concentration of 0.05–0.5 ppm is the optimum for citric acid production. Therefore, at low concentration of Fe, the activity of aconitase will be reduced drastically. Citric acid is responsible for the inhibition of phosphofructokinase (PFK) of EMP pathways which will reduce citric acid production to a great extent. If manganese concentration is increased by 1 ppm, citric acid yield will reduce by 10%. At low concentration of manganese ions, the inhibition of the PFK is reduced drastically due to the formation of ammonium ions which antagonizes inhibition due to citric acid. Mn is a co-factor for isocitrate dehydrogenase. Therefore the Fe and Mn present in cane or beet molasses are to be removed in order to increase citric acid formation in the fermentation medium. These metal ions are removed by boiling with a chelating agent, such as potassium ferrocyanide, for 30 mins. Iron and manganese present in the cane molasses will be precipitated out. These are separated by sedimentation methods. Supernatant is the clear liquid which is used for the preparation of the production medium. An adequate aeration is required (DO concentration of the fermentation broth 20–25% of saturation value) for good production of citric acid. A special type of stainless steel, SS217, is used as the material of construction of the fermenter.

10.1.5 Contamination Problems

Doubling time of the microbial cell is expressed as in Eq. 10.4. *Saccharomyces cerevisiae*

$$t_d = \frac{\ln 2}{\mu} \tag{10.4}$$

has a very low doubling time as compared to *Aspergillus niger*. Cane molasses and beet molasses are also suitable raw materials for the growth of *S. cerevisiae*. Therefore, *S. cerevisiae* is the major contaminant in the citric acid fermentation process.

10.1.6 Flow Diagram of the Citric Acid Fermentation Process

The flow diagram of the upstream processing and citric acid fermentation processes is shown in the Figure 10.5. Cane molasses is the by-product of the sugar industry. Cane molasses is usually transported with the help of tankers and unloaded in the molasses storage tank (MST) located close to the main gate of the factory. As per the requirement, the molasses are transferred to the molasses measuring tank (MMT) located in the fermentation plant. The molasses are mixed with nutrients as per the requirement. The composition of the medium used for the IV and PF is different because the purpose of the IV is to produce cell mass and that of PF is to produce citric acid. This mixing takes place as shown in Figure 10.5, followed by passing through a medium sterilizer. Foam formation takes place during the fermentation process both in IV and PF due to the formation of micelles with soluble proteins present in the medium and air. Foam formation is controlled with the help of anti-foam oil. Separate anti-foam oil tankers are connected to the IV and PF (Figure 10.4). The citric acid concentration is determined by the titrimetric method on the basis of $NaOH$ consumption to increase the pH to 8.6. A table has been developed which indicates corresponding total acid (TA) concentration. Actual citric acid estimation process takes about 1 d time.

10.1.7 Citric Acid Purification or Downstream Processing

The downstream processing deals with the processes involved for the purification of citric acid (Figure 10.5). After the fermentation process, the fermented broth is harvested in the harvesting tank. The first step of downstream processing is cell mass separation from the fermentation broth. Cell mass is separated by a rotary vacuum filter

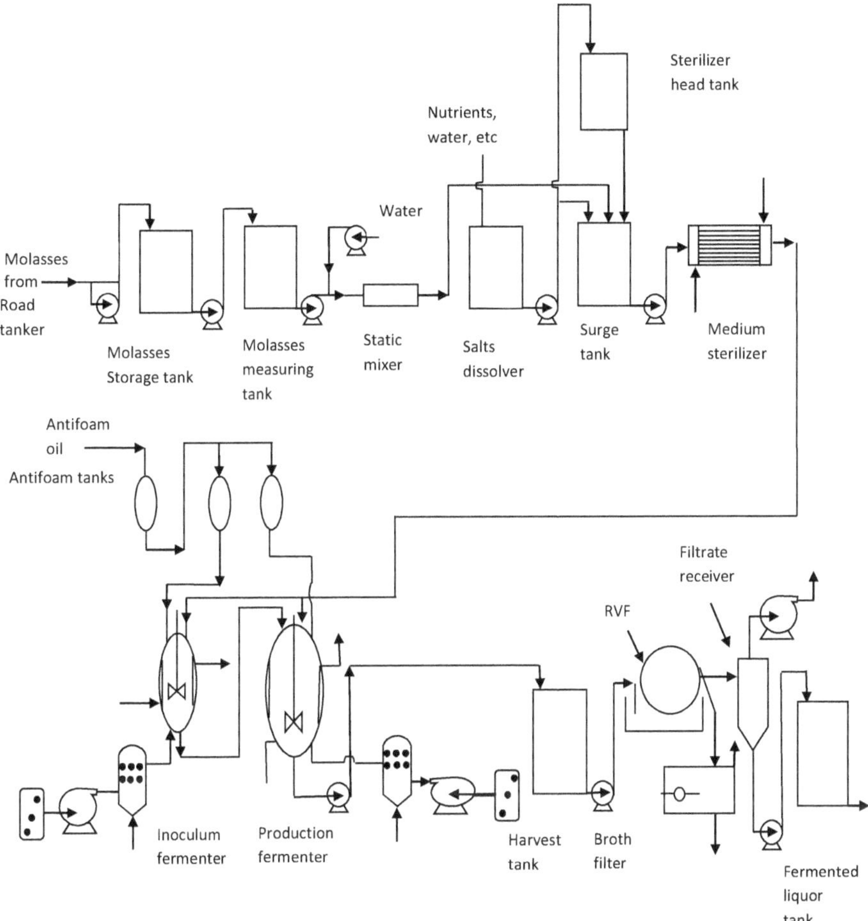

FIGURE 10.4 Flow diagram of the citric acid fermentation process.

(RVF). Cell mass contains 42%w/w chitin which may be used for the production of rough quality paper. This filtered material is first treated with slaked lime, $Ca(OH)_2$ to precipitate out the citric acid in the form of calcium citrate [$Ca_3(C_6H_5O_7)_2 . 4H_2O$]. $Ca(OH)_2$ is produced from quick lime, CaO. Quicklime is produced by heating crushed limestone, $CaCO_3$ to around 1,100 °C in a shaft furnace or rotary kiln. Calcium citrate is separated out by using a pannevis filter. This is followed by hydrolysis of calcium citrate by concentrated sulfuric acid (H_2SO_4) to citric acid and calcium sulphate ($CaSO_4 . 2H_2O$). Calcium sulphate is the raw material for the production of cement. This is separated out by using a gypsum filter. Citric acid concentration after the hydrolysis is around 22%w/v which is concentrated to 60%w/v by passing through the evaporator. This liquid has a dark colour which is partially removed by treating with activated charcoal. After the separation of the charcoal by filtration, the liquid is taken in a crystallizer at around 10 °C to get citric acid in the

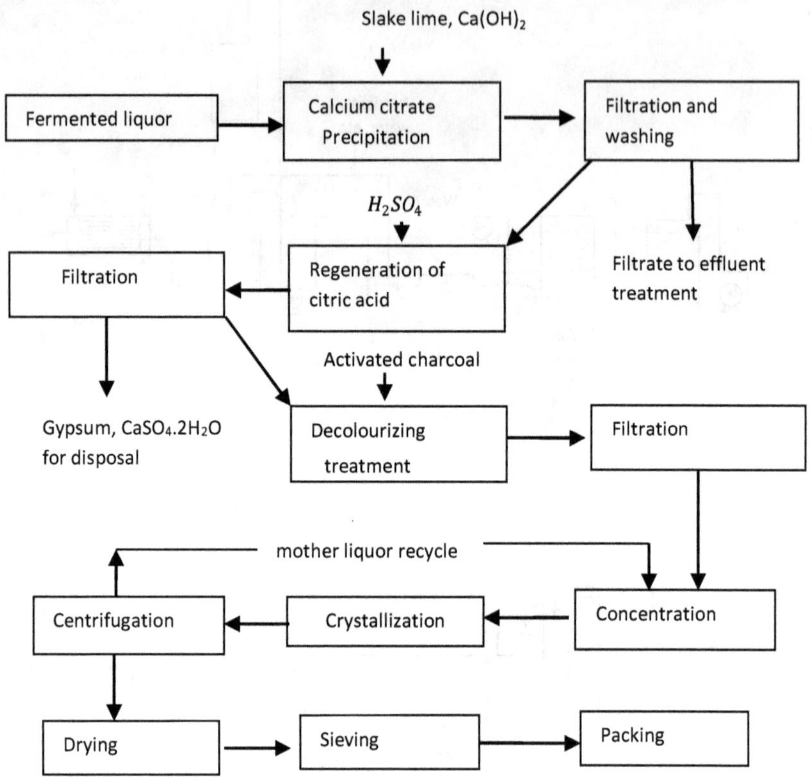

FIGURE 10.5 Block flow diagram for the recovery of citric acid.

form of citric acid monohydrate (CAM, $C_6H_8O_7 . H_2O$). The transition temperature of citric acid is 36.6 °C. CAM is separated by passing through a continuous centrifugation process. The liquid is known as mother liquor which contains about 30%w/v citric acid. This is recycled back to the evaporator. The stoichiometry of the downstream processing processes is shown in Eqs. 10.5, 10.6 and 10.7.

$$\underset{88}{CaCO_3} \xrightarrow{Heat} \underset{56}{CaO} \xrightarrow{H_2O} \underset{74}{Ca(OH)_2} \tag{10.5}$$

$$\underset{420}{C_6H_8O_7.H_2O} + \underset{222}{Ca(OH)_2} \rightarrow \underset{570}{Ca_3(C_6H_5O_7)_2.4H_2O} + \underset{72}{4H_2O} \tag{10.6}$$

$$\underset{570}{Ca_3(C_6H_5O_7)_2.4H_2O} + 3H_2SO_4 + 4H_2O \rightarrow 2C_6H_8O_7.H_2O + \underset{516}{3CaSO_4.2H_2O}$$

(10.7)

The crystals of CAM are passed through a drum dryer followed by a sieving machine to get different sizes of crystals. The selling price of the bigger size crystals is higher compared to the small sizes. They are packed in polythene pouches for marketing.

10.1.8 Production of Citric Acid Anhydrous (CAA)

Citric acid anhydrous (CAA) is produced from CAM using mother liquor containing 30%w/v citric acid. The temperature of the mother liquor is maintained at about 40 ^{0}C (because the transition temperature of citric acid is 36.6 ^{0}C) for the production of citric acid anhydrous. The water molecules present in the CAM will be solubilized in the mother liquor and citric acid anhydrous is produced in the form of solid powder. CAA is separated out through the centrifugation process. This is used in the pharmaceuticals industries.

10.1.9 Calculation of Citric Acid Yield and Productivity

Assuming the volume of unaerated fermenter in batch fermentation process = $200\,m^3$

Initial sugar concentration in the production medium = $200\frac{g}{L} = 200\frac{kg}{m^3}$

Total amount of sugar taken in the medium = $200\frac{kg}{m^3}\times 200\,m^3 = 40000\,kg = 40\,MT$

The following assumptions are made

Batch fermentation time = 140 h and down time =12 h

Total acid (TA) concentration in the fermentation broth =

$$15\,\%\frac{w}{V} = 150\frac{kg}{m^3} = 0.15\frac{MT}{m^3}$$

Citric acid (CA) concentration in the fermentation broth =

$$14.5\%\frac{w}{V} = 145\frac{kg}{m^3} = 0.145\frac{MT}{m^3}$$

$$\text{Total acid yield (TAY)} = \frac{\text{MT of TA}}{\text{MT of sugar input}} = \frac{200\,m^3\times 0.15\frac{MT}{m^3}}{40\,MT}\times 100 = 75\,\%$$

$$\text{Yield on the basis of CAM (CAY)} = \frac{\text{MT of CAM}}{\text{MT of sugar input}} =$$

$$= \frac{200\,m^3\times 0.145\frac{MT}{m^3}}{40\,MT}\times 100 = 72.5\,\%$$

$$\frac{CA}{TA} = \frac{0.145\frac{MT}{m^3}}{0.15\frac{MT}{m^3}} = 0.967$$

$$\text{Productivity} = \frac{\textit{Total amount of CAM produced}}{\textit{Total batch time}} = \frac{0.145\frac{MT}{m^3}\times 200\,m^3}{140\text{ h}+12\text{ h}} = 0.19\frac{MT}{h}$$

In the above calculations, the following matters are to be taken into account:

- The actual volume obtained after emptying the fermentation vessel is less than the initial filled-up volume due to the evaporation of water due to aeration. Since aeration is required in the fermentation process, there is every possibility that some of the water present in the medium gets evaporated out.
- The volume of the cell mass is assumed to be 2%v/v of the fermented filled volume. Therefore, this is to be taken into account.

Problem 10.1: Citric acid is produced from cane molasses (containing 50% w/w sucrose) by *Aspergillus niger* using a batch submerge fermentation process. The capacity of the fermenter is 200 m^3. Average time of fermentation and citric acid yield are 140 *h* and 90% *w/w*, respectively. Compute the following:

(a) Amount of cane molasses required for the fermentation process, assuming initial sugar concentration of $220\frac{g}{L}$
(b) Productivity of the fermenter assuming the idle time of 10 *h*
(c) Percentage utilization of sugar.

SOLUTION

The following data is available:

Capacity of fermenter $= 200\,m^3$

Concentration of sugar in cane molasses $= 50\ \%\frac{w}{V}$

Average time of fermentation $= 140\,h$
Down time or idle time $= 10\,h$
Citric acid yield $= 90\ \%$

(a) Initial sugar concentration in the production medium $= 220\frac{g}{L} = 220\frac{kg}{m^3}$

Total amount of sugar required in the fermentation process $= 220\frac{kg}{m^3} \times 200\,m^3$

$= 44000\,kg = 44\,MT$

$$\therefore \textit{Cane molasses required} = \frac{44\,MT}{0.5} = \mathbf{88\,MT}$$

(b) Citric acid yield = $90\ \%\frac{w}{w} = 0.90$

Total amount of citric acid produced = $220\frac{kg}{m^3}\times 200\ m^3\times 0.9 = 39,600\ kg$

$\therefore$ Productivity = $\frac{39,600\ kg}{140\ h+10\ h} = \mathbf{264\frac{kg}{h}}$

(c) The stoichiometry of the process is shown in Eq. 10.8.

$$\underset{342}{C_{12}H_{22}O_{11}} \rightarrow 2C_6H_{12}O_6 \rightarrow \underset{2\times 210}{2C_6H_8O_7.H_2O} \quad (10.8)$$

From Eq. 10.8, we can write

$420\,kg$ CAM is produced from $342\,kg$ sugar

39,600 kg CAM is produced from $\frac{342\,kg\ \times\ 39,600\,kg}{420\,kg} = 32,246\,kg$ sugar

Sugar utilization = $\frac{32,246\ kg}{44000\ kg}\times 100 = 73.3\%$

Problem 10.2: $200\,m^3$ of citric acid fermentation broth has been transferred from the fermenter to the harvesting tank. The broth contains $10\%w/V$ of cell mass and $12\%w/V$ of citric acid. Find the following:

a) Amount of mycelium produced
b) Quicklime consumption for calcium citrate precipitation process
c) Maximum of amount of gypsum produced ($CaSO_4\ 2H_2O$)
d) Amount of water is to be removed to increase the citric acid concentration from 22% to $60\%w/V$.

SOLUTION

Following data are available

Volume of the fermentation broth = $200\ m^3$

Cell mass concentration = $10\ \%\frac{w}{V} = 100\frac{g}{L} = 100\frac{kg}{m^3}$

Citric acid concentration = $12\ \%\frac{w}{V} = 120\frac{g}{L} = 120\frac{kg}{m^3}$

(a) Amount of cell mass or mycelium produced =

$$100\frac{kg}{m^3}\times 200\ m^3 = 20{,}000\ kg = \mathbf{20\ MT}$$

(b) Amount of citric acid produced $=120\frac{kg}{m^3}\times 200\,m^3 = 24{,}000\,kg$

From Eq. 10.6, we get

$420\,kg$ citric acid requires $222\,kg$ slake lime

$24{,}000\,kg$ citric acid requires $\frac{222\ kg\times 24{,}000\ kg}{420\ kg}$ 12,686 kg slake lime

Again, from Eq. 10.5, we get

$74\,kg$ slake lime is produced from $56\,kg$ quicklime

12,686 kg slake lime is produced from $\frac{56\,kg\times 12{,}686\,kg}{74\,kg}$ = **9,600 kg quicklime**

(c) From Eq. 10.6, we get

$420\,kg$ citric acid produced $570\,kg$ calcium citrate

24,000 kg citric acid produced $\frac{570\,kg\times 24{,}000\,kg}{420\,kg} = 32{,}571\,kg$ calcium citrate

Again, $570\,kg$ calcium citrate produces $516\,kg$ gypsum

$32{,}571\,kg$ calcium citrate produces $\frac{516\,kg\times 32{,}571\,kg}{570\,kg} = \mathbf{29{,}486\,kg}$ **gypsum**

(d) Basis: $100\,m^3$ of citric acid (containing 22 $\%w/V$ citric acid) obtained in the filtrate from the gypsum filter

Assuming $V\,m^3$ of 60 $\%w/V$ citric acid is obtained after passing through the evaporator

We know that $V_1 S_1 = V_2 S_2$

$$\therefore\ 100\,m^3\times 0.22 = V\times 0.6$$

$$V = \frac{100\,m^3\times 0.22}{0.6} = 36.7\,m^3$$

$\therefore$ Amount of water to be evaporated $= 100\,m^3 - 36.7\,m^3 = \mathbf{63.3\,m^3}$

10.2 LACTIC ACID

Lactic acid is an organic compound with the formula, $CH_3CH(OH)COOH$. It is used as a food preservative, curing agent, and flavouring agent. Poly-lactic acid has tremendous potentiality in the medical applications. It is an ingredient in processed foods like cheese, yoghurt, kefir etc. and is also used as a decontaminant during meat processing. Lactic acid is produced commercially by fermentation of carbohydrates such as glucose, sucrose, or lactose, or by chemical synthesis.

10.2.1 Characteristics of Lactic Acid

The empirical formula of lactic acid is $C_3H_6O_3$. There are two forms of lactic acid: L(-) lactic acid and D(-) lactic acid as shown below.

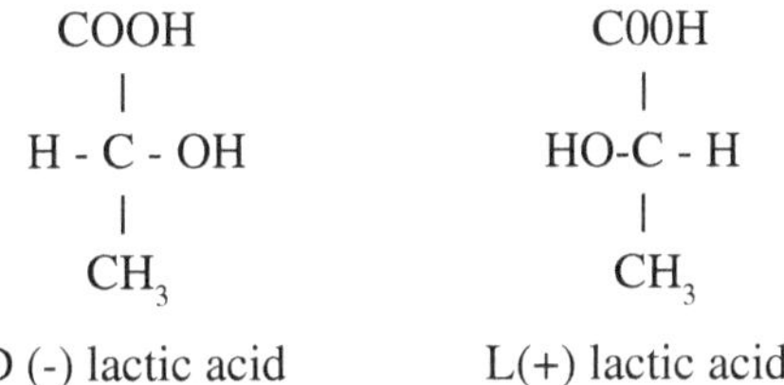

The lactic acid formed during fermentation is usually L (-) form. It is a colourless to yellow syrupy liquid or a white powder. The specific gravity lactic acid is 1.029 and the melting and boiling points are 18 °C and 122 °C. Lactic acid powder is water-soluble. Lactic acid is soluble in water and ethanol. It is corrosive to metals and tissue. Lactic acid is found primarily in sour milk products such as yoghurt, kefir, and cheeses. It is produced by an anaerobic fermentation process. Lactic acid producing bacteria is deliberately added in the baker's yeast fermentation process to extend storage life. The casein in fermented milk is coagulated due to lactic acid.

10.2.2 History of Lactic Acid Fermentation and Biochemical Pathways

In 1780, Scheele isolated and identified lactic acid as the principal acid in sour milk. However, Blondeau first discovered lactic acid as a fermented product in 1847. Schultze (1868) showed the presence of lactic acid bacteria in yeast cultures of distilleries. Delbruck determined the most favourable temperature of lactic acid fermentation. He concluded that relatively high temperature favoured high yield of lactic acid. In 2006, global production of lactic acid reached 275,000 tonnes with an average annual growth of 10%.

Lactic acid production takes place under anaerobic condition. In homofermentative process, lactic acid bacteria usually follows the Embden–Meyerhof Pathway (EMP) to produce lactic acid. Glucose or lactose is converted to pyruvic acid, $[CH_3(CO)\ COOH]$. when passed through the Embden–Meyerhof Pathway. In presence of NADH, pyruvic acid is reduced to lactic acid as shown in Eq. 10.9.

$$\underset{\text{EMP pathways}}{C_6H_{12}O_6 \dashrightarrow 2CH_3(CO)COOH} \xrightarrow[\text{Lactate dehydrogenase}]{2\,NADH} 2CH_3CH(OH)COOH + 2NAD^+ \quad 10.9$$

Homofermentative bacteria convert about 95% of fermentable hexoses to lactic acid. Small amounts of volatile acids and CO_2 are also produced. Heterofermentative bacteria differ from the homofermentative species due to the fact that lactic acid is only one of several principal products formed from sugar. Other compounds include ethyl alcohol, acetic acid, formic acid and carbon dioxide. It follows the pentose phosphate pathway. In the lactic acid fermentation process, lactic acid is mostly produced in the form of calcium lactate. Calcium hydroxide is added during the fermentation process to maintain the pH in the fermentation media and lactic acid is usually converted to calcium lactate.

10.2.3 Lactic Acid Fermentation Process

Lactic acid bacteria (LAB) are a diverse group of bacteria capable of lactic acid production. These bacteria are found in cheeses, yoghurts or decomposing plants. They are Gram-positive and non-sporulating. They have the ability to produce lactic acid as a major metabolic end product of carbohydrate fermentation. They are anaerobic but also tolerate aerated environments and can survive at high acid (pH 3–6) and high ethanol concentrations. The *Lactobacillales* can be divided into different genera, such as *Lactobacillus, Leuconostoc, Pediococcus, Lactococcus and Streptococcus* etc. High lactic acid producing bacteria such as *Lactobacillus delbrueckii* favour temperatures of 44–45 ^{0}C. Examples of homofermentative bacteria are *L. casae, L. delbrueekii, L. bulgaricus etc.* An example of heterofermentative bacteria is *L. acidophilus*.

In the baker's yeast fermentation process, lactic acid bacteria are added into the fermentation for the following reasons:

- They act as a preservative.
- Most yeast utilize lactic acid as a C-source.
- Most of the contamination to yeast is caused by the alkalinity of the medium. To avoid undesirable bacteria grow such as putrefying bacteria and to prevent contamination in the fermentation process, lactic acid bacteria are used. Therefore, aseptic fermentation is relaxed.
- They are used to acidulate to inhibit the development of butyric acid bacteria in the manufacture of yeast.

Lactic acid bacteria require complex medium for their growth. They fail to grow in medium containing any lack of amino acid. They utilize B-vitamins. They are now produced commercially from corn sugar, molasses and whey. The sugar in the production medium is normally adjusted to a concentration of 5 to 20%w/v depending on the nature of the raw material and the conditions of the process. Usually yeast extract or malt extract is used, which contains all the amino acids required for their growth.

During fermentation, this should yield lactic acid as one of the major products. Lactic acid bacteria are usually microaerophilic or anaerobic in nature. The fermentation proceeds best when the pH is on the acid side of neutrality. Calcium hydroxide is used as a neutralizing agent and to adjust pH 5.5–6.5. Certain growth factors are essential for certain lactic acid bacteria. Riboflavin and at least one other 'activator' e.g. nicotinic acid, are required by certain lactic acid bacteria for their growth and metabolism. Some nutrients such as malt sprouts, corn-steep liquor, and thin grain residue to block strap molasses increase lactic acid yield.

Whey is a cheap raw material. It contains about 4.6%w/v lactose. Whey also contains albumin (protein), B-vitamins, and minerals. This medium is suitable for the manufacture of lactic acid. Selection of organisms depends on the temperature of fermentation. *L. delbrueekii, L. bulgaricus* are mostly used for the industrial fermentation process for lactic acid production. Following incubation at 43 °C for 24 h, this starter is added to the main fermentation tank. Fermentation is usually completed in about 2 d. Yields of 85 to 90% lactic acid on the basis of the sugar fermented are obtained in controlled fermenter. Calcium lactate is soluble in water and tends to crystallize out if the medium contains a higher concentration of sugar (about 15%w/v). Noncorrosive material is used for the construction of fermenter. Usually the fermenter is constructed with heavy wood. Modern fermenters are constructed from stainless steel. The nitrogen source in whey is lactalbumin. The volume of inoculum is 10%v/v. *L. bulgaricus* require high temperature (45–50°C) for their growth. This is done for two days. $Ca(OH)_2$ is added in every 6 h intervals to keep the acidity of the medium below 0.6%. The fermentation broth contains lactic acid and calcium lactate. The fermentation broth is heated to 80–95 °C for the coagulation of proteins and is then filtered. The filtrate is then allowed to evaporate to dryness to get Ca-lactate. For the preparation of pure Ca-lactate this is dissolved in water and treated with activated charcoal for decolourization and mixed thoroughly and kept for 15 min and filtered. To prepare lactic acid from Ca-lactate, it is treated with dil. H_2SO_4 to get a dilute solution of lactic acid. $CaSO_4$ is precipitated out. If the dilute solution is heated to high temperature then some chemical changes may occur e.g. formation of lactate anhydride. The block flow diagram of lactic acid fermentation process is shown in Figure 10.6 ((Rohr and Kubicek, 1981; Thakur, 2013; Mettey, 1992).

To prepare 100 kg of lactic acid (80% yield), the following compounds are taken in given amounts: 2500 kg of whey, 100 kg $Ca(OH)_2$, 58 kg H_2SO_4 (100%).

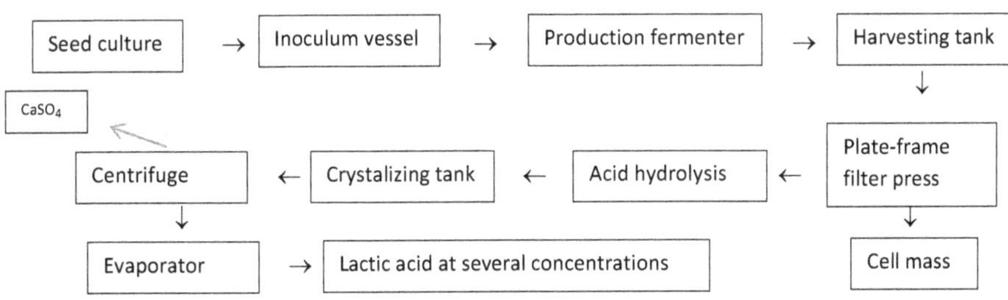

FIGURE 10.6 Block flow diagram of the lactic acid fermentation process.

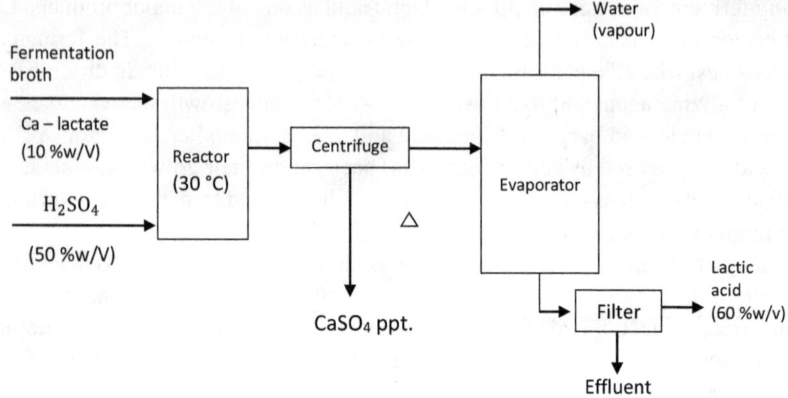

Reaction:

$$\text{Ca} - \text{lactate} + H_2SO_4 \rightarrow \text{Lactic acid} + CaSO_4$$

FIGURE 10.7 Downstream processing of lactic acid production.

Details of the downstream processing of lactic acid production are depicted in Figure 10.7.

10.2.4 Spoilage Reactions Caused by Lactic Acid Bacteria

The following spoilage reactions are caused by lactic acid bacteria:

- In the dairy industry, this can also cause souring in milk.
- In orange juice concentrates, lactic acid bacteria cause spoilage due to off flavour, souring.
- In the sugar refinery industry, *Leuconostoc meseteroids* can grow, which forms high viscous sugar which may interfere in the crystallization of sugar. In the wine-making industry, these bacteria cause ropiness.
- In meat products, this can form turbidity and surface growth, which is undesirable.

10.2.5 Grades and Uses of Lactic Acid

There are four principal grades of lactic acid (LA).

- Crude or technical grade → It contains 22, 44, and 80% LA strength
- Edible → 50–80% LA
- Plastic → 50–80% LA
- U.S.P. (medicinal) → 75–85% LA

The colour of crude grade LA varies from yellow to brown. It is used for deliming of animal skin, and dying of silk and other textiles. Edible-grade LA is available from straw to yellow colour. It is used in jam, jelly, fat extraction etc. Plastic-grade LA is

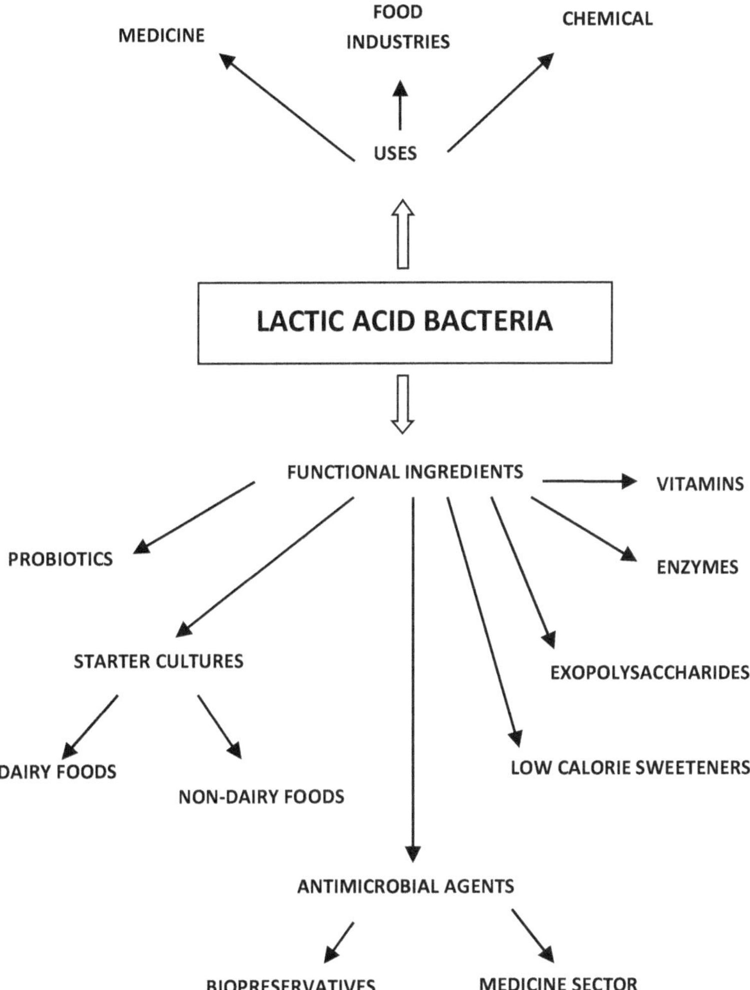

FIGURE 10.8 Different uses of lactic acid producing bacteria.

colourless. It is generally used for the manufacture of plastic. U.S.P. grade LA is used for medicinal purposes. The uses of different lactic acid producing bacteria are shown in Figure 10.8. Poly-lactic acid has tremendous potentiality in the operation theatre. Cheese is a milk product which can be preserved for a long time due to the presence of lactic acid. LA can be used as a chemical. A functional ingredient is a bioactive compound that can be used in the manufacture of functional food products such as probiotics, biopreservatives etc.

10.3 ACETIC ACID OR VINEGAR

Acetic acid or vinegar is one of the world's most popular acidified condiments. Vinegar is a substance arising from the bacteria *Acetobacter* sp. which transforms alcohol to acetic acid. Acetic acid is the main component in vinegar. Vinegar has been very essential not only for food processing as a preservative for fish, meat, and vegetables, but also as an additive in ketchup, spices, sauces, and mustard. Cider vinegar is obtained from fruit juice. Cider vinegar is a highly useful beverage, as it helps to facilitate various types of positive benefits for consumers, such as antidiabetic effects and low blood cholesterol levels by hindering low-density lipoprotein (LDL) oxidation, along with other advantages. Acetic acid is produced in a two-stage fermentation process: alcoholic fermentation by *Saccharomyces cerevisiae* followed by acetic acid fermentation by *Acetobacter aceti.* The conventional process (known as the Orleans process), and the generator process are two kinds of vinegar manufacturing techniques. There are several parameters such as yeast strain, oxygen concentration levels, fermentation temperature, and sugar that affect the rate of production of cider vinegar. Vinegars are highly nutritious, along with sugars, organic acids, amino acids, melanoidins, tetramethylpyrazine, and polyphenols (Chen et al., 2016).

10.3.1 Microorganisms Involved in Vinegar Production

Microbial genera and species such as yeast and acetic acid bacteria are involved in the fermentation process for vinegar production. Environmental conditions and bacterial diversity are key factors in the production of vinegar, and they can influence the performance of the system. The most important group of yeasts helpful in vinegar production are genus *Saccharomyces*, *Hanseniaspora, Kluyveromyces*.

Saccharomycetaceae and other species facilitate fermentation, showing good ethanol tolerance. *S. cerevisiae* is the most important yeast for the fermentation industry for the production of ethanol. Ethanol fermentation processes were discussed in Chapter 9. Yeasts can tolerate high acidity that facilitates their survival in low pH such as fruit juices. The substrate used for yeast metabolism is monosaccharides such as glucose, fructose and mannose. These monosaccharides are metabolized by yeast into two molecules of pyruvate through glycolysis followed by conversion of pyruvate into ethanol and carbon dioxide.

$$C_6H_{12}O_6 \longrightarrow CH_3CH_2OH + 2\,CO_2 + 55\text{ kcals}$$

Acetic acid bacteria (AAB) belong to the family *Acetobacteriaceaea* and *Gluconobacter*. These are Gram positive, are rod-shaped and motile due to the presence of flagella, and are catalase positive and oxidase negative. They display aerobic metabolism with oxygen as electron acceptor. AAB growth has been observed in pH ranges from 5.4 to 6.3. The optimum temperature required for the growth of the AAB cells is 25–30 °C. AAB can grow in substrate such as fruits, flowers and fermented beverages. The stoichiometry of the acetic acid production from ethanol is given below

$$CH_3CH_2OH + O_2 \longrightarrow CH_3COOH + H_2O + 118\text{ kcals}$$

This is a two-step oxidation reaction in the conversion of ethanol to acetic acid. It involves activity of the enzyme alcohol dehydrogenase (ADH) and aldehyde dehydrogenase (ALDH). The first step is carried out by ADH, which is further oxidized to acetic acid by ALDH. The first reaction is exothermic in nature. The second step involves the oxidation of acetaldehyde to acetic acid. The concentration of ethanol in the system affects the AAB. A high concentration of ethanol decreases the activity of the AAB while a lower concentration of ethanol leads to an over-oxidation process.

10.3.2 Methods of Vinegar Production

There are two widely used fermentation processes used for vinegar production. Alcoholic fermentation and acetic acid fermentation are key processes in vinegar production. Alcoholic fermentation is a fast process as compared to acetic acid fermentation. Sugars are converted into ethanol through the metabolic activity of the yeast; usually, *Saccharomyces cerevisiae.* However, in acetic acid fermentation, *Acetobacter* can further oxidize ethanol into acetic acid. Anaerobic conditions are suitable for alcoholic fermentation while acetic acid fermentation is carried out under aerobic conditions.

10.3.2.1 Aerobic Submerged Fermentation

Submerged fermentation (SF) is an aerobic process in which the ethanol is oxidized to acetic acid by AAB under controlled conditions. It is the most common and widely used industrial scale method that improves general fermentation conditions such as aeration, stirring, heating, etc. In this method, the optimum temperature is maintained by a heat exchanger during the fermentation process where bacteria are suspended in the medium. Suitable alcoholic stocks, a continuous aeration system and suitable AAB strains that need a small amount of nutrients for growth are the major necessities for vinegar production in the submerged fermentation process. Submerged fermenters comprise a stainless steel reactor with capacity higher than 10,000 L with a well-fitted aeration supply, cooling system and foam controlling system. Aeration systems play an important role in the SF process. Oxygen transfer depends on the size of the air bubbles (Garcia-Ochoa and Gomez, 2009). The optimum temperature required for the vinegar production is 30°C. There are three steps involved in SF: 1) loading of raw materials and culture into the fermentation medium, 2) fermentation process with complete removal of product from the tank, and 3) remaining part of the media, culture for the next cycle. The acetification process is an exothermal process. Semi-continuous modes of fermentation are usually preferred for SF at industrial level because it is most beneficial for vinegar production, and reduces the risk of substrate inhibition and catabolic repression. In this process, the alcoholic substrate is added after the acetification process has started. The semi-continuous mode of operation is performed in two reactors arranged in parallel. The first bioreactor contains the inoculum resulting from the previous cycle to which wine and other alcoholic products are added. Once the acetifying mass approaches an ethanol content of approximately 2–3%v/v, it is injected into the second bioreactor where it persists until the ethanol is exhausted (0.2–0.3%v/v) and the required acetic acid content is obtained. The time required for the cycle ranges from 18 to 30 h. Fermentation time is dependent upon the bacterial lag phase, the initial concentration of ethanol and

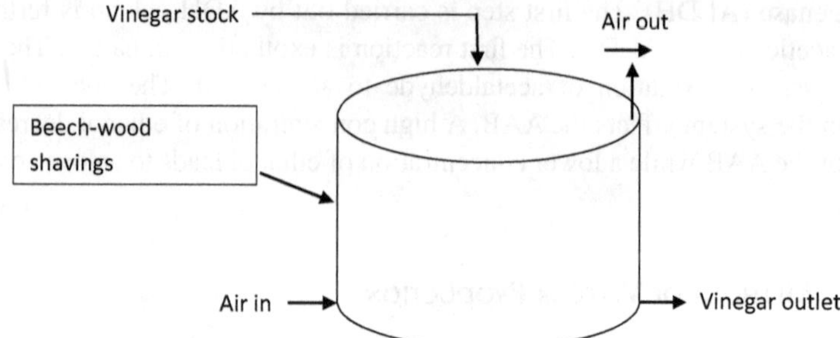

FIGURE 10.9 German or generator process for acetic acid (vinegar) production.

proficiency of the aeration system. Sustaining the microorganisms in the tank is the core goal of the vinegar production system. To maintain the viability of the cells the process has be adjusted under controlled conditions.

The generator method is one of the quick processes used for vinegar production. Sometimes this method is also known as the 'German method' or the 'Trickle method' (Figure 10.9). It is a quick method that produces the final product within days instead of weeks. In this method, the fermentation tank is usually constructed from wood or stainless steel. Partition of the tank is through a screen; a large area of the tank is filled with solid materials almost to the top of the tank and air is inserted and forced in an upward direction and a smaller area of the tank forms a collection chamber which collects the vinegar. A pump circulates the alcohol-acetic acid mixture from the source to the top of the tank, similar to a trickling filter function in wastewater treatment. Heat generated in the tank by the exothermic reaction causes a draft which provides oxygen for the aerobic conversion of alcohol to acetic acid. Cooling water in the heat exchanger regulates the temperature. The optimum temperature for the whole process is 30 °C. Approximately 3 to 7 d are required to complete the vinegar production. The generator method is easy to maintain once the airflow and temperature as well as ethyl alcohol concentration are maintained at the required level. The level of ethyl alcohol does not fall below 0.3–0.5%.

Sugar cane juice is used as the raw material for the production of cane vinegar. It appears in a dark yellow to golden brown colour and has a mellow flavour. In the production process of cane vinegar at industrial level a cleaning step is vital to obtain good quality of cane juice. Contamination should be avoided in order to get a pure-quality final product. Microbial contamination must be removed by a process of filtration and pasteurization. After the cooling process, the culture of *Saccharomyces cerevisiae* is added in the correct proportion for the production of alcohol. Before addition of the culture, cells must complete their lag phase in order to get high yield. Anaerobic conditions should be maintained in order to grow cells in a proper environment. A hydrometer is used to check whether alcoholic fermentation is complete or not. A filtration process is used to removes the yeast cells. Oxygen is very important and its supply should be maintained throughout the acetic acid fermentation process. Sampling is needed in order to check the quality of the vinegar after certain time intervals and, based upon this, adjustment of the process is possible. Final filtration is carried out after the fermentation reaches the desired level. Standardization, quality checking, bottling steps and labelling are the final steps in the manufacturing process.

Various research studies have reported that immobilization provides certain improvements in the fermentation process through increasing biomass, option of reusability, temperature and protection of cells from toxic effects of low pH. Some research groups are trying to develop an effective method using immobilization cells of *Acetobacter aceti* for production of vinegar by using raw material cane juice.

10.3.2.2 Factors Affecting Vinegar Production

pH: The optimum pH required for the growth of acetic acid bacteria to grow ranges from 5.5 to 6.6. Numerous studies state that acetic acid bacteria can still grow at low pH such as 3.0 and some strains also have the ability to survive in the most acidic environments such as pH 2.0. Production of vinegar involves three different strains of acetic acid bacteria on the basis of their survivability in acidic environments such as acetophilic strains (pH 3.5), acetotolerant strains (pH 3.5-6.5), and acetophobic strains (growing at pH higher than 6.5).

Temperature: The optimal temperature required for the growth of acetic acid bacteria ranges from 25°C to 30°C. Thermotolerant bacteria have the ability to survive at a temperature of 40 °C. Some research reports suggest that some acetic acid bacteria can grow at 10°C but have slower growth rate and a low metabolic rate.

Production methods: Production methods of vinegar can also influence the quality and quantity of the final product. As described in the production section, different methods such as the Orleans method, the submerged method and the generator method will affect the quality of the vinegar. The Orleans method is a time-consuming method but it gives a pure quality of vinegar as compared to the submerged process. Therefore the choice of production method will influence the final product.

Chemical composition and quality of vinegar: The quality of vinegars is based upon different factors such as appropriate starter culture, environmental conditions, starting material, method of production and ageing. Assessment of vinegar quality requires determination of the compounds present in the vinegar and their sensory analysis. Existence of aromatic compounds in the vinegar significantly affects the quality of the vinegar. The aroma of vinegar contains many compounds with widespread variation of volatilities, polarities and concentrations ranging from several mg/L to ng/L. To date, various research teams have found 100 different chemical compounds in the aroma of the vinegar such as carbonyl compounds, acetals, lactones, acids, alcohols, ethers and volatile esters. These are all found at different concentrations in the vinegar. One of the research studies states that contact of vinegar with wood enhances the complexity of the aroma. The presence of volatile and non-volatile compounds can be detected by using gas chromatography. The presence of polyphenolic compounds determines the colour and astringency of vinegar. Acetification is an aerobic process in which reaction of phenolic compounds with oxygen determines the colour of the vinegar. For example, oxygen affects polyphenolic compounds at different levels. In submerged fermentation, the flavanol content of vinegar is influenced by oxygen availability while in surface acetification vinegar does not affect phenolic compounds such as aldehyde. The ageing process involves reaction of compounds over time. The most important factor in vinegar production is its taste. Vinegar is difficult to taste because of its strong sensations. The strong presence of acetic acid covers other flavours and some other products which are required for tasting. The qualities used to evaluate vinegar taste include colour, aromatic intensity, woody scent, wine smell and pungent feeling. High-quality vinegar contains large amounts of vanillin, eugenol and benzaldehyde, and the presence of these compounds indicates the quality of the vinegar. Tesfaye et al. (2002) have stated the criteria for evaluation of the vinegar quality. To evaluate the purity of the alcohol it is necessary to test the various compounds present, such as minerals, alcohols, acids, phenolics and other volatile compounds. The second parameter that indicates the quality of vinegar is its aroma, flavour and other organoleptic properties. The main factor that influences the aroma of vinegar is the ethanolic material from which it is made.

10.3.2.3 Types of Vinegar

Vinegar may be classified as natural or artificial. Natural vinegar is produced by a fermentation process. Artificial vinegar is produced from synthetic acetic acid. There are different types of natural vinegar produced and used in various processes. Malt vinegar, apple cider vinegar, wine vinegar, balsamic vinegar, fruit vinegar and many other types of vinegar occur in today's global market. The classification of vinegar is basically dependent upon the raw material used for the production.

1. *Wine vinegar:* Wine vinegar is mostly used in the Mediterranean countries and central Europe and is made from red or white wines. It has lower acidity than white or cider vinegar. It is also called grape vinegar. Acetic acid fermentation of grapes as raw material is used for the production of wine vinegar.
2. *White vinegar:* This is made from distilled alcohol. Also known as 'spirit vinegar' or 'grain vinegar' or 'white distilled vinegar'.
3. *Balsamic vinegar:* This is dark brown in colour with a sweet-sour flavour and is made from the white Trebbiano grape. It is a kind of aromatic, aged vinegar. True balsamic vinegar is aged for 12 to 25 years.
4. *Malt vinegar:* Malt barley is used as substrate for the production of malt vinegar. It is light brown in colour.
5. *Rice vinegar:* Rice white vinegar is mildly acidic in nature and has an uncomplex flavour. The Japanese and Chinese prefer rice vinegar and its varieties for cooking purposes.
6. *Herbal vinegar:* Sometimes herbs and spices are added to the wine or white-distilled vinegars to flavour them. The most popular, tasty and aromatic flavours used are garlic, basil, nutmeg, clove and cinnamon.

Nutrients and bioactive compounds present in the vinegar are the main components in determining the quality of vinegar. Microelements, amino acids, sugars and vitamins are the nutritional components usually involved in vinegars (Chou et al., 2015). These can provide energy, maintain acid-base balance, improve the response of the immune system, and control the metabolism of the cells.

In vinegars, amino acids are derived primarily from the raw materials and microbial degradation of proteins. The existence of both free and nonprotein amino acids have been reported in recent research findings. Amino acids are present in grain and fruit vinegars in high amounts in different kinds of vinegars; e.g. glutamic acid. Alanine concentration is found in the range from 1.59 to 3.51 mg/mL, the maximum being in vinegar aged in Shanxi reported by research groups.

Glucose, fructose, mannitose, arabinose, and xylose are the monosaccharides that are reported in vinegars in several studies. Sucrose, maltose, and mycose are disaccharides that are involved in vinegars (Koyama et al., 2017). In fruit vinegars, fructose and glucose are present and provide their sweet taste. Sugarcane vinegar contains six alditols – arabitol, sorbitol, xylitol, erythritol, ribitol, and inositol and eight monosaccharides – fructose, ribose, xylose, rhamnose, galactopyranose, mannopyranose, arabinopyranose, and glucopyranose according to Li et al. In certain vinegars, polysaccharide macromolecules are present in relatively tiny quantities.

Grain vinegars consist mainly of vitamin B including vitamins B1, B2, niacin acid, and nicotinamide. Vitamins B3, B5, C, and folic acid are present in palm vinegar. The niacin component that is present in vinegar plays a vital role in excreting cholesterol and expanding blood vessels (Zhang et al., 2018).

Vinegar can be used as a copper, arsenic, and chromium extracting tool from chromate copper arsenate (CCA) using wood vinegar. This highlights that wood vinegar can be an option for synthetic chemical substances used to extract metal elements from treated wood waste (Choi et al., 2012). Additionally, to enhance the productivity of solar disinfection systems, i.e. SODIS, vinegar is now used to reducing the pH. The entire findings show that catalyst selection i.e. vinegar was a significant factor in sunlight disinfection, in addition to low pH.

10.3.2.4 Uses of Vinegar

Regular consumption of vinegar may be useful in the prevention of obesity. Acetic acid from vinegar was used as a responsive, specific, and reliable agent to diagnose the development of human cancer cells using cervical cancer cells. Vinegar has high cancer diagnostic ability, as it is affordable and free of adverse side effects compared with the toluidine blue and metachromatic dye usually used for cancer detection. Vinegar is used in head lice, ear infections, nail fungus and scars.

Today vinegar is primarily used in the food processing industry as mentioned below:

i) Vinegar is a salty and sharp liquid used as a preservative for condiments and foods. A food condiment, sprinkled on certain foods such as fish at the table. Vinegar is primarily used to flavour and preserve foods and as an ingredient in salad dressings and marinades.
ii) Vinegar is incredibly beneficial for pickling and preserving vegetables and meat, since it can reduce the pH of food below that which even spore formers may not survive.
iii) Vinegar is one of the most used acidic condiments. It is an essential component of especially famous French sauces, which often contain vinegar.
iv) Approximately 70% of today's vinegar is distributed to various food industries where it used in the manufacturing of sauces, salad dressings, mayonnaise, tomato products, cheese dressings, mustard, and soft drinks.

10.4 CONCLUSIONS

Several biochemical industries are involved in the production of oxychemicals. Citric acid, lactic acid and acetic acid are very important oxychemicals. Cane molasses is found to be a suitable raw material for citric acid production by *Aspergillus niger*. Metal ions play a pivotal role in the citric acid fermentation process. The estimation of product concentration in most of the biochemical industries, such as citric acid, penicillin etc., takes quite long time. Therefore, a quick estimation technique is needed for these industries for the decision-making process for the operation of fermenters. Two forms of citric acid are marketed as citric acid monohydrate (CAM) and citric acid anhydrous (CAA). Citric acid and lactic acid have very important applications in food processing as well as pharmaceutical industries; *L. delbrueekii, L. bulgaricus* are mostly used for the industrial production of lactic acid from whey, which is a by-product of cheese making industry. Vinegar is an acidic liquid product produced from alcoholic fermentation by using yeast and then acetous fermentation by acetic acid bacteria under suitable conditions. It is used to flavour and preserve the foods and also used as ingredient in salad dressing. There are three different methods used for vinegar production. The generator or trickling method, the submerged method or the acetators method, and Orleans method. The submerged method is widely used at commercial level, and produces a high quantity of vinegar with high concentration of acetic acid. The Orleans method is slower but it gives pure vinegar. The generator method is faster compared with the other methods. Vinegar contains various kinds of organic acids, esters, ketones and aldehydes that contribute to the quality of the vinegar. Vinegar also contains certain bioactive compounds including carotenoids, phytosterol, phenolic compounds and vitamins C and E. Vinegar is used as a flavouring agent for food purposes as well as for therapeutic purposes.

REFERENCES

Bailey JE and Ollis DF, Biochemical Engineering Fundamentals, McGraw-Hill, New Delhi, 2010.

Chen H, Chen T, Giudici P, Chen F, Vinegar Functions on Health: Constituents, Sources, and Formation Mechanisms. Comprehensive Reviews in Food Science and Food Safety 15, 1124–1138, 2016.

Choi Y-S, Ahn BJ, Kim G-H, Extraction of chromium, copper, and arsenic from CCA-treated wood by using wood vinegar. Bioresource Technology 120, 328–331, 2012.

Chou C-H, Liu C-W, Yang D-J, Wu Y-HS, Chen Y-C, 2015. Amino acid, mineral, and polyphenolic profiles of black vinegar, and its lipid lowering and antioxidant effects in vivo. Food Chemistry 168, 63–69, 2015.

Das D and Das D. Biochemical Engineering: An Introductory Text Book, Jenny Stanford, Singapore, 2019.

Doran PM, Bioprocess Engineering Principles, Second Edition, Academic press, Waltham, 2012.

Garcia-Ochoa F, Gomez E, Bioreactor scale-up and oxygen transfer rate in microbial processes: An overview. Biotechnology Advances 27, 153–176, 2009.

Koyama M, Ogasawara Y, Endou K, Akano H, Nakajima T, Aoyama T, Nakamura K, Fermentation-induced changes in the concentrations of organic acids, amino acids, sugars, and minerals and superoxide dismutase-like activity in tomato vinegar. International Journal of Food Properties 20, 888–898, 2017.

Mattey M, The production of organic acids, *Critical Reviews in Biotechnology*, 12, 87–132, 1992.

Max B et al., Biotechnological preparation of citric acid, *Brazilian Journal of Microbiology*, 41, 4, 2010.

Moo-Young M, Comprehensive Biotechnology, Pergamon Press, 2019.

Patel AH, Industrial Microbiology, Mac Millan Publishers India, 2012.

Prescott SC and Dunn CG, Industrial Microbiology, McGraw Hill Book, and K O Gakusha, Tokyo, 1959.

Röhr M and Kubicek CP, Regulatory aspects of citric acid fermentation by *Aspergillus niger*. Process Biochem. 16, 34–37, 1981.

Thakur IS, Industrial Biotechnology: Problems and Remedies, I. K. International Publishing House, New Delhi, 2013.

Tesfaye W, Morale MK, Garcia-Parrilla MC, Troncoso AM, Wine vinegar : Technology, authenticity and quality evaluation, Trends in Food Science and Technology, 13:12–21, 2002.

Wittmann C and Liao JC (eds.), Industrial Biotechnology: Products and Processes, Wiley-VCH, 2017.

Zhang Y, Zhou W, Yan J, Liu M, Zhou Y, Shen X, Ma Y, Feng X, Yang J, Li G, A Review of the Extraction and Determination Methods of Thirteen Essential Vitamins to the Human Body: An Update from 2010. Molecules 23, 1484, 2018. https://doi.org/10.3390/molecules23061484

11 Amino Acids and Vitamin Production

Amino acids are the building blocks of proteins. Most proteins are enzymes in nature. Therefore amino acids are very significant for living organisms. Amino acids have a high amount of nitrogen and thus are distinguished from lipids and carbohydrates. On the other hand, vitamins are essential micronutrients required for a wide variety of metabolic processes. These are usually required in trace quantities. Vitamins can be classified into two types: fat soluble vitamins: vitamin K, vitamin E, vitamin D, vitamin A; and water-soluble vitamins: vitamin B, vitamin C. Vitamin B-12 is a part of the B vitamin family (consisting of 8 other vitamins) that performs a key function in healthy brain and nervous control. It is involved in various cellular metabolic processes as well as DNA and fatty acid synthesis. It plays a vital role in the catabolism of threonine, methionine, leucine, isoleucine, and valine. Lack of vitamin B-12 can lead to a variety of deficiency symptoms and overall lethargy and tiredness (Vandamme, 1992).

11.1 AMINO ACID SYNTHESIS VIA FERMENTATION

The basic structure of amino acid is alpha carbon. Four different functional groups are present (such as amine, hydrogen, side chain (R), and carboxyl group).

Amino acids are classified as follows:

Non-polar, aliphatic group: These groups of amino acid are insoluble (hydrophobic) in water because they do not have any charge; e.g. glycine, isoleucine, leucine, and valine.

Non-polar, aromatic group: These groups of amino acid are insoluble (hydrophobic) in water because they do not have charge and they have cyclic structure so these groups of amino acid do not dissolve in water; e.g. tryptophan and tyrosine.

Polar groups: These groups of amino acid are soluble in water (hydrophilic) because they have a charge which helps amino acids to dissolve; e.g. cysteine, serine, and asparagine.

Based on nutrition, amino acids can be classified as follows:

Non-essential amino acids: Amino acids that can be synthesized in the body are called non-essential amino acids. These amino acids are not required in the diet; e.g. alanine, serine, glutamic acid etc.

Essential amino acids: Essential amino acids are not synthesized in the body. These amino acids are required in the diet; e.g. lysine, leucine, methionine, histidine etc.

Nowadays, people in under-developed and over-populated nations are suffering from deficiency of essential amino acids. Therefore, in such nations essential amino acids are fed to domestic animals (feed grade) to improve meat quality. This meat can be consumed by people to overcome deficiency inessential amino acids. Since 2011, many industries have been involved in essential amino acid production via chemical synthesis or through natural resources. Synthesis from these two types of protocols has been successful but is very expensive. Therefore biotechnological approaches have replaced them and are less expensive. This approach also has fewer side effects on humans and animals as compared to chemical synthesis. Non-essential amino acids have important biological applications. Therefore, they are also synthesized by fermentation. Essential amino acids are also synthesized through fermentation. *Corynebacterium glutamicum* is a preferred bacterium used for the synthesis of both essential and non-essential amino acids. As with beverages and baker's yeast, amino acids have been produced via fermentation since the year 2015 (Bongaerts et al., 2001).

11.2 L-GLUTAMIC ACID PRODUCTION

Amino acids are compounds that consist of carbon, hydrogen, oxygen, and nitrogen. They fill in as monomers or building squares and are made out of amino, carboxyl, hydrogen atoms, and a distinctive side chain, all attached to a carbon molecule, the alpha carbon. In an alpha-amino acid, the amino and carboxylate bunches are connected to a similar carbon particle, which is known as the alpha carbon. The different alpha-amino acids contrast depending on the side chain (R gathering) connected to their alpha carbon. All amino acids are known to be active optically except glycine. Optically active mixes can pivot the plane of captivated light either clockwise or counter-clockwise. Optically active aggravates that turn the plane of captured light clockwise are said to be dextrorotatory while those that pivot the plane of captivated light counter-clockwise are said to be levorotatory. Different types of amino acids are found in the body; however, 20 are required for the process of protein synthesis. They generally are of two types i.e. essential as well as non-essential amino acids. Amino acids that can be synthesized by our body and are not essential to be present or supplied through food are called non-essential amino acids such as alanine, arginine, asparagine, aspartic acid, cysteine, glutamic acid, glycine, proline, serine and tyrosine. Amino acids that cannot be synthesized in our body in sufficient quantity and need to be supplied through diet are called essential amino acids such as histidine, leucine, isoleucine etc. (Bender, 2012).

11.2.1 Classification of Amino Acids Based on Their R Group

Chemical nature of R group: Aliphatic; gly, ala, val, leu. Aromatic; phe, tyr, trp. Hydroxyl; ser, thr. Carboxylic; asp, glu. Imino; pro. Sulphur; cys, met.

Production of amino acids: Manufacturing of amino acids started in the year 1908. Several amino acids have been produced extensively throughout the world such as glutamic acid, lysine, and methionine which are considered to be produced globally. There are several methods by which they can be produced, such as the protein hydrolysis method, the chemical synthesis method, the microbiological method, semi-fermentation, use of enzymes or immobilized enzymes, and direct fermentation (Ikeda, 2003).

Glutamic acid is an alpha-amino acid which has two carboxyl group -COOH and one amino acid group. Glutamic acid can also act as an excitatory neurotransmitter and serves as the precursor for the synthesis of the inhibitory gamma-aminobutyric acid (GABA). It can exist in two optical isomers i.e. D (-) and L(+) form. The L form is the form that occurs most widely in nature whereas the D form exists in a few contexts such as in cell walls of bacteria. Glutamic acid is a non-essential amino acid (i.e. the body can synthesize it) which is used for protein synthesis as well as a flavour enhancer in various food products. Glutamic acid was the first amino acid which was discovered and manufactured by a Japanese researcher Kikunae Ikeda of the Tokyo Imperial University in the year 1908. He identified and isolated the brown crystals which were left due to evaporation of kombu broth and thus MSG or monosodium glutamate (flavour enhancer) was found. Professor Ikeda termed the flavour umami. He then patented a method of mass-producing a crystalline salt of glutamic acid, MSG. It is produced on a large scale with an estimated annual production of approximately 1.5 million tons or even more (Bender, 2012).

11.2.1.1 Microorganisms Involved in Glutamic Acid Production

Production of amino acid can be done via various fermentation methods which involve the use of several microbes such as *Corynebacterium glutamicum.* Yeasts and moulds can also be used for the synthesis of amino acids. However, there are several species of bacteria which have highest efficiency for glutamic acid production; a few noteworthy strains are: *Corynebacterium* spp. *(C. glutamicum); Brevibacterium* spp. *(B.divericartum); Microbacterium* spp. *(M. flavum); Arthrobacter* spp. *(A. globiformis)* (Kumagai, 2000).

Glutamic acid bacteria (GAB) shows high glutamate dehydrogenase activity and lacks α-ketoglutarate dehydrogenase activity. These are Gram-positive, non-motile, with non-sporility and they require biotin for their growth.

Particular enzymes required for glutamic acid production are shown in the following reactions for the glutamic acid formation.

Iso-citrate → α-ketoglutarate dehydrogenase → L-Glutamate

(Isocitrate dehydrogenase)   (Glutamate dehydrogenase)

The isocitrate dehydrogenase produces α-ketoglutarate from isocitrate by decarboxylation and dehydrogenation reaction in the citric acid cycle, which is a key precursor in catalysing the amination reaction for the production of L-glutamic acid by glutamate dehydrogenase. In generalized citric acid cycles, there is an association of α-ketoglutarate dehydrogenase complex for the production of α-ketoglutarate but the commercially utilized strains in the glutamic acid production usually lack α-ketoglutarate dehydrogenase activity. However, this interruption, caused due to lack of α-ketoglutarate dehydrogenase activity, is compensated by the synthesis of oxaloacetate which in combination with acetyl co-enzyme produces isocitrate which in turn is utilized for the production of glutamic acid. GAB does not secrete glutamate out of the cell due to the rigid cell wall under normal growth conditions. The permeability of the bacterial cell membrane could be improved through a set of approaches: i. Limiting the development of normal phospholipids using biotin deprived media; ii. Constraining glycerol with glycerol auxotrophs; iii. Reducing oleic acid in oleic acid auxotrophs (for microbes requiring oleic acid for growth); iv. Adding surfactants e.g. Tween-60; v. The addition of penicillin to biotin-rich media (Enei et al., 1989).

11.2.2 Substrate

A wide variety of carbon sources can be used for the fermentation process of glutamic acid production. Glucose and sucrose are two of the most common carbohydrates used as a carbon source. Other carbohydrate sources such as cane and beet molasses or starch hydrolysates can be used as nutritional input in media. Generally, penicillin derivatives and fatty acids derivatives are added in a medium containing molasses (Bender, 2012).

Inorganic sources such as ammonia or ammonium sulphate are generally used (Table 11.1). In the industrial production of glutamic acid, ammonia feeding permits pH control and helps to control the problem of ammonia toxicity. During the process of fermentation pH 7–8 is maintained. Oxygen supply is one of the most important factors required for media optimization. An adequate supply of oxygen is maintained throughout the fermentation process as oxygen deficiency results in the formation of

TABLE 11.1
The Glutamic Acid Yields with Different Carbon Sources

Carbon source	Organism	Yield (g/L)
Sugar beet molasses	*C. glutamicum*	>100
Glucose+ ammonium acetate	*Brevibacterium divaricatum*	100
Ethanol	*Brevibacterium sp.136*	59
Acetate	*B. flavum*	98
	Methylomonasmethylovora	7
Benzoic acid	*Brevibacterium sp.*	80

lactate and succinate. The excess of oxygen might result in the deficiency of ammonium ions which stops the growth of microbes and production of α-ketoglutarate and thus lowers the glutamic acid production. The optimal biotin concentration is dependent on the carbon source used. If the media is glucose, 5 μg/L biotin is required. If the media is acetate-based media then the requirement will be 0.2–1.0 μg/L. Phosphates, vitamins and other necessary supplements are usually provided with corn steep liquor (Bongaerts et al., 2001).

11.2.3 Fermentation Process

A fermentation of glutamic acid from glucose using *Brevibacterium divaricatum* is as follows (Figure 11.1):

Glutamic acid can be produced from different carbon sources (Table 11.1).

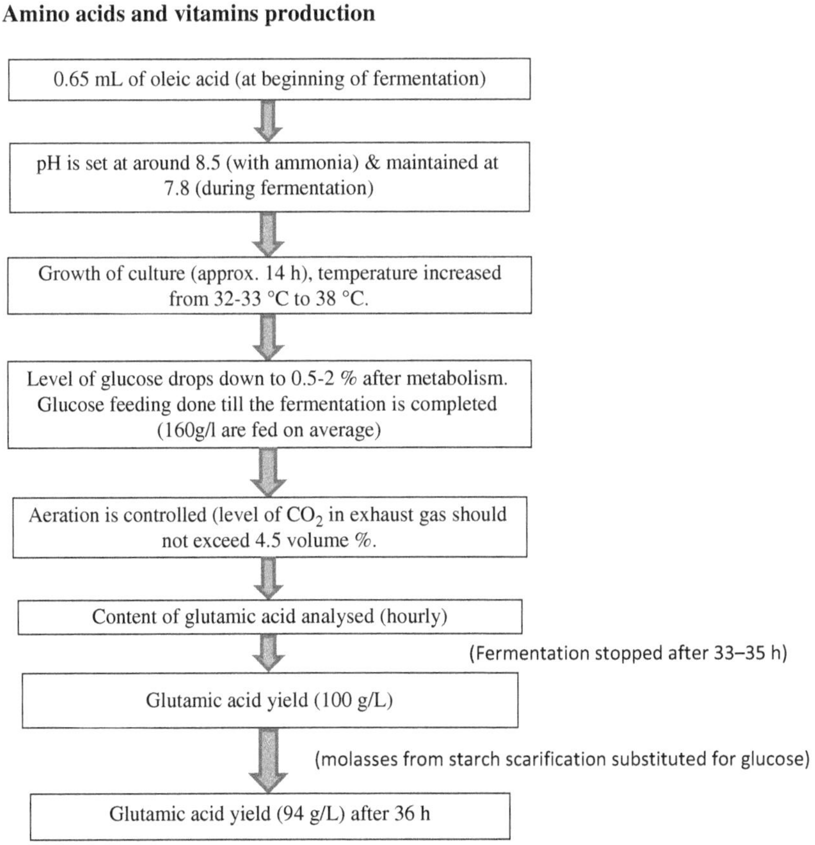

Amino acids and vitamins production

0.65 mL of oleic acid (at beginning of fermentation)

↓

pH is set at around 8.5 (with ammonia) & maintained at 7.8 (during fermentation)

↓

Growth of culture (approx. 14 h), temperature increased from 32-33 °C to 38 °C.

↓

Level of glucose drops down to 0.5-2 % after metabolism. Glucose feeding done till the fermentation is completed (160g/l are fed on average)

↓

Aeration is controlled (level of CO_2 in exhaust gas should not exceed 4.5 volume %.

↓

Content of glutamic acid analysed (hourly)

↓ (Fermentation stopped after 33–35 h)

Glutamic acid yield (100 g/L)

↓ (molasses from starch scarification substituted for glucose)

Glutamic acid yield (94 g/L) after 36 h

FIGURE 11.1 Block diagram for glutamic acid production.

11.2.4 Production of Amino Acids with Auxotrophic Mutants

Auxotrophic mutants belong to those classes of strains which require additional growth supplements which a wild type strain does not require. The first amino acid which was developed from an auxotrophic mutant strain of *Corynebacterium glutamicum* was L-lysine which is used for the synthesis of protein. It is found in various food products such as cheese, yogurt, meat, milk etc. (Enei et al., 1989).

11.2.5 Downstream Processing of Glutamic Acid

After the completion of the process of fermentation, the cells are separated from the broth with the help of various filtration methods (Figure 11.2). Separation of cells can be carried out by using a rotary vacuum filter or filter aids. The filter aids (based on diatomaceous earth) help in increasing the porosity of the filter cake which is directly proportional to the flow rate. There are various extraction methods available to carry out the purification and recovery process of the product (Figure 11.3). Therefore the extraction method to be used depends on the desired purity level of the product.

Two such methods that can be used for purification are chromatography and crystallization. Ion-exchange chromatography can be used for the purification and separation of amino acids such as glutamic acid from the fermentation broth based on their affinity to the ion exchanger. The adsorption of amino acids is based on the type of ion exchange resins. Ion-exchange resins are of two types: i) anion exchange resins and ii) cation exchange resins.

When the amino acids carry positive charge they bind with cation exchange resins (the pH of the solution is less than the isoelectric point of the amino acids) and thus result in the elution of negatively charged amino acids, whereas anion exchange resins are used when the molecule of interest is negatively charged (pH of the solution is higher than the isoelectric point of the amino acids). Crystallization can also be used for recovery of amino acids. Since amino acids are amphoteric (they contain both basic and acidic groups) their solubility is affected by the pH of the solution. Several factors affect the solubility of amino acids, such as temperature, and thus lowering the temperature may result in the recovery of the product (precipitation of amino acids with the help of salts like calcium, ammonium, zinc etc.). After this, an acid or alkali treatment can be given to obtain the free form or acid form of the amino acid (Kinoshita, 1959).

11.2.6 Applications of Glutamic Acid

There are several applications of glutamic acid in food, beverage, and cosmetics production. For food production, it acts as flavour enhancer, and nutritional supplement. For beverage production, it is used in various drinks and wines to enhance the flavour. During cosmetics manufacturing, it is utilized as a hair restorer in the treatment of hair loss and as a wrinkle removal in preventing ageing. Apart from this, glutamic acid is used as intermediate in the manufacturing of various organic chemicals (Bender, 2012).

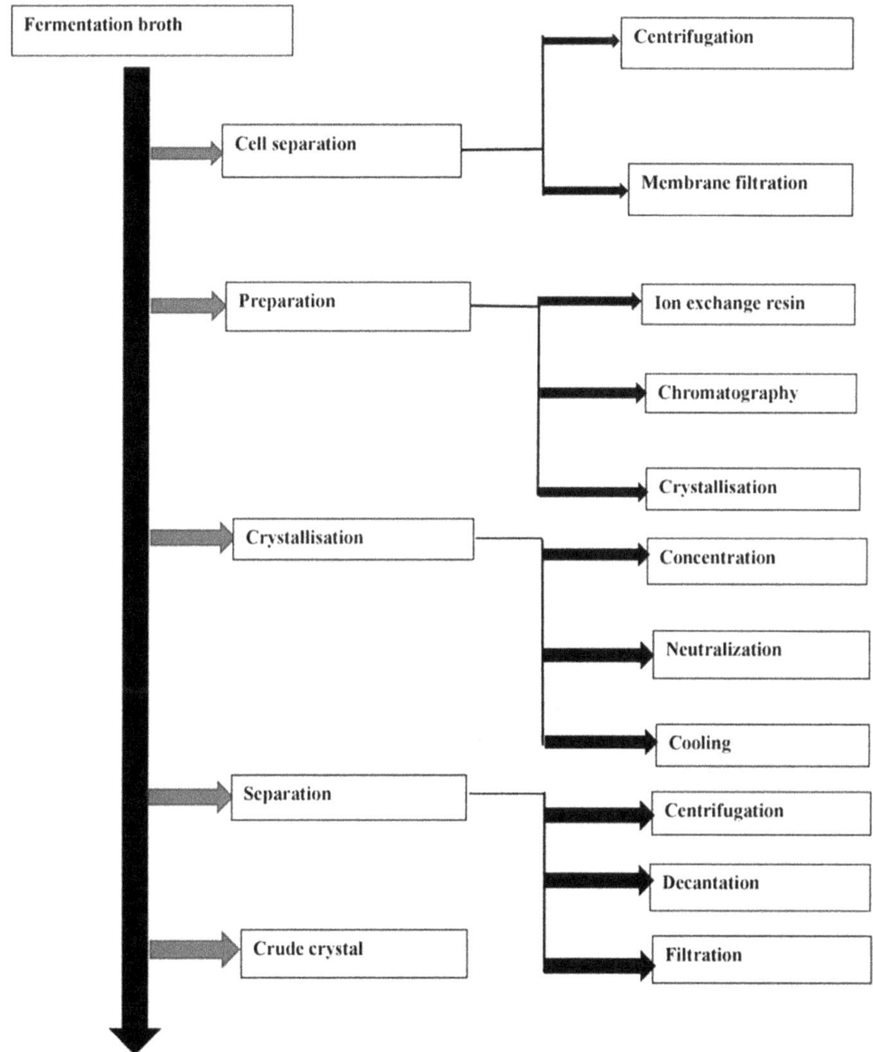

FIGURE 11.2 Downstream processing of glutamic acid production.

11.3 FERMENTATION OF L-LYSINE

L-lysine is known as an alpha amino acid. It also has use in the biosynthesis of proteins. Its contents include an alpha-amino group, and an alpha carboxylic acid group. It is a basic, charged, aliphatic compound. The first-time developer of homoserine auxotroph of the *Micrococcus glutamicus* was Kinoshita et al. in 1958, in which, with the assistance of UV radiation, it was then found that this was able to accumulate a

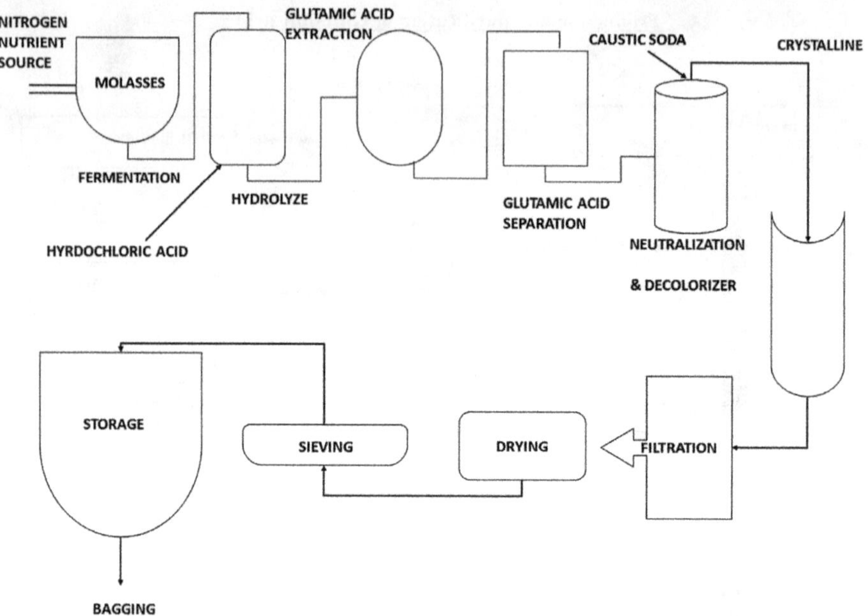

FIGURE 11.3 Process diagram of glutamic acid in the form of monosodium glutamate (MSG).

large amount of the L-lysine which was in the culture broth. Lysine is a building block for protein. It is an essential amino acid because our body cannot make it, so we need to obtain it from food. It is naturally found in food and is used as supplement. The industrial usage of this amino acid includes usage as a supplement in the fields of poultry and swine and all other livestock. It also has usage in pharmaceuticals and in dietary supplements and cosmetics, and exhibits great importance in the economy. It is also used as a food improver. The amino acids are made by fermenting ingredients with other microorganisms. Molasses and other ingredients are added to a specific medium that is used in the cultivation of microorganisms aiding in the complete process of fermentation. This process helps in multiplication of microorganisms and also in the making of amino acids. The enzymes present in the microorganisms help to accelerate the reactions and also help in the breaking down and synthesizing of new substances. There are around 10 to 30 types of enzyme which help fermentation in a series of reactions (Hashimoto, 2017).

11.3.1 Types of L-Lysine Fermentation

Every year 35,000 metric tons of L-lysine is produced. Industrially L-lysine is produced by two different fermentation methods. They are categorized as follows:

(a) *Indirect fermentation:* In this type of fermentation two different types of microorganism are used (dual fermentation). For the initial process of fermentation auxotroph mutant of *E. coli* is used which produces diaminopimelic

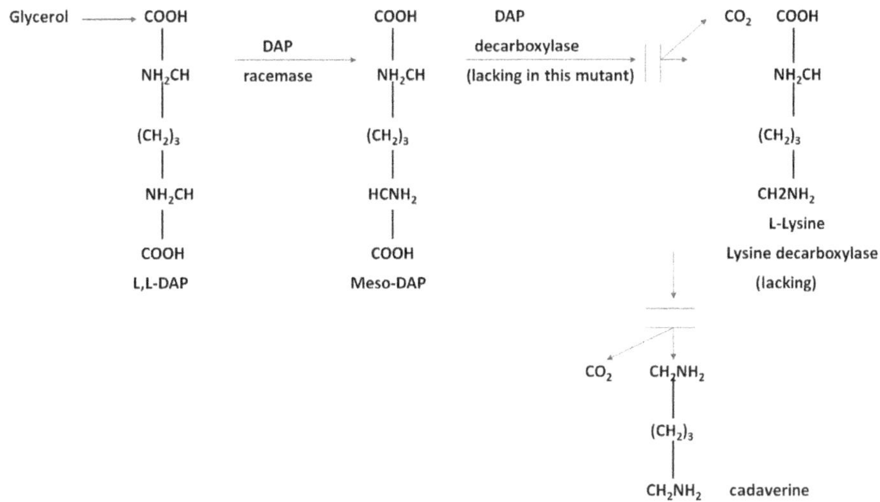

FIGURE 11.4 Indirect fermentation pathway for L-lysine production.

acid or DAP (Figure 11.4). In the second phase, *Aerobacter aerogenes* converts diaminopimelic acid into L-lysine. It has been found that the amount of lysine decarboxylase should be lower in the fermentation process in order to reduce the formation of cadaverine, and thus to increase the level of L-lysine production (Ikeda, 2003).

(b) *Direct Fermentation:* Direct fermentation is a process where L-lysine is produced fermentatively from any substance. For production of L-lysine, this process is used throughout the world. It was discovered that L-lysine is produced from carbohydrates using homoserine or with the use of threonine and added methionine auxotroph of *Corynebacterium glutamicum.* Observing the homoserine auxotroph of *Brevibacterium flavum,* a process was reported which was of the same kind. It was recognized later as a sensitive threonine mutant because growth was inhibited due to the excessive amount of threonine and therefore, by the addition of methionine, the inhibition was released (Figure 11.5). The reason for the phenomenon was because of the feedback of the inhibition of the residual known as homoserine dehydrogenase which is via threonine. In other bacteria, homoserine auxotrophs were found to produce L-lysine but their yields were lower than that from the auxotroph of the *Coryneform* bacteria. There was also the production of fairly large amounts of L-lysine by threonine and leucine auxotrophs which was inferior to homoserine auxotroph. Others were also inferior such as that of the auxotroph of *Corynebacterium glutamicum* and other bacteria (Ikeda, 2003).

For the production of L-lysine, this bacterium is extensively used based on the commercial fermentation process (Table 11.2). There is a need for at least one amino acid or threonine or isoleucine and methionine in the addition of the homoserine in order to fulfil the need for double auxotrophs and growth is much stabilized. There is

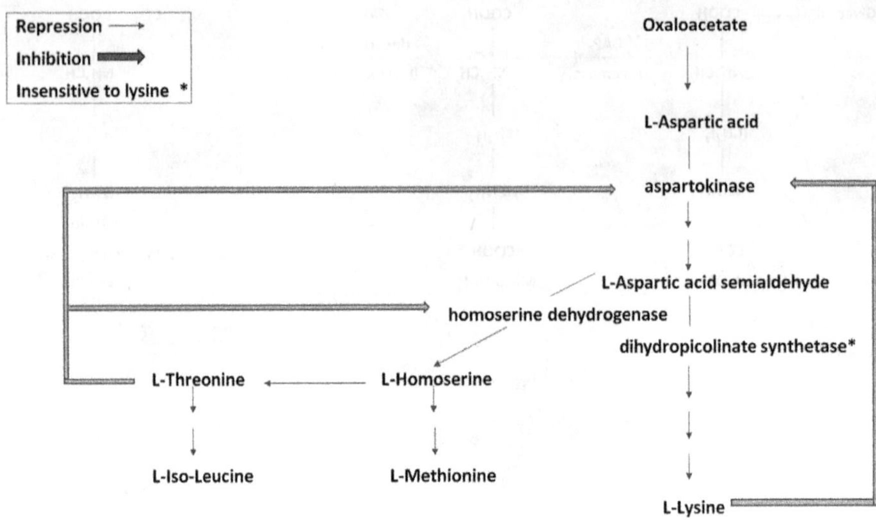

FIGURE 11.5 Direct fermentation pathway for L-lysine production.

a possibility of non-reversal of the culture to a wild type but also of lysine production in higher yields.

Mutagenesis: Mutagenesis of the microorganisms included exposing them to UV light rays or x-rays and also under chemical mutagen treatment. Also, genes can be synthesized and isolated from L-glutamic acid which would produce *Coryneform* bacteria. Then an *in vitro* mutate treatment can be given which would help in the substitution of wild type genes on the chromosomes of the bacteria. Various strains are also used for auxotrophic or resistive properties and mainly to enhance them. Ten enhanced monologues are then used as biocatalysts for the excretion of amino acid. This plays a direct role in the extraction of L-lysine from fermentation. A mutant having both characteristics, which are resistance and ability to produce L-lysine, helps in the extraction (Bongaerts et al., 2001).

Mutagenesis by the help of protoplast fusion: This is the process in which the culturing process takes place where a protoplast fusion mutant strain is cultured and derived by the fusion of protoplast between the strains of *Brevibacterium lactofermentum* and *Corynebacterium glutamicum.* Also 'Recombinant DNA technology' is used these days to enhance the production of L-lysine (Enei et al., 1989).

11.3.2 Medium Preparation for Fermentation

Medium composition plays an important role in the fermentation process along with the importance of the physical parameters such as agitation, aeration rate, pH, temperature, foaming, and dissolved CO_2 rate. All the requirements to prepare a culture medium are necessary to prepare a good medium. The composition is as

TABLE 11.2
The Difference in the Fermentation Process for L-Lysine Production

The Fermentation Process of L- Lysine (Direct fermentation)	Fermentation Process of L-Lysine (Indirect fermentation)
1) Preparation of inoculum A mutant strain of *C. glutamic* is used from the stock solution to produce the inoculum. 2) Preparation of media The seed culture to be prepared includes: glucose: 20 g, peptone: 10 g, meat extract: 5 g, sodium chloride: 2.5 g in 1 L of tap water Seed culture procured is reinoculated for 2nd seed culture in media containing: Molasses: 200 g soy protein hydrolysate 18 g in 1 L tap water. After obtaining first and second media production media is used which includes: Corn-steep liquor, ammonium sulphate, glycerol, $CaCO_3$. 3) Fermentation process The process if followed with certain parameters such as – 28 °C and 60 h. Also, it should be noted there should be biotin concentration more than 30 mg/L in the medium for avoiding accumulation of L-glutamic acid. 4) Harvest and Recovery The same process is employed for harvesting as indirect fermentation.	1) Preparation of inoculum Pure inoculum of both *E. coli* and *A. aerogenes* should be produced from the stock solution. These organisms should lack the ability to produce diaminopimelic acid (DAPA) decarboxylase and lysine decarboxylase respectively in the medium. 2) Preparation of medium The fermentation media should have glycerol, corn-steep liquor as carbon sources, and ammonium hydrogen phosphate, as a nitrogen source. Also, calcium carbonate is necessary. The pH is mostly maintained at 8. 3) Fermentation Process First, the inoculum is added in the fermenter with the medium, the process is carried out for 3 d at 28–30 °C temperature. With a sequence of enzymatic steps during the first stage of fermentation, glycerol is converted into L, L-diaminopimelic acid, which is partially converted into D, L-isomer, and mesodiaminopimelic acid by the action of diaminopimelic acid racemase enzyme. As *E. coli* lacks diaminopimelic acid decarboxylase enzyme, the metabolites formed cannot be converted into L-lysine. In the second half of the process, 1–2 d old culture of A. aerogenes is mixed to the fermentation broth. The microorganism is allowed to grow for one day at 24 °C. After sufficient growth of the microbe, toluene is added to the fermentation broth which causes lysis of cells. Due to this enzyme diaminopimelic acid decarboxylase is liberated in the medium. Thus, L-diaminopimelic acid is converted into meso-diaminopimelic acid by the action of diaminopimelic acid racemase enzyme. Then it is finally converted into L-lysine. 4) Harvest and recovery The L-lysine is obtained in pure form after acidification by anyone of the following separating processes: Precipitation at the isoelectric point, ion exchange chromatography, electrophoresis, extraction with organic solvents.

follows: n-Alkane (C_{14}-C_{18}), ammonium sulphate, $CaCO_3$, K_2HPO_4, KH_2PO_4, $MgSO_4$. $7H_2O$, NaCl, $FeSO_4$. $7H_2O$, $ZnSO_4$. $7H_2O$, $MnSO_4$. $4H_2O$, tap water, pH, steaming (Bender, 2012).

Inoculum production: Pure inoculum of both *E. coli* and *A. aerogenes* is produced from the high-yielding stock culture. The microorganisms which are used should lack the ability to produce diaminopimelic acid (DAPA) decarboxylase and lysine decarboxylase enzymes, respectively, so that DAPA and L-lysine are produced which cannot be further metabolized by the respective organisms. Both organisms should also possess strong, L-diaminopimelic acid racemase activity in order to convert all residual L,L-diaminopimelic acid to meso-diaminopimelic acid. The composition of the medium which is employed for inoculum production is similar to the fermentation medium. The medium is prepared in tap water. The cells of the organisms are separated from the growth medium by centrifugation or sedimentation (Kinoshita, 1959).

Preparation of medium: Both the inoculum and fermentation media contain glycerol, corn-steep liquor as carbon sources, and ammonium hydrogen phosphate which acts as the nitrogen source. Calcium carbonate is also used in the production of the medium. The levels of all of the nutrients are kept low in the inoculum medium. Apart from supplying carbon sources, the corn-steep liquor also provides the L-lysine required for the initial growth of auxotroph of *E. coli*. The pH of the medium is maintained at neutral, at a slightly alkaline level which would be a pH of 8.0 (Kinoshita, 1959).

Fermentation Process: Sufficient quantities of the sterilized medium are fed into the fermenter. Pure and required quantities of the inoculum of *E. coli* which is around 4%v/v are added to the fermenter. The fermentation is carried out for 3 d at 28 to 30 °C range of temperature. The level of L-lysine quantity provided to *E. coli* is very important because the provision of low quantities which are less than optimal results in back mutation and greater quantities would result in the feedback control of lysine biosynthesis, both of which badly affect yield. Through a sequence of enzymatic steps during the first stage of fermentation glycerol is converted into L,L-diaminopimelic acid, which is partially converted into D,L-isomer, and meso-diaminopimelic acid by the action of diaminopimelic acid racemase enzyme. Thus the above-mentioned metabolites accumulate in the fermentation broth because of the auxotrophic *E. coli* which lacks the diaminopimelic acid decarboxylase enzyme. Hence it cannot be converted into L-lysine. The broth contains approximately 40% L,L-isomer, and 60% meso-isomer of diaminopimelic acid. During the second half of the fermentation, 1–2 d old culture of *A. aerogenes* is added to the fermentation broth formed at the end of the first fermentation process. The microorganism is allowed to grow for one day at 24 °C. After sufficient growth occurs toluene is added to the fermentation broth which causes lysis of cells of *A. aerogenes*, due to which the enzyme diaminopimelic acid decarboxylase is liberated into the fermentation broth. By this time most of the L,L-diaminopimelic acid is converted into meso-diaminopimelic acid by the action of diaminopimelic acid racemase enzyme (Ikeda, 2003). The meso-diaminopimelic acid is completely converted into L-lysine by the action of diaminopimelic acid decarboxylase. Some of the important features of lysine fermentation are depicted in Figure 11.6.

Medium composition	-C-source : Glucose, fructose, sucrose, soluble starch -N source: ammonium, nitrate salts, yeast extract, soyabean protein—hydrolysate.
Culture conditions	-pH: 7.0, neutralized by ammonia or urea -T: 28-30 ℃ -High oxygen concentration -Oxygen
Operation mode	-Stirred tank bioreactor -Batch, fed batch -Design with high oxygen transfer
Downstream processing	-Cell separation: centrifuges or ultrafiltration -Concentration: cation exchange resin and elution with alkali -Neutralization: hydrochloric acid -Crystallization: spray dried

FIGURE 11.6 Major steps involved in the fermentation process.

11.3.3 Harvesting and Product Recovery

Mutant strains of *Bacillus licheniformis* are also employed in the production of L-lysine. The strains of the mutants were obtained by the introduction of both analogue resistance and auxotrophy. The medium contains 10% cane molasses. A temperature of 40 °C is suitable for L-lysine production. The sporulation activity which reduces yield can be suppressed by the addition of certain antibiotics like tetracycline and chloramphenicol. These mutants yield approximately 30 mg of L-lysine per mL of carbon source used. L-lysine is also produced by the enzyme process. A racemase mixture of D and L-amino- caprolactum can be transformed by the L-α-aminocaprolactum hydrolase to lysine. Racemase enzyme converts D-α-aminocaprolactum to L-α-aminocaprolactum (Figure 5.5). The L-α amino-caprolactic hydrolase and racemase enzymes are obtained from the bacteria *Achromobacter obae* and yeast *Cryptococcus lauranti* (Enei et al., 1989). L-lysine is obtained in pure form after acidification by anyone of the following separating processes**:** i) Precipitation at the isoelectric point, ii) Ion exchange chromatography, iii) Electrophoresis, iv) Extraction with organic solvents (Nakayama et al., 1978).

Corynebacterium strains are mainly used for the production of L-lysine. This has a series of steps which include: i) Fermentation, ii) Cell separation by centrifugation or ultrafiltration, iii) Product separation and purification, iv) Evaporation, and v) Drying. There is an alternative application of the continuous process which serves as an application in the L-lysine production. There are many advantages that are seen during the development of the continuous fermentation processes in the production of citric acid and gluconic acid. Multiple steps were developed which led to a new biotechnological process; these steps include: i) The proper identification of a suitable biological system, ii) Characterization of a suitable biological system, iii) Bioreactor productivity is increased by systematic media and its optimization, iv) The adaptation of fermentation technology, v) Cell separation using downstream processing by the method of centrifugation or by ultrafiltration, evaporation, separation, or drying (Ikeda, 2003).

11.3.4 Regulation of the Fermentation of L-Lysine

The aims for increase in production of l-lysine include price reduction. Glucose is phosphorylated upon cellular absorption in which it gets converted to glucose-6-phosphate and there is a consumption of phosphoenolpyruvate. However, it is observed that sucrose gets converted to fructose and also glucose-6-phosphate under the process of phosphotransferase system and also the invertase reaction. The Embden–Meyerhof–Parnas process undergoes the catalysis of glucose which is also known as glycolysis and the pentose phosphate pathway. Glucose-6-phosphate isomerase and glucose-6- phosphate dehydrogenase compete for the substrate glucose-6-phosphateduring the glucose catabolism, which results in either fructose-6phosphate or 6-phosphogluconolactone, respectively. It is seen that for the pentose phosphate cycle the oxidative part is the reason where glucose-6-phosphate gets converted into ribulose-5-phosphate under the supply of reduction equivalents in the form of NADPH. There is also an interconversion seen between pentose, hexose, and triode phosphates while proceeding the pentose phosphate cycle. 5-phosphoribosyl-l-pyrophosphate is necessary through nucleoside biosynthesis and acts as a precursor in nucleoside biosynthesis for the production of aromatic amino acids. The NADPH functions like a reduction equivalent in the numerous anabolic biosynthesis. Its assumption is that 4 NADPH molecules are used in the biosynthesis of one lysine molecule. Therefore the carbon flux remains constant. Some enzymes are identified as anaplerotic enzymes, including pyruvate carboxykinase, pyruvate carboxylase and also phosphoenol pyruvate carboxylase, phosphoenol pyruvate carboxylase or also known as PEPC. These enzymes catalyse the reaction by the addition of one mole of CO_2 to the phosphoenol pyruvate and pyruvate. This fulfils the TCA and OAA. This also plays a role in supplying aspartic acid and the subsequent formation of lysine and others. With L-aspartate inflow into the lysine pathway, which is synthesized by oxaloacetate transamination by *C. glutamicum,* this will transform 2,6-dicarboxylate to diaminopimelate to l-lysine intermediate piperdine. There are two different routes to achieve this. There are two points where the carbon flux is regulated. At the first point feedback inhibition of aspartate kinase is observed by monitoring the levels of both L-threonine and L-lysine. At the second point, the level of dihydrodipicolinate synthase is controlled. L-lysine export has also been shown to be an important factor for l-lysine production. Fermentation of L-Lysine is also regulated by type of carbon-source, nitrogen-source, presence of metal ion concentration, dissolve oxygen concentration, temperature, and pH of the microbial medium (Bender, 2012).

Future developments: There is a need to apply a variety of mutant strains for the fermentation of L-lysine using different microorganisms which are obtained by different methods and this overall technology is the first in the list in the biotechnological production of bulk chemicals. This process is the best microbial process and from a futuristic point of view, there is a huge scope for improvements in concentration and yield of L-lysine (Bender, 2012).

11.3.5 Uses of L-Lysine

L-lysine is required for calcium absorption and collagen formation (to improve bone and muscle health). L-lysine is a precursor to the formation of L-carnitine. Carnitine is required for the conversion of fatty acids to energy. It is required for the synthesis of enzymes, antibodies, and some hormones. It is also needed for defence against chronic viral diseases. Deficiency of L-lysine can cause symptoms such as anaemia, hypothyroidism, weakened immune system, weak bones, and muscles etc. L-lysine is available as capsules, powders, tablets and other forms which are used as supplements by athletes and weightlifters.

11.4 SYNTHESIS OF METHIONINE

For the microbial production of methionine, three mutated bacterial strains – *Corynebacterium glutamicum, Escherichia coli,* and *Brevibacterium heali* – have been screened and utilized. The methionine biosynthesis pathway is easier in *C. glutamicum* and *B. heali* compared to *E. coli*. Methionine is an essential amino acid required in diet. Plants contain less protein as compared to meat. Therefore, vegetarians do not get a sufficient amount of methionine from their diet and they may suffer from methionine deficiency. Symptoms of methionine deficiency include depression, free radical formation, Parkinson's, liver deterioration, schizophrenia, and toxaemia. L-methionine is produced by indirect fermentation. Steps to produce L-methionine are i) aspartate conversion into aspartyl phosphate by enzyme aspartate kinase and ATP, ii) aspartyl phosphate conversion into aspartate semi-aldehyde by enzyme aspartyl semi-aldehyde dehydrogenase, and NADPH, iii). aspartate semi-aldehyde converts into homoserine by enzyme homoserine dehydrogenase and NADPH, iv) homoserine conversion into methionine by enzyme methionine dehydrogenase and NADPH (Ikeda, 2003).

Composition of media for *C. glutamicum*, *E. coli* and *B. heali*: 10% glucose and 5% maltose (as carbon source), inorganic salts, peptones (nitrogen source), and vitamin B_7 and B_1; these are all micronutrients required by the bacteria.

11.4.1 Uses of Methionine

Methionine (if in sufficient amounts) functions as an antioxidant and detoxifying agent. It also helps in the removal of heavy materials from the body and histamine from the brain. It also functions as a lipolytic agent (degrades excess fat).

11.5 SYNTHESIS OF TRYPTOPHAN

Genetically modified strains *of C. glutamicum* and *E. coli* are used in tryptophan production. Ten years ago, L-tryptophan production was done via chemical processes. However, due to an increased demand for L-tryptophan, synthesis is now done by enzymatic reactions or fermentation.

11.5.1 L-Tryptophan Fermentation

Microbes commonly utilized for the processing of L-trp are *E. coli, Corynebacterium glutamicum, Brevibacterium flavum, Bacillus subtilis,* yeast etc. The formation of L-trp is somewhat unique for varied microorganisms. In *Escherichia coli*, the synthesis of L-trp involves three parts: the core biochemical pathway, the common pathway for aromatic amino acids, and the branch pathway for L-trp. The key biochemical pathway applies to both erythritol4-phosphate (E4P) and phosphoenolpyruvate (PEP) in the glucose pentose (HMP) (DHHP) pathway to glycolysis (EMP). The typical path corresponds to the phase of shikimic acid (SHIK) to branched acid (CHA) from its start of DAHP. The remaining direction from CHA to the moiety of L-Trp is referred to as the branch pathway of L-Trp. Currently, most L-Trp metabolic research work concentrates on the typical pathway and the L-Trp branch pathway transformation (Bang et al., 1983).

11.5.2 Composition of Media for *E. Coli* and *C. Glutamicum*

For seed culture preparation of agar slants contains 0.5% NaCl, glucose (20 g/L), peptone (5 g/L), yeast extract (1 g/L) and all micronutrients. For fermentation, 30% molasses, 0.7% corn-steep liquor and micronutrients are required. Mutated strains of *C. glutamicum* and *E. coli* can produce a high amount of L-tryptophan (Bang et al., 1983).

11.5.3 Uses of L-Tryptophan

Sufficient amounts of L-tryptophan are required to prevent depression and insomnia, prevent sleep apnoea, anxiety, and facial pain, prevent menstrual dysphoric disorder, prevent hyperactivity disorder and Tourette's syndrome (Bang et al., 1983). Table 11.3 summarizes the microbial production of L-lysine, L-methionine, and L-tryptophan.

TABLE 11.3
Different L-Lysine, L-Methionine, and L-Tryptophan Production

	L-Lysine	**L-Methionine**	**L-Tryptophan**
Microbes Used	***C. glutamicum***	***C. glutamicum, E. coli* and *B. heali***	***C. glutamicum* and *E. coli***
Method of synthesis:	Chemical, indirect, and direct fermentation.	Chemical, indirect, and direct fermentation.	Chemical, indirect, and direct fermentation.
Carbon source used:	Starch and glucose.	Molasses.	Starch and molasses.
Nitrogen source:	Ammonium and urea.	Ammonium and creatinine.	Ammonium and urea.

11.6 MICROBIAL PRODUCTION OF VITAMIN B12

Vitamin B-12 is the most structurally complex and largest of all B vitamins. The structure consists of a corrin ring with a central cobalt atom with four pyrrole units. Cobalt is linked to the nitrogen of the four pyrroles. The empirical formula is of cyanocobalamin ($C_{63}H_{90}N_{14}O_{14}PCO$)

Types of vitamin B12: Cyanocobalamin: Deep pink, associated with octahedral cobalt(II) complexes, use to treat pernicious anaemia (Hugenholtz and Smid, 2002).

Hydroxocobalamin denoted by B-12a: Used as an antidote for cyanide poisoning because of its high affinity for cyanide ions. Supplied typically in water solution for injection (Hugenholtz and Smid, 2002).

Methylcobalamin: Appears as dark red crystals, present in human plasma and cell cytoplasm.

In methylcobalamin, cyanide is replaced by a methyl group. Used in the treatment of peripheral neuropathy (Hugenholtz and Smid, 2002).

Chemical Properties of B12: Vitamin B-12 is very stable at high temperatures and at pH 4.5 to 5.0. However, vitamin activity is lost in a strongly acidic and alkaline environment. It can be rapidly degraded in light. It is soluble in water, ethanol, and methanol (Hugenholtz and Smid, 2002).

Sources of B12: Vitamin B-12 is not found in vegetables, fruits, and other foods of plant origin. Meat, poultry, fish, shellfish, eggs, milk and milk products, and liver are good sources from animals (Hugenholtz and Smid, 2002).

11.6.1 Sources and Dietary Recommendations

The National Institute of Health (NIH) advises 2–4 micrograms (μg) of vitamin B-12 per day for teenagers and adults over 14 years of age, 2–6 μg among pregnant women and 2–8 μg for women who are lactating. Vitamin B-12 is usually found in meats so people who do not consume meat, such as vegans, are recommended to receive vitamin B-12 in the form of supplements. (Hugenholtz and Smid, 2002).

Deficiency Symptoms: Deficiency of vitamin B-12 can lead to permanent and serious damage, especially to the nervous system and brain. Slightly lower than normal vitamin B-12 levels might cause symptoms of deficiency such as anxiety, confusion, memory problems, and fatigue. Other symptoms include constipation, sexual problems/infertility, Loss of appetite/weight loss, muscle pain, hearing and vision problems. Deficiency can lead to Alzheimer's/dementia, learning disability in children, autoimmune disorders, cardiovascular disorders, cancer, megaloblastic anaemia, and neurological disorder such as glossitis (Hugenholtz and Smid, 2002).

11.6.2 Biosynthetic Pathway of B-12 and Microbial Fermentation

Vitamin B-12 is synthesized naturally by microbes. The B-12 biosynthetic pathway is comparable to chlorophyll and haemoglobin biosynthesis. A variety of microorganisms are used for the production of B-12. These include: *Acetobacterium, Aerobacter, Flavobacterium, Agrobacterium, Serratia, Alcaligenes, Azotobacter, Bacillus, Clostridium, Proteus, Pseudomonas, Corynebacterium, Propionibacterium, Lactobacillus, Micromonospora, Nocardia, Protaminobacter, Rhizobium, Salmonella,*

Streptomyces, Streptococcus, Xanthomonas, and Mycobacterium. The most commonly used microorganisms are *Bacillus megaterium, Propionibacterium freudenreichii, Streptomyces olivaceus*, and *Pseudomonas denitrificans*. Mutant strains can produce 50,000 times more B-12 than wild types. Mutations are carried out by UV rays or X-rays. Chemical agents like N-methyl-N'-nitrosoguanidine, nitrosoethylurea, ethyleneimine, dimethylsulfate, and mustard gas can also be used (Vandamme, 1992). A technique of protoplast fusion between *Rhodopseudomonas spheroides* and *Protaminobacter rubber* resulted in a hybrid strain called *Rhodopseudomonas protamicus.* This new strain provided vitamin B-12 up to 135 mg/L.

Microbial production of B-12 can be carried out via extraction as a by-product fermentation of streptomycin, aureomycin, chloramphenicol, or neomycin. Yield is very low in this process. Extraction is as end-product fermentation. This is used commercially.

Overview of steps involved in microbial production of B-12: i) Media formulation and sterilization ii) Preparation of starter culture, iii) Fermentation, iv) Recovery of fermented product.

Production of Vitamin B-12 uses *Propionibacterium sp: Propionibacterium freudenreichii* and *Propionibacterium shermanii*, and their mutant strains are commonly used for vitamin B-12 production.

Media preparation: Corn steep glucose, inverted molasses, soybean meal, beet molasses, alcohols (methanol, ethanol, isopropanol), and hydrocarbons (alkanes, decane, hexadecane). Ammonium salts such as ammonium phosphate and ammonium hydroxide. Buffering/neutralizing agents and salts (ferrous, magnesium, cobalt salts). The carbon and nitrogen sources can be replaced by lactoserum or skimmed milk. Media is sterilized by autoclaving (Vandamme, 1992).

Preparation of starter culture: The starter culture is prepared by using any of the following microbes: *Bacillus megaterium, Streptomyces olivaceous, Propionibacterium shermanii, Pseudomonas denitrificans, Rhodopseudomonas palustris.*

Fermentation: The sterilized medium is poured into a stirred tank fermenter with a 1% starter inoculum. At first, the anaerobic state is retained in order to enable *Pseudomonas shermanii* to develop 5,6-dimethyl benzimidazole cobalamide (DBC). Sterile air (with continuous stirring for aeration) is pumped into the fermenter at the end of anaerobic fermentation. The aerobic process is continued for 4 d. Along with DBC, a small amount of pseudo vitamin B-12 (5′-deoxyadenosylcobalamin) is also produced during fermentation. These compounds form the immediate precursors of cyanocobalamine (Vandamme, 1992).

Recovery of cyanocobalamine: Vitamin B-12 exists in the form of DBC and pseudo-vitamin inside the microbial cells. The culture broth including the cells is extracted and centrifuged at maximum velocity to produce a condensed cell mass. The cell mass acquired is processed with diluted acid followed by heat shock and is maintained at pH 6.5–8.5 for 30 min at 80–120 °C. This heat treatment releases DBC-precursors pseudo vitamin B-12 complex. Further, it is treated with a cyanide solution to separate the DBC and pseudo vitamin B-12. This releases free cyanocobalamine (Vit B-12) into the solution (Vandamme, 1992).

Downstream processing: In solution, cyanocobalamine is isolated by a column adsorption chromatography using IRC-50 resin. Using a phenolic solvent, the adsorbed cyanocobalamine is eluted out. The solvent portion is evaporated through exposure to atmospheric air which leaves behind cyanocobalamin crystals (Hugenholtz and Smid, 2002).

Production of B-12 using Streptomyces olivaceus:

Production of B-12 using *Streptomyces olivaceus* can yield about 3.3 mg/ L of vitamin B-12 (Vandamme, 1992).

Preparation of starter inoculum: Slant cultures of pure *S. olivaceus* are inoculated in 100–250 mL of inoculum medium. Mechanical shakers are used to aerate the system. This culture in flask is then used to inoculate larger tanks of inoculum. Bennett's agar containing yeast extract, beef extract, casein, glucose and agar dissolved in demineralized water at pH 7.3 is used as starter media (Vandamme, 1992).

The production medium consists of carbohydrate (dextrose), protein and cobalt, and other salts. For maximum yield of cobalamin, cobalt is added. For conversion of other cobalamins to vitamin B-12, cyanide is used (Hugenholtz and Smid, 2002).

Batch or continuous sterilization can be used. For batch sterilization, the medium is heated at 250 °F for an hour. Continuous media is sterilized at 330°F for 13 min by mixing with steam. A temperature of 80 °F in the production tank is satisfactory during fermentation. At the initial phases, pH declines due to the accelerated sugar intake, then increases to around 2 to 4 primarily due to mycelium lysis. The pH with H_2SO_4 and reduction agent Na_2SO_4 is held at 5. The optimum aeration rate is 0.5 volume air/volume medium/min. Agents such as corn oil, soya bean oil, silicones, and lard oil can be used (Hugenholtz and Smid, 2002).

During fermentation, cobalamin is mostly associated with the mycelium. The boiling of the fermentation broth with pH of 5 recovers cobalamin from mycelium. Cobalamin-containing broth is then processed to produce B-12 in crystalline form. Mycelium is removed by filtering the broth and its filter treated with cyanide to aid the conversion of cobalamin to cyanocobalamin. Adsorption of cyanocobalamin from the solution is done by passing it through adsorbing agents packed in a column. Elution of adsorbed cyanocobalamin from the adsorbent is done by an aqueous solution of organic bases or Na-Cyanide and Na-thiocyanate solutions Extraction is carried out by countercurrent distribution between cresol, amyl-phenol, or benzyl alcohol and water or a single extraction into an organic solvent (e.g. phenol). Chromatography using alumina and a final crystallization step is required to obtain crystals of cyanocobalamin (Vandamme, 1992).

Production using *Pseudomonas denitrificans* is a completely aerobic process. It is necessary to add cobalt and 5, 6-dimethyl benz imidazole to the medium. The yield of vitamin B-12 is usually increased by adding betaine to the medium (Vandamme, 1992).

Pseudomonas denitrificans stock culture is lyophilized with skimmed milk. Test tubes containing media of agar, beet molasses, brewer's yeast, manganese, and magnesium salts are incubated for 4 days at 28 °C. Maintenance media with agar is used as seed culture, and incubated at 28°C for 3 days on a rotary shaker. A 5 L fermenter is inoculated with 0.15 L of seed culture and 3.3 L of sterilized media (composed of beet molasses, 5,6-dimethylbenzimidazole, yeast, magnesium, manganese, and ammonium salts, adjusted to pH 7.4) incubated at 29 °C with constant agitation and aeration (Hugenholtz and Smid, 2002).

11.7 L-ASCORBIC ACID PRODUCTION

L-ascorbic acid (Asc), publicly recognized as vitamin C, seems to be an important water-soluble supplement for humans and a few other mammals. Its discovery was related to the disease scurvy. Scurvy, as it is known, is the deficiency of vitamin C most commonly seen in sailors and navy personnel. The sailors used to travel on the sea for months and could not get enough vitamin C via their diet, and hence used to suffer from this disease. Symptoms of this disease seen were bleeding gums, bleeding of mucous membranes, anaemia, and eventually death. Medical practitioners found that a lack of antiscorbutic factor causes scurvy. Ascorbic acid was first isolated from adrenal glands in the year 1928 by AlbertSzent-Gyørgyi, who called it hexuronic acid. 1933 was the first time the chemical structure for Asc was deduced, by Norman Haworth (Figure 11.7). He was awarded a Nobel Prize in 1937 for his work on Asc production (Hashimoto, 2017).

Once Asc was discovered, scientists looked for a method to obtain its pure form. This lead to the development of industrial production process of Asc in the 1930s. Tadeus Reichstein, a Polish chemist, synthesized Asc in bulk and was the first to artificially produce Vitamin C. The Reichstein process is a combination of chemical and bacterial fermentation sequences which is still used nowadays to produce Vitamin C (Hashimoto, 2017). This process patent was bought by a pharmaceutical company, Hoffman-La Roche, which was the first company to manufacture and sell vitamin C – under the brand name of Redoxon – in 1934. Subsequently, two-step microbial fermentation was carried out in China between the 1960s and 1970s.This process was more cost-beneficial as compared to the Reichstein process and produced more than 80% vitamin C (Hashimoto, 2017).

Natural sources of Asc: A large number of animals and plants can synthesize Asc. In plant sources, Asc production depends upon plant variety, soil conditions, the climate where it is grown, storage conditions, and preparation method. Fruits and vegetables are the richest sources of Asc. e.g. Kakadu plum and camucamu fruit contain the highest concentration of Vitamin C. In animal sources, Asc is mostly present in the liver and least in the muscles (Hashimoto, 2017).

Citrus fruits such as orange, grapefruit, lime and lemon are an abundant source of Asc. Many non-citrus fruits such as strawberries, pineapple, kiwi fruit, blueberries,

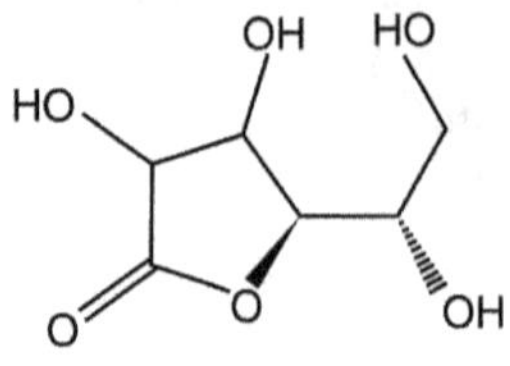

FIGURE 11.7 Vitamin C (ascorbic acid) deduced by Norman Haworth.

watermelon, and apple are good sources of Asc. Vitamin C is also found in many vegetables like broccoli, kale, spinach and carrot.

Physiological effects of vitamin C in humans: Ascorbic acid has many physiological roles in humans; these include: i) synthesis of collagen, carnitine and neurotransmitters, ii) synthesis, and catabolism of tyrosine, iii) metabolism of microsome (Hashimoto, 2017).

During metabolism: It functions as a reduction agent, giving electrons and avoiding oxidation in order to hold iron and copper atoms in reduced form. It also acts as a scavenger of several types of reactive oxygen species (Hashimoto, 2017).

Side effects of ascorbic acid: Large doses may cause indigestion and diarrhoea when taken on an empty stomach, and overdoses can also cause iron overload disorders and kidney stones (Hashimoto, 2017).

11.7.1 Physiochemical Properties of Ascorbic Acid

At room temperature, it is a white crystal, odourless with a melting point of 190–192 ^{o}C, water soluble while slightly soluble in ethyl alcohol and glycerine but insoluble in chloroform and ethyl ether, acidic and can easily be oxidized by O_2, Cu^{2+} and Fe^{3+}.

Properties of Asc: Chemical name: i) 2-oxo-L-threo-hexono-1,4-lactone-2,3-enediol; ii) Molecular formula: $C_6H_8O_6$; iii) Molecular weight: 176.13g.

Methods of assay for detecting Asc: Titrimetric, fluorimetric chemiluminescent gas chromatographic(GC), high-performance liquid chromatography(HPLC), spectrophotometric methods(Hashimoto, 2017).

11.7.2 Biosynthesis of Ascorbic Acid in Mammals and Plants

Asc is produced in both mammals and plants. The glucose acts as precursor for the biosynthesis of Asc (Figure 11.8).

11.7.3 Industrial Fermentation of Ascorbic Acid

There are many pathways for the biosynthesis of L-ascorbic acid such as i) L-sorbose pathway, ii) D-sorbitol pathway, iii) 2-keto-D-gluconicacidpathway, iv) 2,5-diketo-D-gluconic acid pathway, v) D-gluconic pathway. However, the following two pathways are industrially applied: The Reichstein process and the two-step fermentation process (Ikeda, 2003). Strains used for the fermentation of ascorbic acid are *Saccharomyces cerevisiae and Zygosaccharomyces baili.*

11.7.3.1 Reichstein Process

Steps involved in the Reichstein process:

Bioconversion of D-sorbitol to L-sorbose: This is the only step that involves the use of microbes. Acetic acid bacteria are used in this step. They are known to partially oxidize sugars and sugar alcohols e.g. *Gluconobacter* is an acetic acid bacteria. *Gluconobacter oxydans* converts D-sorbitol to L-sorbose with a 100% conversion rate (Ikeda, 2003).

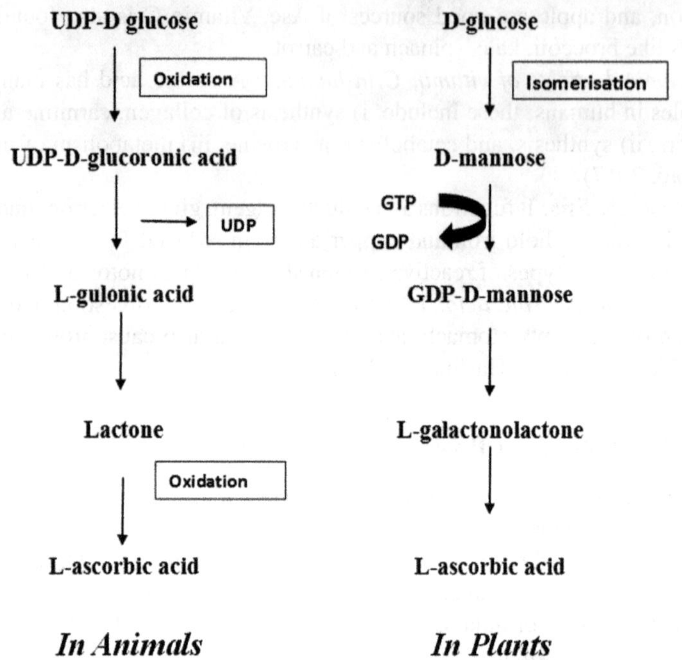

FIGURE 11.8 Pathway for the biosynthesis of L-ascorbic acid in animals and plants.

Enzyme produced: G. oxydans produces enzyme sorbitol dehydrogenase (SLDH). D-sorbitol, D-mannitol, glycerol can be oxidized by this enzyme.

Conversion of L-sorbose to 2-KLG: C2, C3, C4 and C6 of L-sorbose are covered by acetone against oxidation; diacetone L-sorbose would then oxidize the carboxyl group via potassium permanganate formed by diacetone-2-KLG. 2-KLG is produced through hydrolysis and has been lactonized to ascorbic acid.

11.7.3.2 Two-Step Fermentation Process

This process is the improved version of the Reichstein process.

The first step involves conversion of D-sorbitol to L-sorbose:

Microorganisms involved: Many strains convert D-sorbitol to L-sorbose e.g. *G. suboxydans, A. suboxydans* and *G. oxydans. G.oxydans* is a Gram-negative bacterium belonging to the family *Acetobacteraceae*. There are many membrane-bound dehydrogenases present on its surface which lead to incomplete oxidation of sugars, alcohols, and acids. These cause build-up of oxidative products making this strain important for industrial use (Kinoshita, 1959).

Dehydrogenase: G. oxydans contain two membrane-bound SLDHs, pyrroloquinolinequinone-dependent D-sorbitol dehydrogenase (PQQ-SLDH), flavin adenine dinucleotide-dependent D-sorbitol dehydrogenase (FAD-SLDH) and NADP-dependent D-sorbitol dehydrogenase (ADP-SLDH). Among these PQQ-SLDH is now generally believed to convert D-sorbitol to L-sorbose.

Second step fermentation process: L-sorbose to 2-keto-L-Gulonic Acid

Producing microorganisms: Two strains –the companion strain and the conversion strain – are involved in this step.

Companion strain: These are activators such as *P. striata, B. megatarium, B. thurengiensis* that increase the development of the conversion strain and significantly boost the 2-KLG yield. They form spores during cultivation. *B. megaterium* and *B. cereus* seem to be the two major companion strains used throughout the industrial fermentation of ascorbic acid (Kinoshita, 1959).

Conversion strain: K. vulgare is known as the 2-KLG producing strain as it contains the whole enzymes for conversion of L-sorbase to 2-KLG.

Sorbose/ Sorboson dehydrogenase: This enzyme catalysis the conversion of L-sorboson to 2-KLG. The molecular weight of this enzyme is 135 kDa. Optimum enzyme activity occurs at 7.0–9.0 pH.

The companion strain secretes certain metabolites during the fermentation process. Metabolites could be amino acids, proteins, or other substances. In fermentation, extracellular metabolites and cytosols produced as a result of companion strain lyses have been shown to facilitate the propagation of *K. vulgare* (Ikeda, 2003).

11.7.3.3 Applications of Ascorbic Acid

It is utilized in the dairy, beverage, feed and pharmaceutical sectors. A few applications of Asc are as follows (Table 11.4).

11.8 CONCLUSIONS

Essential amino acids like L-lysine, L-methionine, and L-tryptophan are required for good health. Large quantities of essential amino acids can be produced through fermentation techniques. Production of amino acids by fermentation is cheaper and has fewer side effects as compared to chemical production. Microbes such as

TABLE 11.4
Industrial Application of Ascorbic Acid

Fields	Usages
Pharmaceutical industry	Treating scurvy, chronic and viral hepatitis, virus flu, burn of cornea and conjunctiva.
Food industry	Inhibits oxidation of fruits, change of flavour, prevents oxidation of lipids.
Feed industry	Enhances immune function, accelerate growth, for treatment of trauma or burn.
Cosmetic Industry	Promotes collagen formation, enhances skin elasticity.

Corynebacterium glutamicum and *Escherichia coli* can be genetically modified to produce high amounts of amino acid. Therefore, amino acid production by fermentation has a high potential to satisfy the increased demand for amino acids. For ascorbic acid production, the two-step fermentation process is environmentally friendly, cost-effective, and well-established. Batch fermentation is used for the industrial production of Asc. In the two-step fermentation process, recombinant strains are prepared by inserting the gene of interest, which enhances the activity of the enzyme and also stimulates the growth of the strain. Genetic engineering can also be employed to combine the traits of *K. vulgare* and the companion strain which can produce 2-KLG directly from L-sorbose or D-glucose. According to the economic and environmental benefits and even the advancement of modern methods in genetic modification, fermentation is by far the most commonly applied method on an industrial scale. Thus multiple experiments involving mutagenesis and metabolic engineering methods have been implemented with the goal of enhancing amino acid and strain-producing vitamins and thus increasing efficiency and extending the range of fermentation materials and feedstock. It is important to build microbial strains with better production of amino acids and vitamins, and lower development of by-products for reduction in purification costs. Separating technologies, such as nanofiltration membranes, can be integrated within traditional fermenters that combine production and purification in the same unit according to the principles of system intensification. These methods require a range of inexpensive substrates and waste to be processed into high-quality proteins, rich in important amino acids and vitamins. Such characteristics allow the ideal candidates for manufacturing human or animal quality protein products reflecting a revolutionary development for the future.

REFERENCES

Bang W-G, Lang S, Sahm H, Wagner F, Production L-tryptophan by Escherichia coli cells. Biotechnol. Bioeng. 25, 999–1011,1983.

Bender DA, Amino Acid Metabolism. John Wiley, 2012.

Bongaerts J, Krämer M, Müller U, RaevenL, Wubbolts M, Metabolic Engineering for Microbial Production of Aromatic Amino Acids and Derived Compounds. Metab. Eng. 3, 289–300, 2001.

Enei H, Yokozeki K, Akashi K, Recent Progress in Microbial Production of Amino Acids. CRC Press, 1989.

Hashimoto S, Discovery and History of Amino Acid Fermentation, in: Yokota, A., Ikeda, M. (Eds.), Amino Acid Fermentation, Advances in Biochemical Engineering/Biotechnology. Springer Japan, Tokyo, pp. 15–34, 2017.

Hugenholtz J, Smid EJ, Nutraceutical production with food-grade microorganisms. Curr. Opin. Biotechnol. 13, 497–507, 2002.

Ikeda M, Amino Acid Production Processes, In: Faurie R, Thommel J, Bathe B, Debabov VG, Huebner S, Ikeda M, Kimura E, MarxA, Möckel B., Mueller U, Pfefferle W (Eds.), Microbial Production of L-Amino Acids, Advances in Biochemical Engineering/ Biotechnology. Springer, Berlin, Heidelberg, pp. 1–35, 2003.

Kinoshita S, The Production of Amino Acids by Fermentation Processes, in: Umbreit WW (Ed.), Advances in Applied Microbiology. Academic Press, pp. 201–214, 1959.

Kumagai H, Microbial Production of Amino Acids in Japan, in: FiechterA (Ed.), History of Modern Biotechnology I, Advances in Biochemical Engineering/Biotechnology. Springer, Berlin, Heidelberg, pp. 71–85, 2000.
Nakayama K, Araki K, Kase H, Microbial Production of Essential Amino Acids with Corynebacterium Glutamicum Mutants, in: Friedman M (Ed.), Nutritional Improvement of Food and Feed Proteins, Advances in Experimental Medicine and Biology. Springer, Boston, MA, pp. 649–661, 1978.
VandammeEJ, Production of vitamins, coenzymes and related biochemicals by biotechnological processes. J. Chem. Technol. Biotechnol. 53, 313–327, 1992.

12 Penicillin, Cephalosporin, and Streptomycin Production

An antibiotic is a chemical molecule that is usually produced by microorganisms during stress conditions, which can inhibit cell growth or even destroy bacteria or other microorganisms. Antibiotics are basically 'secondary metabolites'. The industrial manufacturing of antibiotics began with penicillin and streptomycin during the Second World War (Quinn, 2012). Metabolites that are not needed for cell growth and development and are the final products of primary metabolites are referred to as secondary metabolites. At the stationary phase of microbial growth secondary metabolites emerge.

Not all antimicrobial substances are antibiotics. There are some anti-microbial substances that are not included in antibiotics. These substances are produced by some higher plants and animals (Nagórska et al., 2007). For example, bacteriocins, which are also produced by microorganisms, are different from typical antibiotics. The bacteriocins are larger macromolecular proteins that can either kill or inhibit the growth of closely related bacteria. Unlike bacteriocins, antibiotics are more diverse in their chemistry and affect even distantly related organisms (Riley, 2011).

12.1 ANTIBIOTICS AND THEIR CLASSIFICATIONS

Antibiotics are usually produced via fermentation. Nowadays, several antibiotics are developed through staged fermentation wherein the high-yield microorganism strains are cultivated under optimum conditions. In some cases, they are produced using semi-synthetic processes where a precursor is produced using fermentation which is later modified by integrating chemical side chains. Antibiotics have been developed for around 80 years, after the development of β-lactams and sulfonamides. Ampicillins are produced via semi-synthetic production (post-production modification of natural antibiotics). Synthetic production of antibiotics are made in the laboratory; e.g. quinoline. Moreover, there are many commercially available antibiotics used in medicine for their anti-microbial action (Table 12.1). There are various mechanisms by which the antibiotics acts on various microbes. These mechanisms are shown in Figure 12.1. Antibacterial drug-target interactions are well studied and typically categorized into three classes: DNA replication inhibition, protein synthesis inhibition, and cell-wall inhibition. The pathways of degradation of bactericidal antibiotics are generally due

TABLE 12.1
The Mode of Action and the Generic Name of the Antibiotics

Mode of action of antibiotics	Generic name/type
Cell wall synthesis inhibitors Beta-lactamase inhibitors	Penicillins, Cephalosporins, Vancomycin Polymyxin, Bacitracin
Protein synthesis inhibitors Inhibit 30s subunit Inhibit 50s subunit	Aminoglycosides(gentamycin), Tetracyclines, Macrolides Chloramphenicol, Clindamycin, Streptogramins
DNA synthesis inhibitors	Fluoroquinolones (ciprofloxacin), Metronidazole
RNA synthesis inhibitors	Rifampin
Mycolic acid synthesis inhibitors	Isoniazid
Folic acid synthesis inhibitors	Sulfonamides, Trimethoprim

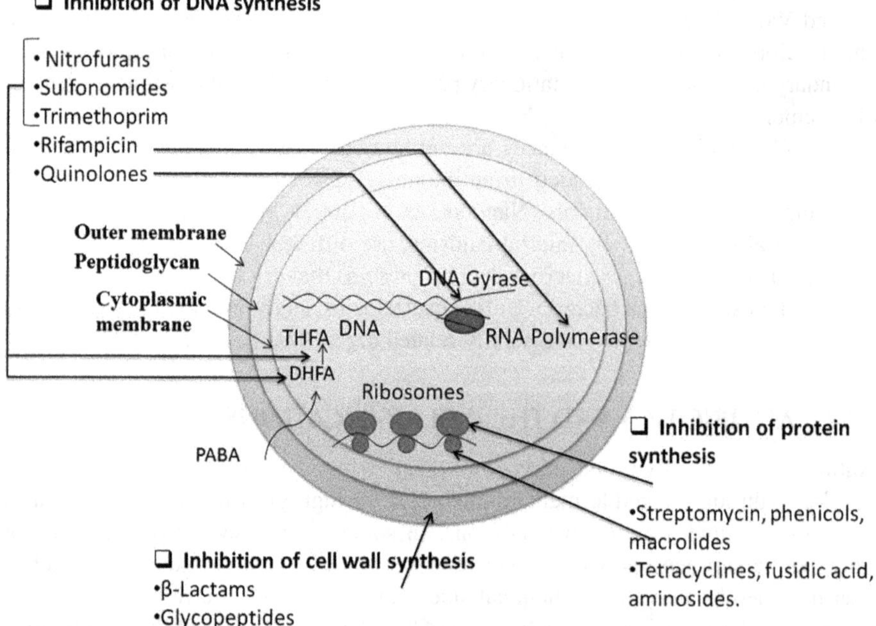

FIGURE 12.1 Mode of action of different antibiotics on bacterial cell components.

to class-specific interactions between the drug targets (Walsh, 2000). Bacteriostatic drugs target the 30S and 50S subunit inhibiting the microbial ribosomal function and thereby inhibiting protein synthesis. The main purpose of this chapter is to discuss the industrial production of various antibiotics. Furthermore, the terminology of antibiotics is particularly ambiguous because it depends on the supplier. The same antibiotic may have as many as 13 separate brand names. Therefore antibiotics are identified by a total of three names: the chemical name, which tends to be lengthy and is seldom used even in science or medical literature; the second is the generic or general name, usually a shorter version of the chemical name or that given by the discoverer; the third is the tradename or trademark that the producer provides in order to differentiate it from the drug by varying the name (Hopwood et al., 1980).

12.1.1 Classification of Antibiotics

Antibiotics are classified in several ways- i) based on mechanism of action; ii) based on the spectrum of activity, and iii) based on the mode of action.

Based on spectrum activity, antibiotics are classified into two classes: the first class consisting of narrow-spectrum antibiotics are more specific and only active against certain groups or strains of bacteria (Kumavath, 2017). Some common narrow-spectrum antibiotics are penicillin G, cloxacillin, vancomycin, bacitracin, and fluxacillin. On the contrary, broad-spectrum antibiotics inhibit a wider range of bacteria e.g. amoxicillin; tetracycline; cephalosporin; chloramphenicol; erythromycin. Narrow-spectrum antibiotics are usually preferred since the effect on other, non-disease causing bacteria is limited (Hopwood et al., 1980).

Furthermore, antibiotics can be classified based on the mode of action: the bacteriostatic antibiotics restrict bacterial growth mostly by altering production of bacterial protein, DNA replication, or even other facets of cellular metabolism in the bacteria. Tetracyclines, sulfonamides, spectinomycin, trimethoprim, chloramphenicol, macrolides and lincosamides are included in this group. Bactericidal antibiotics like cephalosporin; penicillin; erythromycin; aminoglycosides; cotrimoxazole etc. can kill the bacterial population (Williams and Vickers, 1986). There is an ever-increasing demand of antibiotics worldwide (Table 12.2).

12.2 ANTIBIOTIC WITH BETA-LACTAM STRUCTURE

The composition of beta-lactam is not very typical and apart from the antibiotic classes to be listed, it is present only in certain alkaloids and some anti-metabolite toxins, including pachystermines from *Pachystradra terminals* (Chang et al.,

TABLE 12.2
Worldwide Sales of Various Antibiotics (Barber et al., 2004)

Class	Representative drugs	Worldwide sales ($ millions)
Cephalosporin	Cefaclor, cefuroxime	8,446
Penicillins	Amoxicillin, ampicillin	4,413
Fluoroquinolones	Ciprofloxacin, ofloxacin	3,309
Macrolides	Clarithromycin, erythromycin	2,927
Tetracyclines	Minocycline	744
Amino-glycosides	Gentamycin	729
Glycopeptides	Vancomycin	462

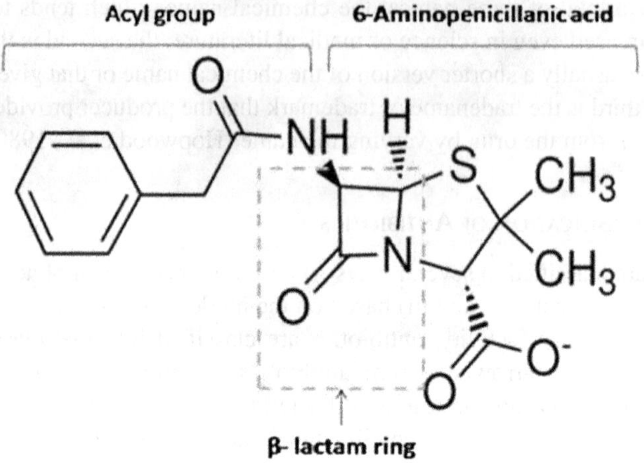

FIGURE 12.2 Structure of penicillin G (β-lactam antibiotics).

2000), *Pseudomonas tabici* producing wild-fire poison (Yi et al., 1990) and even *Streptomyces verticillus* producing anti-tumour antibiotics, phleomycins and bleomycin (Kawano et al., 2000). The commonly used beta-lactam antibiotics are penicillins and cephalosporins whereas some fairly recent members are cephamycins, nocardicins, thienamycins. Apart from nocardicins, the above antibiotics are derivatives of bicyclic core structures wherein the lactam ring unifies with the ring molecule employing a nitrogen atom and a carbon atom (Figure 12.2). This ring compound is 5-membered in penicillins (thiazolidine), thienamycins (pyrroline), and clavulanic acid (oxazolidine). It is 6-membered in cephalosporins and cephamycins (dihydrothiazolidine). Beta-lactam antibiotics prevent the development of the bacterial cell membrane's peptidoglycan. Given the absence of this factor in mammalian cells, beta-lactam antibiotics have low toxicity to mammals (Bruggink et al., 1998).

Mode of action: Penicillin destroys certain classes of bacteria by specifically targeting the transpeptidase activity which has the main role in the final step in the biosynthesis of the cell wall, the peptidoglycan cross-linkage. According to Tipper and Strominger (1965), penicillin acts as a structural analogue for the pentapeptide side chains of peptidoglycan having acyl-D-alanyl-D-alanine terminus. The β-lactam ring of the antibiotic is highly reactive and irreversibly nucleophilically reacts with the active site of the transpeptidase present on the cell wall (Bycroft and Shute, 1985).

The organism utilized for penicillin production: Throughout the time of penicillin development, when the method of surface cultivation was used, a version of the initial culture of *Penicillium notatum* found by Sir Alexander Fleming was used. Once, however, assembly moved to a submerged production, a variety of *Penicillium chrysogenum* NRRL 1951 (after the U.S. Department of Agriculture's Northern Regional Research Laboratory) found in 1943 was implemented. In the submerged community, penicillin yield was up to 250 Oxford Units (1 Oxford Unit = 0.5988 sodium benzylpenicillin) which was ~ 2-3 folds higher in comparison to *Penicillium notatum*. A 'super-strain' X 1612 was produced from the NRRL 1951 variant. Another variant was developed through ultraviolet irradiation of previously produced strain X-1612, resulting in the WISQ 176 strain which was developed by the University of Wisconsin. Moreover, additional ultraviolet irradiation of this newly developed WISQ 176 resulted in the development of another strain BL3-D10, which made solely one 75th of the maximum amount penicillin of WISQ 176, lacking the yellow pigment. Current penicillin manufacturing *P. chrysogenum* strains are far more productive than their parents (Elander, 2003). These are produced through natural selection and mutation utilizing ultraviolet irradiation, x-ray irradiation, or nitrogen mustard application. At present, it is understood that there are several naturally occurring penicillins, viz. penicillins G, X, F, K. Penicillin G (benzylpenicillin) was reported as having been slightly more successful against *Pyogenic cocci.* A higher yield was observed by augmenting the product with phenylacetic acid, analogues (phenylalanine and phenethylamine) of which can be found in corn-steep liquor used to produce penicillin in the U.S. Current penicillin-producing strains are highly unpredictable, and so are most industrial species, and appear to return to low-yielding strains, especially in the case of recurrent agar cultivation. They are thus unremarkably stored in liquid nitrogen at -196 °C or the spores may be freeze-dried. Penicillin has been shown to be created by a large range of organisms as well as the fungi *Aspergillus, Malbranchea, Cephalosporium, Emericellopsis, Paecilomyces, Trichophyton, Anixiopsis, Epidermophyton, Scopulariopsis, Spiroidium* and also the *Actinomycete, Streptomyces.* The sole variety of penicillin created by *Actinomycetes,* however, is penicillin N (with the chemical structure D- (- aminoadipyl)) sometimes accompanied by cephamycins and/or diacetyl-3-0-carbamoylcephalosporin C (Hopwood et al., 1980).

12.3 INDUSTRIAL PRODUCTION OF PENICILLIN

Penicillin, the antibiotic drug, was not extracted until the 1940s (by Florey and Chain), only in time to be employed towards the close of the Second World War. Penicillin became the first significant industrial drug developed by an aerobic fermentation. Penicillin is generated by the *Penicillium chrysogenum,* which requires lactose, other sugars and supply of nitrogen (yeast extract) to grow well enough in the medium. Penicillin, like many antibiotics, is a secondary metabolite and is produced only in a stationary phase. Usually, the fed-batch cycle is used to extend the stationary duration and thereby maximize output. There are three critical steps which are media preparation, inoculum addition along with the important additional parameters which would allow the production of penicillin. Glucose and lactose are both used in the media as a carbon source for diauxic growth of *Penicillium chrysogenum.* High concentrations of glucose can lead to the reduction of penicillin production and thus lactose is added into the media (Birol et al., 2002). Based on the biosynthetic pathway of penicillin production in *Penicillium chrysogenum,* there are some additives highlighted in Figure 12.3 i.e. L-Cysteine and L-Valine which need to be added into the media for optimum production of penicillin. The additional parameters which allow production areas follows; temperature (initial 48 h: 30°C; 25°C maintain throughout the process), pH (6.2 to 6.8 is optimal; above pH 7.3 leads to degradation), 30% dissolved oxygen (DO) concentration; decrease in DO leads to a drastic decrease in penicillin production, air with 1 % CO_2 is essential for germination because decrease in CO_2 reduces germination, high stirring speed increase cell density but lowers penicillin

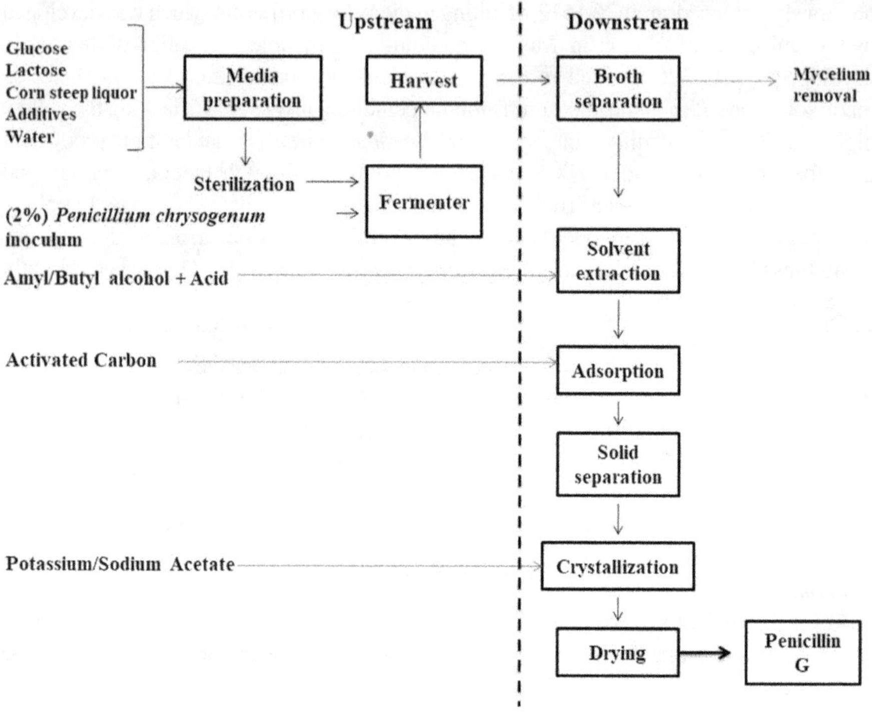

FIGURE 12.3 Block flow diagram (BFD) of penicillin production

production. The development of penicillin is created by introducing surfactants in a process which is very much unclear. However, in recent times it has been found that penicillin yields are higher if adjusted in step with the growth phase, and the temperature is maintained at 25 ° C. Therefore, 30–32 °C is considered optimal for trophophase, and 24 ° C for idiophase. Aeration and agitation are vigorous in keeping the parts of the medium suspended and in keeping yield within the extremely aerobic fungus. The fermentation of the penicillin can be separated into three stages. Throughout this rapid development, the primary process (trophophase) lasts about 40 h in the output of mycelia. The second step (idiophase) has a period of 5 to 7 d; development is decreased as the antibiotic is produced. In the third phase, the sources of carbon and nitrogen are depleted, the production of antibiotics ceases, the mycelia lyse releases ammonia and also the pH increases (Hopwood et al., 1980).

In the downstream process, penicillin is purified in a three-step process; in the first step, solvent extraction, all the penicillin molecules are extracted from the solution with the help of amyl/butyl acetate. Continuous, counter-current, multistage centrifugal extractors (Podbielniak D-36 or Alfa–Laval ABE 216) are used for this purpose. To avoid degradation of penicillin during solvent extraction at low pH, some parameters are optimized, such as filtration time needs to be short and low temperatures are maintained (Birol et al., 2002). For the total recovery of penicillin during solvent extraction, two serially connected extractors are utilized. In the second step, penicillin is adsorbed using activated carbon where penicillin is isolated. In the later stage, solid impurities are separated and then isolated in the form of a salt with the help of potassium or sodium acetate. Crystallization may be conducted by utilizing the solvent or aqueous phase. Some critical parameters need to be adjusted during crystallization such as Na-, K-acetate, and penicillin concentrations, pH, and temperature. Excess amounts of Na or K salts are added to the penicillin-rich solvent before crystallization in an agitated vessel. The crystals are separated by a rotary vacuum filter. The crystals may be washed and pre-dried with anhydrous butyl alcohol to remove some impurities. Crystalline penicillin G or V is sold as an intermediate or converted to 6-APA (6-aminopenicillanic acid), which is used for the production of semi-synthetic penicillins (Winkle and Herwick, 1945).

The production cost of penicillin utilizing glucose as the substrate is \$15/kg. The important cost in penicillin production is raw materials which are 35% along with various utilities. Penicillin manufacturers are trying to minimize production costs by utilizing cheaper raw materials such as starch or molasses and also using genetically engineered strains that could maximize the yield. Approximately 10,000 tons of penicillin V and 26,000 tons of penicillin G are produced worldwide annually. Penicillins are usually used in medical applications and also as intermediates for the production of other semi-synthetic drugs using penicillin acylase. The modifications produced from penicillins are diagrammatically represented in Figure 12.3.

12.3.1 Industrial Production of Cephalosporin

The fermentation process for cephalosporin production is quite similar to that of penicillin production as they share a similar intermediate production pathway. In the upstream process of cephalosporin production, there are three critical steps; media preparation, and inoculum addition along with the important additional parameters which would allow the production of cephalosporin. Glucose and lactose are both used in the media as a carbon source for diauxic growth of *Cephalosporium acremonium.* High concentrations of glucose lead to the reduction of cephalosporin production so lactose is added into the media (Barber et al., 2004). Based on the biosynthetic pathway of cephalosporin production in *Cephalosporium acremonium,* some additives are required i.e. L-cysteine and L-valine which need to be added into the media along with methionine as a source of sulfur for optimum production of cephalosporin. The additional parameters which allow the products are as follows: temperature (initial 48 h: 30°C; 25°C maintain throughout the process), pH (6.2 to 6.8 as optimum; above pH 7.3 leads to degradation), DO 30%; decrease in DO leads to a drastic decrease in cephalosporin production, Air with 1% CO_2 is essential because decrease in CO_2 reduces germination, high stirring speed increase cell density but lowers cephalosporin production (Jermini and Demain, 1989).

In the downstream process, cephalosporin is purified in a three-step process: in the first step, solvent extraction, all the cephalosporin molecules are extracted from the solution with the help of amyl/butyl acetate. In the second step, cephalosporin is adsorbed using activated carbon where cephalosporin is isolated. In the final stage, solid impurities are separated and then isolated in the form of a salt with the help of potassium or sodium acetate (Jermini and Demain, 1989).

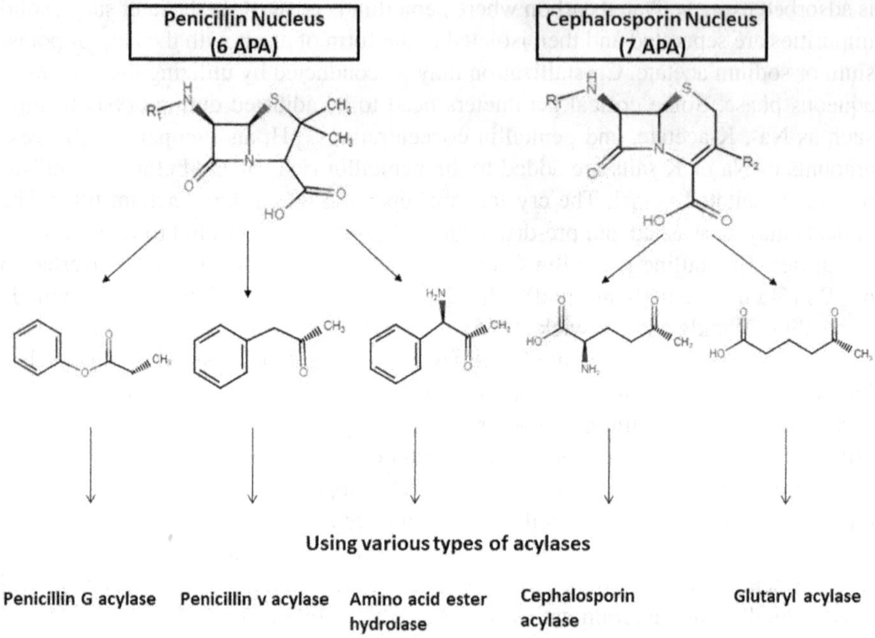

Penicillin Nucleus (6 APA)

Cephalosporin Nucleus (7 APA)

Using various types of acylases

Penicillin G acylase | Penicillin v acylase | Amino acid ester hydrolase | Cephalosporin acylase | Glutaryl acylase

FIGURE 12.4 Modifications of penicillins and cephalosporins using various acylases.

Nowadays, the major interest is 7 ACA (7-aminocephalosporanic acid) which is the structural nucleus present in cephalosporin C and also the precursor for various modifications of cephalosporin. Thus cephalosporin was chemically hydrolysed to produce 7 ACA which was a very tedious process and also time-consuming. Recently, enzymatic hydrolysis has been used for the production of 7 ACA from cephalosporin (Figure 12.4). The two key enzymes responsible for the production of 7 ACA, D-amino acid oxidase and glutaryl amidase are immobilized which leads to efficient production on a large scale (Barber et al., 2004).

12.3.2 Modifications of Penicillin and Cephalosporin

During the 1940s, there were experiments conducted where phenylacetamide derivatives were integrated into a benzylpenicillin molecule (Figure 12.4).

This possibility came into existence by inducing a mould for the production of new antibiotics by chemical induction into antibiotic precursors. These new antibiotics like phenoxymethyl penicillin (Penicillin V) had better acid stability than penicillin G; penicillin O (allythiomethyl penicillin) has a low allergic reaction. In 1959, precursor-starved *P. chrysogenum* fermentation was used for the isolation of 6-amino penicillanic acid which opened the gates to the era of semi-synthetic antibiotics. For the preparation of semi-synthetic penicillins, penicillin G or penicillin V is cleaved producing 6-APA which is later modified by integrating the 6-acyl group chemically or enzymatically with the help of acylases. Mostly bacterial acylases are used as they cleave penicillin G rapidly (Jermini and Demain, 1989).

12.4 OTHER BETA-LACTAM ANTIBIOTICS

Cephamycins (7-Methoxycephalosporins): Cephamycins are cephalosporins with a methoxy group at C-7 having a carbamoxy-loxymethyl function at C-3 and were first produced by *Streptomyceslipmanni,* while cephalosporin was produced by *Streptomycesclavuligenis.* For the production of cephamycins, methionine is necessary for stimulation of the production of cephamycins from penicillin N (Barber et al., 2004).

Nocardicin: Nocardicin is produced by *Nocardia uniformis.* This antibiotic is unique among beta-lactams because it is monocyclic (i.e. no ring coalesces with the ring of beta-lactam). This is highly successful against the Gram-negative *Proteus* compared with Cepham, Cephazolin, and Enteric bacteria, which have minimal efficacy against *Pseudomonas.* However, elevated development exists even during *in vivo* diagnosis with *Pseudomonas.* This is inactive against Gram-positive bacteria including yeasts, including moulds. Norcardicin A and B are significantly distinct in composition and are non-toxic to mammals (Jermini and Demain, 1989).

Clavulanic Acid: In 1976 this antibiotic was first represented using another novel isolation technique. For this procedure, agar plates comprising 10 mcg/ml of penicillin are seeded with the development of *Klebsiella aerogenes* by beta-lactamase. Study experiments did not inhibit cells, although the penicillin added will not have an inhibition zone on the test plate while a diffusible beta-lactamase receptor is present. Clavulanic acid, cephalosporins, and penicillin N are produced using *Streptomyces clavuligenis* strains using the method mentioned above. Nevertheless, clavulanic acid was not found using the traditional approach. This is a weak antibiotic, but it does have broad-spectrum antibacterial action. Nonetheless, if it is used with penicillinase-susceptible antibiotics with greater utility than itself it has a clear possible advantage. It is structurally similar to cephalosporins (Williams and Vickers, 1986).

12.5 AMINOGLYCOSIDE ANTIBIOTICS: STREPTOMYCIN

Streptomycin is an antibiotic (antimycobacterial) drug, first discovered in a class of drugs called aminoglycosides and was the first antibiotic remedy for tuberculosis. It is derived from the actinobacterium *Streptomyces griseus.* Streptomycin is a bactericidal antibiotic. Streptomycin may not be administered orally but must be administered by regular intramuscular injections (Dulaney and Perlman, 1947).

12.5.1 Mode of Action

Streptomycin is a bactericidal antibiotic, most effective against both Gram-negative and Gram-positive bacteria. It has inhibitory action against most of the species of *Mycobacterium* and is referred to as one of the most effective treatments against tuberculosis. Streptomycin is an antibiotic of class aminoglycoside. It destroys bacteria by inhibiting their protein synthesis by binding at the 30S subunit of the ribosome, thereby causing a structural defect interfering with codon-anticodon interaction which causes the production of faulty protein (Luzzatto et al., 1968).

12.5.2 Industrial Production of Streptomycin

In the upstream of streptomycin production, various additives are incorporated into media which include a carbon source, protein source, and some other factors called inducers which may induce the pathway required for the production. For streptomycin production, glucose is used as a carbon source which acts as the best carbon source for providing higher yields of streptomycin as it gives the basic carbon skeleton of streptomycin (Figure 12.5). Along with glucose even fructose, maltose, lactose, galactose, mannitol, xylose and starch can also be utilized as the carbon source. In general, polysaccharides and oligosaccharides offer lower streptomycin yields; corn-steep liquor, casein hydrolysate, yeast extract, and soybean meal etc. act as a protein source; other additive salts such as magnesium, potassium, calcium, boron, molybdenum, sulfate, and chlorides, along with the inducer myo-inositol are added into media with 2%*Streptomyces griseus* inoculum. L-naphthalene acetic acid and phenylactic acid can be added into the media along with proline for better yields. The optimal fermentation temperature is between 25 and 30 ° C and the optimum pH level is between 7.0 and 8.0. However, the highest development rate of streptomycin exists within a pH range of 7.6 to 8.0 (Dulaney and Perlman, 1947).

Streptomycin is an extracellular product thus, in the downstream process, the separation process is appropriate where the mycelium is removed; all streptomycin molecules are captured using hydrophobic interaction chromatography (HIC) after

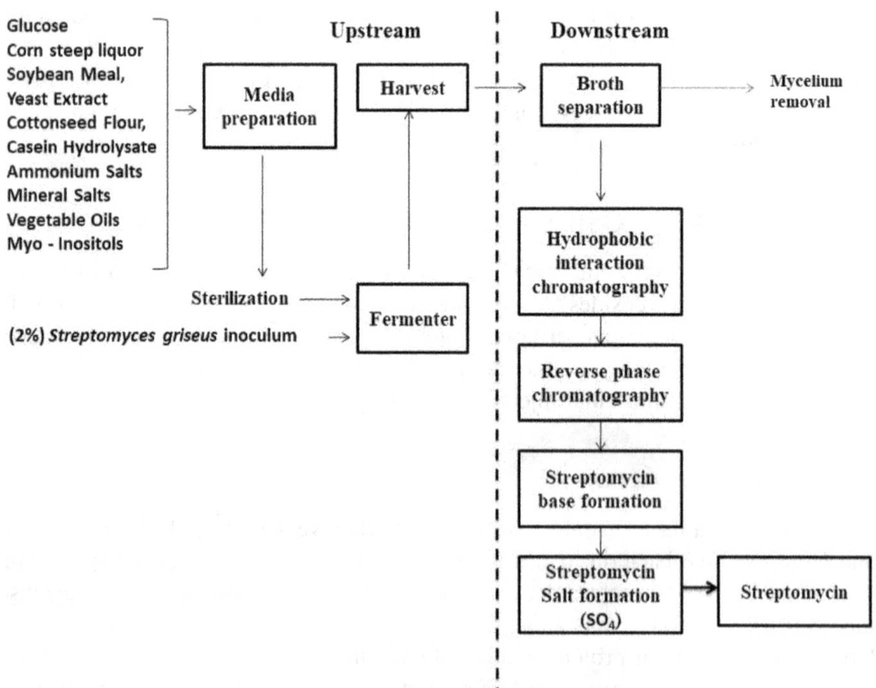

FIGURE 12.5 Block flow diagram (BFD) of streptomycin production.

which there is the isolation of streptomycin using reverse phase chromatography. Streptomycin is usually purified in salt form; streptomycin SO_4 salt formation is conducted. Along with streptomycin, vitamin B12 is also produced which has an altogether different market (Hopwood et al., 1980).

12.6 AMINOGLYCOSIDE ANTIBIOTICS: TETRACYCLINE

Tetracyclines are a group of antibiotics that include tetracycline. These are produced through *Streptomyces* spp. fermentation or by the industrial refining of natural resources. These are components of an octahydro-naphthacene, a group of hydrocarbons containing four annulled six-member rings (Figure 12.6). Tetracycline activity can be classified based on period: examples of short-acting are tetracycline, oxytetracycline; intermediate-acting are demeclocycline, lymecycline, and long-acting are doxycycline, minocycline. Tetracycline is used to treat a wide variety of infections, including acne. It is an antibiotic that works by stopping the growth of bacteria. It is applied in the treatment of infections like septicaemia, endocarditis, and meningitis (Goodman, 1985).

Tetracycline is a broad-spectrum polyketide antibiotic that is normally developed by the *Streptomyces sp.* This antibiotic has a bacteriostatic impact on the bacteria by reversibly binding to the bacterial 30S ribosomal subunit, thus preventing the aminoacyl tRNA binding to the ribosomal receptor region. Tetracycline is actively diffused through the bacterial membrane via porin channels, thus reversibly binding to the 30S ribosomal subunit and stopping tRNA from binding to the mRNA-ribosomal complex and thereby reducing protein synthesis (Goodman, 1985).

12.6.1 Industrial Production of Chlortetracycline

Tetracycline is also produced by the same genus as that of streptomycin but the biosynthetic pathway by which it is produced is quite different. Production media is entirely designed by understanding the biosynthetic pathway. In the upstream of tetracycline production, various additives are incorporated into the media such as the carbon source, starch.

In the biosynthetic pathway of streptomycin and tetracycline, glucose is the carbon source but in the production of tetracycline, starch is used. The market cost of tetracycline is comparatively less than streptomycin. Protein sources such as soybean meal,

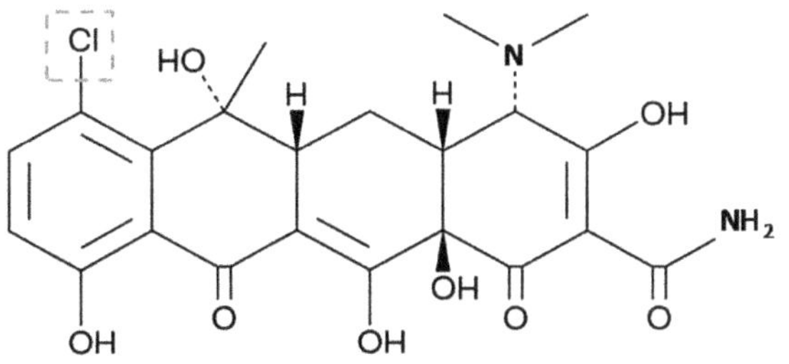

FIGURE 12.6 Structure of chlortetracycline.

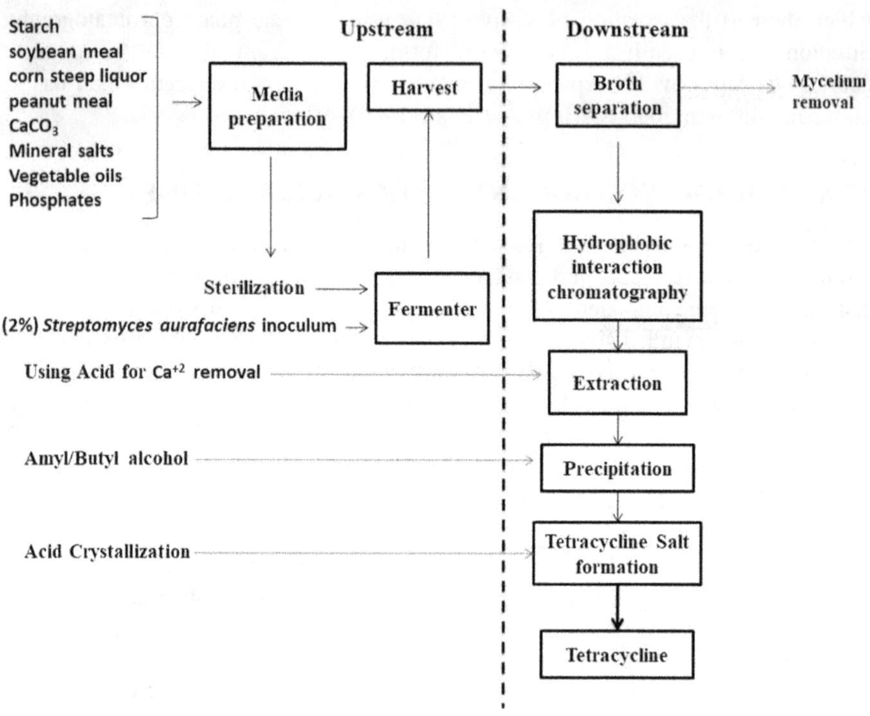

FIGURE 12.7 Block flow diagram (BFD) of tetracycline production.

corn-steep liquor, and peanut meal along with some other additives like mineral salts, vegetable oils, phosphates, and $CaCO_3$ are incorporated to complete the media. It has been reported that $CaCO_3$ enhances tetracycline production (Williams and Vickers, 1986). The *Streptomyces* strains can adapt the initial pH range from 5 up to 7 while the optimum pH is 6.5. Due to the sensitivity of various metabolic activities of *Streptomyces* to initial pH change, evaluation of the optimum levels of initial pH is very important for the overall economic feasibility of the production process (Rogalski, 1985).

In the downstream of the production, as the antibiotic is extracellular the biomass is first separated from the media and then the tetracycline is purified in four different steps. Hydrophobic interaction chromatography (HIC) where complete extraction of tetracycline molecules takes place (Goodman, 1985). As $CaCO_3$ has been added into the media, there are calcium ions in the media which would hamper the purification process thus they are removed using acid extraction (Figure 12.7). The next step of purification is precipitation using amyl/butyl acetate ester. Tetracycline is isolated in the form of salt which is performed using acid crystallization.

12.7 PHARMACOLOGY OF ANTIBIOTICS AND QUALITY CONTROL

After production of each antibiotic, they are subjected to toxicity and quality control tests. Upon purification, the impact of the antibiotic on the usual functioning of the target tissues and organs (its pharmacology) including its potential toxic activity (toxicology) must be checked on a wide number of animals of many types. Besides, successful methods of governance must be established. Once these measures of pharmacology and toxicity are all accomplished, the producer can file an Investigational New Drug Application with the Board of Pharmacy and Poisons. If licensed, the antibiotic may be assessed for toxicity, resistance, absorption and excretion on volunteers (Elander, 2003). If subsequent studies on a limited number of patients are positive, the medication can be included in a wider population, typically in a hundred patients. If all goes well, the medication can also be used in clinical medicine. Such methods, from the moment the antibiotic is detected in the laboratory before the clinical trial is started, typically spread over many years (Williams and Vickers, 1986).

Before distribution, quality assurance is of utmost significance for the development of antibiotics. Since antibiotic processing requires a fermentation cycle, precautions must be taken to ensure that at any stage during development, virtually no contamination is added. To this end, all production machinery and the medium are completely sterilized with steam. The consistency of all the substances is tested periodically during fabrication. Frequent tests of the state of the microorganism community during fermentation are of special significance. They are performed using various chromatography methods. Various physical and chemical properties of the finished product, such as pH, melting point and moisture, are often tested (Hopwood et al., 1980).

12.8 THE NECESSITY FOR NEW ANTIBIOTICS

12.8.1 Antibiotic Resistance

If the concentration of drugs needed to prevent or destroy the microorganism is higher than usual usage, the medication is known to be immune to the microorganism. In certain instances, it has been found that a few microbes can generate a protein that deactivates an antibiotic or inhibits the transfer of the antibiotic to the cell. These microbes are potentially very dangerous for the human population (Mathur and Singh, 2005). Due to antibiotics being used so commonly, various microorganisms develop resistance towards them (Walsh, 2000). This is because of the inherent nature and properties of microorganisms but also because of over-prescription of antibiotics without performing appropriate testing for the infection. This prophylactic use of antibiotics in low doses causes the microbes to become resistant against antibiotics using the ability to transfer genetic material through transformation and conjugation which exhibits the development of resistance. Cross-susceptibility to a single antibiotic often results in susceptibility to certain antibiotics, typically of a specific chemical type, to which the bacteria would not have been exposed. Cross-resistance can exist; for example, in both colistin and polymyxin B, or both clindamycin and lincomycin.

12.8.2 Development of Pathogenic Organisms

In patients who are immunocompromised, non-pathogenic microorganisms which were antecedently standard commensals became pathogens because of the widespread use of antibiotics. Thus *Proteus* sp., *Actinobacte*ria sp., and yeasts have all gained new standing as pathogens particularly in medical care units (Mathur and Singh, 2005).

Mutasynthesis: In this method, a mutant of an antibiotic producer is fed with various other precursors which may lead to the production of new improved varieties of antibiotics. Various antibiotics are being produced using this method, such as puromycin from *Strep. rimogus*, butrosin from *Bacillus circulans*, and ribostamycin from *Strep. Ribosidifacus* (Mathur and Singh, 2005).

12.9 CONCLUSIONS

Antibiotics can eliminate or inhibit the growth of other pathogenic microorganisms and are used in the treatment of external or internal infections. Although certain antibiotics are developed by microorganisms, most of them are now formulated synthetically. Antibiotic misuse, also referred to as antibiotic neglect or overuse of antibiotics, is a severe concern. Misuse or overuse of antibiotics creates severe health effects. This leads to the development of multidrug-resistant bacteria, which can have tolerance to several antibiotics and cause life-threatening infections. Therefore, awareness is important in order to avoid unnecessary use of antibiotics.

REFERENCES

Barber MS, Giesecke U, Reichert A, Minas W, Industrial Enzymatic Production of Cephalosporin-Based β-Lactams, in Brakhage, A.A. (Ed.), Molecular Biotechnology of Fungal Beta-Lactam Antibiotics and Related Peptide Synthetases: -/-, Advances in Biochemical Engineering. Springer, Berlin, Heidelberg, pp. 179–215, 2004.

Birol G, Ündey C, Çinar A, A modular simulation package for fed-batch fermentation: penicillin production. Comput. Chem. Eng. 26, 1553–1565, 2002.

Bruggink A, RoosEC, de Vroom E, Penicillin Acylase in the Industrial Production of β-Lactam Antibiotics. Org. Process Res. Dev. 2, 128–133, 1998.

Bycroft BW, Shute RE, The Molecular Basis for the Mode of Action of Beta-Lactam Antibiotics and Mechanisms of Resistance. Pharm. Res. 2, 3–14, 1985.

Chang LC, Bhat KPL, Fong HHS, Pezzuto JM, Kinghorn AD, Novel Bioactive Steroidal Alkaloids from Pachysandra procumbens. Tetrahedron 56, 3133–3138, 2000.

Dulaney EL, Perlman D, Observations on Streptomyces griseus-I. Chemical Changes Occurring During Submerged Streptomycin Fermentations. Bull. Torrey Bot. Club 74, 504–511, 1947.

Elander RP, Industrial production of β-lactam antibiotics. Appl. Microbiol. Biotechnol. 61, 385–392, 2003.

Goodman JJ, Fermentation and Mutational Development of the Tetracyclines, in: Hlavka, J.J., Boothe, J.H. (Eds.), The Tetracyclines, Handbook of Experimental Pharmacology. Springer, Berlin, Heidelberg, pp. 5–57, 1985.

Hopwood DA, Chater KF, Hugo WB, Holliday R, Hartley BS, Brenner S, Hartley BS, Rodgers PJ, Fresh approaches to antibiotic production. Philos. Trans. R. Soc. Lond. B Biol. Sci. 290, 313–328, 1980.

Jermini MFG, Demain AL, Solid state fermentation for cephalosporin production by *Streptomyces clavuligerus* and *Cephalosporium acremonium*. Experientia 45, 1061–1065, 1989.

KawanoY, Kumagai T, Muta K, Matoba Y, Davies J, Sugiyama M, The 1.5 Å crystal structure of a bleomycin resistance determinant from bleomycin-producing Streptomyces verticillus1 1Edited by I. A. Wilson. J. Mol. Biol. 295, 915–925, 2000.

Kumavath R, Antibacterial Agents.BoD – Books on Demand, 2017.

Luzzatto L, Apirion D, Schlessinger D, Mechanism of action of streptomycin in E. coli: interruption of the ribosome cycle at the initiation of protein synthesis. Proc. Natl. Acad. Sci. USA. 60, 873–880, 1968.

MathurS, SinghR, Antibiotic resistance in food lactic acid bacteria—a review. Int. J. Food Microbiol. 105, 281–295, 2005.

NagórskaK, BikowskiM, Obuchowski M, Multicellular behaviour and production of a wide variety of toxic substances support usage of Bacillus subtilis as a powerful biocontrol agent. Acta Biochim. Pol. 54, 495–508, 2007.

Quinn R, Rethinking Antibiotic Research and Development: World War II and the Penicillin Collaborative. Am. J. Public Health 103, 426–434, 2012.

Riley MA, Bacteriocin-Mediated Competitive Interactions of Bacterial Populations and Communities, in: Drider, D., Rebuffat, S. (Eds.), Prokaryotic Antimicrobial Peptides: From Genes to Applications. Springer, New York, NY, pp. 13–26, 2011.

Rogalski W, Chemical Modification of the Tetracyclines, in: Hlavka, J.J., Boothe, J.H. (Eds.), The Tetracyclines, Handbook of Experimental Pharmacology. Springer, Berlin, Heidelberg, pp. 179–316, 1985.

WalshC, Molecular mechanisms that confer antibacterial drug resistance. Nature 406, 775–781, 2000.

Williams ST, Vickers JC, The ecology of antibiotic production. Microb. Ecol. 12, 43–52, 1986.

Winkle WV, Herwick RP, Penicillin—a review *. J. Am. Pharm. Assoc. Sci. Ed. 34, 97–109, 1945.

Yi YK, Park EK, Kim JH, Occurrence of angular leaf spot symptom on tobacco plants caused by non-toxin-forming mutant of wild fire pathogen, *Pseudomonas syringae*pv. tabaci in Korea. Korean J. Plant Pathol. 6, 81–85, 1990.

13 Vaccines, Hepatitis B and Insulin Production

13.1 INTRODUCTION TO VACCINE

The human body is susceptible to disease via a number of pathogens; these include microorganisms such as bacteria, virus, viroids, fungi, and protozoa. A defence mechanism is required by the body to fight off the pathogenic organisms and this is where the immune system makes its way into the system. The immune system defends us against germs. Germs are these pathogens themselves. The various parts of the immune system include lymphoid organs such as thymus and bone marrow, the lymphatic system along with the white blood cells, and antibodies and the spleen. Immunity (resistance) is mostly due to WBCs (white blood cells). These are produced in the bone marrow. WBCs interfere with and directly attack pathogens. They keep moving through the blood, as a patrolling system, looking for foreign particles (pathogens); when they come across one, they do their work. Lymphocytes such as the natural killer cells, T-cells, B-cells, are various kinds of white blood cell. The entire immune system is essential for resistance against pathogens. The lymphatic system is a transport system that assists in immunity by housing, carrying and transporting all the essential organs and cells to counter a microbe invasion. The toxins produced by the microbes are countered by antibodies; these recognize the pathogenic antigen and mark them for destruction. They themselves do not neutralize the toxin but mark it and then white blood cells along with other proteins come into action and neutralize the threat.

13.1.1 Innate Immunity

Innate or non-specific defensive systems are the first line of protection against all harmful microorganisms. They can work irrespective of the type of microbe. They include the following:

(a) *Anatomical barriers*: This includes mechanical barriers including the skin that physically keeps microorganisms out while mucous membranes of the respiratory and urinogenital tract of the food canal traps microorganisms. In fact, mucous membranes comprise a natural collection or flora of microorganisms that keep foreign organisms out.

(b) *Physiological barriers*: The structure of the human body protects these pathogens. Therefore, the elevated temperature of the human body in the form of fever along with the acidic aspect of the stomach, holds away certain pathogens. Chemical mediators, such as lysozyme present in tears, have been shown to break bacterial cell walls.
(c) *Phagocytosis and endocytosis*: WBCs eliminate and digest the pathogen, while specialized cells engulf and break down foreign particles.
(d) *Inflammatory reactions*: Tissue injury and inflammation cause leakage of serum vascular fluid proteins containing antibacterial agent and inflow of white blood cells contributing to pus development.

We can classify immunity in humans into two categories: I. Naturally acquired; II. Artificially acquired. These are categorized based on where our body has acquired the ability to fight off pathogens.

Active immunity (both naturally and artificially acquired) is the production of antibodies by the body by itself in the event of a pathogenic invasion. A memory is stored and when in future the same pathogen infects again, the body has memory and it helps in the quick destruction of the toxins. In passive immunity antibodies are acquired by a different source; this is a one-time use kind of situation because no memory is stored and when the same pathogen infects again the body does not have antibodies ready because there is no information on how to neutralize the toxin. These can be acquired both naturally and artificially as shown below.

Naturally Acquired		Artificially Acquired	
Active	**Passive**	**Active**	**Passive**
Infections caused by pathogens for the first time; body produces antibodies for that specific pathogen.	Antibodies transferred to a foetus from its mother either by colostrums or to a new born by breast feeding.	Vaccination, live attenuated virus is introduced into the body	Antibodies injected into body by serums

13.2 TYPES OF VACCINES

A vaccine is an antigenic preparation used to produce active immunity to a disease, in order to prevent or reduce the effects of infection by any natural or 'wild' pathogen (Wittmann and Liao, 2016). Different types of vaccine are available: 1) Inactivated; 2) Attenuated; 3) Toxoid; 4) Subunit; 5) Conjugate (Figure 13.1).

13.2.1 Killed/Inactivated Vaccines

Bacterial vaccines are killed vaccines, whereas the viral vaccines are those among the inactivated vaccines. Typhoid is one of the examples of killed vaccines; it was the first one to be produced and was used among English troops in the late 19th century. Inactivated vaccines such as polio and hepatitis A are currently in circulation in the

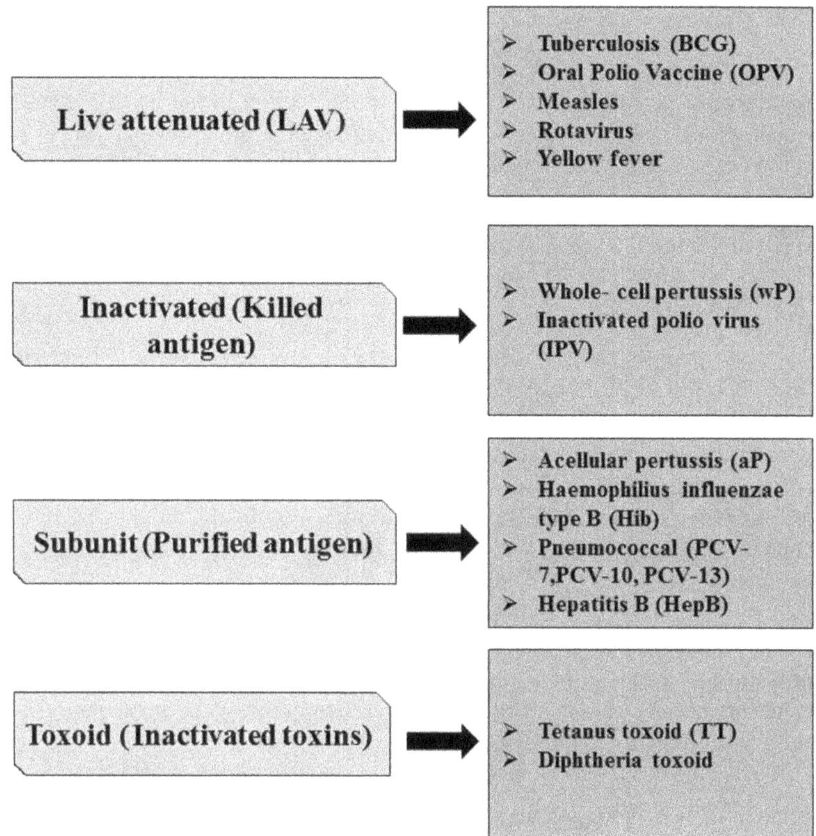

FIGURE 13.1 Different types of commercially available vaccines.

UK, and the most common killed vaccine in many countries is the whole cell pertussis vaccine. The adaptive immune response that the killed/inactivated vaccine generates is directed against a much broader range of antigens (Ellis and Brodeur, 2012). When injected with the vaccine, whole organism phagocytosis occurs. Hepatitis A is an example of an inactivated vaccine. Its protective efficacy is more than 90%. The vaccine is formalin inactivated, the cell culture is adapted and it is a strain of HAV. When injected it gives rise to neutralizing antibodies.

13.2.2 Attenuated Vaccines

The vaccine is made of a culture that has been modified in such a way that it reduces the pathogenicity of the organism, but still retains the original antigens that are virulent. The method for reducing the virulence of the living pathogen is by passaging them through hosts different from usual (Germanier, 2012). In some cases, the non-virulent strains of the pathogen may also be used to create vaccines. Occasionally it is observed that the attenuated pathogens might regain their full virulence and cause disease in the victim's body; an example would be OPV (Sabin Oral Polio Vaccine).

13.2.3 Toxoid

Inactivated bacterial exotoxins are toxoids. The toxins are inactivated by treatment. Toxins, themselves harmless, trigger antibody production in the body (Powell and Newman, 2012). In some diseases the bacteria liberate a protein which proves to be the cause of the disease and not the bacteria themselves, and to denature the protein, the toxin is exposed to formaldehyde. Now some epitopes on the protein molecule are retained, which provokes antibody production.

13.2.4 Subunit

This is a more recent advancement in the field of vaccine production. Modern advancements bring safer, more effective and cheaper vaccines. The subunit vaccine is a type in which instead of a living microorganism or a dead microorganism, only the antigens or the epitopes which can induce the antigen production are used. Sub-unit vaccines are also known as surface molecule vaccines, because the materials that induce antibody production are located at the surface of the microorganism. The advantages in using this kind of vaccine are that they are safer and there is less antigenic competition since only a few components of the microorganism are included in the vaccine production (Ellis and Brodeur, 2012). They have a disadvantage, this being that these vaccines require strong adjuvants; the compounds that are administered along with vaccines to increase their immunogenicity are known as adjuvants. These adjuvants often induce tissue reactions. The duration of the immunity that the vaccine provides is shorter than that of live vaccines.

13.2.5 Conjugate Vaccines

These are similar to subunit vaccines but instead of surface molecules a part of the microorganism is used to make the vaccine. The capsules of these bacteria are recognized by an adult's immune system but the infant or a baby's immune system does not recognize it. Therefore to counter this issue a protein from a different microorganism is linked to this capsule; this makes the baby's immune system respond to the combined vaccine and produce antibodies, and initiates an immune response against the disease-causing organism (Allan et al., 1978).

13.3 STAGES INVOLVED IN VACCINE PRODUCTION

There are several stages involved in vaccine production (Table 13.1). They are as follows:

Inactivation – Antigens are prepared; Purification – Antigen purification is carried out; Formulation – Vaccine is produced

Inactivation: Generation of the antigen: Viruses are grown so that they can be used to extract the antigens. Primary cells such as chicken embryos or fertilized eggs or cell lines are used to grow the cultures. Bioreactors are used to grow bacteria. A growth medium specifically used to enhance the production of

TABLE 13.1
Some of the vaccines and their culture media along with the excipients in them

Vaccine	Culture Media	Excipients
BCG	Synthetic or semisynthetic	Asparagine, citric acid, lactose, glycerine, iron ammonium citrate, magnesium sulphate, potassium phosphate
DT vaccine	Synthetic or semisynthetic	Aluminium potassium sulphate, bovine extract, formaldehyde, thimerosal
Hepatitis A vaccine (Havrix)	Human diploid tissue culture (MRC-5)	Aluminium hydroxide, amino acid supplement, formalin, MRC-5 cellular protein, neomycin sulphate, phosphate buffers, polysorbate 20
Hepatitis B vaccine (Engerix-B)	Yeast or yeast extract	Aluminium hydroxide, phosphate buffers, yeast protein
Influenza vaccine (Afluria)	Chicken embryo	Beta-propiolactone, calcium chloride, dibasic sodium phosphate, egg protein, monobasic potassium phosphate, monobasic sodium phosphate, neomycin sulphate, polymyxin B, potassium chloride, sodium taurodeoxychoalate, thimerosal (multi-dose vials only)
Polio vaccine (IPV – Ipol)	Vero (monkey kidney) cell culture, Medium 199	Calf serum protein, formaldehyde, neomycin, 2-phenoxyethanol, polymyxin B, streptomycin
Rabies vaccine (Imovax)	Human diploid tissue culture (MRC-5)	Albumin, MRC-5 cells, neomycin sulphate, phenol
Typhoid vaccine (inactivated – Typhim VI)	Synthetic or semisynthetic	Disodium phosphate, monosodium phosphate, phenol, polydimethylsiloxane, hexadecyltrimethylammonium bromide
Yellow fever vaccine (YF-Vax)	Chicken embryo	Egg protein, gelatine, sorbitol

antigens is used in these bioreactors (Germanier, 2012). Yeasts, bacteria and cell culture are used to grow recombinant proteins. Releasing and the isolation of the antigen: after the generation of the antigen the release of it from the culture is necessary. The antigen will be isolated from the proteins and other parts of the site of production that are still present in the bacteria and viruses.

Purification: The antigen will be purified to obtain a much better and higher quality and higher purity; protein purification comes into play here.

Additional components: In this step, as the name suggests, additional components are added to the vaccine to improve its efficiency. Addition of adjuvants for example is done in this step, a material which enhances the vaccine's immune response. Stabilizers are added and then the vaccine is formulated for prolonged storage or multi-dose vial usage. With vaccines that contain multiple ingredients along with different antigens, the process becomes tricky and time consuming. All components are thoroughly and uniformly mixed and packaged safely in a syringe or vial (Brown et al., 1993).

Packaging: Once all the above steps are completed along with formulation, the final product is ready, and the vaccine is safely packed and sealed with sterile stoppers. The products are packed and stored for further distribution.

13.4 INTRODUCTION TO RECOMBINANT PROTEIN

Amino acids are the building blocks of protein, which forms the main constituent in our human body, and are responsible for natural metabolism. They are formed in the cytoplasm in the cells, after the translation of mRNA molecules which migrate from nucleus to the cytoplasm. This forms the basis of the Central Dogma of Molecular Biology (Burnett, 1983). Proteins are omnipresent, and have various vital roles to play throughout the body, such as that they work as enzymes to catalyse a reaction or work as receptors for cell signalling, or antibodies in immune responses, while most of them are structural molecules.

Currently, proteins are available in two forms: native form and recombinant form (r-protein). Native proteins are extracted directly from microorganisms, although the drawback lies in the fact that the quantity obtained is extremely low. However, modern techniques use genetically engineered microbes, which are available in large quantity, and can be used in a variety of sectors (Powell and Newman, 2012). Selection and isolation of genes of interest from source organism using various methods is followed by cloning of the desired gene into an expression vector, which leads to efficient expression of the recombinant protein, due to inducible promoters and expression signals.

The synthesis of enzymes of microbial origin can be optimized by well-characterized genomes, judicious selection of host strain, cost-effectiveness, as well as construction of proper recombinants to ensure that there is production only of the desired recombinant protein (r-protein) and the final product is free from unwanted metabolites and proteins. The final products are passed to testing using the current Good Manufacturing Practice (cGMP) guidelines to test for purity and uniformity. Large-scale anaerobic microbial fermentation first employed the use of recombinant protein to produce acetone, butanol and citric acid in the early 1900s. In 1972, Berg, Cohen and Boyer discovered recombinant DNA which changed the face of several industries such as textile, leather, food, paper, plastics, polymers and even the medical and diagnostic area. The production of therapeutic enzymes was another major advancement in this field, as they are produced in adequate quantity, with improved properties and superior quality. These enzymes can be used as oncolytics, thrombolytics and anti-coagulants, and are known as biopharmaceuticals. It is anticipated that up to 50% of all drugs will be biopharmaceuticals in the coming 5 to 10 years. In this chapter, the basics of r-protein production are explained with schematic representation, along with its application in various fields, and future prospects are discussed (Robinson, 2016).

13.4.1 Recombinant Protein or R-Protein

Proteins are present in four forms in nature: primary, secondary, tertiary and quaternary, out of which the tertiary structure is responsible for most of the functions in the human body. DNA is present in the nucleus which is transcribed to form mRNA.

This molecule migrates from nucleus to cytoplasm, carrying the genetic information present in DNA, and are translated into proteins. Recombinant proteins (r-protein) are those molecules which have been coded by recombinant DNA (rDNA), which is a clone of genes of interest into the desirable vector. This hybrid vector is then inserted into the host organism, mostly bacteria, which continues to divide and eventually leads to the formation of r-protein. This technique has been used to produce therapeutic hormones and enzymes, known as biopharmaceuticals, for daily and medical uses. Proteins help the body to build, repair and maintain body tissues; they are important for metabolism and maintaining good health. They help to reduce the risk of heart disease and lower blood pressure (Allan et al., 1978). However, this natural protein is found in limited amounts in the diets of most people or is difficult to absorb for people with genetic defects (for example: Hartnup disease, lysinuric protein intolerance). Hence, the need is to synthesize genetically engineered proteins which might fulfil the daily requirements of individuals.

Our body relies on protein production by using transcription and later translation, but in vitro this can be achieved by Recombinant DNA Technology (RDT), which will produce recombinant proteins (Brown et al., 1993). Plasmids are small circular DNA molecules which are present on bacteria which can transmit to daughter cells by methods such as cell division, binary fission or conjugation. Scientists have devised a novel method which uses Restriction Enzymes (RE) to cleave the insert as well as the plasmid, which is used to introduce the protein-coding gene into the plasmid, growing them in culture to maximize the amount of desirable genes (Figure 13.2). The target genes are then introduced into host cells or vectors which possess specialized machinery for transcription and translation, which represents an expression system.

There are certain disadvantages when relying on naturally occurring proteins. They exist in extremely low concentrations in tissues or blood samples of plants or

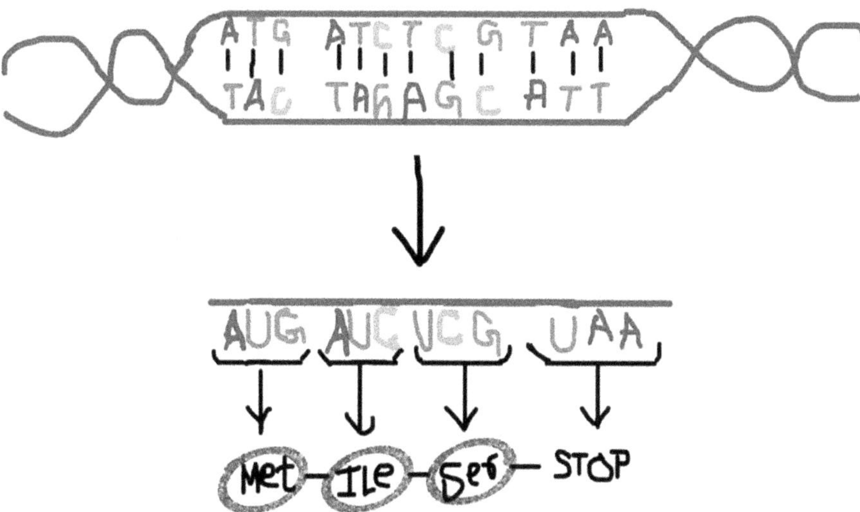

FIGURE 13.2 Central Dogma of Molecular Biology proposed by Francis Crick.

animals, and also the isolation and purification costs are not-economical (Germanier, 2012). Hence, relying on genetic engineering for large-scale production and use of various types of recombinant proteins has proved to be more successful.

The use of Restriction Endonucleases or Restriction Enzymes (RE) in genetic engineering is prominent as it can cleave genes at specific locations. The same RE is used to cleave and isolate the target gene of interest, for example, a human gene located on the chromosome of humans as well as the plasmid. RE produces two ends – the blunt end and the sticky end. Sticky ends produce cohesive bonds, and the type of RE which produces such ends are preferable. There are four different types of restriction enzymes; type I, II, III and IV, among them, type II enzymes are the most specific to both the site of recognition and restriction. The DNA which has the code for the protein of interest located on the chromosome of the source organism is obtained by cutting the portion using restriction enzymes and is then isolated out of the organism. The portion of the plasmid of the bacteria, in which the cut DNA is to be inserted, is also cleaved using the same restriction enzyme as the DNA template was cut; this leads to the formation of open ends. The DNA template and plasmid ends are ligated using DNA ligase. The plasmid which consists of the gene of interest is then incorporated into the bacteria, which are then cultured, and manufactured at large scale producing the product of the inserted gene. The protein is then isolated from the culture, purified and processed to make it available as a commercial and therapeutic product (Ellis and Brodeur, 2012). Isolation and purification of the desired protein is carried out for medicinal and industrial use (Figure 13.3).

13.5 TYPES AND CLASSES OF MICROBIAL RECOMBINANT PROTEINS

Manufacturing the r-protein in microbial system is transforming science. The biggest advantage of recombinant DNA technology is the development of the microbial proteins used for human therapeutics such as hormones, growth factors and antibodies. The advantages of microorganisms over available sources is that they are easy to handle, they have high division rate and the production yield is high. Microorganisms produce a large number of essential products such as carbohydrates polymers (macromolecules), nuclei acids, protein, and small molecules. The recombinant proteins have many applications, such as that they are used in molecular biology laboratories and some are used in research projects. The vectors used for the production of r-proteins are divided into two classes based on the size of the recombinant protein:

Prokaryotic system: *E. coli* is used as the source of the expression of the proteins because this method is the most easy and quick. There are limitations of the proteins as well such as that large proteins, disulphide bond rich proteins and the proteins requiring post-translational modifications cannot be expressed in the model organism.

Eukaryotic system: The best suitable vector for this system is yeast (*Saccharomyces cerevisiae* and *Pichia pastoris*) because of its higher yield and lower cost of production. The limitations of the prokaryotic system have been solved; it can produce high molecular weight proteins larger than 50 kDa, and removal of single sequence and glycosylation can be carried out.

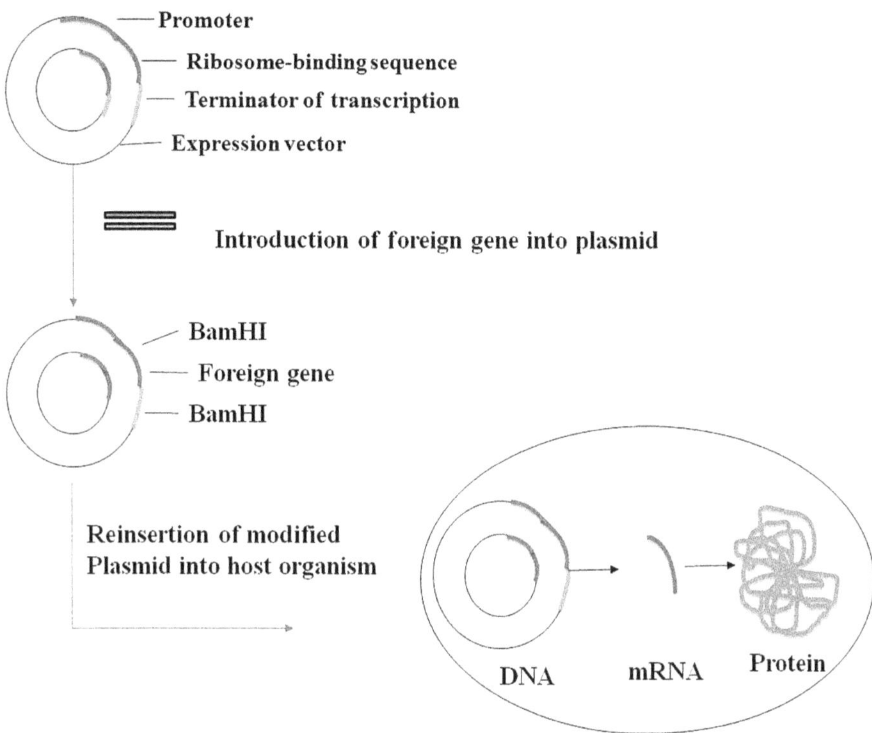

FIGURE 13.3 Flow chart of steps involved in recombinant protein production.

13.5.1 Classes of Microbial Recombinant Protein

On the basis of their applications, proteins can be categorized into some significant classes such as antibodies, enzymes, antibiotics, proteins, vaccines, food supplements and bioactive compounds. These classes are defined below:

Microbial enzymes: Proteins are available in two categories: enzymatic and non-enzymatic. That means that most of the enzymes available are proteinaceous in nature. The production of enzymes for use as drugs is a vital aspect of pharmaceutical companies (Powell and Newman, 2012) e.g. *Aspergillus oryzae* has been used to produce β-galactosidase, *Bacillus brevis* is used for α-amylase production and *E. coli* to produce alkaline phosphatase.

Antibodies: Efficient production systems is required for the synthesis of recombinant antibodies due to the surge in demand in medical, diagnostic and research sectors (Ellis and Brodeur, 2012) e.g. *Bacillus megaterium* plays a role in the production of fibrin-specific single-chain antibody (scFv and scFab fragments).

Antibiotics: Different microorganisms have been used to obtain antibiotics which are necessary for the treatment of various infections. Penicillin G acylase is produced from *B. subtilis.* The isopenicillin N synthetase ('CyClaSe') gene of *Cephalosporium acremonium* has been inserted into *E. coli.* Cyclase genes of

Penicillium chrysogenum and *Streptomyces clavuligerus* have also been cloned in the *E. coli* system (Brown et al., 1993). The expandase/hydroxylase gene of *C. acremonium* has been cloned in *E. coli.*

Microbial proteins: Protease, amylase, xylanase, trypsin, growth hormone and asparaginase are few microbial proteins which have industrial and commercial application on a daily basis (Siber et al., 2008).

Microbial vaccines: Formulation of recombinant protein subunit vaccines is done using protein antigens, which have been manufactured in heterologous host cells. Several host cells are available for this purpose, ranging from *Escherichia coli* to mammalian cell lines e.g. 11 permitted vaccines against Hepatitis B virus and one against human papilloma virus (HPV) have been formulated using *Saccharomyces cerevisiae*. In both cases, the recombinant protein forms highly immunogenic virus-like particles (Powell and Newman, 2012).

Food supplements: β-carotene, astaxanthin, and C-phycocyanin (C-PC) are some of the products of recombinant proteins. β-carotene can be used as a food colouring agent as well as a source of pro-vitamin A. The carotenoid astaxanthin has potential applications in the food and feed industries (Hendriksen, 2012).

Bioactive compounds: Actinomycetes microorganisms play a major role in the manufacturing process of antibacterial agents and other bioactive compounds such as drugs and natural pigments of industrial importance. Some of the antimicrobial compounds from various Actinomycetes species are Abyssomicin (Antibacterial and antitumor) from *Verrucosispora* sp., Actinoflavoside (Antifungal) from *Streptomyces* sp., Analogs-metacycloprodigiosin (Anticancer) from *Saccharopolyspora* sp., Essramycin (anti-inflammatory) from *Streptomyces* sp., etc. (Germanier, 2012).

13.6 HEPATITIS B VACCINE

Vaccines are antigenic preparations when injected into the bloodstream, and stimulate the immune system to generate antibodies that defend the body from subsequent infections. They normally contain an inactivated form of the infectious agent as the antigenic material. Usually heating or related treatment is used for formulating vaccines, so that the virus particles become attenuated and are no longer infective (Germanier, 2012). The inactivation process must be extremely efficient and complete because even one live virus particle could cause infection. Many virus particles for the manufacture of vaccines are obtained from tissue cultures. However, a few of them do not grow in tissue cultures – such as the Hepatitis B virus.

Hepatitis B is a widely spread human disease which mainly affects the liver, causing chronic hepatitis, cirrhosis, and liver cancer. The hepatitis B virus is a 42 nm particle, called the Dane particle. It consists of a core containing a viral genome (DNA) surrounded by a phospholipid envelope carrying surface antigens. Infection with hepatitis B virus produces Dane particles and 22-nm sized particles. The latter contain surface antigens which are more immunogenic. It is, however, very difficult to grow hepatitis B virus in mammalian cell cultures and produce surface antigens.

RDT can overcome the glitches in the production of vaccines (Figure 13.4). The important discovery has been that isolated components of the virus such as proteins of

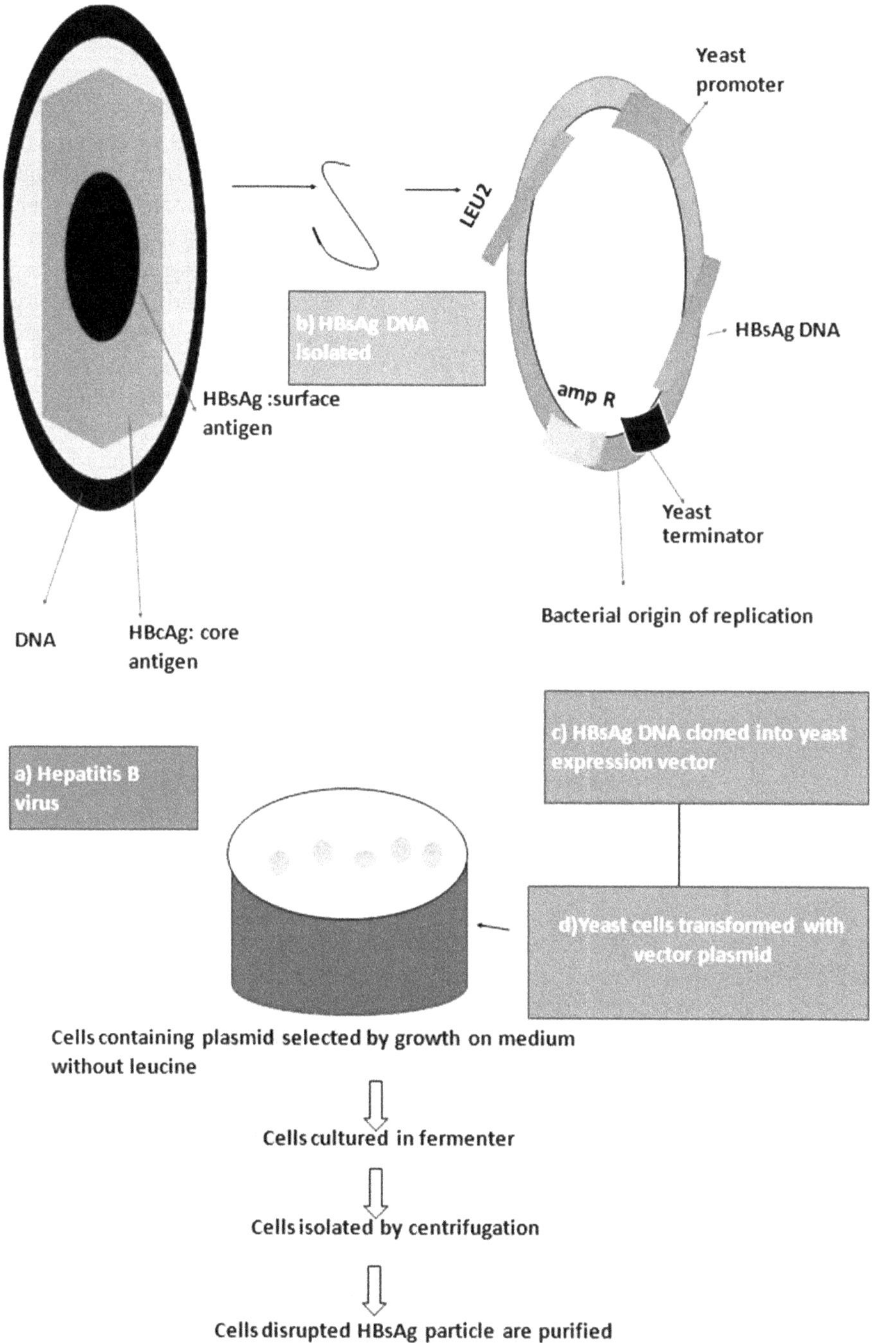

FIGURE 13.4 Schematics of Hepatitis B subunit production using rDNA technology with yeast vector.

the virus coat, can also aid as antigens to provoke the synthesis of virus-specific antibodies. The antibodies bound to the virus after infection deactivate the virus. Based on this principle, the production of recombinant vaccines is a two-step process: the first involves cloning of insert in expression vectors and the second is the production of recombinant vaccinia viruses. The hepatitis B virus (HBV) causes infection of the liver and is transmitted by blood transfusion, inadequately sterilized syringes and needles and by sexual contact (Ellis and Brodeur, 2012). Vaccines are prepared from inactivated small surface antigen proteins (HbsAg) found in the serum of carriers. Viruses cannot be grown in tissue culture. The production of anti-hepatitis B vaccine is precarious as it involves working with very large quantities of the virus. The vaccines require large quantities of serum, which makes them very expensive (Chisari, 1996).

The infectious hepatitis B virus particles have two proteins: a core protein associated with the virus genome and a coat protein or S protein envelope. The core protein is not useful as a vaccine, but is used for diagnosing the disease (Powell and Newman, 2012). The HBV coat has the hepatitis B surface antigen (HbsAg), which is found in the form of aggregates in the blood of patients. The coat protein is not synthesized in large quantities in *E. coli*, but has been cloned in yeast using a vector based on a 2 μm circle. The 3.2kb HBV genome contains the HbsAg gene. Initial cloning is done in prokaryote (e.g. *E. coli)* using a shuttle vector which can replicate on both bacteria as well as yeast (eukaryotic cell). Yeast is used because it is easy to culture and grows rapidly. Yeast vectors are based on a plasmid called the two-micron plasmid (2 μm circle) which inhabits yeast cells (Robinson, 2016).

Yeast-bacterial shuttle vectors have origin of replication (ori) of bacteria from pBR322 and a strong yeast promoter (Robinson, 2016). The HbsAg surface antigen coding sequence is isolated from the HBV genome and cloned into a yeast expression vector. The vector has a strong promoter (from the alcohol dehydrogenase I gene) and a transcription terminator. It contains both bacterial and yeast replication origins and selection markers (bacteria Ampr and yeast LEU2). Yeast cells are transfected with the vector plasmid. Cells containing the plasmid are selected by growth on a medium lacking leucine. They are then cultured in a fermenter where they grow to high densities. Media composition and fermentation conditions are aimed at increasing the retention of plasmids. At the end of fermentation, the HBsAg is harvested by centrifugation followed by lysing the yeast cells. It is separated by hydrophobic interaction and size-exclusion chromatography. The resulting HBsAg is assembled into 22-nm diameter lipoprotein particles. The HBsAg is further purified up to greater than 99% by a series of physical and chemical methods. The purified protein is treated in a phosphate buffer with formaldehyde, sterile filtered and then co-precipitated with aluminium hydroxide to form a bulk vaccine (Figure 13.5). The final product is adsorbed to aluminium hydroxide which acts as an adjuvant. Lot-to-lot consistency is thoroughly accessed before marketing the product. Vaccine quality is decided by the following factors: Potency; Sterility; Purity; Immunological response; Stability; Protection and longevity.

The vaccine contains thimerosal, a mercury-derived preservative at 1:20,000 or 50 μg/mL concentration as well as aluminium at less than 50 mg/mL of vaccine. The vaccine is either sold as a monovalent vaccine or is combined with other antigens

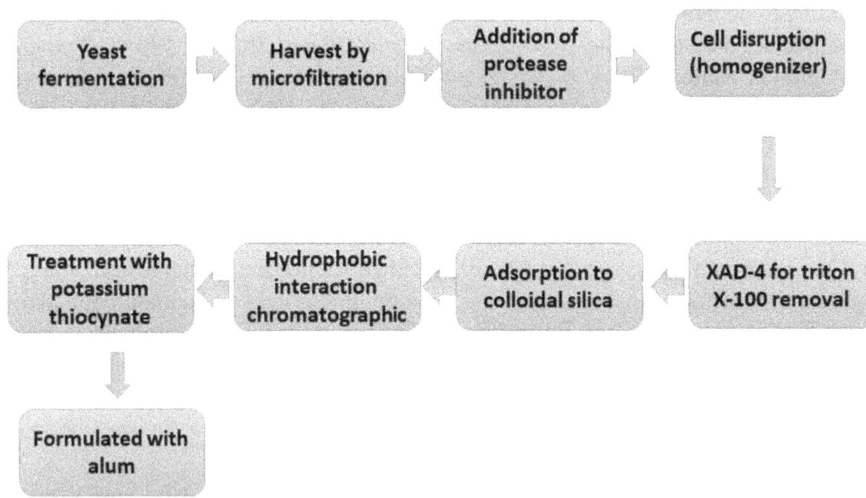

FIGURE 13.5 Downstream processing steps for the production of Hepatitis B vaccine.

such as diphtheria. The first recombinant protein vaccine for human use, licensed by Merck in the US in 1986, was Recombivax HB1 for prevention of hepatitis B infection. The protein produced from yeast is marketed as a vaccine for the subunit HBV. Popular products available are Recombivax HB (Merck); Engerix-B developed by GSK; Elovac B developed by Human Biologicals Institute, Indian Immunologicals Limited Division; Genevac B (Serum Institute). Shanvac B is India's first genetically engineered vaccine (Guni) against HBV developed by Shantha Biotechnics Pvt. Ltd in 1997. India is the fourth country to develop this highly advanced vaccine.

Market trends of Hepatitis B vaccine: The global demand for hepatitis therapeutics is estimated to hit USD 25.8 billion by 2025. A few factors projected to drive demand over the forecast period include the-number of people infected with hepatitis, increased alcohol and drug use and enhanced accessibility to hepatitis drugs. About 400 million individuals worldwide are afflicted with at least one type of hepatitis, based on WHO figures, and about 1.4 billion die annually because of the disease (Wittmann and Liao, 2016). Few state and private organizations are currently participating in the vaccine campaign to avoid hepatitis. HCV leads to more severe complications as compared to hepatitis A and B. According to WHO reports, 150 million people are infected by hepatitis C infection per year worldwide. In addition, most patients infected with HCV develop liver cancer or cirrhosis of the liver, and about 700,000 people die from hepatitis C infection and related disorders. Key players include Gilead, Johnson & Johnson, Merck & Co. Inc., Bristol-Myers Squibb Company, and AbbVie Inc.

13.7 INSULIN PRODUCTION FOR HUMAN USE

Human insulin, or humulin, was the first licensed drug produced through genetic engineering. An essential hormone that controls the absorption of sugar, insulin is formed in the pancreas by a limited number of cells and secreted into the bloodstream.

A failure to generate insulin leads to diabetes, but regular insulin doses are adequate to counteract or at least allay the disease's harmful effects (Burnett, 1983).

A protein named 'isletin' was isolated from the Islets of Langerhans in the pancreas by Banting and Best and was injected into a dog which had diabetes, thus curing the dog. It was from then onwards that the importance of insulin (later named) to cure diabetes was known, but it had some drawbacks. Before the advent of genetic engineering, insulin was extracted from bovine or porcine pancreas, which led to bloodshed of those innocent animals and other difficulties. Among many such difficulties was the immune response by the individual against the animal insulin, though both insulins are of similar chemical composition. It also led to inflammation in many patients. The method of extraction from animals made it difficult to obtain in large amounts to meet the needs of patients. Thus the concept of manufacturing insulin *in vitro* was understood by scientists. Due to the knowledge that genes have the code for all proteins, scientists could easily formulate the procedure for the production of proteins using RDT.

In 1955, Sanger first determined the sequence of insulin and also found that the peptide chain bound all the amino acids and keeps it together in a chain. In 1979, human insulin was able to be produced using RDT in bulk. The process is to cut the genomic DNA that codes for the insulin gene using restriction enzymes and inserting into a suitable vector, allowing the vector to grow under suitable conditions and then extracting, processing and purifying proteins for therapeutic usage. During the 1920s, before the recombinant molecule was produced, insulin was obtained for the treatment of diabetes mainly from the pancreases of farm animals like pigs and cows. Although this hormone is clinically active in mammals, the sequence of amino acids is never similar to the human molecule. It will also take a couple of months for all the livestock to mature and be able slaughtered for their pancreas. A few patients have, therefore, developed antibodies to insulin injection, occasionally leading to severe inflammatory responses. Since recombinant human insulin is the same as the natural product, immunogenicity really should not be a concern. Bioengineered insulin became the first significant biotechnological product. The new recombinant DNA technique was used for a more efficient approach (Wittmann and Liao, 2016). In 1978, a modified variant of the human insulin gene was constructed and implanted in *E.coli* at the laboratory of Herbert Boyer in the University of California, San Francisco. Eli Lilly Company received clearance for its genetically modified insulin in 1982.

Insulin comprises of two amino acid chains: the acidic A chain of peptide consisting of 21 amino acids and the basic B chain of peptide containing 30 amino acids. The dual polypeptide chains (A, B) attached through disulphide bond and thus two DNA fragments are sliced out of the genome and inserted into two separate vectors which are later ligated together following purification of each strand to render the protein usable (Figure 13.5). Once the A and B chains are synthesized, the 30 amino acid C peptide chain is then fused together to create a form known as pro-insulin. Pro-insulin is cleaved enzymatically to generate insulin (Ellis and Brodeur, 2012).

Expression of human insulin in E. coli: The first recombinant insulin was developed through the expression of both A and B chains independently and then by reconstructing them into a complete insulin molecule (Figure 13.6). DNA fragments

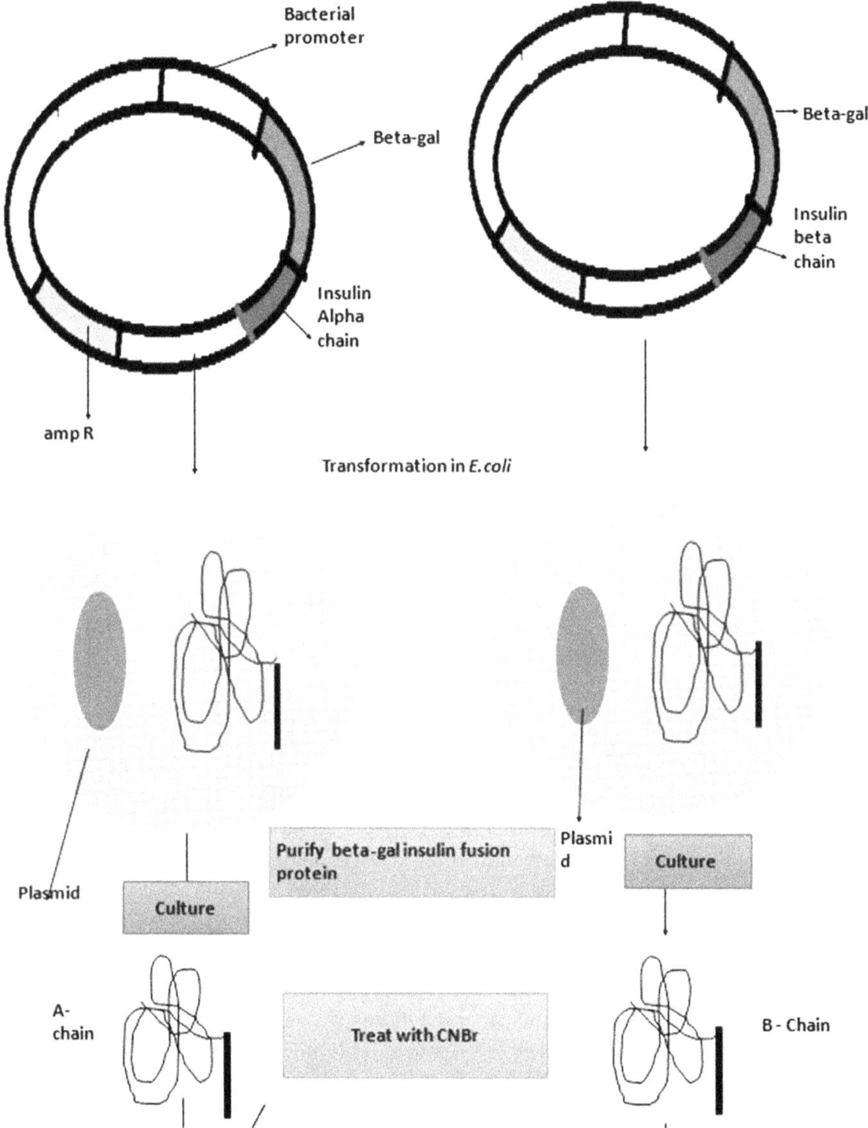

FIGURE 13.6 Human insulin production using rDNA technology.

coding both the insulin chains were generated by adding two complementary oligonucleotides which had been chemically fabricated. Each strand was linked to a bacterial expression vector so that the carboxy terminus of the enzyme β-galactosidase (β-gal) would be fused when the insulin chain is translated.

The expression vectors were transformed into *E. coli* and β-gal-insulin fusion proteins were amassed in bacterial cells which were harvested. By transfection, plasmids are inserted to the *E.coli* cells with the help of ligases. The bacteria containing

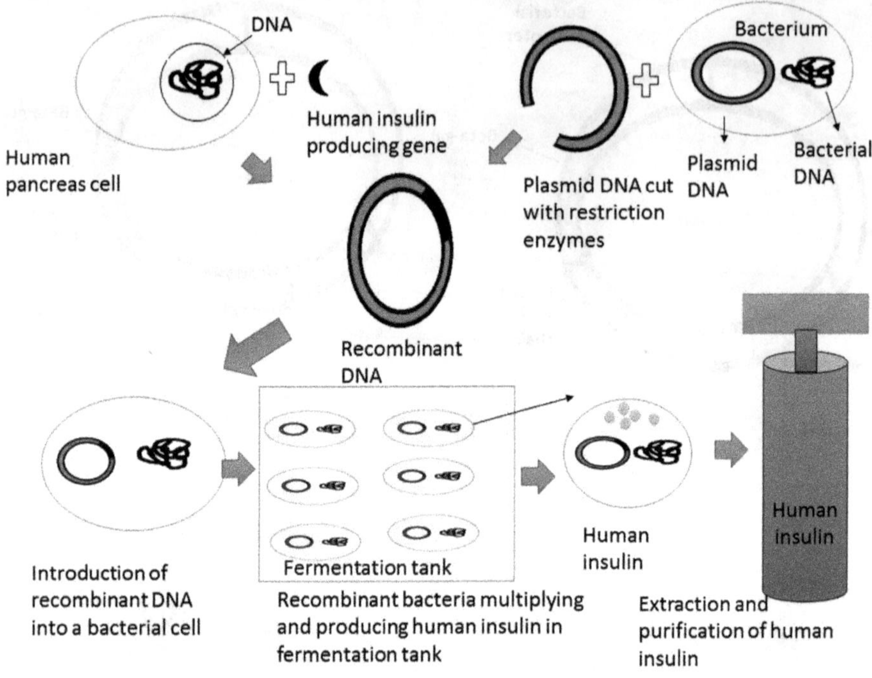

FIGURE 13.7 Humulin production steps using genetically engineered *E. coli*.

the plasmids which can synthesize insulin then undergo a fermentation process. Fermentation is carried out at optimal temperatures under suitable conditions. Care is taken to avoid any contamination with foreign bacteria. β-gal-insulin fusion proteins from these cells were then purified. Insulin coding DNA was modified to begin with a methionine codon. This design was established to eliminate the β-gal portion from the insulin polypeptide. Chemical treatment of this fusion protein with cyanobromide (CNBr) was carried out, resulting in peptide bond cleavage after all methionines. Natural insulin peptides have been obtained in this way. Because β-gal contains other methionine recurrences, CNBr has been treated with many small peptides. The A and B chains were separated and then combined together to produce active recombinant insulin (Figure 13.7).

DNA extraction is done from the cells by the addition of lysozome that digests the outer layer of the cell wall. A detergent mixture further separates DNA from the cell membrane (Figure 13.8). The bacterium's DNA is then treated with cyanogen bromide, a reagent that splits protein chains at the methionine residues. This separates the insulin chains from the rest of the DNA. The two chains are then mixed together and joined by disulfide bonds through the reduction–reoxidation reaction. The composition and purity of the insulin batches was evaluated against DSP. Normally High Performance Liquid Chromatography (HPLC) is utilized to assess insulin purity. Other isolation methods, such as X-ray crystallography, gel filtration and amino acid sampling, are still being used.

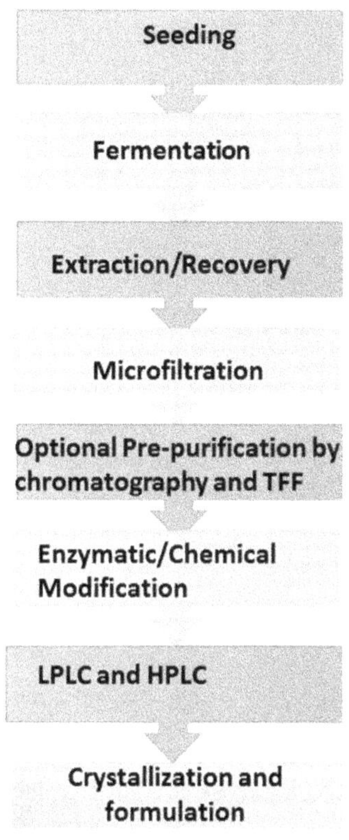

FIGURE 13.8 Steps involved in the downstream processing of insulin.

Manufacturing for human insulin complies with the protocols of the National Institutes of Health procedures (Powell and Newman, 2012).

Market trends of human insulin or humulin: In 2019, the worldwide demand for human insulin hit a size of US$34.8 trillion. In the future, the industry is projected to hit a size of US$58.3 trillion by 2025, experiencing a compound annual growth rate of about 9 percent between 2020 and 2025 (Wittmann and Liao, 2016). One of the main factors driving the growth of the industry is the increasing incidence of obesity and diabetes around the world. A larger number of people suffer from chronic lifestyle diseases attributable to sedentary lifestyles, rising geriatric demographics and poor eating patterns, thus increasing the need for insulin therapeutics. In addition, due to their high quality and cost-effectiveness, the demand for biosimilar drugs offers a boost to market growth. Various developments in technology, such as the production of pen instruments and safety pen needles for delivering HI in the body, have had a positive effect on consumer growth. These protection pens ensure reduced injury, pain and illnesses by blood-borne pathogen transfer, relative to regular needles, some of the major players being B. Braun Melsungen AG, Biocon, Eli Lilly & Company, Julphar, Novo Nordisk, Pfizer, Sanofi, Sedico, Wockhardt, Ypsomed Holding, etc.

13.8 CONCLUSIONS

The protein code carried by the recombinant DNA is called r-protein. Recombinant means that it is a combination of a plasmid and the DNA of interest through the process called genetic engineering. The plasmid is donated by the bacteria, in recent advances. Introduction of the DNA into the plasmid of a bacteria helps in expressing the product of the inserted DNA by the bacteria. This technique is used to produce many enzymes, hormones, antigens, antibodies, and products used for therapeutics that come under the category of r-protein. R-protein has tremendous application in the market because each protein has its own function to play in the lifestyle of people. A large part of the population is under-producing the naturally occurring essential proteins which are important for proper functioning of an individual. Mostly, the naturally synthesized proteins are either not produced, produced less or are not in a form usable by the individual because of faulty genetic expression. Thus such genetic engineering techniques come as a gift to individuals who require such proteins urgently. Hepatitis B vaccine and Humulin have proved to be a boon for the entire human race.

REFERENCES

Allan WH, Lancaster JE, Toth B, Newcastle disease vaccines, their production and use. Newctle. Dis. Vaccines Their Prod. Use, Food and Agriculture Organization of the United Nations, Rome, 1978.

Brown F, Dougan G, Hoey EM, Martin SJ, Rima BK, Trudgett A, Vaccine design. Vaccine Des, 1993.

Burnett JP, Commercial Production of Recombinant DNA-Derived Products, In: Inouye M (Ed.), Experimental Manipulation of Gene Expression. Academic Press, pp. 259–277, 1983.

Chisari FV, Hepatitis B Virus Transgenic Mice: Models of Viral Immunobiology and Pathogenesis, in: Chisari, Francis V, Oldstone MBA (Eds.), Transgenic Models of Human Viral and Immunological Disease, Current Topics in Microbiology and Immunology. Springer, Berlin, Heidelberg, pp. 149–173, 1996.

Ellis RW, Brodeur BR, New Bacterial Vaccines. Springer Science & Business Media.

Germanier R, Bacterial Vaccines. Academic Press, 2012.

Hendriksen CFM, Laboratory animals in vaccine production and control: Replacement, reduction and refinement. Springer Science & Business Media, 2012.

Powell MF, Newman MJ, Vaccine Design: The Subunit and Adjuvant Approach. Springer, 2012.

Robinson JM, Vaccine Production: Main Steps and Considerations , In: Bloom BR, Lambert P-H (Eds.), The Vaccine Book (Second Edition). Academic Press, pp. 77–96, 2016.

Siber GR, Klugman KP, Mäkelä PH, Pneumococcal vaccines: the impact of conjugate vaccine. Pneumococcal Vaccines Impact Conjug. Vaccine, 2008.

Wittmann C, Liao JC, Industrial Biotechnology: Microorganisms. John Wiley, 2016.

14 Alpha-Amylase, Protease, Lipase, and High-Fructose Corn Syrup Production

14.1 INTRODUCTION

Enzymes are biological catalysts that enhance the rate of reaction and play a vital role in the case of biological reactions. Catalysts increase the rate of otherwise slow or imperceptible reactions without undergoing any net change in their own structure. They are highly specific and selective and hence the chosen reaction can be catalysed with a particular enzyme to exclude side-reactions thereby eliminating undesirable by-products. This eventually reduces the expenditure associated with purification during downstream processing. In addition, the product is usually generated in sterile conditions.

Chemical catalysts used in industry were very complicated in nature, with the added disadvantage that processes carried out with the help of these chemical catalysts often required high temperature, pressure, and moderate specificity. Enzymes proved to be an efficient substitute for chemical catalysts. Enzymes work in milder conditions compared with chemical catalysts. High specificity and rapid reaction rates are the unique characteristics of these enzymes. Enzymes have been applied widely in various sectors of industry, such as textiles, paper, food, cosmetics, pharmaceuticals, and many others (Figure 14.1).

Enzymes have been used in starch saccharification, beer production, disorders of digestive systems, and also in the production of cheese from milk. Table 14.1 shows wide applications of enzymes in commercial food processing. Microbes are used for the production of exocellular enzymes. With the advent of recombinant technology, it is possible to produce the desired enzymes in high quantities on a large scale with less cost. Many different microbes have been used to produce a variety of industrially important enzymes (Krishna, 2011).

In addition to this, a significant number of enzymes are used for analytical (research), medical or diagnosis purposes. Table 14.2 shows a few of the important enzymes and their function; these are used frequently for analytical purposes.

14.1.1 Enzyme Classification

Enzymes have a wide range of activity and are very specific to the substrate. Enzymes are classified based on their enzyme catalysing reactions. The EC (enzyme

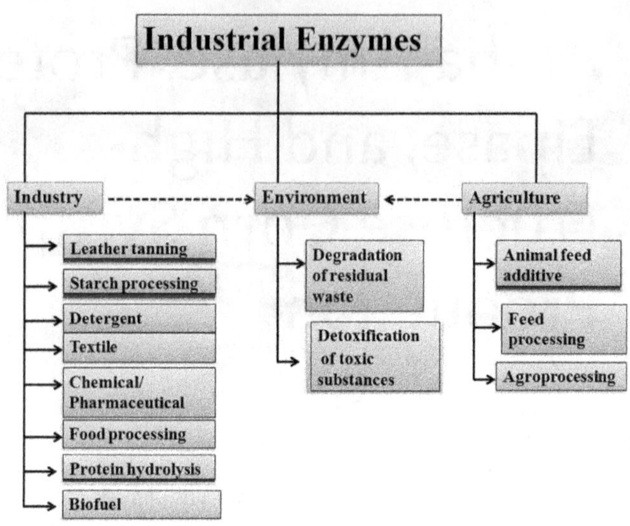

FIGURE 14.1 Application of enzymes in industrial, environmental and agricultural aspects.

TABLE 14.1
Application of Important Commercial Enzymes for Food Processing

Enzyme	Source	Applications
Starch processing α-Amylase	*Bacillus licheniformis*, *Aspergillus* spp.	Starch liquefactions, alcohol production
β-Amylase	Plant (malt)	Maltose production, alcohol production
lucoamylase	*Aspergillus* spp.	Starch saccharification, brewing, baking
Glucose isomerase	*Bacillus coagulans*	High fructose corn syrup sweeteners
Invertase	*Saccharomyces cerevisiae*	Invert sugar, sugar confectionery
Pullanase	*Klebsiella* spp.	Debranching of starch, brewing
Dairy processing Rennet	Stomach of calves	Cheese manufacture (milk coagulation)
Microbial rennet	*Mucor miehei*	Cheese manufacture (milk coagulation)
Lipase/ esterase	Fungal, bacterial, animal	Cheese ripening, milk fat modification, sausage ripening
Protease/peptidase	*Aspergillus Niger*	Cheese ripening
Lactase	*Kluyveromyces, Aspergillus* spp.	Lactose hydrolysis
Catalase	Bovine liver, *Pyrobaculum calidifontis*	Milk sterilization, bleaching
Fruit/vegetable processing Pectinase	*Aspergillus* spp. Trichoderma, Aspergillus spp.	Extraction/clarification of fruit juices
Cellulase		Fruit and vegetable processing

TABLE 14.2
Enzymes Used for Analytical Purpose

Enzyme	Applications
Nucleases	Cleave the phosphodiester bond between two nucleotide subunits. Used in rDNA technology. Remove one nucleotide unit at a time from either 3' or 5' end. Used in rDNA technology.
Ligases	Ligase enzyme used to bind to piece of DNA by phosphodiester linkage between the two strands. It acts as a biological glue and can bind both blunt ends and sticky ends. Used in rDNA technology.
Reverse transcriptase enzyme	Converts RNA template into DNA strand that is complementary to RNA strand. Used in rDNA technology.
Alkaline phosphatase	Remove the phosphate group present in 3 prime ends of DNA. Used in molecular biology research.
Polynucleotide kinase	Obtain from *E.coli* and add phosphate group in 5 prime end of DNA molecule. Used in molecular biology research.
Terminal deoxy-nucliotidyle transferase	Has the capacity to add one or more nucleotides in 3 prime end of DNA molecule. Used in molecular biology research.
Lysozyme	Attached with peptidoglyan and hydrolysed the glycosidic bond between N- acetylmuramic acid and N-acetyleglucosamine. Used in molecular biology research.
Angiotensin converting enzyme (ACE)	ACE help in the conversion of angiotensinI to angiotensin II by removing dipeptide. Contract blood vessel thus increase blood pressure.
Acetylcholin esterase	Catalysed the breakdown of neurotransmitter acetylcholine and acetate group. Helps in the termination of synaptic transmission.
Neucleotidase	Catalyse the hydrolysis of nucleotide into nucleoside. Convert adenosine monophosphate and guanosine monophosphate to adenosine and guanosine.
Kinases	Catalyse the transfer of phosphate group from ATP to specific substrate. Used in biochemistry research.
Trypsin	Hydrolysed the polypeptide bond into smaller peptide.
Hexokinase and Glucose-6-phosphate dehydrogenase	These enzymes are used in the measurement of glucose body.
Glucose oxidase	Diagnosis of glucose level in body. Used as a part of glucose biosensor to determine blood glucose concentration.
Bromelain	Used for the treatment of indigestion and inflammation.
Rennin	Rennin used in the therapy and helps in the digestion of milk protein.
Papain	Help in the meat protein and loosen necrotic tissue in wounds.
Catalase	Catalyse the decomposition of hydrogen peroxide to water and oxygen and prevent the formation of CO_2 bubbles in the blood.
Asparaginase	Used in the treatment of the lymphocytic leukaemia and convert asparagine into aspartate.
Cholesterol Oxidase	Used as a part of calorimetric biosensor to determine blood cholesterol concentration.

commission) number is a numerical classification scheme for enzymes, based on the chemical reactions they catalyse. Most of enzyme name ends with '-ase'; except for pepsin, rennin, and trypsin (Krishna, 2011).

14.2 SOURCES OF ENZYMES AND THEIR APPLICATIONS

Enzymes such as amylase, glucanases, proteases, beta-glucanases, amyloglucosidase, pullulanases, acetolactate decarboxylase, cellulases, pectinases, trypsin, rennin, lipases, lactases, mannanase, catalase, xylanases, ligninases, esterase, laccases, peroxidases, and transglutaminases, nitrile hydratase, D-amino acid oxidase, glutaric acid acylase, penicillin acylase, ammonia-lyase, humulin restriction enzymes, DNA ligase, and polymerases are used in multiple industries such as alcohol and beverage, fruit drinks, baby food, food processing, dairy, detergents, textiles, paper and pulp, rubber, oil and petroleum, pharmaceuticals, biopolymers, and molecular biology (Souza and Magalhães, 2010).

14.2.1 General Therapeutic Application of Enzymes

The main reason for using therapeutic enzymes is because of their higher efficiency. These enzymes are widely used in therapeutics like oncolytics, thrombolytics, anticoagulants, and are also used as a replacement for metabolic deficiencies. These enzymes can function well at relatively lower concentrations (Table 14.2). Therapeutic enzymes are used in treating cancer, damaged tissues, and treating infectious diseases. Some of the widely used enzymes in therapeutics are hyaluronidase, asparaginase, lysozymes, trypsin, collagenase, glutaminase, ribonuclease, streptokinase etc. These enzymes can be derived from bacteria, fungus, and algae (Souza and Magalhães, 2010).

14.3 SAFETY CONCERNS

The recent development in biotechnology has involved enzymes in multiple industries such as food, detergents, textiles, medicine, brewing etc. Hence usages must be monitored in order to avoid air-born enzymatic allergens and immunogens that cause sickness. Multiple organizations are involved in developing standards, and monitoring and regulating the usage of enzymes. Some of these are FCC, JECFA and AMFEP in Europe, and ETA has developed enzyme usage regulations in the USA (Krishna, 2011).

14.4 APPLICATION OF ENZYMES

14.4.1 Amylase

The first enzyme to be discovered and isolated, amylase catalyses the conversion of starch into sugar. It is found as different forms of amylase present in nature which possess industrial importance e.g. α-amylase, β-amylase etc. Amylase has wide

TABLE 14.3
Amylases and Their Applications in Their Respective Fields (Souza and Magalhães, 2010)

Industry	Application
Starch industry	Hydrolysis of starch in the starch liquefaction process.
Paper industry	90% of liquid detergent consists of amylase.
Detergent industry	Used to remove starchy stains such as chocolate, potato, gravies, custard, etc., in laundry.
Food processing industry	Enhances the fermentation rate inside dough in the bakery industry via hydrolysing the starch into smaller dextrin. Also used in the brewing industry, cake production, producing digestive aids, and processed fruit juice production.
Textile industry	Desizing of fabric is carried out by amylase after adding sizing agents such as starch. It acts as a strengthening agent in the threads.
Medicine	Acute inflammation of the pancreas, perforated peptic ulcer, strangulation ileus, torsion of an ovarian cyst, macroamylasemia, and mumps are caused by a higher concentration of amylase.

applications in industry (Table 14.3). Different types of amylase enzyme are used commercially.

α-amylase: this basically helps in hydrolysing the alpha bonds of large alpha-linked polysaccharides, such as starch and glycogen, yielding glucose and maltose. Major forms of alpha-amylase are found in human saliva (human α amylase) and pancreatic juice which help in the breaking down of 1,4 glycosidic linkages of starch. It is also found in seeds as a food reserve. α-amylase is used in ethanol fermentation by converting starch into oligosaccharides. Termamyl, an isoform obtained from *B. licheniforms* is used in detergents (Souza and Magalhães, 2010).

β-amylase: Beta amylase is found in plants, bacteria, and fungi species; it is not present in animal tissue; it helps in the hydrolysis of α 1,4- glycosidic linkages. It is also called 1,4-D-glucan maltohydrolase, glycogenase, or saccharogen amylase. In plants, beta-amylase reduces starch to maltose during fruit ripening. It is also present in seeds in an inactive form. Microbes also produce β-amylase to reduce extracellular starch. Both, α- and β-amylases are used in the production of beer and liquor, to reduce starch into fermentable sugars (Souza and Magalhães, 2010).

γ-amylase: Gamma amylase helps in the hydrolysis of (1-6) glycosidic linkages and the last(1-4) glycosidic linkages at the nonreducing end of amylose and amylopectin. It is also called glucan1,4--glucosidase; amyloglucosidase; exo-1, 4-glucosidase; glucoamylase; lysosomal-glucosidase; 1,4-D-glucan glucohydrolase. It is mainly used in the food industry, pharmaceuticals, the cosmetic industry and is also used in agriculture and environmental engineering (Souza and Magalhães, 2010).

Fungal amylase: *Aspergillus niger and Aspergillus oryzae* are filamentous fungi, producing extracellular enzymes including alpha-amylase. These fungal species are used in the large-scale production of amylase. The fungal amylase produced is more accepted as

GRAS – generally recognized as safe – because it greatly avoids bacterial contamination as it is acid-tolerant (pH 3). *Aspergillus oryae* is mainly used to produce amylase as well as in the production of citric acid and acetic acid (Souza and Magalhães, 2010).

Bacterial amylase: Bacterial amylase is mainly used due to its thermostable property and ability to produce in large quantities in a bioreactor. Industrially, the bacterial amylase has optimum functionality in high salt conditions as well as stability at higher temperatures. *Bacillus subtilis, Bacillus stearothermophilus, Bacillus licheniformis, and Bacillus amyloliquefaciens Chromohalobacter* sp., *Halobacillus sp., Haloarculahispanica, Halomonasmeridiana, and Bacillus dipsosauri* are used to produce alpha-amylase (Krishna, 2011).

14.4.2 Lipases

The hydrolysis of fat is carried out by lipases. Lipases perform an essential role in the digestion, transport, and processing of dietary lipids. Human pancreatic lipases convert triglycerides into monoglycerides and two fatty acids. Lipases are also produced from bacteria and fungus. The application of lipases is wide and they can easily be produced in large quantity; hence it is an industrially very demanding enzyme (Table 14.4).

TABLE 14.4
Lipases and Their Application in Various Fields

Industry	Application
Food industry	Used in biolipolysis i.e. removal of fats in processed meat and fish. Utilized to food to add flavours. Applied in fermentation step in sausage preparation.
Textile industry	Used in desizing cotton and denim fabrics. Used in the removal of size lubricants.
Detergent industry	Lipases are supplemented to the detergent mixture to enhance its efficiency.
Diagnosis	The concentration of lipids in the body can be diagnosed in certain conditions such as pancreatitis where lipase is applied.
Medical application	Used as digestive aids. Used in the treatment of malignant tumours. HGL(Human Gastric Lipase) used for enzyme substitution therapy. Applied in reducing serum cholesterol level.
Cosmetics	Water-soluble retinol are made from immobilized lipases; one of the major ingredients in topical anti-obesity cream.
Biosensors	Immobilized lipids along with glucose oxidase are used in blood cholesterol and triglycerides determination.
Biodegradation	Lipases degrade biodegradable polymer such as poly trimethylene succinate, which is used as biosensors. Castor oil degradation is carried out lipases produced by *Pseudomonas aeruginosa.* Used in ecorestoration by degrading oil spilled in the coastal environment by soil microbial lipases.

Bacterial lipases are mostly used in the detergent industry. The lipases produced by the bacteria are extracellular, and hence can easily be produced on a large scale. *Achromobacter, Alcaligenes, Arthrobacter, Bacillus, Burkholderia, Chromobacterium, Enterococcus, Corynebacterium and Pseudomonas* are used primarily to produce lipases. *Bacillus subtilis, Bacillus licheniformis, Bacillus amyloliquefaciens, Serratiamarcescens, Pseudomonas aeruginosa, and Staphylococcus aureus* produce lipases which are used in various industries such as oil mills, soap, dairy, and slaughterhouses (Treichel et al., 2010).

Lipase-producing fungi are *Mucor, Candida, Penicillium, Rhizopus, Geotrichum, Rhizomucor, Aspergillus, Humicola,* and *Rhizopus*. Extracellular thermostable lipase is produced by thermophilic *Mucor pusillus, Rhizopus homothallicus,* and *Aspergillus terreus*. *Mucor* sp. produces an extracellular, thermostable, inducible, and alkaliphilic lipase (Treichel et al., 2010).

14.5 PRODUCTION OF ENZYMES

Most of the microbial enzymes produced at industrial scale are exocellular in nature. The steps involved in enzyme production are: i. microbial culture, media preparation and propagation (upstream); ii. fermentation (midstream) and iii. separation and purification of enzyme followed by formulation (downstream). Figure 14.2 depicts the steps involved in enzyme production.

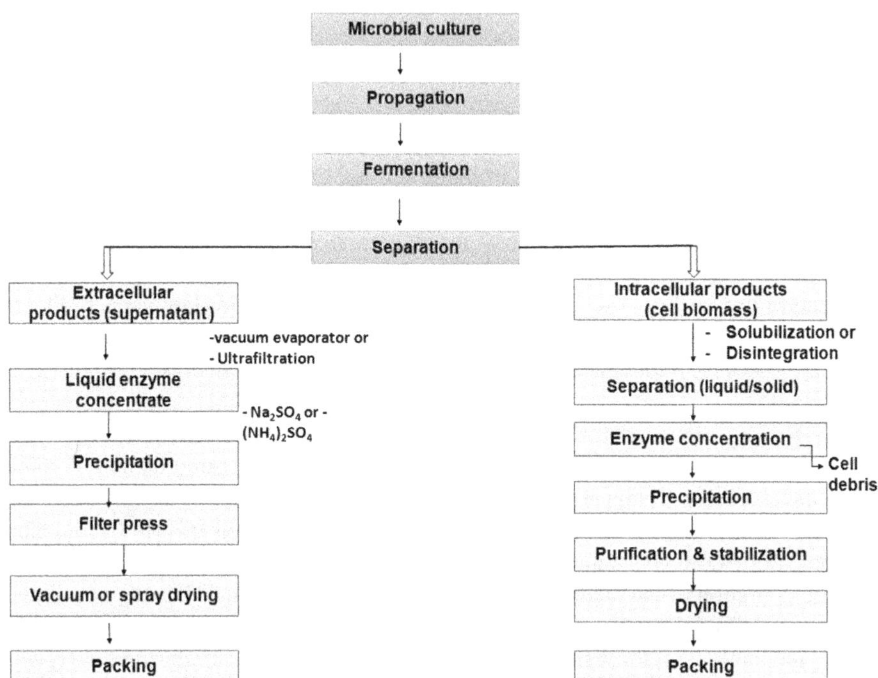

FIGURE 14.2 Flow chart showing the steps involved in enzyme production.

In general, two types of fermentation technique are employed for the industrial production of microbially derived enzymes.

14.5.1 Submerged Fermentation

For the processing of microbial produced enzymes, submerged liquid fermentation is traditionally used. Submerged fermentation entails the microorganism being submerged in an aqueous solution that contains all the nutrients necessary for development. Parameters such as temperature, pH, oxygen consumption and carbon dioxide formation are measured and controlled to optimize the fermentation process. In the harvesting of enzymes from fermentation media, insoluble materials, such as microbial cells, must be removed. This is usually achieved by centrifugation. As most industrial enzymes are extracellular (secreted into the external world by cells), once the biomass has been removed, they remain in the fermented broth. The enzymes in the residual broth are then concentrated according to their intended use through evaporation, membrane filtration, or crystallization. When pure enzyme preparations are needed, they are typically chromatographically isolated by gel or ion exchange. Several forms of submerged fermenter are known and can be grouped in a variety of ways: form or structure, aerated or anaerobic, batch or continuous. The aerated stirred tank batch fermenter is the most widely used type of fermenter.

14.5.2 Solid-State Fermentation (SSF)

SSF is used for the production of enzymes from microorganisms under conditions of low moisture content for growth. The medium used for SSF is usually a solid substrate (e.g. rice bran, wheat bran, or grain), which requires no processing. In order to optimize water activity requirements, which are of major importance for growth, it is necessary to take into account the water sorption properties of the solid substrate during fermentation. In view of the low water content, fewer problems due to contamination are observed. The power requirements are lower than for submerged fermentation. However, factors such as inadequate mixing, limitation of nutrient diffusion, metabolic heat accumulation, and ineffective process control limit the wider application of SSF. The SSF technique is generally applicable for low-value products with less monitoring and control. There exists a potential for conducting SSF on inert substrate supports impregnated with defined media for the production of high-value products.

SSF has higher volumetric productivity and the energy requirements are very low. It is easier to meet aeration requirements as it resembles the natural habitat of some fungi and bacteria. Temperature control is typically not difficult, in order that organisms are exposed to a constant temperature throughout their growth cycles. The availability of O_2 to the biomass can be controlled reasonably well at a particular level of saturation of the medium. Temperatures can rise to values that are well above the optimum for growth due to inadequate removal of waste metabolic heat. In other words, the temperature to which the organism is exposed can vary during the growth cycle; O_2 is typically freely available at the surface of the particle, however, there may be severe restrictions in the supply of O_2 to a significant proportion of the biomass that is within a biofilm at the surface or penetrating into the particle.

Production by the bacterial submerged state method requires media optimization and includes 5% starch, ammonium nitrate, sodium citrate, $MgSO_4.7H_2SO_4$., $CaCO_3.2H_2O$, peptone, yeast extract etc. Phosphate concentration in the medium has a regulatory effect on enzyme production by *B. amyloliquefaciens*. A high phosphate concentration promotes maltose uptake and microbial growth, while a high maltose uptake rate suppresses enzyme biosynthesis due to a catabolite repression effect. Thus the phosphate concentration should be optimized in the culture medium.

The inoculum is prepared by the shake flask culture method. The inoculum is then grown in small fermenters before final fermentation. The optimum pH for the production of amylase is 6–8. Maximal enzyme production occurs at a relatively low temperature of about 27–30 °C, but it depends on the bacterial strain used. Thermophilic bacteria such as *Thermonospora* sp. produce maximum enzymes when the temperature is 53 °C. Aeration is done in a range of 0.8–1 vvm for 48 h. The enzyme fermentation rate is usually very low during the exponential phase of growth but just before the rate of growth decreases and spore formation begins, amylase production increases (Figure 14.1).

The production of fungal alpha-amylase is mostly carried out by SSF using wheat bran as substrate. *Aspergillus oryzae* is considered the best microorganism for amylase production. For media preparation, wheat bran is moistened and steamed for 1–2 h to extract soluble starch. (8% w/V) starch, sodium nitrate, magnesium sulphate, KCL, $FeSO_4$, and malt extract are added to prepare an optimized medium. Fungal spores are produced in solid media to use as inoculum. The temperature is maintained at 28–30°C. The duration of fermentation is usually 3–5 d. Fungal moulds are grown in specially designed trays equipped with perforated covers. High rates of aeration are provided to encourage growth (Figure 14.3).

The cell mass present in the fermentation broth is separated by means of filtration or centrifugation. For fungal amylase, filtration alone is sufficient to separate

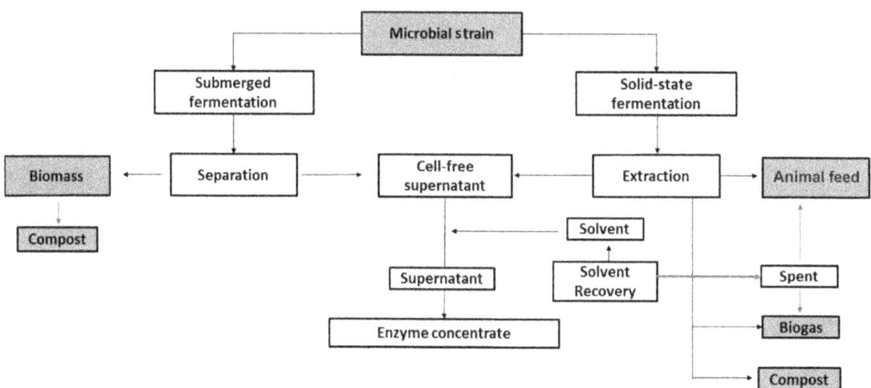

FIGURE 14.3 Block flow chart for enzyme production.

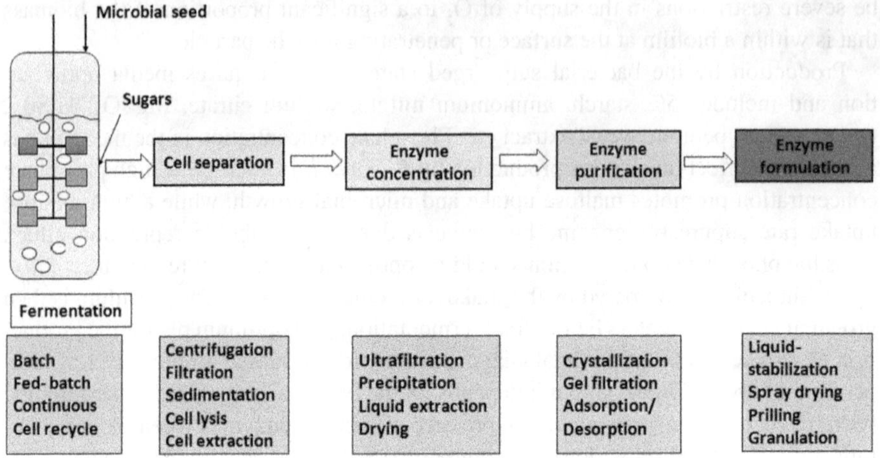

FIGURE 14.4 Downstream processing of the enzyme production.

solid from liquid, while for bacterial amylase filtration and centrifugation are carried out. Figure 14.4 shows the general steps involved in the downstream processing of enzyme production.

The primary task of formulation is to minimize losses in enzymatic activity during transport, storage and use. Secondary purposes include prevention of microbial contamination, avoidance of precipitation or haze formation, minimizing formation of sensitizing dust or aerosols and improving colour and odour by preventing denaturation, catalytic-site deactivation and proteolysis i.e. 'prevent unfolding' by altering the protein's environment so as to induce a compact protein structure. There are several ways to accomplish this. Different forms of granulation and immobilization are employed for enzyme formulation, tabulated in Table 14.4.

14.6 PROTEASES: PRODUCTION AND APPLICATIONS

Proteases are enzymes that degrade proteins. Proteolytic enzymes – or proteases or peptidases – are found in living organisms and play a vital role in the growth of cells, cell differentiation and signalling, metabolism, and many other processes. Proteases are an important enzyme which have been widely used for many decades. These enzymes catalyse the breakdown of proteins into small polypeptides or various single amino acids. There is an worldwide demand for protease owing to its applications in industry. These environmentally friendly enzymes are much better than chemical catalysts. Protease can be produced by microorganisms in optimum conditions (Li et al., 2017).

14.6.1 Proteases and Their Types

Proteases are naturally present in all organisms and share a gene content of 1–5%. These enzymes have been involved in several physiochemical reactions taking place

TABLE 14.5
Enzyme Formulations Based on Their Applications

Enzyme formulation type	Application
T-granulates	Physical strength and minimum dust. High shear granulation and coating techniques. Detergent industry.
Micro granulates	Fluidized bed drying for finer particle size distribution and safety (non-dusting) in food industry
Liquid formulations	Liquid product formulated and stabilized with polyols like glycerol, sorbitol, MPG, sugar, salts to decrease water activity
CT-Granulates	Coated-Tough for heat sensitive enzymes to prevent denaturation. Feed industry.
BG/SG granulates	Smaller particle size, easy incorporation into flour, safety. Spray drying and fluidized bed drying. Bakery industry.
Immobilized enzyme	High productivity at low cost. Enzyme immobilized on a carrier or in a matrix, enhancing stability and preventing leakage into substrate during application. Starch, oil & fat industry.

in the body, ranging from digestion of food to complicated cascades which are mainly regulated by the proteases. These proteases are classified based on the mechanism of catalysis (Krishna, 2011).

Serine proteases are also called serine endopeptidases, which have an active site of an enzyme containing a serine residue. These serine proteases have an efficient role in functioning in the body, ranging from blood clotting, providing immunity, and inflammation to help the digestive system to digest food in both eukaryotes and prokaryotes. Serine protease usually attacks the active site with serine on the peptide bond. Serine proteases have endo- or exopeptidase activity. Most of the serine proteases hydrolyse amides and esters. These enzymes have been efficiently expressed in mammalian, insect, yeast, and bacterial expression systems. Serine proteases are mainly alkaline in nature and active at pH 7.0–11.0 (Krishna, 2011).

Aspartate proteases are also called aspartyl proteases and are protoleytic enzymes. They are present in plant viruses, plants, and vertebrate organisms. These proteases cleave the dipeptide bonds which have hydrophobic residues and beta-methylene groups. These enzymes belong to the pepsin family and have a role in maintaining overall health which includes blood pressure and digestion of food (Li et al., 2017).

When the cascade of the protease reactions takes place apoptosis change happens in mammalian cells which undergo programmed cell death. This cascade particularly involves aspartate-specific cysteine proteases. This kind of protease includes a mechanism of catalysis which has the nucleophilic cysteine thiol in a catalytic step (Li et al., 2017).

These types of enzyme have secreted or transmembrane enzyme families which process and degrade various proteins. Metalloproteinases are characterized as secreted or membrane-anchored metalloproteinases. These classes of enzyme consist of Zn^{2+} or Ca^{2+} at their active site. Metalloprotein metal ions generally interact with a water molecule and enhance reactivity (Brouta et al., 2001).

In the case of threonine proteases, there is in particular N-terminal threonine which performs the function of catalysis at every site. This is often seen to attack at the lumenal surface (Li et al., 2017).

14.6.2 Production of Protease

Proteases at the industrial production scale are obtained mostly from microorganisms. Microbial protease production is as common as amylase or lipase production. The steps involved in the microbial process of protease are shown in Figure 14.5.

Microbes require an environment where they can receive proper nutrition whether in the form of macronutrients consisting of protein, lipid and carbohydrates or in the form of micronutrients (Anwar and Saleemuddin, 1998). It has been observed that the serine protease is predominantly produced by the bacterial species such as *Pseudoalteromonas, Microbulbifer, Shewanella, Photobacterium, Psychrobacter, Photobacterium, Vibrio, Bacillus, and Halobacillus,* while another class of protease, the cysteine proteases, were found to be produced by the fungal microbes *Sporotrichum pulverulentum* and *Aspergillus oryzae*. Metalloproteases are produced by several microorganisms such as *Bacillus cereus, Eupenicillium javanicum, Microsporumcanis, B. megaterium, Streptomyces griseus,* and *Bacillus subtilis.* There are various microbes efficient at producing proteases but the *Bacillus* strain turns out to be the most efficient producer of protease enzymes. Literature review suggests that the alkaline protease is produced by the bacteria *Bacillus* species which is isolated from *Pernaviridis.* Furthermore, *Bacillus subtilis, Bacillus amyloliquefaciens,* and *Bacillus proreolyticus*, have also been isolated for production of alkaline proteases (Li et al., 2017). Meanwhile a wide range of fungal species such as *Fusarium*

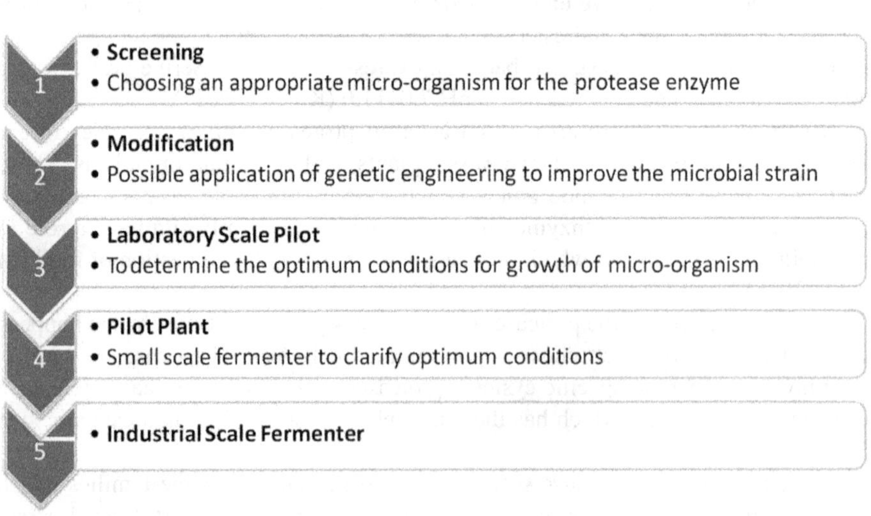

1
• Screening
• Choosing an appropriate micro-organism for the protease enzyme
2
• Modification
• Possible application of genetic engineering to improve the microbial strain
3
• Laboratory Scale Pilot
• To determine the optimum conditions for growth of micro-organism
4
• Pilot Plant
• Small scale fermenter to clarify optimum conditions
5
• Industrial Scale Fermenter

FIGURE 14.5 Microbial process of protease production.

graminarum, Chrysosporium keratinophilum, Penicillium griseofulvin, Aspergillus niger, Aspergillus flavus, Aspergillus melleu and *Scedosporium apiospermum* have been reported to be a good source of serine proteases. *Scedosporium apiospermum* and *Penicillium griseofulvum* are also considered good sources of proteases. Proteases isolated from fungal sources are in high industrial demand because of their stability, substrate specificity, and broad diversity. In a few cases, fungal proteases are preferred over bacterial protease as the mycelium is easily separated.

Most of the commercial proteases are obtained from the genus *Bacillus.* Neutral proteases are obtained with bacteria growing at the pH range 5–8 and can also tolerate low heat. The bacterial neutral proteases are known for their high affinity for hydrophobic amino acid pairs. Alkaline proteases produced by bacteria have high activity at alkaline pH. The optimum temperature for the production of these enzymes is 60 °C which also makes them an ideal component for the detergent industry. The protease enzyme produced from *Pseudomonas aeruginosa* is produced via the submerged fermentation. When the *Bacillus* is used for the production of proteases, both SSF and submerged fermentation are employed. However, 90 percent of proteases are produced by submerged fermentation (Li et al., 2017). The advantages of using these types of fermentation are the reduction rate of contamination, low investment, and a high level of conversion from the substrate to products. The submerged protease produced by using nutrients like fructose or lactose or by molasses (as carbon source) and corn steep liquor (as nitrogen source) (Anwar and Saleemuddin, 1998) has the advantage of simple media, a lack of complex machinery, and economical investment. In the past few years various industrial or agricultural by-products, for example, wheat bran, red gram husk, and sugarcane bagasse have been utilized as substrates having low cost in the case of SSF for protease production. The efficient production of the enzyme depends on various important physicochemical factors which include the concentration of substrate, pH, agitation, time of incubation, and temperature incubation. The ideal process control parameters depend on the source of microbes, the desirable end product, the fermentation method used, and many other factors (Li et al., 2017). Temperature is a vital parameter that has to be applied in protease production and it depends on microbes for having the maximum production and cell growth. The optimum or ideal temperature for the production of protease produced from the bacterial species of *Bacillus* has shown to be variable. The ideal temperature for protease production by different *Bacillus* species has been reported by several researchers as 45 °C for *Bacillus thermoruber,* 50 °C for *Bacillus licheniformis,* 70 °C for *B. subtilis,* 50 °C–70 °C for marine *Bacillus* sp. The fungi which can produce protease are *A. niger, Aspergillus fumigatus,* and *Penicilliu mitalicum A. flavus*, while their optimum temperature lies between 30 and 60°C (Anwar and Saleemuddin, 1998).

As with temperature, the pH also affects the biochemical reaction. For both *B. licheniformis* and *B. subtilis*, the optimum media pH was found to be 9.0, while in case of bacteria *Bacillus clausii,* optimum pH was found to be 12. The fungal proteases producing *strain P. italicum* and *A. niger* require acidic pH for protease production. In contrast, the fungus *Aspergillus versicolor* requires pH of 9.0 for the production of protease (Li et al., 2017).

The time required for ideal protease production by bacteria or fungi can be as long as 48 h to 9 d: for *P. aeruginosa* 20 h, *B. subtilis* DM-04 24 h, *Bacillus* species 72 h. The fungal species had 72 h incubation time for *A. niger* MTCC 281 but 96 h for *P. godlewskii* SBSS 25 (Li et al., 2017).

Substrate type and concentration play a major role in the protease formation. Wheat straw, barley, coffee pulp, sugarcane bagasse, copra waste, and grapes are used as feedstock or substrates. These type of substrate require pretreatment that includes washing and milling. Selection of a suitable substrate depends on two important factors; namely, cost and availability of the substrate. Deoiled cake of Jatropha is reported to have a good growth of bacteria and the substrate transformation is carried out by the microbe *P. aeruginosa* PseA (Pınar Çalık et al., 2001).

14.6.3 Industrial Applications of Proteases

There is a global demand for protease enzymes and they are widely used in many applications. Proteases are largely used in the detergent, leather, pharmaceutical, and food industries. Proteases are used in detergent formulations to enhance washing and to clean contact lenses effectively. They have also been applied to spot removal in a green manner fora shorter duration of soaking in detergents and less agitation. In various detergents, different enzymes have been added and have proved effective in washing for household purposes. Proteases also have wide usage in the leather industry and are used to improve the quality of leather in an eco-friendly manner and save both time and energy. In the food industry, proteases are used to modify the properties of proteins in food thus increasing solubility, flavour, digestibility, and nutritional value. Protease enzymes are also used in processing medicinal dietary products, juice fortification, and soft drinks. Peptides are applied in manufacturing cheese, hydrolysing peptide bonds to produce macro peptides and casein. In addition, proteases are used in tenderization of meat and produce soy products and soy sauce. In bakeries, proteases facilitate maintaining uniformity of dough, and uphold gluten strength, flavour, and texture in bread. Fungal and bacterial proteases have contributed to synthesizing therapeutic agents and curing diseases such as clot-dissolving and cancer, and are an anti-inflammatory agent. Protease obtained from *Aspergillus oryzae* is used as an agent for curing lytic enzyme deficiency. A few proteases have also been reported to have antimicrobial properties (Anwar and Saleemuddin, 1998). Protease proves itself to bean industrially important enzyme keeping the environment clean. Recently, researchers have focused more on the microbial production of this category of enzyme (Krishna, 2011).

14.7 PRODUCTION OF HIGH-FRUCTOSE CORN SYRUP

High-fructose corn syrup (HFCS) is a natural sweetener also known as glucose-fructose syrup which is mainly used in desserts, beverages, and other sweet foods. Previously, enzymes were not available commercially and hence the production of syrups was not possible. The only syrup available until 1935 was 42 DE (dextrose equivalent) acid converted corn syrup. In 1940, enzymes became commercially available and corn syrup was produced from corn starch enzymatically. A major achievement was the production of glucose isomerase which converts glucose to its sweeter isomer fructose. Fructose is approximately 1.7 times sweeter than glucose. The HFCS are differentiated based on their fructose content. The first HFCS was produced in 1967 and contained 15% fructose. Later, there was improvement in the yield processes and 42% and 55% fructose-containing HFCS were produced. Three major HFCS products differ by their fructose content; 42%, 55%, and 90%. Various quantities of fructose are used in products; for example, 42% fructose is mainly used in most food products which have liquid sweeteners. Higher levels of fructose content such as 55% are used in soft drinks and 90% is used in jams and jellies as a low-calorie sweetener (Parker et al., 2010).

14.7.1 Production Process

The production of HFCS is an enzymatic process. There are several varieties of corn but only the dent corn variety is used as a source of corn syrup. It is a common variety mainly grown in the United States (Li et al., 2017). Other materials used during the processing of corn into corn syrup are sulphur dioxide, hydrochloric acid, various enzymes, and water. A block flow diagram of HFCS production is shown in Figure. 14.6.

Dried and cleaned corn kernels are placed in large stainless-steel tanks called steep tanks. Warm water with sulphur dioxide is added to the tank , forming a sulphurous acid solution; this process is used to soften the kernels, which makes it easy to remove starch from them. The softened kernels are then further processed to remove the inner portion of the kernel, called the germ, which contains most of the corn oil. The germ is separated and pumped onto a series of screens and washed several times to remove the starch.

The slurry formed is composed of starch, proteins, and fibre. This slurry is further processed to remove starch from the fibre. The starch and protein mixture is called mill starch, and is centrifuged. Owing to the difference in specific gravity the heavier starch can be separated from the lighter proteins. The starch is diluted with water before being washed and filtered to remove the remaining protein. It is again centrifuged in order to obtain pure starch. The total and soluble protein contents of the starch slurry should be lower than 0.3% and 0.03% respectively.

Saccharification is the process of breaking starch into sugars in the presence of enzymes. Saccharification of liquefied starch slurry is achieved by the enzyme glucoamylase which produces more dextrose from the branched chains of the starch. There are certain conditions for this step of saccharification; the temperature should be 60 °C, the pH should be adjusted to 4.3, and a holding time of around 65 to 75 h is required. The feed which undergoes saccharification contains 30 to 35% dry substance and 1:1 of glucoamylase solution per ton of dry weight of starch. The product obtained is 94 to 96% dextrose, 2 to 3% maltose, 1 to 2% higher saccharides, and 30 to 35% dry substance. To obtain 42% fructose HFCS, the saccharification process needs to produce 94–96% dextrose, which requires controlled environmental conditions. The dextrose syrup produced by the process of liquefaction and saccharification needs to be refined for the removal of ash, proteins, and metal ions; otherwise these may cause interference with the process of

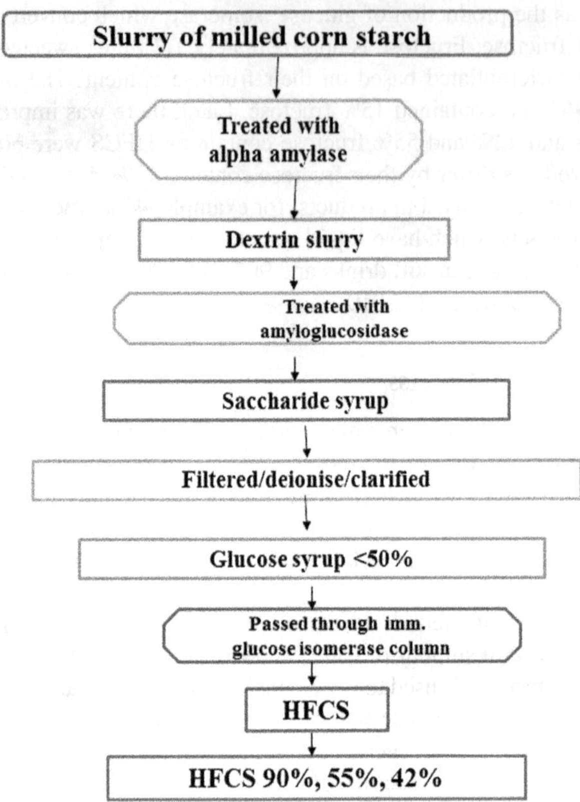

FIGURE 14.6 Industrial production of HFCS.

isomerization. This process of refining is carried out on rotary precoat vacuum filters. Next, they are passed through several more checks and polish filters to remove traces of particles. The syrup is then decolorized with the help of activated carbon and deionized by ion exchange systems (Figure 14.7). The decolorized and deionized dextrose syrup is evaporated to concentrated dextrose and Mg^{+2} ions are added to activate the isomerase.

The isomerization process of glucose to fructose takes place in a packed column of an immobilized enzyme; glucose is converted to fructose (Figure 14.8). The required conditions are temperature of 55–65°C, pH 7.5 to 8, and holding time should be 0.5 to 4 h. The optimum temperature is 60 °C; temperatures above this can cause inactivation of enzymes and temperatures lower than 55°C may result in slower production of fructose and also cause microbial contamination (Li et al., 2017). The optimum pH for enzymatic activity is 8. The process of enzymatic isomerization of glucose to fructose is reversible.

The fructose concentration in the effluent is expected to be half the amount of dextrose in the feed; e.g. if the feed contains 96% of dextrose, the fructose will be 48%. However, exact equilibrium is not reached in the given holding time and the effluent

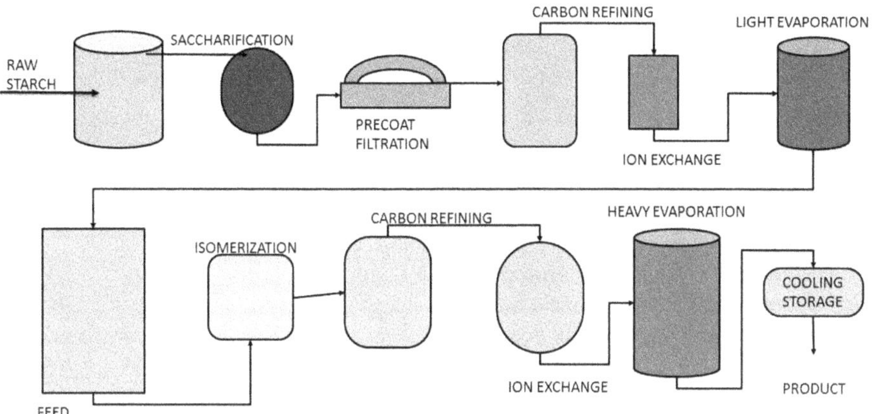

FIGURE 14.7 Flow diagram of high-fructose corn syrup production.

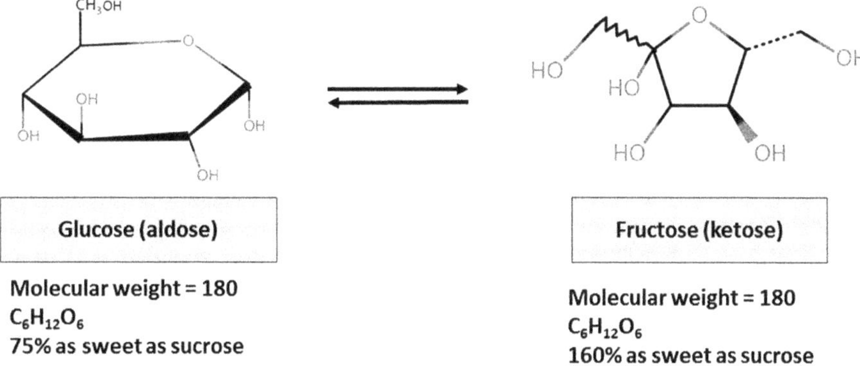

FIGURE 14.8 Conversion of glucose to fructose by glucose isomerase.

contains 42% fructose. The activity of enzymes drops after a particular time period and their half-life is 70 to 120 d. Therefore, at the later stages, the feed flow rate should be lowered to obtain 42% fructose. Usually, series and parallel configurations of immobilized enzyme columns are used to compensate for activity loss in enzymes. Variations in temperature, feed pH, and flow rate affect the uniformity of product quality. Polarimeters helps in obtaining constant fructose levels. The columns must be replaced two or three times a year. Enzyme cost constitutes most of the operating costs. Reduction in isomerization cost can be brought about effectively by improving enzyme stability and activity. The color and ions of the HFCS produced by dextrose isomerization can be removed by carbon treatment and ion exchange respectively. The 42% refined HFCS is evaporated for shipment, yielding 71% solids.

The HFCS obtained after isomerization contains 42% fructose, 52% of dextrose, and 6% oligosaccharides. The fructose in 42% syrup must be concentrated to obtain

55% and 90% fructose syrups. Fructose forms a complex with certain cations like calcium, which is used for the concentration of fructose in HFCS.

There are two types of process for the enrichment of fructose from 42% syrup:

1) Utilization of an inorganic resin for the selective molecular adsorption of fructose.
2) Chromatographic fractionation using organic resins.

The elution of fructose from the column is done using deionized and deoxygenated water. Enrichment is carried out in a column packed with low cross-linked fine-mesh polystyrene sulphonate-Ca cation exchange resin. This enriched syrup contains 90% fructose and is called Very Enriched Fructose Corn Syrup (VEFCS). This is blended with 42% fructose syrup to obtain the desired fructose content, such as 55%. The effluent from isomerization may be recycled back to the feed solution to obtain 42% fructose syrup in the effluent of the isomerization column (Parker et al., 2010). The raffinate stream which is rich in oligosaccharides is recycled back to the saccharification step. The water which is used as an eluent during enrichment of the syrup should be minimized to increase the solid content of the syrup.

HFCS has replaced sucrose as a sweetener with low calories, to a large extent. The price per pound of HFCS was 10 cents lower than sucrose in 1981. Recent developments in HFCS technology have increased this difference further, leading to a further replacement of sucrose and glucose by HFCS as a low-calorie sweetener in carbonated drinks, canned fruits, ice cream, and bakery products over the last two decades. Consumption of low-calorie sweetener is expected to approach 130 lb/capita per annum (Parker et al., 2010). HFCS has several advantages; application of HFCS add more flavour for both sweet and spicy flavours such as baked goods, fruit fillings, tomato products, canned fruit, and beverages. Furthermore, it enhances moisture control, retards spoilage, enhances texture and extends product freshness in baked goods, granola, spaghetti sauce, ketchup and condiments, breakfast and cereal bars.

14.7.2 Public Health Concerns

The excess use of HFCS may be a concern to health for the public. There are two major health-related issues caused due to HFCS: i. Role in metabolic syndromes: obesity, diabetes, and other cardiovascular diseases. Several studies have shown that increased consumption of HFCS causes an increase in obesity and cardiovascular disease; ii. Mercury contamination: HFCS contain traces of mercury. Its production uses caustic soda, which involves mercury cells, and mercury is toxic to the neurological cells and may be harmful (White, 2008).

14.7.3 Safety Concerns

Recent developments in biotechnology have involved enzymes in multiple industries such as food, detergents, textiles, medicine, brewing etc. Therefore usages are to be monitored accordingly to avoid airborne enzymatic allergens and immunogens that cause sickness. Multiple organizations have been monitoring and developing standards in the usage of enzymes. Some such organizations are FCC, JECFA, and AMFEP in Europe, and ETA in the USA has developed enzyme usage regulations.

The global market size for enzymes was projected at USD 9.9 billion in 2019 and is projected to rise from 2020 to 2027 at a compound annual growth rate (CAGR) of 7.1 per cent. It is expected that rising demand from end-use industries such as food and beverage, biofuels, animal feed, and home cleaning will boost market growth over the forecast period. Increasing demand for high-quality foodstuffs coupled with natural taste and flavour has led to consumer growth over the past decade. This pattern indicates the need for the expansion of processed and flavoured foods within the product category of industrial enzymes. Apart from alpha amylase, cellulase, lactase, amylase, pectinase, and mannanases are the prominent carbohydrases used in several end-use industries such as food and beverage, animal feed, and pharmaceuticals. Key players – including Novozymes, DuPont, and DSM – together represent over 75% of the market share.

14.8 CONCLUSIONS

Enzymes are widely studied and are used in multiple fields. The present chapter summarizes the important usage of the most commonly used enzymes in different fields and broad-spectrum usage of enzymes. Enzymes are manufactured industrially and sold for multiple purposes. Various types of microorganism are used to produce enzymes and the specificity of the enzymes produced by these organisms is heavily researched. The enzyme industry is one of the busiest and most profitable industries.

REFERENCES

Anwar A, Saleemuddin M, Alkaline proteases: A review. Bioresour. Technol. 64, 175–183, 1998.

Brouta F, Descamps F, Fett T, Losson B, Gerday C, Mignon B, Purification and characterization of a 43.5 kDa keratinolytic metalloprotease from Microsporum canis. Med. Mycol. 39, 269–275, 2001.

Krishna PN, Enzyme Technology: Pacemaker of Biotechnology. PHI Learning, 2011.

Li Y, Wu C, Zhou M, Wang ET, Zhang Z, Liu W, Ning J, Xie Z, Diversity of Cultivable Protease-Producing Bacteria in Laizhou Bay Sediments, Bohai Sea, China. Front. Microbiol. 8, 2017.

Parker K, Salas M, Nwosu VC, High fructose corn syrup: Production, uses and public health concerns. Biotechnol. Mol. Biol. Rev. 5, 71–78, 2010.

Çalık P, Çalık G, Özdamar TH, Bioprocess development for serine alkaline protease production: a review. Rev. Chem. Eng. 17, 1–62, 2001.

Souza P, Magalhães P, Application of microbial α-amylase in industry - A review. Braz. J. Microbiol. 41, 850–861, 2010.

Treichel H, de Oliveira D, Mazutti MA, Di Luccio M, Oliveira JV, A Review on Microbial Lipases Production. Food Bioprocess Technol. 3, 182–196, 2010.

White JS, Straight talk about high-fructose corn syrup: what it is and what it ain't. Am. J. Clin. Nutr. 88, 1716S–1721S, 2008.

15 Fermented Milk Products

Yoghurt, cheese, and other fermented milk products have been consumed for over a millennium. The belief that this is beneficial to health is as old as the products themselves. Fermented milk products are rich in protein, vitamins, and minerals. Yet only since the 1930s have scientific communities been backing up various beneficial claims about fermented milk products. Several research papers have stated that desirable bacteria in milk suppress other disease-causing bacteria in the intestine of human beings. These observations have paved the way for studies on the potential benefits of lactic cultures and fermented products to benefit humankind. Over the past few years, a lot of studies have been concentrated on producing fermented milk products that improve health and have other therapeutic properties. As defined by the International Dairy Federation fermented milk is a milk product prepared from milk – skimmed or not – with specific cultures (Widyastuti et al., 2014). The microflora are kept alive until sale to the consumers and may not contain any pathogenic germs. In this chapter, the production of cheese, yoghurt, and probiotic products are discussed in detail.

15.1 CHEESE

Cheese can be defined as a consolidated curd of milk solids in which milk fat is entrapped by coagulated casein. It is a popular proteinaceous dairy product that is produced by coagulation of the milk protein casein and is available in a wide range of texture and flavours. Cheese is a universal name given to a group of milk-based fermented products consumed all over the world. Several hundred varieties of cheese can be manufactured according to the particular combination of salt, temperature, pH, and culture used. Cheese originated from the weather conditions of the Middle-East many years ago. Being an age-old food item, production started around 8,000 years ago in the 'Fertile Crescent' area surrounding the Euphrates and Tigris rivers of the Middle-East. The 'Agricultural Revolution' that took place in this region helped people to recognize the nutritive value of milk. Milk functions as a good source of nutrition for microorganisms, especially bacteria, favouring their growth and leading to contamination in warm climates (Gorbach, 1990).

Curdled milk was observed when milk was placed in a pouch made from a goat stomach. A scientific study revealed that the enzyme renin present in the goat stomach pouch curdled the milk, and additionally the temperature of these countries was favourable for enzymatic activity to occur. Curdling of the milk due to fermentation of milk sugars followed by a swaying motion would have resulted in the separation of whey and buttermilk which is then further salted, producing high protein-containing nutritious food items i.e. cheese. Cheeses having strong flavour are mainly made from sheep milk or goat milk because of the presence of a large number of fatty acids, caproic acid, caprylic and capric acid. Based on these properties, the treatment of milk, the method of preparation, the conditions to be maintained such as temperature, the properties of coagulum, and the choice of locality, there are many cheese types (Gurr, 1987).

15.1.1 Classification

Cheese types can be classified based on their consistency (soft cheese, hard cheese, semi-hard or semi-soft cheese, extra-hard cheese and fresh cheese), milk used (sheep, buffalo, goat, cow), their manufacture (sour milk cheese, ultrafiltration, rennet), interior (eyes and moulds), the content of fat and surface (soft, hard, with smear, moulds) (Figure 15.1). Additionally, they also differ in certain bioactive components and flavours created during different stages of ripening when the main components of cheese – lactose, fat, and proteins – are broken down by lipolysis, proteolysis, and fermentation. Thus the variety of cheeses on the market is enormous, which is also reflected in the variability in the composition of the different types of cheese. In most countries with a high consumption of milk products, the majority is produced from cow's milk. Milk mainly consists of fat, protein, and water as well as vitamins, minerals, and trace elements. Lactose is rarely present (Gorbach, 1990).

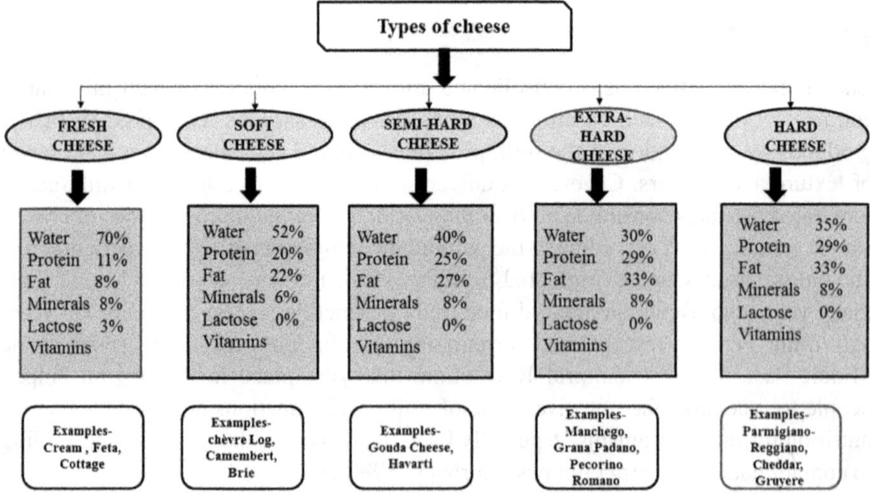

FIGURE 15.1 Classification of cheese.

15.1.2 Nutrition

Lactose: Approximately 70% of the universal population is lactose intolerant because of which consumption of dairy products can lead to various symptoms such as flatulence, nausea, diarrhoea, abdominal cramps, bloating etc. However, the onset of the cheese-ripening process results in the partial washout of lactose along with whey and fermentation of the remainder into alpha-hydroxy acid followed by ethyl alcohol, carbon dioxide, acetaldehyde, 2,3- butanedione and ethanoic acid. The fact that all cheese types – except fresh and a few soft cheese types – are lactose-free is an advantage against lactose intolerance, allowing consumption of these cheeses and hence contributing to a healthy diet (Gorbach, 1990).

Protein: Cheese being an important source of amino acids and proteins is reported to produce all the essential amino acids along with bioactive small peptides which are unique amino acid sequences originally found within bigger protein molecules. These, when bound to proteins, remain in a dormant state. However, in their active form they are involved in a wide range of biological activities such as anti-cancer, anti-thrombotic, anti-caries, anti-inflammatory, and cholesterol-reducing activities (Gurr, 1987).

Vitamins and minerals: Milk and dairy products contain various quantities of vitamins and minerals, the most important being calcium. One third to one half of the recommended daily intake of 1.2 g calcium is supplied by 50 g of hard and semi-hard cheese. It is also a good source of zinc, magnesium, and phosphorus. A portion of hard cheese contains 15% of the daily recommended intake of vitamin A, over 20% of vitamin B6, 10% of vitamin B2, and over 40% of vitamin B12 and the energy content of 10% of a regular diet (Gurr, 1987).

Fat: Fat is one of the most important elements of cheese and constitutes 20% to 35% of the dry weight. It is reported that on average cheese constitutes 60% of saturated fatty acids with the most common being myristic acid (9.8% fat), stearic acid (8% fat) and palmitic acid (26% fat), and 4.6% poly-unsaturated fatty acids especially oleic acid (16.5% fat) and 2.35% monounsaturated fatty acids. Saturated fatty acids (SFA) play a crucial role by regulating genes and proteins and expression and ensuring bioavailability (BA) of PUFA (Buttriss, 1997).

15.1.3 Cheese Processing

Even though cheese making was initiated 8,000 years ago, the advance of technology has led to a better insight into the raw materials that can be used for the production of consistent and better-quality cheese. These include heat killing of pathogens, and the addition of starter cultures and rennet to produce good quality cheese. A guide for the production of good quality cheese includes steps that can be incorporated into cheese of the best quality and longer shelf life. Numerous elements are required for the production of good quality cheese. The source, nature, and standardization of milk play a vital role in determining the quality of cheese produced. Cow milk is known to have increased plasmin activity due to its high somatic cell count in comparison to goat milk which results in better quality of cheese obtained from goat milk than obtained from cow milk. Standardization of milk on the other hand avoids excess fat and casein losses in whey.

This can be carried out via three methods: adding skimmed milk powder or liquid or by removing cream. The starter culture is pure lactic acid bacteria added for flavour development and to lower pH during processing. These include fast-acid, adjunct, and genetically modified starters which can be used in liquid, dried, or lyophilized form.

Colouring is carried out occasionally to change the original milk colour and whiten the curd.

The quality of curd can be improved and growth of microbes interfering with the ripening and maturation of cheese is avoided by the addition of chemicals such as sodium nitrate and calcium chloride. Chymosin is the major enzyme used in cheese processing. Other coagulants include mucorpusillus, cryophonectriaprassitica etc. These coagulating enzymes or rennet hydrolyse k-casein to para-k-casein within curd which is hydrophobic in nature and macro peptide which is hydrophilic in nature (Siso, 1996).

15.1.4 Major Steps Involved in Cheese Production

The objective of cheese making is to obtain the optimum cheese composition for moisture, acidity (pH), fat, protein, and minerals (especially calcium) and to establish the correct structure of the cheese at microscopic level (Johnson, 2017).

The various steps involved in cheese production are

- Milk is standardized
- Inoculation of starter culture
- Addition of rennet for coagulum formation
- Shrinkage of curd (removal of whey)
- Salting of curd and pressing into shape
- Ripening of cheese

15.1.5 Standardization of Milk

The nature of cheese is directly affected by quality of milk being used. For example, cheese made from milk containing fewer fats becomes hard and leathery. If milk contains more fats, then the 'spreadability' property of cheese increases. Protein and fat content in milk both affect the quality of cheese. Therefore, the fat/protein ratio is adjusted to obtain consistent quality (Johnson, 2017).

15.1.6 Pretreatment of Milk

The milk used for cheese processing is obtained from goats, cows, or sheep. The milk is subjected to pasteurization at 75 °C for 15 min to eliminate pathogens and extend shelf life. The process is carried out in order to deactivate the organisms and enzymes responsible for spoilage of milk. For the production of cream or cottage cheese, high-temperature short-time pasteurization or low-temperature long-time pasteurization is used (Johnson, 2017). Pasteurization gives better control for growing desirable strains in order to obtain good texture and quality cheese. This is followed by the standardization of milk by varying the fat and casein ratio. Sometimes milk is homogenized by pressuring it at high pressure through small pores to decrease the size of fat globules present in milk to be used in the production of soft cheese. Finally milk is standardized ready for inoculation of the starter culture.

15.1.7 Inoculation of Starter Culture

In the past, people were used naturally occurring bacteria to produce lactic acid. Nowadays, dairy technologists use artificial starter cultures for inoculation. A starter culture is defined as a preparation of living microorganisms which is deliberately used to assist the beginning of fermentation, producing specific changes in chemical composition and the sensorial properties of the substrate in order to obtain a more homogenous product. The starter culture is a mixture of living entities (microorganisms) that actively takes part in the fermentation process and produces specific changes in the composition of certain chemicals and important properties of the substrate to obtain a more homogenous product. The *Lactococcus lactis* strain is mainly used for cheese preparation at 40 °C temperature and at relatively high temperature, strains such as *Streptococcus thermophiles*, *Lactobacillus bulgaricus*, *L. helveticas* are used. Lactic acid lowers the pH of milk which leads to coagulation of casein. The surface of casein micelles has a negative charge which keeps them away from each other; lactic acid provides H^+ ions which neutralizes the negative charge present on the surface of casein micelles and helps to form clumps together. Thus providing favourable pH helps for the action of the enzyme rennin, which forms curd from casein. Low pH inhibits the growth of proteolytic and undesirable bacteria (Siso, 1996).

However, using lactic acid directly for cheese production has disadvantages such as i) Attack of bacteriophage on lactic acid bacteria may lead to damage of the bacteria (this can be avoided by choosing bacteriophage resistant strains of lactic acid bacteria (Figure 15.2), ii) Lactic acid bacteria are Gram-positive bacteria and susceptible specifically to penicillin which is an antibiotic, and inhibits the growth of

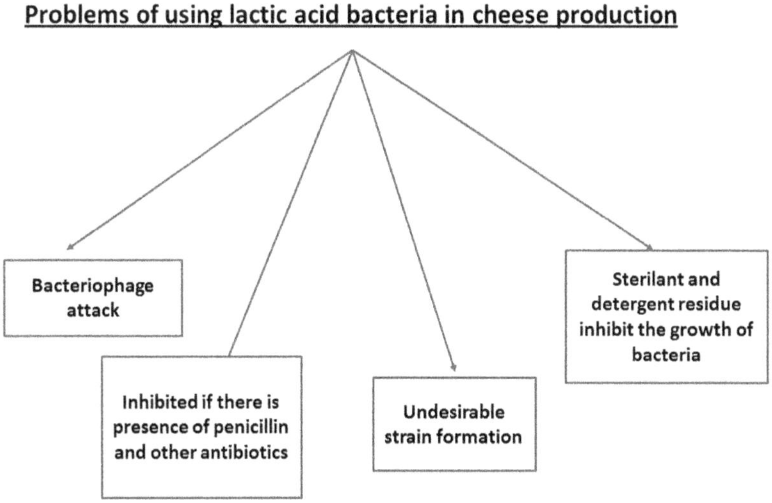

FIGURE 15.2 Challenges of using lactic acid bacteria in cheese production.

bacteria, iii) If due to mutation formation of undesirable strains takes place then they may produce undesirable by-products such as gas and flavour, and can produce antibiotics against other strains of lactic acid bacteria, iv) In the case where some sterilants or detergents remain in small amounts in equipment used for cheese production, it can lead to inhibition of bacterial growth (Gorbach, 1990).

15.1.8 Addition of Rennet for Coagulum Formation

Rennet/chymosin is an enzyme required for coagulum formation derived from the fourth stomach, abomasum, or vell of freshly slaughtered milk-fed calves. 94% rennin + 6% pepsin is found in the extract of young calves; comparatively older cow extract contains 40% rennin + 60% pepsin (Figure 15.3). Pepsin leads to hydrolysis of the formed coagulum and resulting in a low yield of cheese (Johnson, 2017).

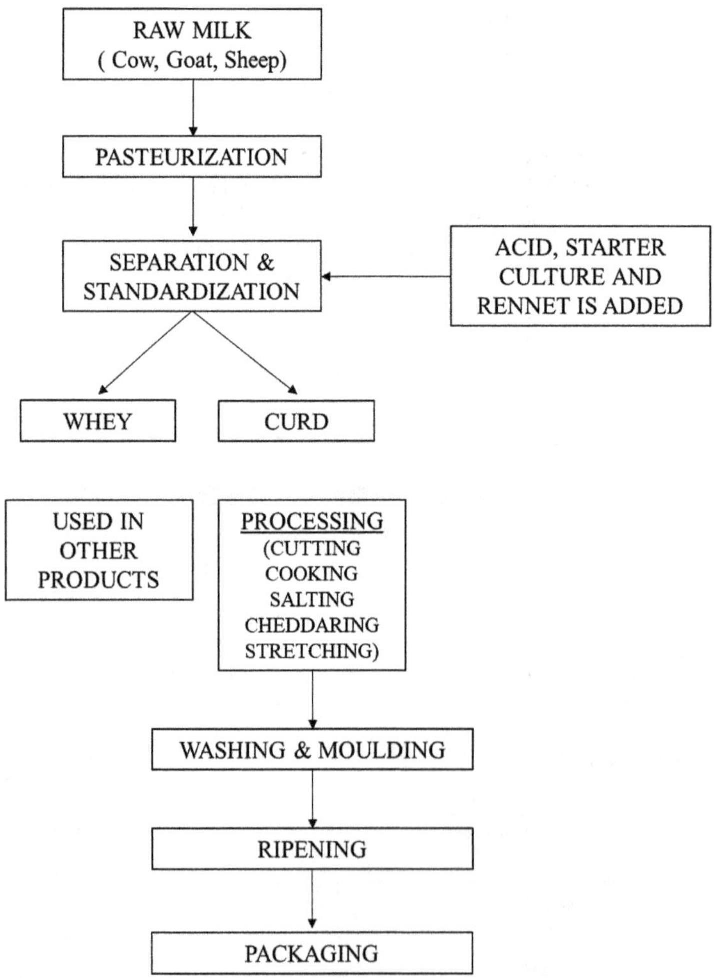

FIGURE 15.3 Flowchart of cheese manufacturing.

Lactic acid bacteria lowers the pH of milk from 6.8–7 to 5.5 by producing lactic acid, adding rennet hydrolyses of k-casein to release para-k-casein and k-casein macropeptide, calcium and phosphate ions having nonpolar nature dissolve in the water phase and para-k-casein remains part of casein micelles. These now bind together to form the curd following the removal of carbohydrates with the k-casein macropeptide and the exposure of binding surfaces. Curd having entrapped most of the bacteria, fat, and other particulate matter is removed. The remaining liquid part – known as whey – contains proteins, lactalbumin, globulin, and yellow-green riboflavin (vitamin B2). As the milk is made of two proteins, casein and whey, the whey protein can be separated from the casein in the milk or formed as a by-product of cheese making (Figure 15.4). The process of cooking curds helps them expel whey, firm up and reach the desired acid levels. Whey is used as a drink. Whey is also used for single-cell protein production.

Some factors that affect rennet coagulation are temperature history of milk, pH of milk during coagulation, temperature of milk during coagulation, Ca^{++} ion content of milk, and casein content of milk. Extract from animal source is expensive so microbial sources are used (Table 15.1).

15.1.9 Shrinking of Curd (Removing of Whey)

The process of shrinking of cheese is initiated by removal of whey after curd formation, then after heating in order to lower moisture content, the size of the cheese is reduced by cutting, pressure is applied, and the pH is lowered. Scalding in cheese production is done to make hard cheese in which curd is cut into small cubes and the temperature is raised to approximately 50 °C. The types of cheese produced by these methods are Parmesan, Emmental, and Gruyere, etc. (Siso, 1996).

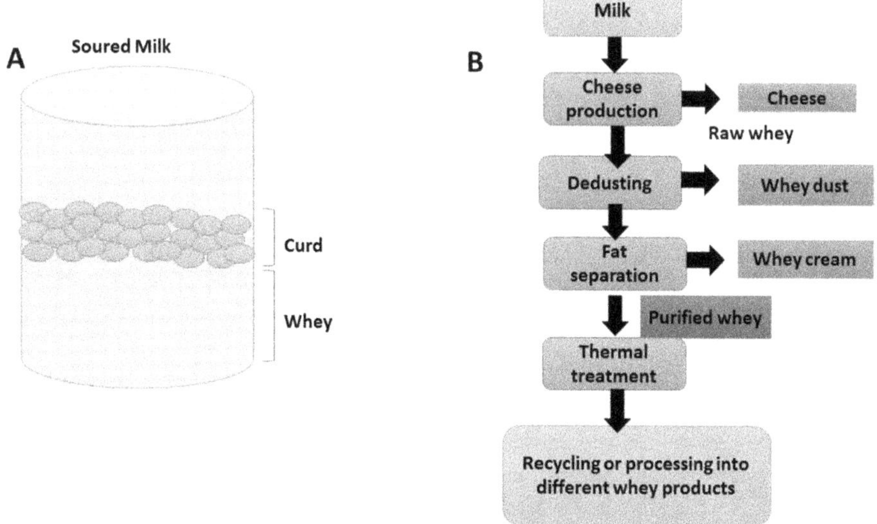

FIGURE 15.4 A. Whey production and B. Steps involved in whey production.

TABLE 15.1
Names of the Rennet Produced from Microorganisms

Microbes	Type of rennet used
Mucor miehei	Harmilase
Mucor miehei	Rermilase
Mucor miehei	Fromase
Mucor pusillus	Emposase
Mucor pusillus	Meito
Endothia parasitica	Suparen
Endothia parasitica	Surd curd
Bacillus subtilis	Mikrozyme

15.1.10 Salting of Curd and Pressing into Shape

Salt controls the acidity and moisture of cheese. Salting is a process of adding salt to formed curd. Salt gives taste to the cheese also acts as an inhibitor to the growth of undesirable proteolytic bacteria. Curd is shaped before going for maturation (Johnson, 2017).

15.1.11 Ripening of Cheese

Maturation or ripening of cheese is a slow biochemical as well as microbiological process in which curd or raw cheese becomes final cheese. Starter cultures bring different flavours by producing different enzymes and also take participating in fermentation (Johnson, 2017).

There are also cheeses that do not undergo a ripening process. These are known as unripened/fresh cheese. In that case, the cheese is cut, packaged and distributed for sale after draining. Bocconcini is an unripened cheese.

Pasta filata is an Italian term for cheeses made with curd that are heated in hot whey and mechanically stretched before being pressed into moulds. The resulting cheeses are more elastic. Examples are Fior di Latte, Caciocavallo, Mozzarella and Bocconcini.

Ageing cheese: The longer a cheese ages, the more flavour will develop. Most cheese are highly acidic, and therefore will kill any harmful bacteria while allowing flavour-imparting moulds to remain. Some moulds are only found in very specific regions, or even in certain caves, meaning that a wide range of cheeses can be produced all over the world.

Blue cheese is a general classification of cheeses that have had cultures of the mould Penicillium added so that the final product is spotted or veined throughout with blue, or blue-grey mould and carries a distinct smell, either from that or various specially cultivated bacteria. Some blue cheeses are injected with spores before the curds form, and others have spores mixed in with the curds after they form. Blue cheeses are typically aged in a temperature-controlled environment such as a cave. Blue cheese can be eaten by itself or can be spread, crumbled or melted into or over foods.

Cheddar cheese is a relatively hard, off-white (or orange if spices such as annatto are added), sometimes sharp-tasting (i.e. bitter), natural cheese. Cheddar is the most popular type of cheese in the UK, accounting for 51% of the country's £1.9 billion annual cheese market.

Roquefort is a sheep milk blue cheese from the south of France, and is one of the world's best known blue cheeses. The cheese is white, tangy, crumbly and slightly moist, with distinctive veins of blue mould. It has characteristic odour and flavour with a notable taste of butyric acid. The blue veins provide a sharp tang. It has no rind. The exterior is edible and slightly salty.

15.2 YOGHURT

Yoghurt is a type of fermented food that is achieved by fermenting milk to preserve nutrients. The sugar present in the milk is fermented to lactic acid which converts milk into curd. In India, it is commonly known as 'Dahi'. The word 'Yoghurt' is a Turkish-derived word meaning coagulated. The origin of yoghurt is not clear because it is found in ancient Indian literature as well as ancient European literature. Yoghurt with honey has been considered the food of gods, and has also helped in indigestion and acidity problems since ancient times. A wide variety of yoghurts is available in the market from fruit-flavoured to Greek yoghurt. Yoghurt is mainly made from cow milk; buffalo milk and goat milk can also be used. The nutritional quality of yoghurt depends on the type of milk used for yoghurt production (Gurr, 1987).

Yoghurt has a relatively good nutritional value which makes it so special. Many dieticians provide yoghurt in medication for people having digestion and acidity problems. Yoghurt has calcium, protein, other minerals, and a wide range of vitamins; only levels of vitamin B1 and pantothenic acid are decreased as they are utilized by the bacterial strains used for producing yoghurt. Folic acid is high in quantity in comparison with milk as bacterial strains produce folic acid.

15.2.1 Microorganisms for Yoghurt Production

The *S. thermophilus and L. bulgaricus* strains are mainly responsible for the formation of yoghurt. These microorganisms should roughly be equal (1:1) in the numbers; otherwise it will result in change texture, flavour, and odour of the yoghurt. *L. bulgaricus* gives the characteristic flavour and odour to the yoghurt. These cultures ferment the lactose sugar and convert it into lactic acid. Production of lactic acid leads to decrease in the pH of milk. Raw yoghurt is formed due to the low pH. The surfaces of casein micelles have negative charge which keeps them away from each other, and lactic acid provides H^+ ions which neutralize the negative charge present on the surface of the casein micelle and help to form clumps and form raw yoghurt (Gorbach, 1990). Low pH inhibits the growth of proteolytic and undesirable bacteria.

15.2.2 Procedure for Yoghurt Production

Commercial yoghurt is produced via the following steps:

Milk is standardized: Milk components are standardized. All components are adjusted in order to get the desired type of yoghurt e.g. fat levels may be reduced or increased to 1–2%. Stabilizers are added if required. These help in improving the consistency and firmness of yoghurt; e.g. carrageenan, gelatins, and gums (Alm, 1982).

Homogenization of milk: Milk is homogenized to break bigger fat molecules into smaller ones in order to maintain uniformity throughout the milk. Pressure and velocity of emulsion are adjusted in order to get the desired size of fat molecules (Widyastuti et al., 2014).

Pasteurization of milk: Pasteurization is done to avoid growth of undesirable microbes, mainly at 80 °–85 °C for 30–35 min or at 90 °C–95 °C for 5–6 min.

Cooling to incubation temperature: Pasteurized milk is then cooled to incubation temperature (45 °C), in order to set standard conditions for growth starter cultures (Widyastuti et al., 2014).

Addition of starter culture: A starter culture is added in order grow the desired microbes, which ultimately ferment lactose into lactic acid and form raw yoghurt (Widyastuti et al., 2014).

Packaging: Packaging is done before incubation. After incubation, cooling (3°C–4°C) is done to stop the growth of the starter culture. In Stirred Yoghurt: - Incubation is done before packaging. After incubation cooling (3 °C–4 °C), stirring, cooling and pumping are done. Sundae Style Yoghurt: Fruit syrup is blended with the milk before adding the starter culture. Swiss Style Yoghurt: Fruit syrup is added after incubation before packaging of the yoghurt (Widyastuti et al., 2014).

15.3 PROBIOTIC FOODS

The word 'Probiotics' is derived from two Greek words –'pro' and 'bios' which means for life. Probiotics have been used in the human diet since very ancient times. For example, cheese and other fermented milk products were one of the favourite foods for Greeks and Romans. In those times they were greatly recommended for children and convalescents. The human intestine can serve as home for various microbes, thus providing it with an adequate quantity of microbes in food as probiotics can serve our body in many ways and greatly improve digestion. The word 'Probiotic' was first used for the substance produced by one living organism that stimulates the growth of another, and was later revised to mean an organism or substance that contributes to intestinal microbial balance. Some more precise definitions state that probiotics are any products which after ingestion have positive effects in various pathological conditions inside the body. Despite all these definitions, the most suitable one for describing probiotics to a nonprofessional is that provided by the International Scientific Association for Prebiotics and Probiotics. It states, 'live microorganisms which, when administered in adequate amounts, confer a health benefit on the host'. As probiotic foods contain living cells, quality regulations and checking need to be very strict, since the slightest contamination from other harmful microbes or an overdose of healthy microbes can have harmful effects on gastro-intestinal systems. Thus probiotics are strictly checked in terms of tolerance to food additives, inhibition of adhesion of pathogenic bacteria, stability in the food matrix, evaluation to antibiotics, resistance to pancreatic secretion and gastric acidity and total safety for the host. All these criteria decide the selection of the strain to be used in probiotics (Somashekaraiah et al., 2019).

15.3.1 Characteristic Lactic Acid Bacteria for Probiotics

In all probiotic products, the commonly used bacteria are lactic acid bacteria such as *Streptococcus, Lactobacillus,* and *Bifidobacterium.* Lactic acid bacteria are the cluster of bacteria which produce lactic acid, are non spore-forming, having low percentage G+C (Guanine + Cytosine), Gram positive rods and cocci having similar physiology, biochemical and genetic properties. Lactic acid producing bacteria consist of 12 genera in the phylum *Firmicutes* and order *Lactobacillus* (Somashekaraiah et al., 2019).

Five sub-clusters are: *Streptococcus–Lactococcus* branch (Family Lactobacillaceae), *Lactobacillus* (Family Streptococcaceae), *Lactobacillus–Pediococcus* (Family Streptococcaceae), *Oenococcus* and *Leuconostoc–Weisella* (Family Leuconostocaceae), *Carnobacterium, Aerococcus, Enterococcus, Tetragenococcus, Vagococcus* (Families Carnobacteriaceae, Aerococcaceae, Enterococcaceae).

Some strains of *Enerococcus* can cause infections, and *Carnobacterium* are also not desirable, since they are considered to be spoilage organisms as fermented products of meat. Species such as *Aerococcus, Vagococcus* and *Weisella* are not generally found in foods, and their significance is not clear.

15.3.2 Probiotic Dairy Products

Probiotic food has a very promising market as many people are making their lifestyle more healthy. Among all probiotic foods, probiotic dairy products are consumed in large quantities and on a daily basis as a part of our diet; these are actually fermented milk products. In many countries they are used to improve the nutritional value of milk, such as improving gut microbiota and providing immunomodulatory effects. Fermented dairy products are also commonly used due to their taste and aroma. Nowadays, such fermented dairy products are still made by using traditional methods. The main action on lactose in dairy products is carried out by lactic acid producing bacteria. Some probiotic milk products are fermented milks – Acidophilus milk, Bifidus milk, Acidophilus-Bifidus milk, Mil-Mil, Yoghurt, Yakult, Kefir, Cheese, Koumiss-Kumiss, Miru-Miru etc. (Somashekaraiah et al., 2019).

There are many probiotic milk products which are now produced and marketed commercially; for production of such products various engineered or selected microbial strains are used as starter cultures. In the selection process, several factors are considered such as; the strain must be viable, and should be present in high amounts at the time of consumption (it should contain at least 6–7 cfu/g of viable bacteria at the time of consumption). Some general selection criteria can be listed as: a) The strain used must be listed or reported in the scientific literature, b) The health benefits claimed should be backed up by concrete proof, c) The strain should form a colony in the gastrointestinal track and should be able to regulate microbial balance, d) It should have resistance against low pH and bile salts to be viable in gastrointestinal conditions, e) It should prevent the growth of other pathogens with its anti-microbial activity, f) It should not have antibiotic resistance and should be safe to be consumed, g) It should be amenable to commercialization, g) It can be stored after manufacturing. After the strain is selected, it is optimized for maximum benefits. One of the key factors in the production of fermented milk products is the exposure to oxygen during storage; this may reduce the viability and functionality of probiotic strains, as they possesses anaerobic metabolisms (Somashekaraiah et al., 2019).

*Probiotic microorganisms used in production of functional dairy products: The m*ain source for the isolation of probiotic microorganisms is the human gastrointestinal system. Widely used species are *Bifidobacterium* and *Lactobacillus;* other species used are *S. thermophilus, E. faecium, E. faecalis, Leuconostoc mesenteroides, Pediococcus acidilactici, E. coli, Propionibacterium freundenreichii, Sporolactobacillus inulinus*, and yeasts such as *Saccharomyces cerevisiae* and *Saccharomyces boulardii*. In the yoghurt industry recent interest for probiotics has increased as a result of *Lactobacillus acidophilus, Bifidobacterium, Lactobacillus casei, Lactobacillus rhamnosus,* and *Lactobacillus reuteri*. Sometimes combinations of organisms are used, mainly for better results and product quality. In combination with a species named *Lactobacillus delbrueckii, L. bulgaricus* is mostly used because of its ability to produce aromatic substances such as acetaldehyde and some acids (Gorbach, 1990).

Composition of milk: The fluid which comes from the mammary glands of mammals to feed young offspring is known as milk. This complex liquid contains various components, among which the main ones are lactose, proteins, enzymes, minerals, fat, and vitamins, which vary from breed to breed (Table 15.2). Skimmed milk is obtained when fats are removed from the milk, and when casein is separated from it, then whey is obtained in cheese making (Buttriss, 1997).

Two types of milk protein are present: casein and whey protein, of which casein is 85% of total milk –consisting of carbohydrate, phosphorus and glycol-phospho-protein.

TABLE 15.2
Different Types of Milk and Their Nutritional Components in %w/w

Animal	Water	Protein	Ash	Lactose	Fat
Ass	89.0	2.0	0.5	6.0	2.5
Camel	87.1	3.7	0.9	4.1	4.2
Reindeer	63.3	10.3	1.4	2.5	22.5
Cow	87.6	3.3	0.6	4.7	3.8
Mare	89	2.6	0.7	6.2	1.5
Goat	87	3.3	0.6	4.6	4.5
Sheep	81.6	5.6	0.9	4.4	7.5
Buffalo	82.1	4.2	0.8	4.9	8.0

Casein is of four types: α, β, kappa and gamma based on the electric charges present. Of whey proteins: 6 % are β-lactoglobulin, 22% α-lactalbumin and 10% immunoglobulins (Gorbach, 1990). Lactose is found in milk, having low sweetening ability and low water solubility; it is a dissacharide of glucose and galactose. Fats with saturated and unsaturated fatty acids are present, of which some have very low molecular weight with 10 or fewer carbon atoms; these consist of one molecule of glycerol and three molecules of fatty acids (Ruiz Rodríguez et al., 2019). Milk contains enzymes such as proteases, esterases, carbohydrates, oxidases, and reductases. Calcium is the major mineral in milk but milk also has phosphorus, sodium, manganese, potassium, sulphates, and chlorides in considerable quantities. In the production of yoghurt, unpasteurized milk is used with the species of bacteria mentioned above. The production of lactic acid lowers the pH to 4–5, giving a sour taste and causing coagulation of milk proteins (Buttriss, 1997). Various fermented milk products are consumed throughout the world indifferent forms (Table 15.3). The following are a few noteworthy examples:

TABLE 15.3
Different Fermented Milk Products, Organisms Used, and Country

Name	Country of origin	Description	Cultures
Yoghurt	Asia/ Balkans	Acidic, set or stirred, characteristic Aroma	*S. thermophilus, Lb. bulgaricus*
Kafir	Caucasus	Stirred beverage, creamy consistency, characteristic taste	*Lc. lactis, Lc. cremoris, Lb. kefir, Lb.casei,Lb.*
Acidophilus milk	USA	Set, stirred or liquid, mild flavour	*Lb. acidophilus*
Lassi	India	Sour milk drink diluted with water consumed salted, spicy or sweet	*Lactococcus spp., Lactobacillus spp., Leuconostoc*
Dahi	India	Set, stirred or liquid beverage	*S. thermophilus, Lb. bulgaricus, Lc. diacetylactis,*
Kumiss	Mongolia	Frothy beverage, acidic, refreshing taste	*Lb. bulgaticus, Lb. acidophilus, yeasts*
Villi	Finland	Viscous stirred product, mildly Sour	*Lc. Lactis, Lc. Cremoris, Lc. diacetylactis*
Leben	Middle East	Set or stirred product, pleasant taste and aroma	*S. thermophilus, Lb. bulgaricus, Lb. acidophilus*
Filmjilk	Sweden	Viscous stirred beverage, clean acid taste	*Lc. Lactis, Lc. Cremoris, Lc. diacetylactis*

Acidophilus milk: *L. acidophilus.* When used to ferment milk, the product is known as acidophilus milk. Milk is first heated at 95 °C (to eliminate competition for the strain) and is homogenized and then cooled till it reaches 37 °C. The starter culture consists of pure acidophilus bacteria with 2–5% concentration and incubated for 12–24 h. After this incubation, milk is again cooled to 5 °C and then kept in cold conditions. For production of unfermented acidophilus milk, *L. acidophilus* culture is directly added to cold milk at 5 °C and kept under cold conditions. Acidophilus milk is beneficial for diarrhoea and intestinal gas problems and this strain in the intestine prevents the activity of other harmful gas-forming microorganisms (Gorbach, 1990).

Bifidus milk and Acidophilus-Bifidus milk: These were first produced in 1948 in Germany, and have a spicy and acidic aroma. Nowadays, cultures of *Bifidobacterium bifidum* and *Bifidobacterium longum* are inoculated at a 10% ratio to the cold milk with pH. After production, 1 g of milk contains nearly 10–100 million cells. It is very effective in the treatment of gastrointestinal and liver disease as it is easily digestible. The bacteria forms L(+) lactic acid inside the body. First, the milk is pasteurized at high temperature and kept for some time and then 10% culture is inoculated in it and the milk is left for incubation at 37 °C until it coagulates and the pH value becomes 4.3 and 4.7. Then the product is packed and stored in cool conditions having 108–109 cfu/mL (Buttriss, 1997).

Mil-Mil: This was originated in Japan by Yakult Honsha. It is a mixture of *Bifidobacteriumbifidum, L. acidophilus* and *Bifidobacterium breve*. This product is rich in small amounts of glucose, fructose and carrot juice and rich in vitamin A (Buttriss, 1997).

Yakult: This is a probiotic dairy product with the strain *L. casei*, discovered by Dr. Shirota in 1935, which is resistant to duodenal and gastric juices and can form antimicrobial substances in the small intestine and also helps macrophages in their activities. The commercial product has been on the market since 1955 from the Yakult Honsha Company (Buttriss, 1997).

Kefir: This was originated in Turkey, and means 'good food'. Kefir is a slightly acidic dairy product produced from kefir grains by specific mixtures of yeast and bacteria. It is naturally carbonated. It is an important part of human diet in North America, North Europe, Japan, Russia, Asia and the Middle East. Due to the action of *Bifidobacterium bifidum* it is very suitable for 6-month-old or older infants. It has been found to be very effective in the treatment of infantile intestinal disease. It has gel-like colonies of casein and microbes in symbiosis. Any type of milk can be used in the production of kefir (Buttriss, 1997).

Koumiss/Kumiss: This has many different names in different countries but the compositions are similar. The composition of this drink is 2% alcohol, 2–4% milk sugar, 0.5–1.5% lactic acid, and 2% fat giving it a taste like sour ayran. For treatment of tuberculosis, pneumonitis, asthma, gynaecological disease, and cardiovascular disease it is highly recommended and has various other benefits such as increasing energy and it can be used for weight gain (Buttriss, 1997).

Acidophilin: Via use of high-quality starter cultures in cow milk, this is produced with an acidic taste. Starter cultures primarily consist of *L. acidophilus* and

L. delbrueckii. In some cases the culture used for kefir is also inoculated with this culture in the ratio 1:1:1 to homogenized milk at 18–25 °C. It is acidophilic with around 0.67–0.72% acidity but the acidity and antimicrobial activity are less than acidophilus milk. The final product has 97% *L.lactis* ssp., 2% *L. acidophilus*, and 1% yeast.

Miru-Miru: This is produced in Japan via a combination of *L. casei, L. acidophilus* and *Bifidobacterium breve*(Buttriss, 1997).

15.4 ADVANTAGES AND DISADVANTAGES OF FERMENTED FOOD

15.4.1 Advantages

The presence of desirable, non-pathogenic bacteria helps in inhibiting the growth of undesirable bacteria and spoilage of food. Some fermented products have low pH and most of the spoilage-causing bacteria cannot grow in lower pH. During the process of fermentation some bacteria – e.g. LAB (lactic acid bacteria) – synthesize vitamins and minerals, produce biologically active peptides, and enzymes like peptidase and proteinase, and help in removing some non-nutrients. For example, exopolysaccharides exhibit a prebiotic effect, and bacteriocins show anti-microbial effects etc. (Gorbach, 1990). The fermentation process helps with the breakdown of complex components of a food into a simpler one, which decreases stress from the body in digesting such complex food. For example, lactose sugar from the milk is converted into glucose plus galaclose.

As the process breaks complex components of food into simpler one it takes less time to cook food. Microflora in humans in different regions have coevolved over millions of years. The microflora of the gut help in digestion of food by producing enzymes, vitamins, and folic acid etc. Some bacteria produce antibiotics which help in boosting immunity. Fermented food provides probiotics that help in digestion, and also dietary fibres which are useful for the growth of already present microflora (Gorbach, 1990).

15.4.2 Disadvantages

Histamine is present in quantity in almost all fermented foods e.g. histamine is present in fermented milk products, fermented non-vegetable items (meat, sausages), fermented vegetables, fermented grains etc. and also in fermented liquids such as wine, beer, and vinegar. Histamine is one of the bioactive chemical messengers from biogenic amines class, and regulates most of the vital functions of the body. The human body requires amounts of histamine for brain function, digestive function, and to respond to allergic reactions etc. If the amount of histamine has risen beyond its critical value, this can lead to acute symptoms such as skin problem (itching, redness, swelling), acidity, gastric reflux, stomachache, respiratory problems, cough, over nasal secretion, hypotension, blood pressure problems, headaches, etc. and chronic symptoms such as chronic fatigue, painful menstruation, panic disorder, depression etc. (Gorbach, 1990).

15.5 CONCLUSIONS

The major classifications of cheese and its types have been discussed briefly along with their nutritional information. The complete process of cheese production including the major steps such as milk standardization, pre-treatment, incubation of starter culture, the addition of rennet, salting etc. has been discussed in detail. A detailed study of yoghurt, the microorganisms in yoghurt, and the complete commercial process of yoghurt production such as homogenization, pasteurization, cooling etc. has been given. Probiotic foods and the characteristic lactic acid bacteria present in probiotics such as the *Streptococcus–Lactococcus* branch (Family Lactobacillaceae), *Lactobacillus* (Family Streptococcaceae) and probiotics in the form of different products such as acidophilus milk, bifidus milk, acidophilus-bifidus milk, mil-mil, yoghurt, yakult, kefir, cheese, koumiss/kumiss, miru-miru etc. and various compositions of milk have all been reviewed. Lastly, the advantages and disadvantages of different fermented foods were discussed.

REFERENCES

Alm L, Effect of Fermentation on Lactose, Glucose, and Galactose Content in Milk and Suitability of Fermented Milk Products for Lactose Intolerant Individuals. J. Dairy Sci. 65, 346–352, 1982.

Buttriss J, Nutritional properties of fermented milk products. Int. J. Dairy Technol. 50, 21–27, 1997.

Gorbach SL, Lactic Acid Bacteria and Human Health. Ann. Med. 22, 37–41, 1990.

Gurr MI, Nutritional aspects of fermented milk products. FEMS Microbiol. Rev. 3, 337–342, 1987.

Johnson ME, A 100-Year Review: Cheese production and quality. J. Dairy Sci. 100, 9952–9965,2017.

Ruiz Rodríguez LG, Mohamed F, Bleckwedel J, Medina R, De Vuyst L, Hebert EM, Mozzi F, Diversity and Functional Properties of Lactic Acid Bacteria Isolated From Wild Fruits and Flowers Present in Northern Argentina. Front. Microbiol. 10, 2019.

Siso MIG, The biotechnological utilization of cheese whey: A review. Bioresour. Technol. 57, 1–11, 1996.

Somashekaraiah R, Shruthi B, Deepthi BV, Sreenivasa MY, Probiotic Properties of Lactic Acid Bacteria Isolated From Neera: A Naturally Fermenting Coconut Palm Nectar. Front. Microbiol. 10, 2019.

Widyastuti Y, Rohmatussolihat R, Febrisiantosa A, The Role of Lactic Acid Bacteria in Milk Fermentation. Food Nutr. Sci., 5, 434–442, 2014.

16 Single-Cell Protein Production

16.1 INTRODUCTION

The concept of 'Single Cell Protein' (SCP) was invented by Professor Wilson of the Massachusetts Technology Institute. Single Cell Protein refers to protein derived from microbes. It has been identified that protein malnutrition is generally a much more serious problem than with most other foodstuffs (nutrients). Starting in the 1950s and 1960s, there have been rising concerns about food deficiency in less-developed countries, mainly due to their population explosion. Because of this issue, alternative and non-traditional kinds of food sources were aimed for. The drawbacks of traditional protein sources were noted. These include: A) In the case of plants, crop yield decreases due to adverse weather conditions. B) In the case of fish, the time requirements to replenish stock are greater. C) In the case of agricultural fields, the availability of land is less. The concept of SCP has subsequently been embraced.

In contrast, SCP production has a range of advantages: (a) SCP can be produced at any time of the year, (b) Microbes grow more quickly than animals or plants. However, some drawbacks have also been noticed in SCP production. Less industrialized countries mostly suffer from protein malnutrition because of weaker financial efforts to generate high-yield mass production by fermentation. However, the use of portable fermenters and recovery methods which do not involve specialized methods can overcome this weakness. Another disadvantage of SCP is that microorganisms contains high amounts of RNA and that their ingestion could result in accumulation of uric acid, and the formation of kidney stones and gout.

16.1.1 Substrates for Single-Cell Protein Production

SCP processing has a broad range of substrates, including hydrocarbons, alcohols, and waste from multiple sources.

1. *Hydrocarbons:*
   The Zobell rule addresses the stages of utilization of hydrocarbon by microbes, and the descriptions are as follows:
   A) Aliphatic hydrocarbons in several generations can be utilized by the yeast strains as substrate. Other hydrocarbon classes, including aromatics, can be oxidized, but are usually not integrated effectively.

B) n-alkanes are generally not incorporated but can be oxidized, with a chain length shorter than n-nonane (Reihani and Khosravi-Darani, 2019). Yield factors increase with the chain length of n-nonane, and the oxidation rate reduces.
C) Saturated hydrocarbons are more easily degraded than unsaturated hydrocarbons.
D) Straight-chain carbon compounds deteriorate more easily than branched-chain hydrocarbons.

Gaseous hydrocarbons: Methane is reported to be the most common substrate for SCP production as compared to all gaseous hydrocarbons. Methane is a prevalent natural gas which is related to petroleum sources ('casinghead gas'). Natural gas is abundant throughout the globe and is quite affordable. Perhaps the main benefit, in contrast with liquid hydrocarbons, is the lack of residual hydrocarbons in the single-cell protein produced from it. The benefits of utilizing methane are faster growth rate, better yield ratio, higher contamination tolerance to efficiency and less foam output. Single-cell protein production from methane has used a mixed population of microorganisms and continuous cultures.

Liquid hydrocarbon: Crude petroleum is the primary source of liquid hydrocarbons. These hydrocarbons were initially investigated as a source of vitamins and lipids for microbes. These studies were later extended to feeding whole paraffin-grown bacterial and yeast cells into rats in the late 1940s. The interesting decision to commercially produce paraffin cells was for 'dewaxing' i.e. removing higher n-alkanes from crude fractions of petroleum. The composition of crude petroleum is highly variable according to country. Most crude petroleum oils, however, consist of 90–95 per cent saturated hydrocarbons (Ritala et al., 2017a). Crude oil is first distilled using atmospheric pressure during petroleum refining, a method called 'topping'. After topping, the materials left contain significant amounts of natural alkanes with carbon atoms larger than C8 ranging from C10–C18. For the production of SCP, various petroleum hydrocarbons are used.

Initial development in SCP was done by British Petroleum (BP) in 1973, which held the most patents. Soon afterwards, many more petroleum businesses and government authorities around the world developed pilot plants and got involved in the research. The material from the distillation column in the BP plant in Scotland was transferred through a molecular sieve to allow only the easily incorporated portion into the fermenter under aseptic conditions of the microorganism n-paraffins (especially 97.5%–99% n-C10-C33 boiling at 30–33°C). Also, in another plant situated at Lavera Industrial Biotechnology South of France, untreated gas oil was utilized in non-aseptic conditions. Approximately 10 percent of the gas oil was used, while the remaining 90 percent went back to the refinery (Abbott and Clamen, 1973). Solvent extraction methods were utilized to eliminate the last traces of yeast cream oil *(Candida tropicalis)* including the lipids content of 0.5 tons of yeast. By 1963, the technique provided five tons of dry weight of yeast per day. In respect of yeast composition, there was a slight difference between the two methods. In economic terms, the marginal benefit contributed to the Lavera cycle (dewaxing process). Additionally, there were fewer expectations that SCP's high petroleum predictions would be fulfilled. Indeed, many proposals for the development of stage fermenters revealed by various oil companies were dismissed (Royer and Nakas, 1983).

Even when there was progress in utilizing n-paraffin and gas oil for SCP production, research was initiated for substituting these substrates for petroleum-based substrates such as methanol and ethanol.

2. Alcohols:

Methanol: Methanol is considered to be a substrate for SCP since it is highly water-soluble, thus avoiding the three-form (water-paraffin-cell) transfer issue inherent in the usage of paraffin. Methanol is easily accessible from a range of methane to naphtha of hydrocarbon sources. Many microorganisms are unable to use methanol as a substrate, so a pure culture can be maintained without much effort. Oil firms in Italy and USA have taken notice of the use of methanol as an SCP substratum. One of the advanced projects among all of these countries was the Imperial Chemical Industries (ICI) project which utilized the bacteria *Methylophilus methylotropha* for SCP production using the loop fermenter ('pressure cycle fermenter'). More than 20 species have been investigated for growth on methanol from the genera *Hansenula, (Hansenulapolymorpha Pichia, Torulopsis and Candida)*(Reihani and Khosravi-Darani, 2019).

Ethanol: Ethanol to SCP is costly in the case of the catalytic process (Ritala et al., 2017a). Additionally, it was developed and marketed by the American Amoco Oil Company in a facility that eventually produced 15 million lbs annually as a flavour-enhancing agent of baked bread, pizza, sauces, etc., mainly utilizing *Candida utilis. Candida acidothermophilum* is a new strain found by Mitsubishi Oil in Japan which is more beneficial than *Candida utilis* because it grows at a lower pH value and temperature, and utilizing impurified ethanol helped to reduce expenses by reducing aseptic conditions and cooling requirements. The output from pilot plants is 100 tons per annum (Tusé and Miller, 1984). In Spain, the *Hansenulaanomala* strain of yeast has been utilized. Furthermore, SCP production by utilizing ethanol (substrate) has been achieved in Russia, the Czech Republic and in Switzerland, utilizing *Acinetobacter caloaceticum* instead of yeast. It is primarily aimed at human consumption and thus it is necessary to decrease the content of nucleic acid.

3. Waste Material:

Petroleum prices have continued to rise in recent years. Consequently, it is doubtful that substrates based on petroleum, such as chemically modified methanol and ethanol, gas oil etc., will be used in long-term processes. Several proposals have already been stopped. However, this is not the end for SCP production, since more focus is being given to substrates obtained from natural products such as plants that can restore energy through a photosynthesis process. Usually, though, these are obtained from other sources as waste materials (Royer and Nakas, 1983).Various research reports are available for SCP production but in an unmanaged way. They may be categorized accurately and are as follows:

Plant/wood wastes: These materials consist of cellulose. Cellulose is crystalline and highly tolerant to fermentation, and therefore preliminary treatment is required. The tolerance is even higher when lignin is available as it generally prevents cellulose from direct degradation. For this reason, waste from different processes such as sulfite pulping can necessarily break lignocellulosic material down before the process initiates SCP production (Reihani and Khosravi-Darani, 2019). Cellulose-containing plant wastes need pre-treatment, after which waste products can then be treated either by a chemical process or by microbes.

There is a significant amount of cellulose-containing agricultural waste around the world at present. Generally, this is cost-effective and harmless. Nevertheless, cellulose, and other content present in waste material, varies and thus there is a need to keep account of the expense of collecting and storing them. Developing countries, especially in tropical regions, are assumed to have huge quantities of plant material but there are obstacles such as lack of workforce, poor finance or managerial expertise, and poor availability of fermentation equipment. Some research specifically directed at developing countries is encouraging. One such high point of the process is the use of labour-intensive techniques, using fermentation equipment and other equipment made from relatively inexpensive materials by Tate and Lyle Ltd, the British manufacturer of sugar. Many developing countries of Africa/Asia and South America use these techniques and manufacture SCP from plant waste locally.

Starch: As an SCP producing substrate, starch waste materials such as rice, potatoes, or cassava etc., can be used because starch hydrolysis can be conducted easily by microbial cells or enzymes. The Swedish Sugar Corporation created the Symba methodin, where the symbiotic association is between *Endomycopsis fibuligera* and *Candida utilis. Endomycopsis fibuligera* hydrolyses starch and converts it to glucose and maltose with alpha and beta amylases, after which these sugars can be used for growth by *Candidautilis*.

Dairy waste: Whey is generated as a by-product during the processing of proteins (and fat) for manufacturing cheese. Whey is a lactose-rich liquid that can be collected from cheese manufacturers in concentrated form and diluted to provide the ideal concentration of lactose. A high-quality edible food yeast *Saccharomyces fragilis* can be grown in whey-containing broth. SCP or alcohol may be produced by this process (Tusé and Miller, 1984).The authors suggested the production of SCP and alcohol within anaerobic conditions to avoid the expense of aeration.

Chemical waste: In chemical waste, a carbon source is usually present in adequate quantity. Therefore, *C. Lipolutica or Trichosporoncutaneum* can be treated with oxanone water to produce SCP. Various substrates such as lemon, palm, coconut coffee waste etc. are examined in order to produce SCP. It is important to know the specific pretreatment required and also the appropriate organisms which can grow in waste substances for SCP production (Ritala et al., 2017a).

16.1.2 Microorganisms Used in SCP Production

Microorganism used for the production of SCP should have various basic characteristics:

16.1.2.1 Non-Pathogenic and Non-Toxic

Large-scale production of microorganisms is harmful to animals or plants and may pose a major health hazard. Microorganisms do not secrete or excrete harmful or carcinogenic substances (Nangul and Bhatia, 2020).

16.1.2.2 Protein Quality and Content

Not only is protein content in species important, but they should also contain many amino acids that are required for human beings.

16.1.2.3 Digestibility and Organoleptic Properties

Microorganisms should be digestible as well as having reasonable flavour and smell.

16.1.2.4 Growth Rate

They must quickly grow into an easily accessible, inexpensive medium.

16.1.2.5 Ability to Adapt to Drastic Conditions

Environmental factors that are antagonistic to contaminants are sometimes beneficial for removing pollutants and thus for reducing production costs. Accordingly, strains which grow at low pH or high temperatures are most typically selected.

At present, for SCP production bacteria, actinomycetes, fungi (especially mould and yeast) are used. Liquid hydrocarbon substrates, and gaseous hydrocarbons such as methane, propane, butane etc. can be degraded by bacteria. n-paraffins, petroleum gas, diesel and alcohols can be degraded by bacteria as well as by yeast. Microbes can simply utilize sugar compounds from carbohydrates e.g. starch. Fungi have an advantage for use as a culture because they have lower RNA content, which can easily be isolated.

16.1.3 Use of Autotrophic Microbes for Substrate Degradation

Photosynthetic bacteria and algae are autotrophic organisms. Major work on SCP production appears to be restricted to algae. The expectations of SCP cannot be fulfilled because photosynthetic bacteria only grow in the absence of oxygen. To establish and maintain anaerobic conditions is difficult.

The total production of the world's oceans and seas of about 550×10^9 tons is extremely high versus all human beings living at about 100 tons a year (Nangul and Bhatia, 2020). Algae are rich in protein content. For economic profit, in marine water algae growth needs to be at least 250 mg/L. However, the protein content is only 3 mg/L. Therefore the main aim is to produce algae in sufficient quantity for economic viability (Tusé and Miller, 1984).The advantages of algae cultivation are: relatively low capital expenditure; can easily be digested by ruminants and other animals; the potential of combining waste disposal with algae production; the availability of sunlight and mild temperatures across the year; the tropics should be the location for algae cultivation.

16.1.4 Dietary Quality of SCP

Microbial cells content decides the nutritional value of SCP, particularly proteins, amino acids, vitamins, and minerals. Further, this depends on the growth conditions of microbes. For the amino acid content of proteins, the FAO has defined guideline values. Based on these, SCP produced from bacteria and yeasts are methionine deficient. Even in moulds, glycine and methionine are poor. This SCP quality can be improved by the addition of small quantities of animal protein (Ritala et al., 2017b).

16.1.5 Safety Issues Associated with SCP Consumption

Governments worried about the carcinogenic chemicals present in SCP produced by using oil as a substrate as well as other disadvantages, such as limited sources and non-renewability. For these reasons companies shifted to non-oil substrate utilization (Nangul and Bhatia, 2020). However, research has been submitted in support of SCP, and SCP will probably be officially accepted eventually.

As well as cancer production and pathogenicity from oil-related products, another problem in SCP production is the intake of highly nucleic acid. The enzyme uricase that oxidizes uric acid into a soluble and excretable compound called allantoin has been lost by humans. The main problem is the deposition of nucleic acid inside the body. When humans consume nucleic acid, the nucleases in pancreatic juice break apart and are transformed by intestinal juices into nucleosides before absorption (Tusé and Miller, 1984). Guanine and adenine are transformed to uric acid, which is not convertible to soluble and excretable allantoin, as previously mentioned. When a food contains a high amount of nucleic acid then that results in an unexpectedly large level of uric acid in the blood plasma. Uricates can be accumulated into a different part of the body, inside kidneys and joints, because of the lower solubility of uric acid; this causes diseases like kidney stone and gout. In 1970, in addition to the amount in the regular diet for adults, the PAG collaborators for SCP set an upper limit of 2 g of nucleic acid each day (Suman et al., 2015). The amount of nucleic acids in the whole cell content of different microbes is as mentioned below (Table 16.1).

Growth rate and cell physiology:

In continuous cultures, growth rate will be higher, and this relates to a higher RNA/protein ratio. Therefore, it can easily be said that the growth rate is directly proportional to RNA content. Hence growth level is lowered as a way to suppress

TABLE 16.1
Microbes for SCP Production and their Nucleic Acid Content

Microorganisms	Nucleic acid content
Moulds	2.5–6%
Algae	4–6%
Yeast	6–11%
Bacteria	Up to 16%

nucleic acid. As per the concept of SCP, the biomass should be higher. Thus the method may have only limited importance. Bases such as NaOH or KOH are quickly dissolved, hydrolysing RNA. The extraction of RNA can also be achieved by hot 10% sodium chloride. Cells should be disrupted before the addition of NaOH or KOH. Sometimes protein needs to extract, purified and concentrated.

Bovine, heat-stable pancreatic juice RNAase was used to hydrolyse yeast RNAs at 80 °C, where cells were highly active at temperature. Using heat shock and chemical treatments, RNAase enzyme is activated and the RNA concentration inside the yeast cell becomes lower (Moreira, n.d.).

SCP Market Trends: The global market value of single cell protein extracts and other standard sources was valued at USD 12,685.7 million in 2016. Rising demand for food and feed that has low fat concentration and optimum amino acid composition are the main factors driving business expansion. It is expected that technical advances in the processing of proteins extracted from a single cell will accelerate lucrative growth in the industry. BlueBiotech Int GmbH is a micro-algal biotechnology corporation which has been developing significant quantities of *Spirulina* and *Chlorella* for more than 10 years. Similarly, with distribution in the U.S. and 30 other nations, Cyanotech Corporation is one of the world's top manufacturers of *Spirulina*. The largest share in 2016 was purchased by North America. Regional development is driven by the adoption of advanced biotechnology techniques, combined with the involvement of large companies operating in this field. U.S.-based firms such as Cyanotech Corporation, Nucelis, and KnipBio have played a significant role in the global market for SCPs.

16.2 YEAST PRODUCTION

Yeast has been used since ancient times mostly for the conversion of fruit juices to harmless drinks as well as for fermenting grains for leavening of dough. Over that period time, the concept of yeast was not clear. Yeast is produced in large quantities by weight as compared to all microbes. Furthermore, it is produced in all six continents. World production of yeast in 1977 appears to be much lower in the Middle East, Asia, and African countries. Six leading companies in the United States produce baker's yeast. These companies are Universal Foods (Red Star Yeast), Fleischmanns, Gist-brocades, Lallemand (American Yeast), Man-day and Columbia. These companies are also the main food companies in the world. Theycomprise13 manufacturing plants. Initial attention is on baker's yeast, and then on food and fodder yeasts. However, for now, the main aim is to produce SCP, and for that purpose food and fodder yeast production is the focus (Reed and Nagodawithana, 1990).

16.2.1 Baker's Yeast Manufacture

Yeasts utilized for bread making is a primitive art, although the mechanism of raising of dough was not originally understood by mankind. It is interesting to compare a history of yeast with industry development. According to ancient records, it was probably developed by a mixture of yeasts and bacteria producing lactic acid. A small productive piece of dough was considered as the inoculum for the next batch to maintain continuous culture. This practice was discontinued and utilized only for a special sour bread from San Francisco, California, USA. Bakery yeasts were collected from wine making and brewing from around the Middle Ages. However, the yeast quality was diverse, and the product was bitter in the case of yeast achieved from beer due to the obvious hops in the beer. In the19th century Pasteur explained the nature of yeast (1855 to 1857). The Vienna method, initiated before 1860, in which cereal mash was used for the manufacture of anaerobic alcohol, was used for good yeast concentration and it can be considered the first significant innovation of baker's yeast technology. Yeasts and alcohol were produced simultaneously in one process even if the quantities were less than optimum (Reed and Nagodawithana, 1990). Pasteur's work eventually resulted in greater aeration, generating more cells and less alcohol. Because, during the World War I, food shortage was prominent, the use of substrate changed to molasses, supported by ammonia and phosphate. Baker's yeast technology took the next big step in its growth by introducing fed-batch or gradual addition of nutrients at the beginning of the process instead of classical batch methods (introduction of all nutrients). The core of this so-called Zulaut system is still used in the manufacture of bakery yeast, which prevents excess molasses of sugar which could lead to the production of alcohol. Because of the highly active aeration, current methods of baker's yeast do not permit alcohol generation. Furthermore, yield grew from 3 percent in the mid-19th century to 13 percent early this century to today's production of over 50 per cent dry yeast weight.

16.2.1.1 Use of Yeast Strains and Techniques

Non-sporulating "torula" yeast were often used for baking, but *Saccharomyces cerevisiae* strains are currently preferred. Occasionally, two strains of baker's yeasts were present: one highly active but with minor effects during preservation, and the other with poor activity but being maintained in storage. A successful breeding system was conducted, in order to develop new strains by using the above strains. The yeast strains for modern fast-growing dough with the following traditional and innovative physiological characteristics are as follows: A) Capable of increasing the population at room temperatures of about 20–25 °C; B) In anaerobic conditions can grow and synthesize enzymes and coenzymes; C) Generate enormous volume of CO_2 in flour dough, instead of alcohol; D) Capacity to withstand autolysis when held at 20 °C; E) The potential for high glycolytic action; F) The capacity to adapt quickly to a naive substrate; G) High invertase and other enzyme activity, to hydrolyse sugar to glucose and fructose; H) Available in farmed flour dough instead of alcohol; I) Capacity to withstand the osmotic effect of dough salts and sugars; J) High productivity, i.e. high yield in terms of dry weight; substrate. In western countries (the Chorleywood Bread Process) introducing complex plants and computerized techniques also placed new demands on routinely used yeasts. These novel proposals involve fermentation of complex sugars, high starting fermentation capacity, rapidly adaptation to maltose fermentation and rapid reassembly in an active dry environment. Additionally, in Eastern European countries, baker's yeast is particularly unprocessed. The yeast is produced from grain mash and fermented for the distillation of alcohol, by the Vienna method (Reed and Nagodawithana, 1990).

Saccharomyces cerevisiae specifically used as a baking strain are suitable for mutating, and thus safe storage is important. The common storage method is storing in liquid nitrogen and an oil culture method where sterile oil is set over a slant of yeast and cooled at 4 °C. However, freeze-drying is not widely applicable as it leads to loss of feasibility and a tendency to mutation in many yeasts.

16.2.1.2 Manufacturing Plant

Molasses are the prime substrate utilized to make baker's yeast. Another sugar-containing substrate e.g. corn steep liquor is used when molasses are not obtainable or are extremely costly. Also, for both alcohol and baker's yeast manufacturing, sulfite liquor has been used in the Soviet Union. Ethanol has not been used on a larger scale but only in experimental studies. When beet and cane molasses are simultaneously provided, they are treated separately: clarified, adjusted for pH, and sterilized. The nutritional deficiency of one substrate is maintained by the addition of some amount by another substrate to level up. Cane sugar has notable contents of biotin, pantothenic acid, magnesium, and calcium, whereas beet molasses are considerably high in nitrogen. Despite this, the molasses concentration is not constant and varies according to the geographic area. If only one type of molasses is present, adequate nutrients are added in order to maintain nutrient quality.

The inert coloured material arising from colloidal agents needs to be removed from molasses since ultimately these particles may contribute to an unwanted yeast colour. To remove this colour, precipitation with aluminium or calcium phosphates or polyelectrolyte flocculants such as alginates and polyacrylamides may lead to clarification. The clarification also contributes to the reduction of foaming. Subsequent sterilization by heating at around 100–110 °C for nearly an hour after the pH of 6–8 is maintained to avoid caramelization of the sugar. The mixed molasses are supplemented with phosphorus, ammonium and lesser quantities of magnesium, potassium, zinc, and thiamine. At intervals, antifoam is added.

Fermenter to produce baker's yeast are made from stainless steel. Commercial fermenters (i.e., the final fermenter) range from about 75 to 225 cubic metres. The medium occupies about 75 percent of the space, leaving the remaining area for foaming. Due to the high initial and operational costs, traditional stirred-tank fermenters with agitator baffles and sparging are not often used in yeast production. However, for baker's yeast, spargers are normally used. Fermenters are aerated using spargers which are arranged in such a way that a large amount of air passes through each unit: about one volume of air/volume of broth/minute. Different types of sparger are accessible. The anaerobic conditions are set when oxygen falls below 0.2 ppm, and alcohol is produced. Water, mineral nutrients, yeasts, and mixed molasses with 1 percent of glucose are mixed. The quantity of mixed molasses incorporated is calculated and does not increase by more than 0.1 per cent of the total sugar in the fermenter. During fermentation, molasses are added at intervals as they are utilized by the yeast, and the concentration is more than 0.1 per cent. The pH is kept at pH 4–6 by adding alkali, and by cooling the temperature to 30 ° C. Experiments are used to calculate the addition of the quantity of molasses at regular intervals.

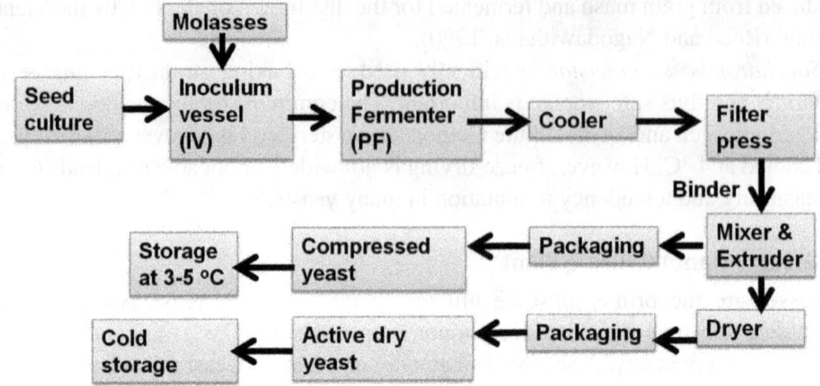

FIGURE 16.1 Block flow diagram of baker's yeast production.

Several other advanced fermenters are constructed with automatic sensors and self-adjusting machinery for temperature pH, oxygenation, sugar, etc. Significant amounts of heat emerge, and a very essential factor is fermenter cooling. 1 lb dry wt. of yeast would require 4.3 lb of molasses, 0.9 lb of ammonia, 0.3 lb of $NH_4H_2PO_4$, 1.1 lb of $(NH_4)_2SO_4$ and 60 lb of air. Continuous fermentation is commonly used for feed yeast by utilization of sulfite liquor but not for manufacture of baker's yeast (Figure 16.1).

Based on yeast inoculation, fermentation time in commercial fermentation or output fermentation varies between 10 and 20 h. Cell production ranges from 3.5 to 5 percent of dry weight. Aeration is conducted for 30–60 min at the final stage of the feeding cycles in some operations in order to permit usage of left-over nutrients. In addition, the budding cells divide in such a way that most cells 'rest' at the start of the budding process. This means that cells divide 'synchronously' when growth restarts. Furthermore, the fermentation broth is cooled, and cells are extracted by centrifugation and are washed by resuspending in water and centrifuged until they are lighter in colour (Reed and Nagodawithana, 1990). It is also concentrated using a rotating vacuum filter or by a filter press. Often the Mautner method is used during vacuum filtration to guarantee friable dry milk. This latter procedure involves utilization of 0.2–0.6 %w/v of NaCl, before filtration, that causes osmosis resulting in cell shrinking. Excess salt is extracted by spraying water, and during the filtration the cells swell again. The resulting product contains 28–30 percent of dry weight (Figure 16.2). The yeast can be converted into a compressed yeast or a dry active yeast. It can be transformed into dried yeast for human or animal feeding.

Baker's yeasts are packed in the form of moist yeasts or in the form of dried active yeast.

Compressed yeast: After-harvesting the yeast product, it is combined with fine particles of ice, starch, fungal inhibitors, and vegetable oils (e.g. glyceryl monostearate); all these ingredients contribute to stability. Then it is condensed into small blocks (1–5 lbs) for domestic use or large blocks(up to 50 lbs), preserved at -7 to 0 °C and transferred into refrigerated containers (Reed and Nagodawithana, 1990).

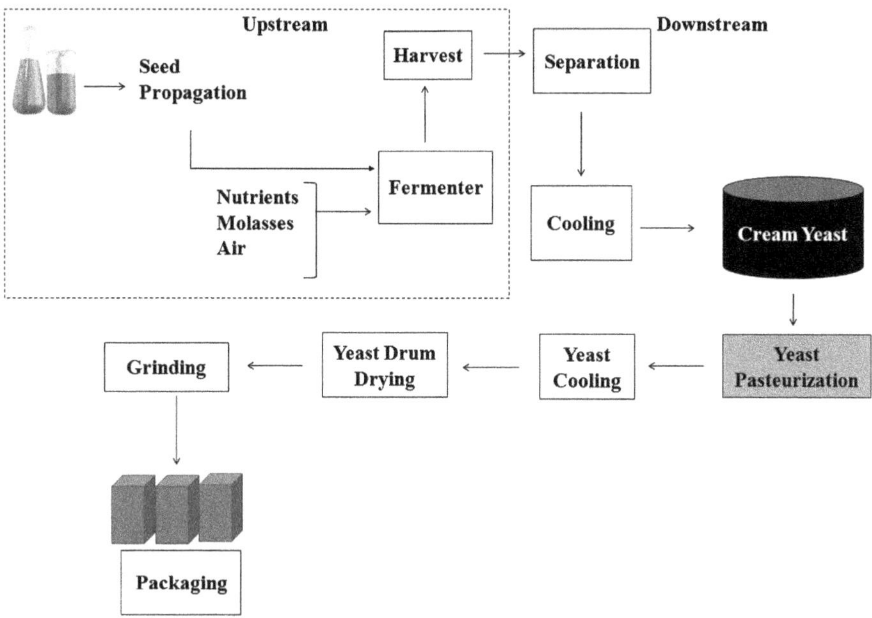

FIGURE 16.2 Recovery of bakers' yeast.

Active dry yeast: This is resistant and also can be used in areas and regions where cooling is inaccessible. For several developing countries baker's yeast in the form of active dry yeast is traded internationally. Special strains are applicable and dry environments can be used for the productive development of dry yeast. It has been noticed that these perform better than standard strains of dry yeast when exposed to various treatments. These treatments involve increasing the temperature to 36°C towards the end of fermentation, adding alcohol-based spent broth (which results from centrifugation or finished fermentation of the yeast), and synchronizing budding by alternating feeding and starvation. The concept behind this addition is not clear (Reed and Nagodawithana, 1990).

By filter pressing methods, about 30–38% yeast cream can be removed through a panel in order to develop a continuous thread-like structure. Next, these thread-like structures are cut finely and dried using several driers, such as tray driers, rotary drum driers, or fluid bed driers. The result shows a moisture value of around 8 percent and can be sold in tins filled with nitrogen. Antioxidants can also sometimes be added in order to maintain shelf life (Reed and Nagodawithana, 1990).

16.3 FOOD YEAST

As food yeasts are consumed by humans, government departments, professional organizations, and producers enforce high rules on these products. The standards of IUPAC maybe cited as a reference here since they are the most accurate. The IUPAC demands non-extraction, fat content not greater than 20 percent, food yeast belonging to the *Cryptococcaceae* family should not contain any inert additives, and should be free of *Salmonella*. There are certain disadvantages such as the excessive RNA content of yeasts which the human kidneys cannot remove, thus making it toxic for the body (Touzi et al., 1982).

Saccharomyces cerevisiae, Saccharomyces carlbergiensis, Saccharomyces fragilis, Candida utilis and Candida tropicalis are yeasts used as food yeasts. Lactose can be used for whey fermentation. Lactose is typically utilized only by *Saccharomyces fragilis* (imperfect stage of *Candida pseudotropicalis*). To produce food yeast, ethanol is a suitable substrate; it is only used by *Saccharomyces fragilis* and *Candida utilis,* the most flexible of all yeasts, and can utilize a broader variety of sources of carbon and nitrogen. Thus it is mainly used in preparations for food yeast (Touzi et al., 1982).

16.3.1 Substrates to Produce Food Yeast Production

Molasses, sulphite liquor, wood hydrolysate, and whey are the most widely used substrates. Other unusual substrates which can be recommended are different kinds of hydrocarbon, alcohol, and waste etc.

Molasses: After separating from the spent liquor by centrifugation, baker's yeast is produced using molasses as explained above and may be dried to produce food yeast. Drum-drying, spray-drying, or fluid-drying of the bed can only decrease the humidity to about 5 percent(Touzi et al., 1982). Thus *Candida utilis* is grown fed-batch in Waldhof fermenters in Taiwan. The fed-batch process is also used in South Africa, using molasses as a substrate. Utilization of *Candida utilis* along with food yeasts has currently been proposed in Cuba and Eastern Europe in the continuous culture of molasses.

Sulfite liquor: The incentive to generate sulfite liquor food and fodder yeast from sulfite liquor also helps in pollution reduction if waste contains fermentable substrates. The application of continuous fermentation is also attractive because sulfite is produced continuously by pulp mills (Figure 16.3).

In most cases, a Waldhof fermenter is used to produce yeasts with the help of continuous methods. Liquors are usually mixed from multiple sources. Sulfite compounds are either excluded by lime precipitation, aeration (steam stripping) or steam flow. The pH is modified with ammonia from an approximate pH of 2 to 5.5 (Touzi et al., 1982). Continuous supply of ammonium, phosphate, and potassium is necessary. *Candida utilis* is a versatile and tough yeast normally used to prevent the addition of biotin. The yeast is harvested and recycled continuously by simultaneously extracting the effluent liquor. Centrifugation concentrates the cells from approximately 1% to 8%. To extract lignosulphinic acid, it is normally washed with water for dilution, as well as being centrifuged. The development of material from sulphide and alcohol compounds is usually poor and can be improved by introducing new carbon compounds such as acetate acid and ethanol. Moreover, with the addition of H_2SO_4, the liquid can be hydrolysed to increase sugar content from 4% to around 24%.

Manufacture of food yeast from whey: Whey is regarded as the effluent of draining milk from the coagulum while manufacturing cheese. It comprises roughly 4% sugar (lactose), 1% minerals and some quantity of lactic acid, which allows the milk protein to coagulate. Whey is a waste product in countries that produce a lot of cheese, but

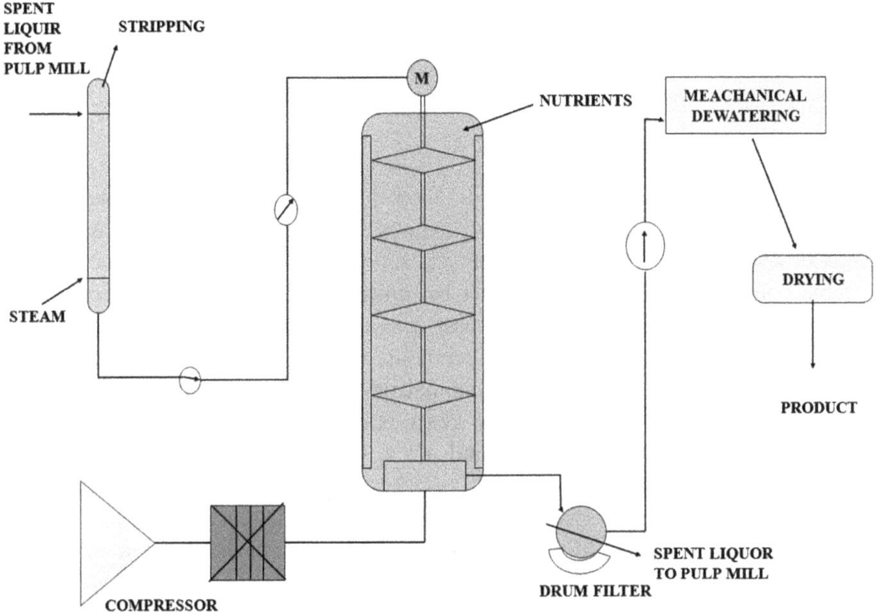

FIGURE 16.3 Flow chart for the production of fodder yeast from sulphite liquor.

it can be converted into alcohol or yeast. Lactose is metabolized by very few yeasts (Touzi et al., 1982). These are *Saccharomyces lactis, Kluyveromyces fragilis,* and *Candida pseudotropicalis.* The whey is diluted, mixed with ammonia, phosphate, minerals, yeast extract and then pasteurized for approximately 45 min at 80 °C and incubated at a temperature of 30 °C, and pH 4.5. In the United States, principally *K. fragilis* is used in continuous fermentation and automatic monitoring of sugar, pH, and minerals. The yeast is recovered by centrifugation along with drum or spray drying (Touzi et al., 1982).

16.4 FEED YEAST

Feed yeasts are like food yeasts. The only variation is that feed yeasts possess fewer rigid standards. Thus feed yeasts are typically used for animals, and are produced by drying out the entire fermented broth, mostly without washing. Thousands of tons of yeast are restored yearly in breweries across the globe for conversion into feed yeast; these need to be 'debittered' of hop resins by continuous rinsing with dilute alkali until the bitter taste is removed. Then they are slightly acidified to about pH 5.5. Cells are retrieved by spray and centrifugation or drum-dried (Touzi et al., 1982).

Alcohol yeast: This type of yeast is used in beer brewing, wineries, and commercial alcohol distilleries. The approaches to baker's yeasts are generally similar to those already mentioned (O'Leary et al., 1977). Contamination is monitored on a plate from a lyophilized vial or tube. A single colony (or preferably a distinct spore by micromanipulation) is selected and increased consecutively in quantity. The yeasts used are specially selected strains for breweries, and are as follows: *Saccharomyces cerevisiae, Saccharomyces uvarumcarlbergensis S. uvarum* (Kourkoutas et al., 2001).

16.5 YEAST PRODUCTS

Different products of yeast are used in food, pharmaceuticals, and related industries. In certain regions of the world, a famous example is marketed as 'Marmite'. The extracts can be collected with autolysis and then with or without harvesting the soluble compound from the autolysate by spray or drum drying. The extract can also be obtained by hydrolysing the yeast cells using an acid solution. The sodium hydroxide is neutralized, filtered, and decoloured with charcoal. Flavouring components such as sodium glutamate and animal or vegetable protein extracts are normally used to enhance yeast products. Yeast extracts are consumed as a nutritional content for pharmaceutical reasons and are a source of Vitamin B12 (Reed and Nagodawithana, 1990).

Market trends of baker's yeast: By 2024, the global demand for baker's yeast is expected to hit USD 2.84 billion, rising at a compound annual growth rate of 5.1 percent during the 2019–2024 forecast period. The rising demand from end-user industries, such as food and beverages, is contributing significantly to the growth of the yeast industry. There is a large demand for baker's yeast, as it is widely used for improving the consistency of bakery products. In terms of supply chain, product selection, networks of distribution, and customer tastes, the European bakery market is well known. In the indulgence areas, such as desserts, pastries, and biscuits, novelty and new product growth are gradually witnessed. Growing demand for processed foods, combined with rising awareness of health-conscious goods, is a major driver of growth in the European region's bakery yeast industry, which, in turn, is driving growth in the market under research. With some global players, such as Associated British Foods PLC (ABF), Lesaffre International, Lallemand Inc., and numerous regional players, mainly from the European region, the global bakery yeast market is moderately concentrated.

16.6 SPIRULINA

Spirulina is a blue-green alga which is multicellular and also filamentous and belongs to the genus *Spirulina* and *Arthrospira* and has 15 species. Out of all the species, *Arthrospira platensis* holds primary status and is also the most widely available. The main role of *Spirulina* is associated with single cell proteins (SCPs). The cells of *Spirulina* are very beneficial for deriving these SCPs. The field of biochemical engineering has always focused on developing methods for the better processing of food. *Spirulina* cells are better than others because of their higher content of protein, amino acids, nucleic acids, essential minerals, vitamins, iron, beta-carotene, essential fatty acids (EFA) etc. and this makes the cells of *Spirulina* best suited for food processing. γ-linolenic acid (GLA) as EFA, β-carotene, linoleic acid, arachidonic acid vitamin B12, iron, calcium, phosphorus, and phycocyanin, a pigment-protein complex are found only in *Spirulina*. This species is found in lake Texcoco, Chad and

Niger, Nakuru, Elementeita, Kilotes, Aranguadi and also along the Great Rift Valley. These are the largest reservoirs of these species. *Spirulina* is also used as a feed supplement in aquaculture, aquarium, and poultry industries. Itis used as food, a dietary supplement, and as a nutritional supplement and natural health product. Some of its components are also used as food colouring agents in Japan (Shimamatsu, 2004). Certain cosmetics also use *Spirulina* as the active ingredient. Much research is involved in upgrading the quality of wild strains to increase their effectiveness and many industries are involved in its marketing and production.

16.6.1 Production and Harvesting

Spirulina is either produced naturally or in laboratory conditions. It is naturally produced in many parts of the world by directly harvesting from natural sources where the species of *Spirulina* dwells, but in the laboratory, there are some specific condition under which *Spirulina* is harvested. Some of the processes which are used to harvest *Spirulina* are nutritional media, and light regimes etc. *Spirulina* acts as an ideal candidate for replacing traditional food items since it can be harnessed and cultivated easily and is sustainable. There are many processes and approaches to harvesting *Spirulina.* Most commercial production is based on an approach where shallow raceways are used on which *Spirulina* strains are mixed in paddle wheels (Figure 16.4). For mass production, there are two types of process; one is open raceway ponds with concrete linings, and the other is a shallow tunnels. The paddles can be large or small and are used to stir the mixture.

In Mexico, *Spirulina maxima* has been harvested in a semi-tropical environment which was 2200 m above sea level in Lake Texcoco. In semi-natural harvesting of *Spirulina* strains, harvesting is done at day and night time and the alga biomass doubles over a three to four day time lapse. Bangladesh uses a dynamic approach

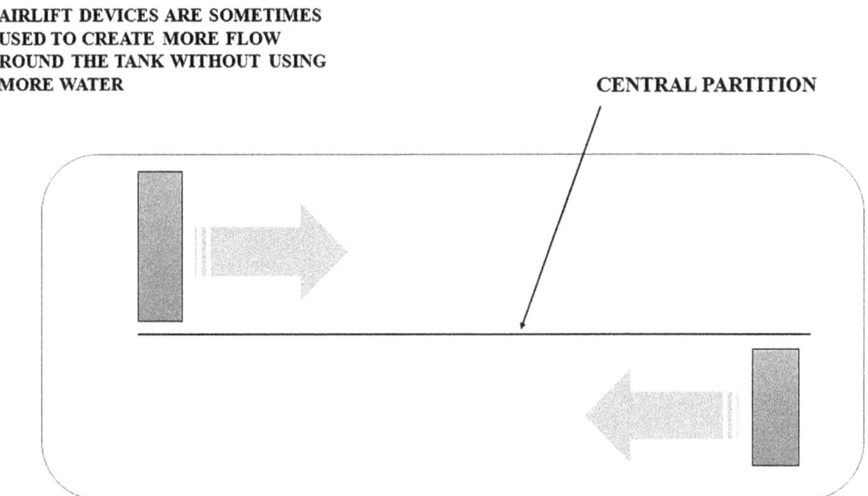

FIGURE 16.4 Raceway pond for bulk spirulina production.

to harvesting *Spirulina* in heterotrophic mode, where it is cultured in wastewater effluents of sugar mills, poultry or fertilizer factories, and also urban waste. One basic approach to harvesting *Spirulina* is by carrying it out in ditches which are unlined and where the flow of water is low, where harvesting is done by using a cloth and the biomass eventually gets dehydrated under the sun. In another system, a digester is fitted which processes wastes and sewage and produces biogas. Then the excluded algal biomass is harvested, which is done with a cloth, and that then produces a quantity of *Spirulina* which is used as a food supplement by farmers and the poor (Hülsen et al., 2018). In an approach to production and harvesting of *Spirulina* named 'mud-pot', a specific medium is used which contains biogas slurry, 2–3 g of salt and *Spirulina* culture. These mud pots are then buried under the soil and filled with water and mixed with medium, and more pure culture of *Spirulina* is added. This medium is stirred many times per day and then within four days the culture becomes mature. This can then be harvested by using a clean cloth (Shimamatsu, 2004; Das, 2015).

For effective production of *Spirulina*, new mediums are being formulated such as RM6, which is a combination of various chemicals and elements and this increases growth of the cultures. Digesters are also used to produce and harvest *Spirulina* strains. These digesters produce biogas, which is filtered, and that gives a gross amount of *Spirulina*. One major limitation in harvesting *Spirulina* at small scale is its high cost in terms of inorganic nutrients (Figure 16.5). Harnessing essential nutrients from water effluents is an excellent alternative to overcome this limitation. The steps included in harvesting and processing *Spirulina* are:

Firstly, a nylon filter is used as a filter in the water body entrance. In the pre-concentration step, the algal biomass is washed in order to lower salt content. In the next step, the interstitial water is removed and then the recovered biomass is neutralized. Then a grinder is used for disintegration. Packing is done and the product is stored (Andrade et al., 2018).

16.6.2 Composition of Spirulina

An ideal cell of *Spirulina* has following components:

Vitamins: Vitamin $B_{1,}$ Vitamin $B_{2,}$ Vitamin $B_{3,}$ Vitamin B_6, Vitamin $B_{9,}$ Vitamin $B_{12,}$ Vitamin C, Vitamin E and Vitamin D.

Protein: 55%–70% protein; amount calculated by dry weight. Contains all essential amino acids.

Minerals: Contains calcium, potassium, copper, chromium, iron, manganese, magnesium, selenium, phosphorus, zinc and sodium.

Fatty acids: Contains high amount of polyunsaturated fatty acids or PUFA, γ-linolenic acid (ALA), linoleic acid (LA), eicosapentaenoic acid (EPA), steariodonic acid (SDA), arachidonic acid (AA) and docosahexaenoic acid (DHA).

Photosynthetic pigments: Contains chlorophyll a, beta-carotene, xanthophyll, echinenone, zeaxanthin, myxoxanthophyll, diatoxanthin, canthaxanthin, beta-cryptoxanthin, oscillaxanthin, 3-hydroxyechinenone, allophycocyanin and c-phycocyanin. This is a brief composition of *Spirulina* cells; experiments have been conducted to get a more detailed composition of these cells. Depending on

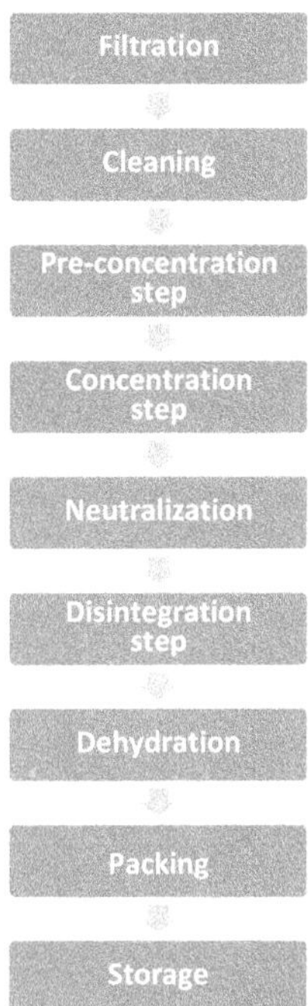

FIGURE 16.5 Flow process diagram of Spirulina recovery.

the origination of the species the composition changes (Saranraj and Sivasakthi, 2014). Table 16.2 shows the approximate composition of *Spirulina* cells deriving from France, India, Thailand, Malaysia and Bangladesh.

16.6.3 Advantages of Spirulina as SCP

Spirulina is used as a source of protein and dietary supplement extensively due to its major advantages, which we have seen. Many experiments and chemical trials have been done in order to test usage of *Spirulina*. It is even being used by farmers and financially deprived people due to its cost effectiveness and higher nutritional composition. Ingesting *Spirulina* as a dietary supplement has many advantages, such as helping in boosting immunity. The nutritional content of *Spirulina* is very high in terms of proteins and other macronutrients and hence this proves very worthwhile for poor people to have as a cost-effective food supplement. When *Spirulina* cells are converted into powder, it can be used to made a variety of food products which are good commercially and for people's welfare (Andrade et al., 2018). *Spirulina* can grow in all three modes: autotrophically, heterothropically and mixotrophically (Figure 16.6).

In the scope of agriculture, *Spirulina* proves to be a very good fertilizer for plants when used in combination with other fertilizers. It is also used in the poultry business as a protein supplement to get better yield. *Spirulina* is also used as a colourant

TABLE 16.2
Approximate Composition of *Spirulina* in % Dry Matter from 4 Originating Places (Saranraj and Sivasakthi, 2014)

Constituent	Malaysia	France	Bangladesh	Thailand
Protein	61	65	60	55–70
Lipid	6	4	7	6
Carbohydrate	14	19	-	-
Fibre	-	3	-	6
Moisture	6	-	9	5
Ash	9	3	11	3–6
NFE (Nitrogen free extract)	4	-	17	15–20
Amino acids (g/100g)-				
Lysine	4.63	-	-	2.60–3.30
Phenylamine	4.10	-	-	2.60–3.30
Tyrosine	3.42	-	-	2.60–3.30
Leucine	8.37	-	-	5.90–6.50
Methionine	2.75	-	-	1.30–2.00
Glutamic acid	7.04	-	-	7.30–9.50
Aspartic acid	5.37	-	-	5.20–6.00
Tryptophan	2.98	-	-	1.00–1.60
Cystine	0.6	-	-	0.50–0.70
Serine	3.84	-	-	-
Arginine	4.94	-	-	-
Histidine	2.81	-	-	-
Threonine	3.35	-	-	-
Proline	4.11	-	-	-
Valine	4.01	-	-	-
Isoleucine	3.85	-	-	-
Alanine	10.81	-	-	-
Glycine	6.66	-	-	-

and hence this alga is effective in a variety of commercial sectors. When seen in marine life, this alga is used as a nutritional supplement for aquatic organisms to get a better yield. It is cheaper than other feeding supplements for fish and other animals (Andrade et al., 2018).

As stated earlier, no more than 15 species of *Spirulina* have been discovered to date. These are *Spirulina platensis*, *S. platensis* Geitler, *S. platensis* NIES-39, *S. platensis* (Nordstedt) Geitler, *S. subsalsa* Oersted, *S. maxima*, *S. subsalsa* fo. *versicolor* (Cohn) Koster, *A. maxima*, *S. subsalsa* Oersted ex Gomont, *S. lonar*, *S. major* Kutzing, *Arthrospira fusiformis* (Voronichin), *S. laxa* G.M. Smith, *S. nodosa*, *A. jenneri* (Kutzing) Stitz, *S. labyrinthiformis*, *S. laxissima* and *S. princeps* West & West.

Future and revolutionary aspects in Spirulina: *Spirulina* is already in use in commercial applications. The largest commercial producers of *Spirulina* are located in USA, Thailand, India, Taiwan, China, Bangladesh, Pakistan, Myanmar, Greece, and Chile. The average cost of *Spirulina* dietary product in India ranges

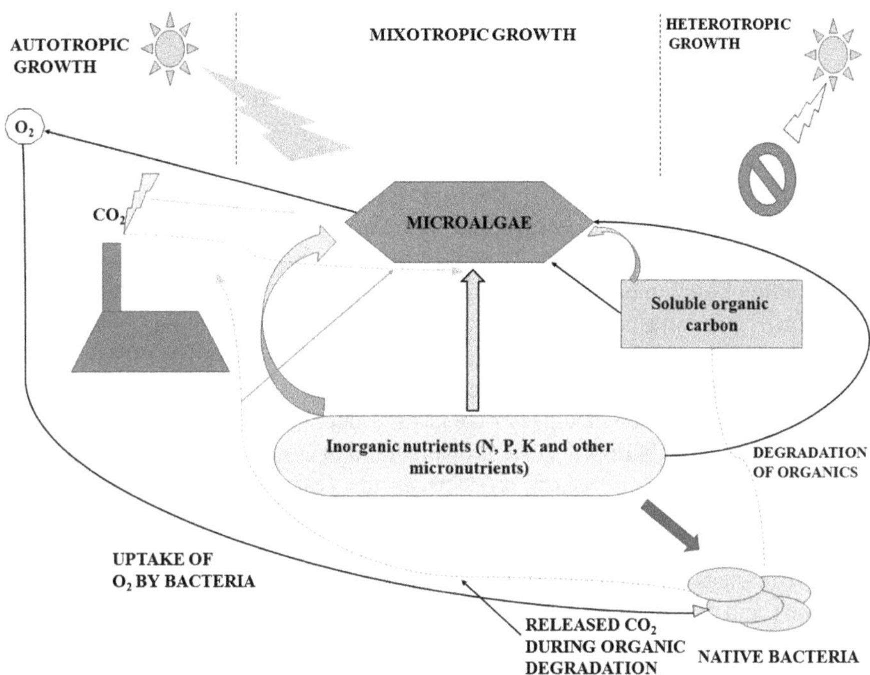

FIGURE 16.6 Different mode of growth for Spirulina sp. for bulk production.

from Rs.600–1,000/kg of product. However, research has been done to manipulate the genes of the *Spirulina* species in order to get a better yield and nutritional content. Plant growth restrictors are being used on *Spirulina* to enhance its growth. One example would be the usage of 6-BA in combination with NAA for increasing the growth rate of *Spirulina*. Others are Gibberellic acid, 2,4-D, IAA and IBA. Other approaches include adapting the growth of *Spirulina* in specific environments to induce growth of this alga on particular platforms. There is large scope for *Spirulina* to be used as an SCP and as an excellent food supplement.

Market trends of Spirulina: According to MRFR analysis, the global *Spirulina* market is expected to grow at a CAGR of 8.02 percent from 2018 to 2023. *Spirulina* is a blue-green micro-alga that is multicellular and filamentous. It is used as a human food substitute and to resolve malnutrition because it is a rich source of protein, vitamins, minerals, carotenoids, and antioxidants. The worldwide demand for *Spirulina* is split into North America, Europe, Asia Pacific, and the rest of the world. Among these, due to rapid economic development, increasing urbanization, and growing middle-class population, Asia Pacific is expected to maintain its supremacy over the forecast era. Cyanotech Corporation (Hawaii), DIC Corporation (Japan), E.I.D. Parry Limited (India), Tianjin Norland Biotech Co. Ltd (China), Jiangxi Alga Biotech (China), Hydrolina Biotech Private Limited (India), and Synergy Natural Products Pty Ltd (Australia)are the major industry participants in the global market for *Spirulina*.

16.7 CONCLUSIONS

The present chapter deals with the production and downstream processing of baker's yeast and *Spirulina*. A need for protein has always guided the production of SCP processes, and this continues to be a significant factor in the development of both old and modern processes. Yeast cells and *Spirulina* are the two most common SCPs produced industrially due to high demand. Baker's yeast is commonly used as a leavening agent in baking bread and bakery products. Baker's yeast is classified on the basis of its moisture contents. Dry yeast forms are good choices for longer-term storage, often lasting more than a year at room temperatures. Genetically stable high-fermenting power yeasts are used in industries which can prevent autolysis. Proper mixture of beet molasses and cane molasses(60:40) is the best nutrient for the development of yeast cells. Yeast extracts have been essential ingredients in savoury tastes as well as in fermentation media over the last few decades. The growth of the baker's yeast market is directly linked to the growing trend in consumption of processed and fast foodstuffs, especially bakery products. In the years 2004–2005, the European and Asian regions produced 51 million tonnes of bakery products, priced at US$107 billion. As a result of the growing global trend, China is currently one of the most attractive markets for baker's yeast, as its demand continues to expand as the population grows and the demand for bakery products shifts.

Spirulina is used as food, as a dietary supplement, as a nutritional supplement and as a natural health product. *Spirulina* is produced by factories at large scale in raceway ponds adopting mechanized agitation, filtering and drying systems with relatively high capital investment. Proper and quick drying are essential features of high-quality *Spirulina* production. For economic reasons, the dryer of choice in large-scale *Spirulina* production facilities is the spray dryer. Sun drying is one of the most economical systems for small projects. The recent goal to acquire biofuels as well as algae by-products is expected to lead to advances in production, extraction and processing of algae products that can support producers of *Spirulina*.

REFERENCES

Abbott BJ, Clamen A, The relationship of substrate, growth rate, and maintenance coefficient to single cell protein production. Biotechnol. Bioeng. 15, 117–127, 1973.

Andrade LM, Andrade CJ, Dias M, Nascimento CAO, Mendes MA, Chlorella and Spirulina Microalgae as Sources of Functional Foods. Nutraceuticals Food Suppl. 45–58, 2018.

Chiattoni LM, Nutritional evaluation of single-cell protein produced by Spirulina platensis. Afr. J. Food Sci., 2011.

Das D, Algal Biorefinery: An Integrated Approach, Capital Pub. Co, New Delhi and Springer, Switzerland, 2015.

Hülsen T, Hsieh K, Lu Y, Tait S, Batstone DJ, Simultaneous treatment and single cell protein production from agri-industrial wastewaters using purple phototrophic bacteria or microalgae – A comparison. Bioresour. Technol. 254, 214–223, 2018.

Kourkoutas Y, Komaitis M, Koutinas AA, Kanellaki M, Wine Production Using Yeast Immobilized on Apple Pieces at Low and Room Temperatures. J. Agric. Food Chem. 49, 1417–1425, 2001.

Nangul A, Bhatia R, Microorganisms: a marvelous source of single cell proteins. J. Microbiol. Biotechnol. Food Sci. 10, 15–18, 2020.

O'Leary VS, Green R, Sullivan BC, Holsinger VH, Alcohol production by selected yeast strains in lactase-hydrolyzed acid whey. Biotechnol. Bioeng. 19, 1019–1035, 1977.

Reed G, Nagodawithana TW, Baker's Yeast Production, In: Reed G, Nagodawithana TW (Eds.), Yeast Technology. Springer Netherlands, Dordrecht, pp. 261–314, 1990.

Reihani SFS, Khosravi-Darani K, Influencing factors on single-cell protein production by submerged fermentation: A review. Electron. J. Biotechnol. 37, 34–40,2019.

Ritala A, Häkkinen ST, Toivari M, Wiebe MG, Single Cell Protein—State-of-the-Art, Industrial Landscape and Patents 2001–2016. Front. Microbiol. 8, 2017a.

Ritala A, Häkkinen ST, Toivari M, Wiebe MG, Single Cell Protein—State-of-the-Art, Industrial Landscape and Patents 2001–2016. Front. Microbiol. 8, 2017b.

Royer JC, Nakas JP, Potential Substrates for Single Cell Protein Production, In: Côté WA (Ed.), Biomass Utilization, NATO Advanced Science Institutes Series. Springer, Boston, MA, pp. 443–459, 1983.

Saranraj P, Sivasakthi S, Spirulina platensis–food for future: a review. Asian J. Pharm. Sci. Technol. 4, 26–33,2014.

Shimamatsu H, Mass production of Spirulina, an edible microalga. Hydrobiologia 512, 39–44, 2004.

Suman G, Nupur M, Anuradha S, Pradeep B, Single Cell Protein Production: A Review 12, 2015.

Touzi A, Prebois JP, Moulin G, Deschamps F, Galzy P, Production of food yeast from starchy substrates. Eur. J. Appl. Microbiol. Biotechnol. 15, 232–236,1982.

Tusé D, Miller MW, Single-cell protein: Current status and future prospects. C R C Crit. Rev. Food Sci. Nutr. 19, 273–325, 1984.

17 Biodiesel, Power Alcohol and Butanol Production

In some areas of the world, petroleum-based resources are restricted and reserved. Day by day, fossil fuel resources are reducing. The global economy is highly dependent on services and goods for export, but on the other hand, transport is mainly dependent on the supply of energy from petroleum-based fuels. The use of fossil fuels has caused global climate change. Thus fossil fuels should be replaced with sustainable and green sources of energy in order to minimize carbon dioxide and greenhouse gas emissions. Through biological systems, biofuels (bioalcohol, biogas, and biodiesel) can be produced which are considered to be efficient alternative sources of energy. Recently, biodiesel has been the only renewable power transporter that can replace diesel fuel explicitly in compression ignition engines, and is receiving rising global attention. Biodiesel is biodegradable, clean with an excellent flashpoint, non-toxic, sustainable, has better viscosity, has fossil-fuel-like calorific value, and is renewable. Biodiesel can be chemically defined as long-chain fatty acid monoalkyl esters. Biodiesel can be derived from a number of inexpensive renewable lipid feedstocks such as vegetable oils or animal fats, algal oils, microbial oils, and waste oils. It comprises esters of short-chain alcohols; primarily, methanol or ethanol. Biodiesels have considerably large higher heating values (HHVs). In gasoline, 46 MJ/kg HHVs are found, which are considerably higher than biodiesel HHVs at 39–41 MJ/kg, petroleum at 42 MJ/kg, and petrodiesel at 43 MJ/kg, but only slightly better than coal at 32–37 MJ/kg. Biodiesel can be mixed with diesel at all ratios to form a biodiesel mixture.

These days, with the help of government and regulatory frameworks through opportunities and prescription of volumetric needs, several nations have emphasized production and use of renewable fuels such as biodiesel. To reduce exchangeable carcinogens and pollutant outcomes, biodiesel was found effective. Consequently, edible vegetable oils such as soybean, rapeseed, and palm oils are being used as auxiliary feedstock for the manufacture of biodiesel in the twenty-first century. The use of edible oils as an energy source, however, has attracted numerous questions from environmental and non-governmental organizations. Second-generation biodiesel has been used as an enticing substitute feedstock for the biodiesel market. For biodiesel production, multiple microbes have also been considered to be beneficial: *Rhodosporidium spp.*, *Rhodococcus spp.*, and *Cryptococcus spp.* Microalgae are being examined for ability to accumulate high oil content as a substitute for vegetable oils for the production of biodiesel. For biodiesel production, microalgae as a feedstock have been extensively reviewed. Microalgae grow at an extraordinarily fast rate; about 100 times faster than terrestrial plants, and they can double their biomass in less than one day. Biodiesel derived from microalgae is among the third-generation biofuels which open up completely new features of renewable energy sources. Microalgae biofuels have been positioned worldwide as among the main areas of research that can bring tremendous benefits to humans and the environment. Pyrolysis, transesterification, and the supercritical fluid process these various methodologies have been developed for biodiesel production. To generate yield of biodiesel and glycerol from oil, transesterification is the most commonly used technique among all these processes.

17.1 BIODIESEL

Biodiesel is classified by ASTM International as a combination of long-chain fatty acid monoalkylic esters produced by renewable resources for use alone or blended with diesel. Biodiesel supplemented with diesel fuel is indicated as 'Bx', where 'x' is the percentage of biodiesel in the blend. For instance, 'B20' indicates a blend with 20% biodiesel and 80% diesel fuel. All biodiesel fuels are of similar renewable origin, developed through photosynthesis conversion of solar to chemical energy that segregates them from slightly earlier photosynthesis. The term 'biodiesel' is applied to long chains of monoalkyl esters of fatty acids derived from waste oils, edible oils, and non-edible oils generated by methanol and catalyst transesterification process of triglycerides. Methanol is usually used for the production of biodiesel due to its low cost and easy availability. Biodiesel production depends on solar energy and is the basis of renewable economic development. The primary concern for renewable fuel is a land struggle for fuel and food development. For the development of biodiesel, a scientific move is taken that involves the optimum production process, enhancement of quantity and quality for biodiesel, the development of feedstocks, and carbon-neutral economic systems (Mata et al., 2010).

17.1.1 First-Generation Biodiesel

Soybean oil, corn oil, olive oil, mustard oil, coconut oil, rice bran oil, palm oil, and rapeseed oil are some examples of edible feedstocks utilized for the production of first-generation of biodiesels. Usage of edible feedstuffs for biodiesel production is very recent and at the start of the biodiesel era (Leung et al., 2010). A major aspect of first-generation feedstock is the accessibility of crops and this has a comparatively simple conversion process. The problem of food source restriction is the major drawback in the usage of these feedstocks, and raises the value of food products. The obstacles behind biodiesel production through edible feedstock are adaptability to high prices, restricted area of cultivation, and environmental conditions. These limitations have forced customers to move to alternative sources for the production of biodiesel.

17.1.2 Second-Generation Biodiesel

Biodiesels of the second-generation utilize non-edible feedstock, such as nag champa oil, *Calophyllumino phyllum* oil, Jatropha oil, *Mahua indica* oil, rubber seed oil, neem oil etc. Because of the limitations of utilizing first-generation feedstock, researchers started to work on non-edible feedstock. The main benefits of utilizing second-generation biodiesel are reduced costs of production, elimination of food scarcity, eco-friendliness, and less land being required for agriculture. Such oils provide the key advantages of using second generation biodiesel. There is no need to rely on food plants, and there is no need for agricultural land only. Second-generation fuels have certain drawbacks such as plant yield, where output decreases for major non-edible plants such as Karanja oil, Jatropha oil, and Jojoba oil. These foodstuffs may grow in insignificant fields. Hence there is pressure on agricultural territories to farm non-edible crops. This significantly affects society's economic growth and food production. Researchers are trying to pay attention to specific alternative solutions that are economically viable and simply available to a large extent, in order to defeat the social and economic problems of non-edible oil. Additional alcohol necessity is another disadvantage for biodiesels of the second generation (Crabbe et al., 2001).

17.1.3 Third-Generation Biodiesel

Third-generation biodiesel is the term used for biodiesel produced from microalgae and waste oils. A higher rate of growth and efficiency, higher oil proportion, less competition for agricultural land, less food chain impact, and low greenhouse effect are the main advantages of biodiesel of the third generation (Ma and Hanna, 1999). Fish oil, waste cooking oil, animal fats, and microalgae etc. are the main sources for the third generation of biodiesel. The need for enormous sums of money, the problem of large-scale development, oil extraction complications, and the need for sunlight are the key drawbacks. All of these viable third-generation biodiesel resources defeat the problems of initial-generation feedstocks affecting flexibility, availability, economic feasibility, environmental parameters, and the food chain. For biodiesel of the third generation using cooking oil, waste oils, waste animal tallow oil, waste fish oil is also considered. This also decreases the waste-handling plant load and water pollution reduces. Beef, pork, goats, and poultry are types of animal fats which currently have growing potential for production of biodiesel.

17.1.4 Fourth-Generation Biodiesel

In the fourth generation of biodiesels, photobiological solar fuels and electro-fuels are utilized. Solar biofuels are generated using raw resources from solar energy for conversion into biodiesel. This conversion approach is a major area of research. The raw resources are accessible globally, are limitless, and inexpensive. Similarly, a combination of photovoltaic or inorganic water-splitting catalysts with the production of metabolically modified microbial fuel is a growing strategy for effective management and processing of liquid fuel (Crabbe et al., 2001).

17.2 FEEDSTOCK FOR THE PRODUCTION OF BIODIESEL

Vegetables, microbial oil, animal fats, and algae are some examples of feedstock which can be used for biodiesel extraction (Table 17.1). Feedstock selection that affects different factors such as price, composition, yields, and biodiesel purity, is an essential aspect of biodiesel production. To categorize biodiesel into edible and non-edible, and waste-based sources, feedstocks are essential factors for accessibility and kinds of origin. The choice of feedstock for the production of biodiesel depends on regions (Leung et al., 2010). The economic dimension and availability of the nation in question are considered before feedstock selection. Canola oil is used as a feedstock in Canada, and soybean oil is used in Brazil and the USA. Coconut and palm oils are used as biodiesel feedstock in Indonesia and Malaysia, and the UK uses rapeseed oils. However, Karanja and Jatropha are considered to be possible feedstocks for biodiesel in India. Rapeseed oil, mustard oil, sunflower oil, and soybean oil were historically used as biodiesel feedstock, but their use as biodiesel feedstock declined due to adverse effects on food plants. The use of edible oils as feedstocks for biodiesel has a major problem because of the impact on the food chain. It is assumed from different studies that the use of non-edible oil as a biodiesel feedstock has many advantages such as low sulfur content, low content of aromatics, biodegradability, no impact on the food chain, and accessibility.

17.2.1 FEEDSTOCK FOR FIRST-GENERATION BIODIESEL

Arecaceae (Palm oil): Malaysia and Indonesia were the two largest palm oil producing nations over the last decade. Nigeria and Brazil have strong potential for developing palm oil. In Europe, palm biodiesel oil demands are rapidly increasing. Palm oil has several advantages such as maximal oil yields for each hectare and are especially economic, in comparison with several other edible oils (Ma and Hanna, 1999). The

TABLE 17.1
Various Feedstocks; Percentage Gained of Oil Material

Various Feedstocks (Edible, Non-edible, and Animal Fats)	Oil (% gained)
Glycine max (Soybean)	15-20%
Arecaceae (Palm oil)	30-60%
Cocos nucifera (Coconut)	63-65%
Brassica napus (Rapeseed)	38-46%
Linum usitatissimum (Linseed)	40-44%
Jatropha curcas (Jatropha)	30-40%
Hevea brasiliensis (Rubber seed)	53.70-68.40%
Azadirachta indica (Neem)	20-30%
Ricinus communis (Castor)	45-50%
Algae	30-70%
Microalgae	30-70%
Chicken waste (Broiler)	30-70%

height of palm trees varies from 10 to 5 m. Monounsaturated fatty acids and high concentrations of saturated medium-chains are present in palm oil. It contains a low percentage of stearic acid 3–6 %, 9–12 % linoleic acid, and a high percentage of oleic acid 36-44 % and palmitic acid 39–48 %. Since palm oil has high saturated fatty acid content, this problem is associated with palm oil in the production of alkali catalysed biodiesel. However, by pursuing an acid-catalysed pre-esterification procedure, this issue was eliminated.

Glycine max (Soybean): Soybean is the world's leading grown oilseed crop. Soybean oil is a very popular source of biodiesel production in the USA. The height of the soybean tree varies from 0.5 to 1.2 m. Soybean generates less oil yield per hectare compared with other such sources. Soybean can fix nitrogen, and that is why it can grow in both tropical and temperate environments. Soya even regenerates nitrogen in the soil. Soybean contains a high content of oleic acid at 20–30%, and linoleic acid at 50–60%, withlow contents of linolenic acid at 5–11 % and palmitic acid at 6–10% (Crabbe et al., 2001).

Cocos nucifera (coconut): Coconut is one of the feedstock sources for biodiesel production. In the Philippines, this feedstock has been approved for biodiesel production. Coconut oil is a triglyceride that has a significant proportion of saturated fatty acids of 86%; polyunsaturated fatty acids are 2%, and monounsaturated fats are 6%, which is low in percentage. Lauric acid 45%, palmitic acid 8%, and 17% myristic acid are present. It has monounsaturated fatty acid in lower amount as oleic acid, and it also has polyunsaturated fatty acid as linoleic acid. Coconut oil is a rich source for biodiesel (Crabbe et al., 2001).

17.2.2 Feedstock for Second-Generation Biodiesel

Madhuca long folia (Mahua): India is the largest producer country of Mahua. This belongs to the family Sapotaceae and is an evergreen tree. For the Mahua plant, cultivation in a warm and moist atmosphere is the best environmental climate. Around 20 to 200 kg of seeds are produced annually from a single Mahua plant (Crabbe et al., 2001). Approximately 20 m is the maximum height of the Mahua plant. In the Mahua plant, linoleic acid is 8.9–18.3%, stearic acid 20.0–25.1%, palmitic acid 16.0–28.2%, and oleic acid 41–51%; this unsaturated acid present in large amounts.

Gossypium (Cottonseed): Europe, China, and the United States are the major producer nations of the cotton crop. *Gossypium hirsutum* and *Gossypium herbaceum* are the primary cotton plant species, and are used for production of cottonseed oil. The height of cotton plants is 1.2 m. This oil has different types of material that are non-glyceride, such as carbohydrates, phospholipids, sterols, resins, linked pigments, and gossypol (Leung et al., 2010). A 17–25% amount of oil is present in cotton plant seeds. 11.67–20.1% palmitic acid, 19.2–23.26% oleic acid, and 55.2–55.5% linoleic acid are present in cottonseed oil.

Jatropha curcas (Jatropha): Jatropha is an oleaginous plant grown in semi-arid, marginal field regions. The average plant height is 5–7 m, and Jatropha plant species are related to the Euphorbiaceae family. India, USA, Brazil, Mexico, Africa, Argentina, Paraguay, and Bolivia are home to the Jatropha crop. About 100–150 cm of rainfall annually is needed for Jatropha plants. In India, the Jatropha tree has been identified as one of the essential sources for biodiesel (Ma and Hanna, 1999). Approximately, 20–60% oil content is present in Jatropha plant seeds. Saturated components such as stearic acid are in lower amounts at 7.1-7.4 %, and palmitic acid is present in higher amounts at 13.6–15.1 %. Unsaturated components such as oleic acid 34.3–44.7% and linoleic acid 31.4–43.2% are also present in Jatropha.

Nicotiana tabacum (Tobacco): North America, South America, Russia, Turkey, and India are the major tobacco-producing countries. Due to tobacco's primary features, which are very much similar to vegetable oil's chemical and physical characteristics, it is regarded as a potential source of biodiesel production. In tobacco, unsaturated fatty acids such as linoleic acid 69.49–75.58% and 35–49% of oil are present in the seeds of the tobacco plant(Leung et al., 2010).

Heveabrasiliensis (Rubber seed): Brazil is the largest manufacturing country of rubber seed, while Malaysia, India, Thailand, and Indonesia are also production nations. The height of the rubber tree is about 34 m. Non-frost climate and heavy rainfall are important for the growth of the rubber plant. It has brown oil of 40–50% by weight of kernel or copra, and the seed has 50–60% oil content. A large amount of unsaturated fatty acids such as linolenic acid 16.3%, oleic acid 24.6%, and linoleic acid 39.6% are present in rubber seed oil (Crabbe et al., 2001).

17.2.3 Feedstock for Third-Generation Biodiesel

Microalgae: Microalgae use sunlight to convert carbon dioxide into well-graded bioactive compounds. Algae cultivation is a category of aquaculture used to grow other materials derived from algae (Crabbe et al., 2001). With the potential of producing large quantities of lipids, these are aquatic plants with one cell ideal for biodiesel production. Microalgae have become progressively important as a more sustainable alternative, because of the significantly greater problem of the greenhouse effect, partially due to the combustion of fossil fuels, and rising petroleum prices. In response, many organizations have become interested in algal fuel development. The greater capital investment needed to convert algae into biofuels is the most significant limitation. The extraction of oil, waste treatment, fuel processing, and the development of algal oil from the harvesting of one algal species are additional challenge since mixed algae crops are found to produce more oil.

Fats of animals: Animal fats are the co-products of the fishing and meat industries. They can be obtained from cattle, hog, fish, and chicken. Because of low retail prices, co-products from animal fats are at present mainly used for biodiesel production (Leung et al., 2010). For the manufacture of biodiesel, lard, beef mutton residues, and yellow fat residues after omega-3 fatty acids can be used. However, higher saturated fatty acids from animal waste fats and free fatty acids are required for the complex development of methods. Strong oxidation stability, high calorific value, and shorter inflammation may cause these advantages obtained from animal waste fats with lower saturated fatty acids. Large volumes of saturated fatty acids create a low-temperature issue performance. Because of the large volumes of saturated fatty acids, biodiesel derived from animal fats is less suitable in cold countries (Ma and Hanna, 1999).

Waste oil material: There is a wide variety of waste oils easily accessible for biodiesel production. By consuming substances of certain kinds that would otherwise have to be discarded, waste oils usually have additional ecological capacity and are affordable. The non-edible sector, waste oil from the food industry, and restaurants and households are the three main categories into which waste oils can be divided. Biodiesel development uses waste oil from soybean, coconut, palm oils, other edible oils, and rapeseed to produce biodiesel. There are also many co-products produced in food factories that can be used (Table 17.1). To generate biodiesel through the pyrolysis process waste oils from the non-food industry can be used such as e.g. waste tyre oil, waste plastic oil, etc. (Crabbe et al., 2001).

17.2.4 Feedstock for Fourth-Generation Biodiesel

Important factors behind the production of biodiesel feedstocks of the fourth generation are inexhaustibility, lower price, easy accessibility, and high energy quality yield. Two main technologies that use photosynthetic water split into its constituents by the use of solar energy are photosynthesis by artificial methods and direct processing of solar biodiesel production. With the help of development in the technology for storing biomass, biodiesel feedstocks of the second and third generation have enhanced the capacity of photon-to-fuel conversion efficiency (PFCE). To produce high-quality fuels with better yield, solar photobiological production system harvests solar energy and uses it. Microbes allow the continuous collection of fuel in a photobioreactor. In an optimal situation, biomass production stops whenever the environment is moved in order to direct the development of photobiological solar fuel. A solution to this may be the immobilization of algae and cyanobacteria. Compared to the normal biomass-harvesting system, PFCE is larger in the photobiological fuel manufacturing process (Crabbe et al., 2001).

17.3 PRODUCTION OF BIODIESEL

Biodiesel is made by transesterification of vegetable oils or animal fats and alcohol. This chemical reaction turns an ester (vegetable oil or animal fat) into a fatty acid mixture consisting of esters of the oil (or fat). Biodiesel is derived from fatty acid methyl ester (FAME) mixture cleansing. To speed up the reaction, a catalyst is used. Transesterification can be basic, acidic or enzymatic according to the catalyst used.

By the transesterification of triacylglycerols (TAGs) with alcohols, biodiesel is produced. (Leung et al., 2010). With a few bacteria, various sources of biomass hydrolysate possessing free sugars have been efficiently used to conduct fermentation and accumulate TAG. Cultures are segregated from supernatant upon completion of production and are subjected to extraction of oil. Perle pounding, sonic shock, and electric shock are methods that are recorded for the extraction of oil. *In vivo, in vitro*, and semi-*in vivo/in vitro* are various methods developed for biodiesel production. TAG, alcohol, and enzymes i.e. lipase are required for *in vivo* production of biodiesel that is produced by utilizing microbes. TAG and alcohol are added externally for *in vitro* systems, and the bacteria produce the lipase that is necessary for transesterification. TAG and alcohol are added externally in semi *in vivo* and *in vitro* processes.

Transesterification is a three-stage reversible reaction which occurs step by step. In the initial step of transesterification, triglycerides in oil are converted into diglycerides and glycerides into monoglycerides and glycerol. In the transesterification process, to support the reaction, a sufficient quantity in a 3:1 ratio of alcohol to oil molar ratio is usually maintained. To drive the equilibrium towards the product side, an excess quantity of alcohol is usually added. The two layers formed are the top layer of the biodiesel phase and the bottom layer of the glycerol-rich phase; the reaction stream is divided at the end of the trans-esterification process (Leung et al., 2010). Karanja and Mahua hybrid varieties are used in a 50:50 quantity ratio to produce biodiesel derived from non-edible feedstocks. In an alkaline transesterification process, an esterification reaction is carried out by using acid, which decreases the desired limit of the quantity of free fatty acid (FFA). The oils can be converted into fatty acids with methyl ester. The catalyst for the esterification reaction is sulfuric acid. The transesterification method is facilitated by the addition of potassium hydroxide (KOH) and methanol as catalysts.

To generate biodiesel from Jatropha, catalytic and non-catalytic processes are used. In the production of biodiesel, it is assumed that an alkaline catalyst and transesterification of the two phases are more appropriate if Jatropha oil has a minimum of 1% of FFA content. For the transesterification process, studies of the production of biodiesel using raw Jatropha curcas oil use one rate of alkali catalyst. 5.5:1, 6:1, 6.75:1, 7.5:1, and 8:1: these different molar ratios of biodiesel at distinct outputs were measured by the researchers. From 50 °C to 70 °C, they measured varying responses of temperature. At 80.5 °C, maximum yield was obtained. Due to the higher fatty acid composition, the researchers noticed that the yield in crude Jatropha curcas oil was poor. The kinematic viscosity had decreased, and the various parameters were found to be comparable to the American Society for Testing Materials (ASTM) requirements; they also observed this after the transesterification process (Ma and Hanna, 1999).

For the production of biodiesel, the usage of various catalysed transesterification reactions has been investigated. Different processes, such as membrane reactor, microwave, and ultrasonic irradiation, reactive absorption, and distillation pillar greatly affect the final transformation and, in particular, the product's yield and quality. Calcium oxide was studied, using the transesterification process for the manufacture of biodiesel, as a catalyst for obtuse horn shell-palm oil waste. The physical and chemical properties were tested for methyl ester and palm oil. Some significant factors influencing the performance of methyl ester as a biodiesel from waste cooking oil analyses include the volume of catalyst, the reaction temperature to the performance of calcined scallop shells catalyst usage and the proportion of methanol to oil in moles. During waste cooking, the oil used as a heterogeneous catalyst for biodiesel processing utilizing nanocomposite doped zinc oxide was examined. By using the copper-doped zinc oxide nanocomposite, biodiesel yields are increased (Leung et al., 2010).

17.3.1 Steps for Production of Biodiesel

Microalgae, waste oils, animal fats, and oils from plant feedstocks are used for biodiesel production. Primarily oil crops depend on the crop species from which the biodiesel comes. Oil crops from first- and second-generation feedstocks are low compared to feedstocks of third-generation oils. Firstly, using various major strategies such as micro-emulsification, transesterification, dilution, pyrolysis etc. oil is produced from seeds or algae biomass, then that oil is converted to biodiesel for biodiesel development (Figure 17.1). For commercial production of biodiesel, the transesterification process is among the most economical processes with high biodiesel yield, and thus it is the most versatile approach (Crabbe et al., 2001).

Oil extraction from different feedstocks: The extraction of oil from seeds of oil crops is the first stage for biodiesel development. Based on facilities and production quantity, there are two techniques for oil extraction (Leung et al., 2010). The first is small-scale pressing, and the second is large-scale or commercial pressing. Pressing, the first step at small scale, is washing the oilseeds and pressing the seeds mechanically at maximum 40 °C. Removal of suspended contaminants with the help of filtration processes is the next step. As a by-product, press cakes are formed and are rich in protein, so can be used as protein fodder. To prevent the growth of microbes and proliferation of proteins, the maximum temperature required is higher than 80 °C for production on a large scale. The requirement for a particular quantity of water is important since a high-water amount creates a problem for solvent diffusion, whereas a lower quantity increase compactness. Hexane is usually used as a solvent, which can eliminate oil at a temperature of 80 °C. At the end of the process, the hexane and oil mixture is collected which is termed grist and miscella extraction. The next step is to compress seeds at an 80 °C temperature after maintaining the temperature and water quantity. Approximately 75% of the total volume of oil can be extracted.

Refining of oil: To remove undesirable materials such as tocopherols, phosphatides, free fatty acids, and dyes a refining process is used. These materials impact additional processing steps and also affect oil's shelf life. The refining process is affected by the physicochemical properties of oil and kind of feedstock used. The preliminary step in refining purification is degumming, in which phosphatide removal is performed. The two strategies used to remove phosphatides are acid and water degumming (Mata et al., 2010).

The next stage of refining is neutralization or de-acidification (Figure 17.1). This is an essential step for edible oils of the first generation, as it avoids the production

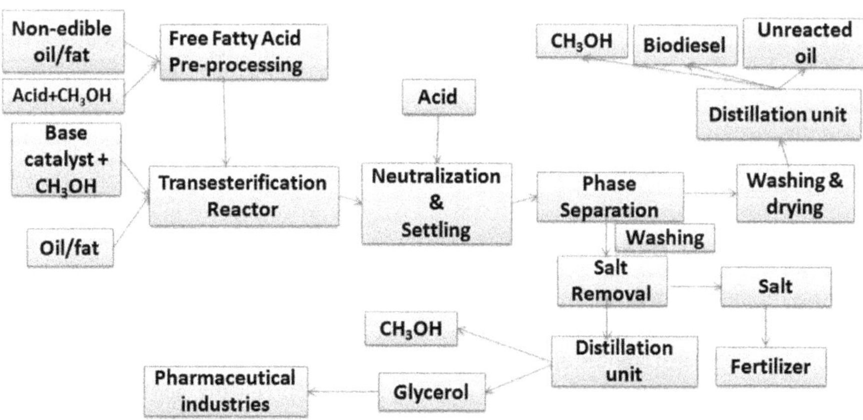

FIGURE 17.1 Steps involved in biodiesel production.

of rancid flavours of FFA. Separation of these types of composites is beneficial not only for edible oil but also for fuel production, as these components directly influence transesterification efficiency and stabilization of storage. Distillation, elimination of pigments, and elimination of odours are approaches included in deacidification by several solvents such as ethanol, alkali neutralization, and propane. Colourants are eliminated by bleaching, which is the next step in refining. Silica gel and activated carbon are the absorbing materials used in this strategy. The elimination of odorous elements such as aldehydes and ketones is refined by deodorization, which is the next stage. Moisture content is removed by dehydration, which is the last stage of refinement. Water removal is carried out via a low-pressure distillation process (Crabbe et al., 2001).

17.3.2 Refined Oil to Fuel Conversion Methodologies

Pyrolysis: For the production of the biodiesel, pyrolysis is one method which can be used. This method is influenced by the difference in components of the chemical composition. The biodiesel produced via pyrolysis process has a calorific value equivalent to diesel fuel; however the viscosity, pour point, and flashpoint of biodiesel produced via pyrolysis are lower than diesel. Through this procedure, the amounts of biodiesel produced have a sufficient quantity of sulfur and water, but also have enough ash and carbon content. High installation price is the main drawback of this method (Crabbe et al., 2001).

Micro emulsification: By carrying out microemulsion, the concerns about viscosity of vegetable oil can be removed. This is the balanced, consistent, and isotropic combination of water, oil, and surfactant. Co-solvent, alcohol, cetane improver, and surfactant are applied to carry out the microemulsion process of oils. By creating micro-emulsions with butanol, hexanol, and octanol, these can reach the highest viscosity requirements for fuel. The microemulsion process is simple. Some of the problems with micro emulsification are stability, less volatility, and high viscosity (Ma and Hanna, 1999).

Dilution: Dilution is the technique through which solute quantity is reduced in solution by raising solvent quantity. To dilute oils, ethanol and diesel maybe used as solvent. Reduction in oil viscosity and density is the outcome of this method. The boiling point valuation for diesel fuel is greater than ethanol, and that is why ethanol can encourage advancement of the combustion procedure across an unburned blend spray. The dilution method is simple, but carbon deposition in the engine cylinder and incomplete burning occur (Crabbe et al., 2001).

17.3.3 Process of Transesterification

From an economic perspective, biodiesel produced from the transesterification reaction has properties similar to diesel fuel. This procedure is beneficial for commercial production. Glycerol and esters are generated at the transesterification phase, while triglyceride reacts with alcohol (Leung et al., 2010). Three fatty acids, i.e. oils and organic fats, are linked to the triglycerides and have a glycerin molecule. By hydrolysing the triglycerides, free fatty acids are produced. Thereafter these free fatty acids

TABLE 17.2
Production of Biodiesel from Feedstocks of Oil

Oil Type	Catalyst used	Type of alcohol	The yield of biodiesel (%)
Mahua oil	KOH	Methanol	95.75
Palm oil (of waste obtuse horn shell)	CaO	Methanol	86.70
Jatropha curcas oil (crude)	KOH	Methanol	80.00
Palm kernel oil	NaOH	Methanol	95.5
Karanja oil	H_2SO_4	Methanol	98.6
Animal tallow oil	NaOH	Methanol	-
Canola oil	KOH or H_2SO_4	Methanol	90-95 or 80
Waste cooking oil	Calcined scallop shell	Methanol	86
Waste cooking oil	Copper doped zinc oxide nano composite	Methanol	97.71

react with alcohol and form esters or biodiesels, and glycerol. In the transesterification process, the end products are segregated due to high-weight glycerol settling down, and biodiesel settling at the top. To prevent reverse operation, the separation process should be very fast. Usually methanol and ethanol are used in the transesterification process. The method of transesterification is called methanolysis when methanol is used to react with free fatty acids. Through the methanolysis process, heat is applied to the oil mixture at 80–90% and methanol at10–20% and catalyst at lower quantity (Table 17.2). The fatty acid methyl ester (FAME) biodiesel is produced after the procedure. If the ethanol is used for a free fatty acid reaction, then this transesterification of the mechanism is called ethanolysis. The fatty acid ethyl ester (FAEE) biodiesel is produced through the ethanolysis procedure (Crabbe et al., 2001). Biodiesel produced from the transesterification method has many advantages as compared with pure plant oil (PPO). In diesel engines, biodiesel can be used with few engine differences as its features are quite comparable to fossil diesel.

17.3.3.1 Advantages of Biodiesel

There are several advantages of using biodiesel as fuel source, as follows:

1. Simple to use – No alteration of the vehicle or the required fuelling equipment.
2. Energy, performance, and economy: Proven power generation, performance, and cost-efficiency have made biodiesel a valuable fuel.
3. Environmental effect: Biodiesel helps increase health by reducing CO_2 emissions that mitigate the impact of climate change.
4. Biodiesel is better to treat because it is less harmful than petroleum and relatively easier to store.
5. Biodiesel limits the use of unnecessary oils.
6. Biodiesel can be produced from a large number of oil-containing feedstocks (Figure 17.2).

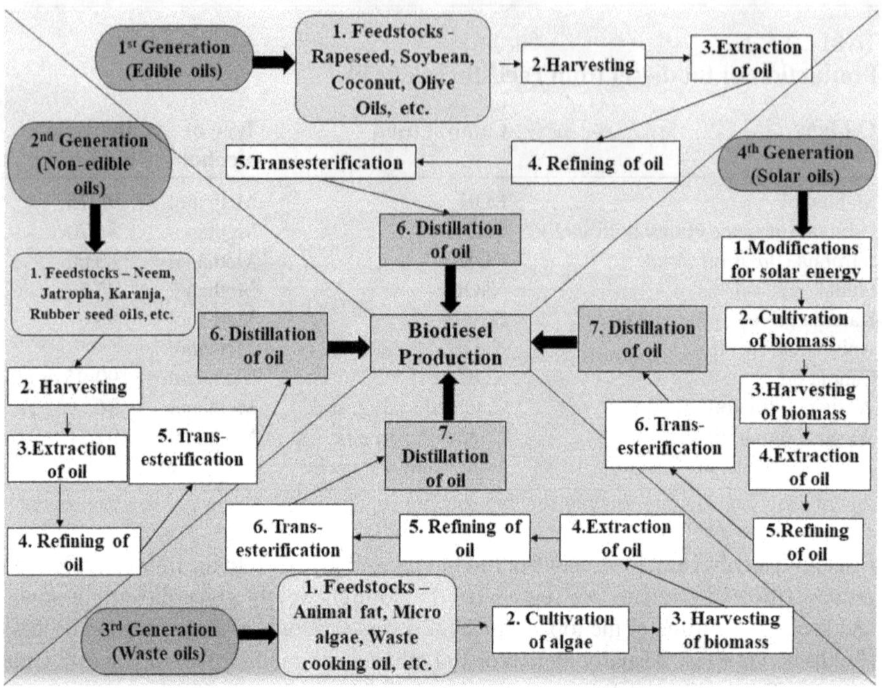

FIGURE 17.2 Production process of biodiesels.

17.3.3.2 Disadvantages of Biodiesel

Nevertheless, a few bottlenecks limit its commercial use, which need to be resolved.

1. At present, biodiesel fuel probably costs around one and a half times more than petroleum diesel fuel.
2. Production of biodiesel fuel from soya crops requires energy, plus the sowing, fertilizing, and harvesting energy.
3. Another drawback of biodiesel fuel is that certain biodiesels can cause damage to engines.
4. When biodiesel removes pollutants from the engine, this dust can collect in it, and clogging arises in the fuel filter. Thus filters should be changed frequently.
5. Biodiesel fuel distribution systems need to be improved, which is another drawback of biodiesel fuel.

17.6 BIOBUTANOL

Butanol ($C_4H_{10}O$) or butyl alcohol is an alcohol which serves the purpose of a solvent at various scales and is also a fuel. Butanol that is a product of a process involving biological mass is termed as biobutanol. Biobutanol is produced by microbial fermentation. It is made from a range of sugars, starch and certain organic matter of the same type as that of ethanol. It can also be produced by fermenting lignocellulosic biomass;

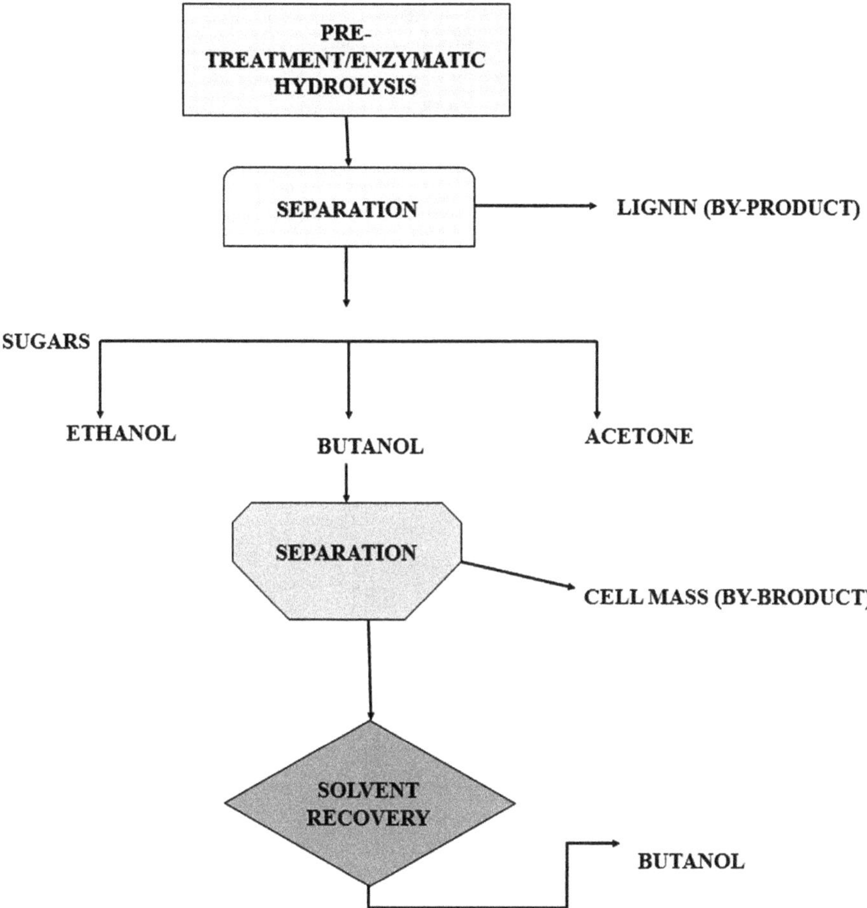

FIGURE 17.3 Butanol fermentation from lignocellulosic materials needs pretreatment.

nevertheless chemical pretreatment or enzymatic hydrolysis is required prior to its use as a substrate (Figure 17.3). The processing of biobutanol production is currently more expensive when compared to processing of ethanol so it has not yet been commercialized on a bigger scale. Biobutanol has considerable merits when compared to ethanol, which is why there is a lot of research going on regarding its development (Bharathiraja et al., 2017).

17.6.1 Properties

Biobutanol is an alcohol which is colourless and is an considered eco-friendly fuel (García et al., 2011). Biobutanol has greater miscibility with diesel and greater energy density when compared with bioethanol. It has lower vapour pressure, which gives it an edge over ethanol when it comes compatibility with conventional fuels.

The structure of butanol provides it with resistance against aqueous contamination, so that it can be transported in pipelines. On the other hand, ethanol absorbs water and causes corrosion in pipelines and hence its transportation is done by means of road transport and on a smaller scale. Since it is mixable at a refinery, it makes biobutanol better that ethanol as its distribution and transportation become a lot more convenient. Its calorific value is 29.2 MJ/L; Self-ignition – 340 °C; Melting point –89.5 °C; Boiling point – 117.2 °C; Flash point – 36 °C

17.6.2 Production

Anaerobic fermentation is amongst the oldest techniques that are employed to produce butanol on an industrial scale. The interest in butanol was an outcome of the scarcity of rubber supply in the early 20th century due to price inflation. Back then, butanol was a raw material for producing butadiene, which was one of the raw materials used to produce synthetic rubber. However, now biobutanol is looked on as a potential substitute fuel (Jang et al., 2012). Butanol production may be achieved by various chemical processes. Butanol can also be produced by the process of fermentation by the *Clostridium* bacterial genera. This fermentation is carried out in the absence of air – that is anaerobically – the final product butanol is called biobutanol.

Biobutanol is produced by fermenting carbohydrates in a process better known as ABE fermentation since its major chemical products are **A**cetone, **B**utanol and **E**thanol (Jang and Lee, 2015). The ratio in which these are typically yielded is 3:6:1. The final concentration of butanol is 3%.

The fermentation process uses microbial activity of bacteria usually from the Gram-positive, sporulating, rod shaped, anaerobic genus *Clostridium. Clostridium acetobutylicum* is the bacterial species that is most commonly used for fermentation. Various strains of clostridia such as *C. acetobutylicum*, *C. saccharobutylicum*, *C. beijerinckii*, *C. aurantibutyricum*, *C. saccharoperbutylacetonicum*, and *C. carboxidivorans* are used for the production of butanol by fermentation. *Clostridium acetobutilicum* is most commonly used for butanol production. Another species called *C. cellulolyticum* utilizes cellulosic biomass in the fermentation of the biobutanol. Another strain known as *C. ljungdahlii* can utilize hydrogen and carbon monoxide present in syngas as a substrate (Jang and Lee, 2015).

In the course of industrial production of biobutanol, using a fermentation process, there are three factors to be taken into consideration:

1) The profitability of the entirety of fermentation.
2) The expense of the recovery of biobutanol.
3) The amount of toxicity in the obtained product.

In spite of these, the fermentation that involves crop products is not economically very sustainable, primarily because of their high prices as they have high demand in the food industry. Therefore for butanol production materials like agricultural waste such as grass, stalks, spoiled grains fruits, leaves, straws etc. are used (Bharathiraja et al., 2017). The algae are not resource intensive and are also economically sustainable. *Chlorella* is an example of micro algae that have about 30–40 percent sugar content.

The high content of sugar in such micro-algal species greatly improves the process of butanol fermentation. Bacterial species such as *Clostridium acetobutylicum* and *Clostridium beijerinckii* are being worked upon genetically so as to develop resistance against butanol concentration in the broth during fermentation (Jang and Lee, 2015).

17.6.3 Physiology of Butanol Production

The ABE-producing clostridia can use both hexose and pentose sugars. Hexose sugars are metabolized via an enzyme-catalysed pathway known as glycolysis. Pentoses are metabolized through the nonoxidative pentose phosphate pathway. Glycolysis produces two molecules of pyruvate which are further used for ABE fermentation (Jang et al., 2012).

The ABE fermentation is an anaerobic process which proceeds in two phases: the first phase is known as the acidogenesis phase followed by the solventogenesis phase (Figure 17.4). In the acidogenesis phase, *Clostridia* grow and produce mainly acetic and butyric acids. These acids lower the pH of the medium. During this phase CO_2 and H_2 are produced as by-products.

The concentration of undissociated butyric acid triggers cells to enter the solventogenic phase. In the solventogenesis phase, acids are re-assimilated into acetone, butanol and ethanol. A semi-continuous approach is generally used for the industrial production of butanol using fermentation. Each fermentation steps lasts for a period of about 21 d (Jang and Lee, 2015). The fermentation plants typically have several trains of big tanks (up to 400 m^3). Fermentation is carried out using a fed-batch method. Fresh feedstock,

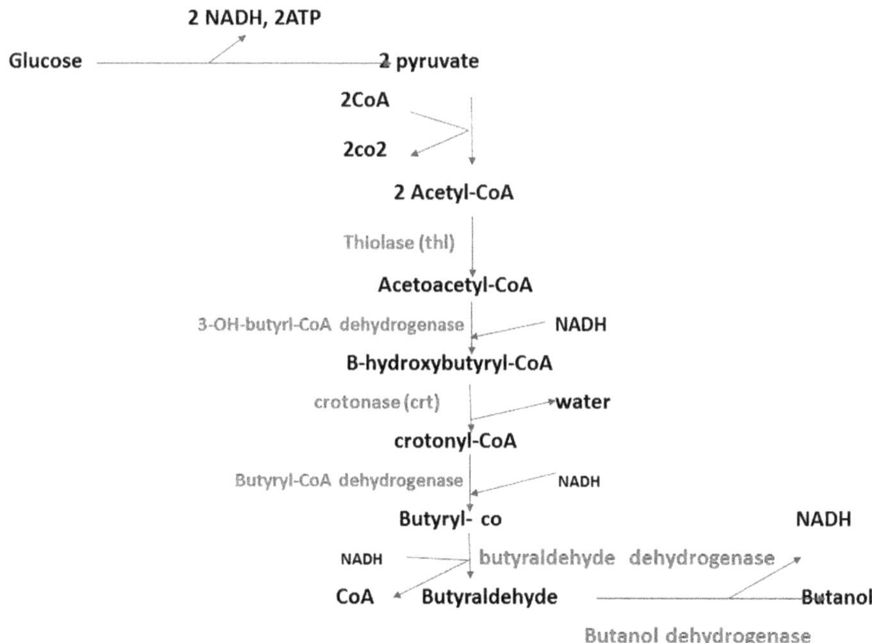

FIGURE 17.4 Butanol biosynthesis pathway.

together with periodic additions of seed culture, flows through the fermenters. This provides sufficient residence time for re-assimilation of the acids produced in the acidogenesis phase to solvents in solventogenesis. Acetone, butanol and ethanol are produced by fermentation in a proportion of 3:6:1 by mass. The final product can then be recovered by distillation based on the boiling points of acetone, ethanol and butanol which are 56 °C, 78 °C and 118 °C, respectively. A new approach of biobutanol production involves the use of metabolically engineered *Clostridia.* The metabolically engineered *Clostridia* immediately enter the solvent-organic phase, avoiding the acidogenic phase. The use of metabolically engineered *Clostridia* decreases production of acids, thereby increasing the yield of biobutanol (Jang and Lee, 2015).

17.6.4 Biobutanol Recovery

There are certain techniques that are used to eliminate the toxicity of butanol to the bacterial cells. Distillation is one of the traditional methods that are employed for product recovery (Figure 17.5).

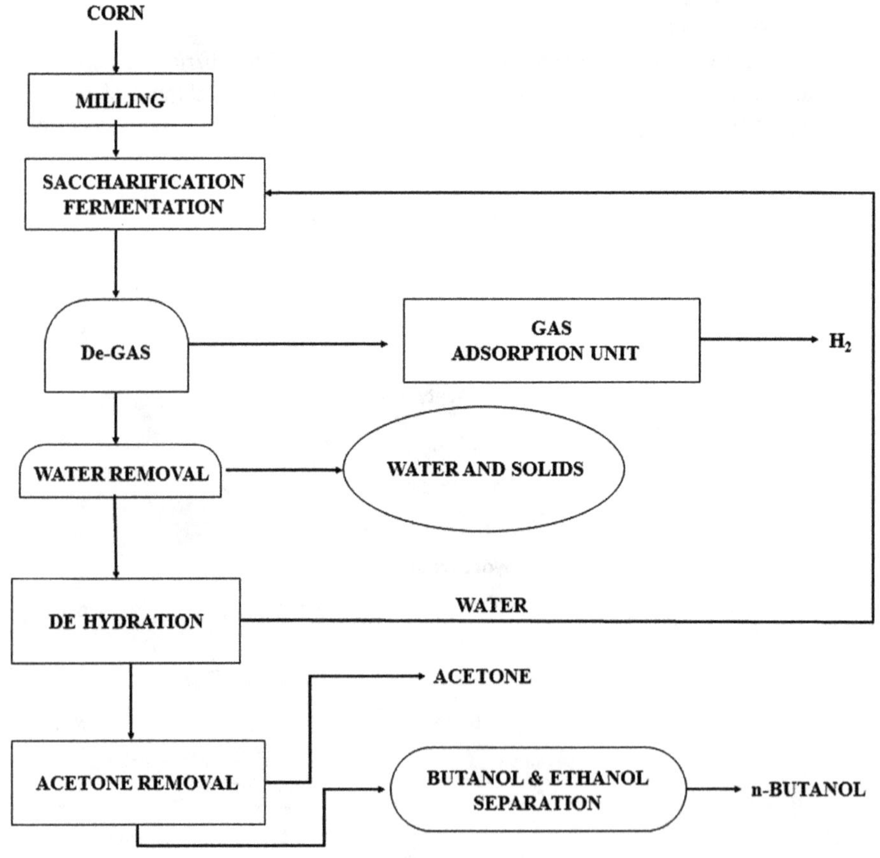

FIGURE 17.5 Block flow diagram of butanol fermentation.

Butanol has a boiling point of 117.2 °C, which is higher than that of water. Because of this, the process becomes energy intensive, hence making the processing costs much higher, especially when the butanol is present in low concentration in the broth (Bharathiraja et al., 2017). Distillation is not considered because it involves the boiling of water which lower compared to that of the maximum concentration of biobutanol that is just around 3 percent by weight. This is uneconomical even in energic terms. It shows less yield and is an expensive method for biobutanol recovery. As a result, other methods are currently used, such as adsorption, membrane pertraction, extraction, pervaporation, reverse osmosis or 'gas stripping'. Specifically, a lot of attention is being paid to pervaporation, since it simultaneously separates as well as concentrates the product (Jin et al., 2011). Several methods are employed for the recovery of butanol from the fermentation broth, of which gas stripping is considered a promising method. This involves separation of volatile compounds by lowering the pressure, heat, or the use of inert gas. The combination of these techniques has various practical applications. The columns which are used to separate miscible liquids are fractional distillation columns, and similar columns are used to separate out products such as butanol.

17.6.4.1 Butanol Recovery by Pervaporation

Pervaporation is considered to be one of the promising techniques to recover or remove harmful chemicals such as alcohols (acetone, butanol and ethanol) present in the broth for bacteria like *C. acetobutylicum.* Selective transport by diffusion of components through a membrane occurs in this particular technique where vacuum is applied to the side of the permeate. On the side with low pressure, condensation of the vapor is carried out(Bharathiraja et al., 2017). Since this method involves the travelling of organic compounds through a permeable membrane, a membrane with hydrophobic properties is considered over others. For pervaporation methods, polydimethylsiloxane membranes and silicon rubber sheets are mostly used. When a vacuum is applied to generate low pressure on one side of the permeate this becomes expensive, which is a drawback of pervaporation (Jang et al., 2012).

17.6.4.2 Butanol Recovery by Adsorption

Adsorption has been studied in separating biobutanol during the fermentation process from the broth, but the ability of the adsorbing agent is less; thus it fails to be used on a large scale. Silicalite is one of the most commonly used adsorbents for recovering butanol during fermentation. It is a form of silica. It has a structure similar to that of a zeolite and is hydrophobic in nature. It exhibits an ability of selective adsorption of small organic molecules of alcohol containing 1 to 5 carbon atoms from dilute aqueous solutions (Jang et al., 2012). The process of butanol recovery during fermentation by means of adsorbents is restricted to laboratories.

17.6.4.3 Butanol Recovery by Membrane Reactor

Immobilization of microbes in membranes or usage of a membrane reactor is used for the recovery of butanol (Jang and Lee, 2015). Membrane reactor techniques are quite feasible on an industrial scale; however, there are a lot of demerits, such as poor mechanical strength and bad transfer resistance. Leakage of cells from the matrices adds to its disadvantages.

17.6.4.4 Recovery of Butanol by Ionic Liquids

The process of removing butanol from the broth is quite a tedious and tough task. The use of conventional solvents proves to be beneficial, but it involves the risk of handling of solvents that are poisonous, volatile, and harmful with high toxicity (Jang and Lee, 2015). Lately an increase in interest in ionic liquids has been observed; ionic liquids are non-volatile and eco-friendly solvents for several chemical techniques. We can find an answer in ionic liquids to butanol recovery from fermentation broth. These are basically organic salts that are present in liquid form at STP, have very low vapor pressure and are sparingly soluble in water. That is why ionic liquids are considered to be precious in butanol recovery from fermentation broth.

17.7 POWER ALCOHOL

Combating climate change pushes the planet to search for renewable energies and power supplies that are low in emissions. Because traffic is one of the main sources of greenhouse gas, i.e. carbon emissions, replacing fossil fuels with cleaner substitutes such as biofuels is an important way to minimize such emissions. Biofuels provide a solution for lowering traffic carbon emissions where other options, such as converting to hybrid cars, are not an option due to high energy prices or lack of automotive charging networks. Various alcohols are utilized as fuel for internal combustion engines. Methanol, ethanol, propanol, and butanol are aliphatic alcohols that are suitable to be used as fuels as they are generated chemically or biologically. Most methanol is synthesized from natural gas, although it can be synthesized using biomass technology. Ethanol is mostly synthesized from biological material through fermentation processes. There is no chemical difference between biologically synthesized and chemically produced alcohols (Awad et al., 2018). Ethyl alcohol is used to produce electricity. This is called power alcohol; for example, as an additive to motor fuels that serve as a substitute for internal combustion engines. It typically has 80 per cent gasoline and 20 per cent alcohol. In the United States and other countries with ethanol, there have been decades of motor fuel production practice. To achieve greater performance, ethanol has been combined with oils. The process of mixing ethanol began in India in 2001. The Government of India required mixing 5 percent ethanol with petrol. Rather than using commercially generated alcohol, nowadays biomass-prepared ethanol is used for this mixture. This is known to be more environmentally sustainable, and therefore is named biofuel (Aditiya et al., 2016).

17.7.1 Bioethanol as Power Alcohol

Ethanol fuel is also known as ethyl alcohol and is also found in alcoholic beverages. It was utilized as a fuel for the first time in 1978 in Brazil. Conventional ethanol is produced from fermentation of sugars like sugar cane and corn. Enzymes have the potential to replace conventional fermentation strategies (Alvira et al., 2010). Around

2000–2007, world production of ethanol for transportation fuel tripled from 17×10^9 litres (4.5×10^9 U.S. gal; 3.71×10^9 imp gal) to over 52 x 10^9 litres (1.4×10^{10} U.S. gal; 1.1×10^{10} imp gal). Ethanol has a 'gasoline gallon equivalent' value of 1.5 and to offset the energy of 1 volume of gasoline, 1.5 times the amount of ethanol is required. Ethanol-mixed fuel is commonly used in Brazil, the United States, and Europe. Many cars on the road today in the U.S. can operate on blends of up to 10 percent ethanol, and in 2011 ethanol contributed for 10 percent of U.S. domestic fuel supplies. Any flexible-fuel vehicles can use ethanol up to 100 percent. The Brazilian government has declared blending ethanol with gasoline compulsory since 1976, and since 2007 the permissible mix is about 25 percent ethanol and 75 percent gasoline. Recent advances in the production and commercialization of cellulosic ethanol can allay some of these concerns. Cellulosic ethanol promises hope as it can be used to generate ethanol by cellulose fibres, a large and fundamental feature of plant cell walls. Cellulosic ethanol will play a much more important role in future according to the International Energy Agency (Aditiya et al., 2016).

17.7.1.1 Production

Fermentation: Microbial fermentation of the sugar produces ethanol. At the moment microbial fermentation acts only explicitly with sugars. Two main plant elements, starch and cellulose, are also composed of sugars and can be converted to fermentation sugars (Aditiya et al., 2016). At present, only the segments of sugar (e.g. sugar cane) and starch (e.g. corn) can be processed economically. There is a lot of operation in the cellulosic ethanol region, where the cellulose portion of a plant is broken down to sugars and then converted to ethanol (Alvira et al., 2010).

17.7.1.2 Distillation

For ethanol to be utilized as a solvent, the solid yeasts and the remainder of water should be segregated. During fermentation, heat treatment is provided to the mash such that there is ethanol evaporation. This process is defined as distillation which recovers ethanol although its purity is restricted to 95–96%, leading to the incorporation of a low-boiling water-ethanol azeotrope of maximum (95.6%w/w) ethanol (96.5%v/v) and 4.4%w/w (3.5%v/v) water. This combination is called hydrous ethanol, and can be used as a fuel alone, but unlike anhydrous ethanol, hydrous ethanol is not miscible at all concentrations of gasoline, and the water fraction is normally isolated to burn in combination with gasoline in gasoline engines for further treatment (Aditiya et al., 2016).

17.7.1.3 Dehydration

There are three methods of dehydration to extract water from an azeotropic mixture of ethanol/salt. In several early fuel ethanol plants, the first step is called azeotropic distillation and consists of adding benzene or cyclohexane to the mixture. This is a two-phase liquid mixture when diluted. The heavier form, detrimental in the trainer, is stripped from the trainer and recycled into the feed while the lighter process, with the stripping condensate, is recycled to the second board. Another initial approach, termed extractive distillation, would be to attach a ternary component that improves ethanol's relative volatility. With growing attention being paid to energy conservation, several strategies have been suggested to prevent distillation for dehydration entirely. From such methods a new approach has arisen, and most modern ethanol plants have adopted it. The pores of the bed are designed to allow water to adsorb while removing ethanol (Alvira et al., 2010). For a period of time, the pad is regenerated to extract the adsorbed water under pressure or in the flow of inert atmosphere (e.g. N_2). Two beds are also used in such a manner that one can adsorb water while the other is being regenerated. This dehydration technique will save 3,000 BTUs/ gallon (840 kJ/L) of energy relative to earlier aceotropic distillation. Recent work has shown that total dehydration is not always necessary prior to blending with gasoline. Alternatively, the azeotropic mixture can be directly mixed with fuel such that equilibrium in the liquid–liquid process can aid in water removal. With minimal energy consumption, a two-stage counter-current system of mixer-settler tanks will achieve full recovery of ethanol into the fuel process (Aditiya et al., 2016).

17.7.2 Environmental Impact

All biomasses pass through at least some of these steps. They need to be grown, collected, dried, fermented, distilled, and burned. These measures require an infrastructure and funding. The cumulative amount of energy produced in the loop is defined as the balance of energy relative to the energy produced by burning ethanol. The energy balance for sugarcane ethanol produced in Brazil is far more advantageous, through one unit of fossil-fuel energy being needed to produce 8 from ethanol. Estimates of the energy balance are not readily made, hence various conflicting reports on these have been created. One study, for example, reports that the processing of ethanol by sugar cane, that demands a warm climate for efficient growth, yields 8 to 9 units of energy per unit spent compared to maize, which yields only around 1.34 units of fuel oil per unit of energy spent. A 2006 study by the University of California Berkeley showed that the production of corn ethanol requires considerably less petroleum than gasoline, after analysing six separate tests. Throughout fermentation and combustion, carbon dioxide, a greenhouse gas, is released. This is balanced out by plants sucking in more carbon dioxide as they grow to provide biomass. Based on the manufacturing process, ethanol releases fewer greenhouse emissions as compared with gasoline (Aditiya et al., 2016).

Air pollution: Ethanol is a suspended solids-free source of fuel combusted with oxygen to form carbon dioxide, carbon monoxide, hydrogen, and aldehydes identical to conventional unleaded gasoline. Today, the compound MTBE (methyl turt-butyl ether) is being phased out due to pollution of the soil, which makes ethanol a desirable substitute. New processing practices include air emissions from macronutrient fertilizer producers, such as ammonia. A Stanford University study by atmospheric scientists found that in Los Angeles, USA, E85 fuel could increase the chance of death through air pollution by 9 percent relative to gasoline: a very large, urban, car-based metropolis that is the worst case scenario. Ozone levels are rising significantly, thereby increasing polluting gases and aggravating respiratory problems like asthma (Aditiya et al., 2016). Brazil uses large quantities of ethanol biofuel. Gas chromatography tests of ambient air have been performed in São Paulo, Brazil, and compared with Osaka, Japan, which does not burn ethanol fuel.

Change in land use: Agricultural alcohol production requires large-scale cultivation, and this requires significant quantities of planted land. University of Minnesota researchers claim that even if all U.S.-grown corn were utilized to generate ethanol it would deplete 12 percent of current U.S. fuel consumption. There are many fears that deforestation cultivates land for ethanol production, while some have indicated that areas currently supporting trees are not necessarily suitable for growing crops. Farming will in any case require a decrease in soil fertility due to a reduction in organic matter. Reduced water quality and production have intensified pesticide and fertilizer use and subsequent dislocation of urban communities. Advanced technology enables producers and processors to effectively generate the same production with less input. The development of cellulosic ethanol is a recent method, which may mitigate land use and related issues (Tomás-Pejó et al., 2008). In an effort to mitigate conflict amongst food requirements versus fuel needs, cellulosic ethanol can be produced from any plant matter, potentially doubling yield. Instead of using just starch by-products from the grinding of wheat and other crops, the production of cellulosic ethanol maximizes the use of all plant materials such as gluten. This solution would have a lower carbon footprint while the number of energy-intensive fertilizers and fungicides with greater material usage remains the same. The technology for the manufacture of cellulosic ethanol is now in the commercialization stage.

17.7.2.1 Market Trends of Bioethanol

Use of fuel in cars in the environment emits vast amounts of GHGs, causing environmental pollution and global warming. Bioethanol is blended with gasoline which most effectively consumes the fuel-blend and reduces vehicle carbon emission levels. Bioethanol is more affordable than gasoline, and so countries are seeking to use it as an alternate fuel to reduce their reliance on crude oil; thus raising the usage of bioethanol in the final-use transportation sector. The worldwide market for bioethanol is expected to rise from USD 33.7 billion in 2020 to USD 64.8 billion by 2025, with a Compound Annual Growth Rate (CAGR) of 14.0% between 2020 and 2025. The market is driven by enforcing the use in the US and Canada of higher-ethanol blends. Significant suppliers such as Archer Daniel Midland (US), POET LLC (US), Green Plains (US), Valero Energy Company (US), Tereos (France), Raizen (Brazil), Flint Hills Wealth (US), Pacific Ethanol (US), The Andersons Inc. (US), and Sekab Biofuels & Chemicals (Sweden) constitute the worldwide bioethanol industry.

17.7.2.2 Biobutanol Market – Growth, Trends, and Forecast (2020–2025)

The leading producers are Butamax, Green Biologics, Gevo and Cathay. The Asia and Pacific regions such as China, India and Japan, dominate the market. Of these, China accounts for the largest market share in biobutanol consumption. China is the largest producer of paints and coatings in the world. Biobutanol is used in synthesizing coatings, adhesives, resins and textiles. Moreover, due to the growing need for coatings, various paint manufacturers are extending their product ranges. Big adhesive companies are expanding their range of goods in India. Over the projected period, these aforementioned factors are expected to fuel market growth.

17.7.2.3 Biodiesel Market Trends

The global biodiesel market size was estimated to be 28.04 billion in 2016. Central and South America is expected to be the fastest-growing market, growing at a CAGR of 11.0% from 2017 to 2025. However, due to lack of production capability and broad potential for R&D in the selection of feedstock for commodity manufacturing, the demand–supply gap is expected to open up opportunities for new players in the sector. The demand for the product was dominated by the automotive fuel sector, which accounted for over 75% of the overall market. Growing demand for fuel to replace crude oil in commercial vehicles is projected to have a positive effect on the industry. The fuel is favourable to the atmosphere because, relative to conventional diesel, it decreases greenhouse gas impact by emitting lower VOC content.

17.8 CONCLUSIONS

Biodiesel is an effective renewable substitute for conventional fuels. This chapter has described all the four generations of biodiesel regarding feedstocks used in the production of biodiesel, distinct biodiesel production innovations, and biodiesel production estimates. Each type of biodiesel generation has its advantages and drawbacks. The development of biodiesel generation has focused primarily on improving its performance with less environmental deterioration. Biodiesel is made from oil using various methods including catalytic distillation, dilution, micro-emulsion, pyrolysis, and transesterification etc. Transesterification is among the most affordable of these conversion strategies, and biodiesel developed from this approach has similar characteristics to diesel. Specific research resources are available in the fields of production of biodiesel, economic viability, improved performance, and reduced emissions.

The use of biobutanol as fuel is still in the nascent stages and at a research level. The yield of butanol is still on the lower side due to the toxic properties of the organism used i.e. *Clostridium acetobutylicum.* When it comes to the synthesis of biobutanol biologically, one of the best methods that can be deployed is Metabolic Engineering. IL (i.e. Ionic Liquids) is a promising method of butanol extraction or recovery. There are some efficient options: Ionic liquids used *in situ*. Use of membrane contractors; as mentioned earlier, biobutanol is considered an eco-friendly substitute fuel compared to ethanol and others. It also has a positive impact on ecology. Its use as fuel will certainly cause a drop in overall GHG emission levels. The use of bioethanol as a power alcohol is also a promising response to the oil and environmental crisis.

REFERENCES

Aditiya HB, Mahlia TMI, Chong WT, Nur H, Sebayang AH, Second generation bioethanol production: A critical review. Renew. Sustain. Energy Rev. 66, 631–653,2016.

Alvira P, Tomás-Pejó E, Ballesteros M, Negro MJ, Pretreatment technologies for an efficient bioethanol production process based on enzymatic hydrolysis: A review. Bioresour. Technol., Special Issue on Lignocellulosic Bioethanol: Current Status and Perspectives 101, 4851–4861, 2010.

Awad OI, Mamat R, Ali OM, Sidik NAC, Yusaf T, Kadirgama K, Kettner M, Alcohol and ether as alternative fuels in spark ignition engine: A review. Renew. Sustain. Energy Rev. 82, 2586–2605, 2018.

Bharathiraja B, Jayamuthunagai J, Sudharsanaa T, Bharghavi A, Praveenkumar R, Chakravarthy M, Yuvaraj D, Biobutanol – An impending biofuel for future: A review on upstream and downstream processing techniques. Renew. Sustain. Energy Rev. 68, 788–807, 2017.

Crabbe E, Nolasco-Hipolito C, Kobayashi G, Sonomoto K, Ishizaki A, Biodiesel production from crude palm oil and evaluation of butanol extraction and fuel properties. Process Biochem. 37, 65–71, 2001.

García V, Päkkilä J, Ojamo H, Muurinen E, Keiski RL, Challenges in biobutanol production: How to improve the efficiency? Renew. Sustain. Energy Rev. 15, 964–980, 2011.

Jang Y-S, Lee SY, Recent Advances in Biobutanol Production. Ind. Biotechnol. 11, 316–321, 2015.

Jang Y-S, Malaviya A, Cho C, Lee J, Lee SY, Butanol production from renewable biomass by clostridia. Bioresour. Technol. 123, 653–663, 2012.

Jin C, Yao M, Liu H, Lee CF, Ji J, Progress in the production and application of n-butanol as a biofuel. Renew. Sustain. Energy Rev. 15, 4080–4106, 2011.

Leung DYC, Wu X, Leung MKH, A review on biodiesel production using catalyzed transesterification. Appl. Energy 87, 1083–1095, 2010.

Ma F, Hanna MA, Biodiesel production: a review1Journal Series #12109, Agricultural Research Division, Institute of Agriculture and Natural Resources, University of Nebraska–Lincoln.1. Bioresour. Technol. 70, 1–15, 1999.

Mata TM, Martins AA, Caetano, Nidia S, Microalgae for biodiesel production and other applications: A review. Renew. Sustain. Energy Rev. 14, 217–232, 2010.

Tomás-Pejó E, Oliva JM, Ballesteros M, Realistic approach for full-scale bioethanol production from lignocellulose: a review. JSIR, 6711, 2008.

18 Biopesticides and Biopolymers

With limited land resources and a rising human population it is becoming increasingly important to reduce pest-associated crop losses. Farm yield losses from pests are estimated at 10–30%, although there could be substantial additional post-harvest losses. It has been estimated that up to 6% of cereals and pulses, 10% of oilseeds, 18% of fruits and 13% of vegetables are destroyed during processing, handling and storage. Meanwhile, due to environmental and safety issues, many of conventional organic pesticides have been banned from use. Microbial pesticides consisting of bacteria, entomopathogenic fungi or viruses (and often including the metabolites which are produced by bacteria or fungi) seem to be a promising solution to the setbacks of the conventional methods. Biopolymers are the natural polymers that the cells of living organisms create. Biopolymers consist of monomeric groups, which are bonded covalently to form larger molecules. There are three major groups of biopolymers, identified by the monomers used and the biopolymer structure formed: polynucleotides, polypeptides, and polysaccharides. However, this chapter mainly focuses on microbial biopesticides.

18.1 MICROBIAL PESTICIDE PRODUCTION

Pesticides play a huge role in agriculture. In today's world where organisms are developing resistance to most pesticides there is a need to find new techniques other than chemical methods of prevention. In 1976 the United Nations Development Programme (UNDP) along with the World Health Organization (WHO) started a special programme for research and training in tropical diseases in which they studied the diseases malaria, trypanosomiasis, filariasis, leishmaniosis and leprosy. The production of these microbial pesticides has given promising results by increasing crop yield, maintaining the storage life of crops. They are also more effective i.e. give better results than natural methods which include the use of manure; also, microbial pesticides are environmentally friendly, they do not mix in the soil and reduce soil fertility or get to water bodies and cause contamination. They can be categorized into bacterial, viral, fungal and protozoan. The most widely used out of all these classes of insects is *Bacillus thuringiensis* mainly because of its ability to target specific insects.

Alternatives to chemical methods:

1. *Use of predators:* In this method, via the use of predators, insects are eliminated. The best example is use of *Gambusia affinis* to eat mosquito larva.
2. *Genetic modification:* The male insects are made sterile so that mating does not give result to a new generation. There are various means of modifying organisms by chemical method or by irradiation.
3. *Hormones:* Pheromones are secreted by insects which act as sex attractants. Insects are destroyed when attracted.
4. *Pathogen use:* The use of pathogen production of microbial pesticide. The idea first originated via the disease caused by *Bombyx mori* in silkworm.

18.1.1 An Elaborate Discussion of *Bacillus thuringiensis*

Bacillus thuringiensis commonly known as 'bt' is harboured in the gut of caterpillars such as moths and butterflies. They also grow on leaf surfaces, in aquatic environments or on storage products such as crops. They are Gram-positive in nature. There are many subspecies of *Bacillus thuringiensis* but the most commonly used is *Bacillus thuringiensis* subspecies kurstaki (btk), subspecies raelensis (bti) and subspecies aizawa.

Bacillus thuringiensis forms spores in unfavourable conditions which leads to the formation of parasporal crystals of proteinaceous insecticidal beta-endotoxins called cry or crystal protein. It is called cry protein because this protein is encoded by cry genes. It accounts for about 1% of the total agrochemical market in the world. It is sold in the form of mixtures of dried spores and toxic crystals (Butt et al., 1999). The insect ingests this toxin in this form (Figure 18.1). After intake of the crystal a protoxin is released. First the protoxin is converted into an active toxin protein which diffuses through the membrane of the gut. This toxin paralyses the gut and the insect stops feeding, leading to its starvation and ultimately death. Also it is observed that the toxin makes pores in the membrane causing misbalance in the ion gradient which makes the gut swell (Kumar et al., 2019). A few common *Bacillus* used as biopesticide are listed in Table 18.1.

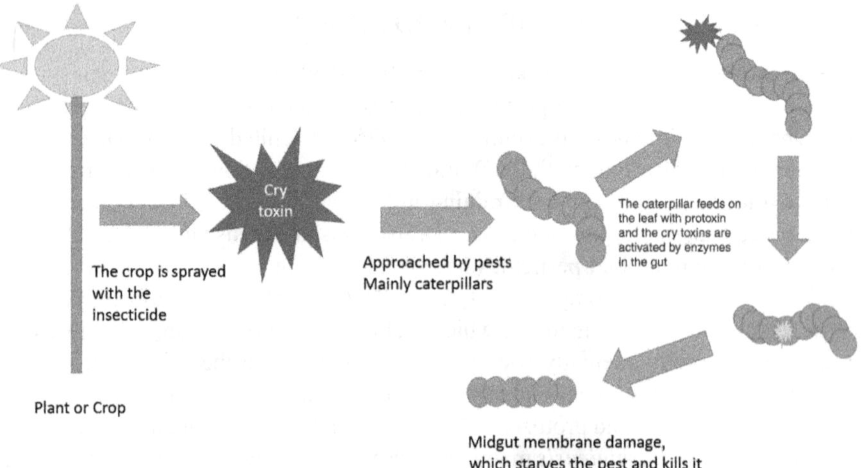
Cry
toxin
The crop is sprayed
with the
insecticide
Approached by pests
Mainly caterpillars
The caterpillar feeds on
the leaf with protoxin
and the cry toxins are
activated by enzymes
in the gut
Plant or Crop
Midgut membrane damage,
which starves the pest and kills it

FIGURE 18.1 Schematics of mode of action of microbial pesticide.

TABLE 18.1
Commercial *Bacillus* Pest Control Agents

Organism	Target pest	Commercial products	Company
B. thuringiensis kurstaki	Caterpillars	Dipel	Abbott
		Thurcide/javelin	Sandoz
		Biobit/foray	Novo nordisk
		Cutlass/condor	Ecogen
		MVP	Mycogen
		Agree	CIBA-CEIGY
		Wormbuster	Bactec
Bt. israelensis	Mosquitoes/blackflies	Acrobe	American
		Taknar	Cyanamid
		Vectobac	Sandoz
		Skeletal	Abbott
		Bactimos	Novo nordisk
Bt. tenebrionis	Fungus gnats	Gnatrol	Abbot
	Colorado potato beetle	Novdor	Novo nordisk
		Trident	Sandoz
		Foil	Ecogen
		m-track	Mycogen
Bt. aizawai	Wax moth	Certain	Sandoz
	Diamondback moth	Florbae	Novo nordisk
		centari	abbott

Other favourable candidates for the production of microbial pesticide production in bacteria include:

1. *Bacillus thuringiensis* subspecies kurstaki: This is commonly known as garden dust or caterpillar killer. It is used to kill lepidopterans.
2. *Bacillus thuringiensis* subspecies israelensis: This is available in the form of granules or liquid. It affects larval stages of dipterans.
3. *Bacillus sphericus*: This is used for eliminating mosquito larvae. Its lethal behaviour depends solely on the cell wall of insect. It takes a much longer time to act on organisms than *Bacillus thuringiensis* but its advantage over *thuringiensis* is that it can remain in a dormant stage in mud or sewage water.
4. *Bacillus papillae*: This lives in soil and is Gram-positive in nature. It is an obligate parasite and hence it is produced in the larva of beetle.

18.1.2 Viruses

A large number of viral species infect insects. Hence viruses are also suitable candidates for producing insecticides. The most common viruses are baculoviruses (rod shaped and containing DNA) which infect about 100 insect species. They are preferred because they are easy to produce in large quantities, they are highly target specific, and they are stable under unfavourable conditions because of inclusion bodies. On the other hand, these viruses are slow in reacting and giving results and are also expensive as compared to chemical methods. They are also not stable under ultraviolet rays of the sun (Butt et al., 1999). The method of action is that after the ingestion of virus by the insect polyhedral inclusion bodies dissolve in the gut which leads to the normal functioning of the virus; this leads to the development of more viruses and after rupture of the cell, viruses are released. The death of the insect takes place after 4 to 5 d of ingestion.

18.1.3 Fungi

These are also a potential source of pesticide production; their use is comparatively lower as they require moisture content i.e. humidity (79–100%) and temperature (27 °C) for spore germination. As these factors can only be provided in places such as Russia they are not widely used. They perform their mode of action by penetrating into the cuticle of insects, gaining access to haemolymphs by which the fungi take up nutrients to grow in haemocoel. They are available in the form of conidial powders. Some examples of fungi used are the commercial mycoinsecticide ‘Boverin’ which is applied to potato beetles. *Culicinomyces sp.* acts specifically on mosquitoes (Butt et al., 1999).

18.1.4 Protozoa

In comparison to the above pesticides, the action of protozoa is slow and may take weeks. They are also very difficult to produce and are produced only *in vivo*. One of the most important features of this is that there is a reduction in the number of offspring produced by the infected insect. Examples of protozoa are *Microsporidia,* which infects a wide range of insects, and *Vavra vilivis* and *Nogema algerae* are found to be effective against mosquitoes. So far no protozoan has been able to be cultured via an industrial method (Berini et al., 2018). Table 18.2 summarizes the role of bacilli, fungi, protozoa and viruses as biopesticides.

18.2 CRITERIA FOR THE SELECTION OF ORGANISMS FOR BIOLOGICAL CONTROL

The following are the criteria for the selection of organisms for biological control: i. They should be target specific i.e. should not affect or kill any other insect. ii. The mechanism of killing should be anticipated. iii. They should not be costlier than chemical methods. iv. The organisms should not be harmful to human populations or any other organisms. v. Killing should be fast enough so that damage is kept minimum. vi. The production of microbial pesticides should be cheap and the final product should be stable under ultraviolet light. vii. One should be able to use them again (Burges, 2012).

TABLE 18.2
The Role of Bacilli, Fungi, Protozoa and Virus as Biopesticides

Pathogen	Product name	Host range	Uses
BACTERIA			
Bacillus thurengeinsis var kurstaki	Bactur, bioworm, catterpillar killer, dipel, futura, javelin,thurcide, topside,worthy attack	Caterpillars (larvae of moths and butterflies)	Effective for foliage-feeding catterpillars. Deactivated rapidly in sunlight, apply in evening and direct some spray on lower surface or leaves.
Bacillus thurengeinsis var israelensis	Aquabee, bactimos, gnatrol, lavaX, mosquito attack, skeetal.	Larvae of aedes and psorophora mosquito, black flies, fungus gnats.	Effective against larvae only, active only if ingested. Culex and anopheles mosquito are not controlled under normal application rates.
Bacillus thurengeinsis var aizawai	Certain	Wax moth caterpillars	Used only for the control of moth infestants in honey bee hives.
FUNGI			
Beauveria bassiana	Botanigard, mycotrol,naturalis	Aphids, fungus gnats, mealy bugs, whiteflies	Effective against several pests, high moisture requirements. Lack of storage longitivity.
Lagenidium giganteum	laginex	Larvae of most pests mosquite species	Remain effective in the environment even in dry conditions. Cannot survive high temperatures.
PROTOZOA			
Nosema locustae	NOLO bait, grasshopper attack	European cornborer caterpillar, grasshopper and mormom crickets.	Useful for rangeland grasshopper control. Active only if ingested. Not recommended for small-scale usage.
VIRUSES			
Gypsy moth nuclear polyhedrosis	Gypchek virus	Gypsy moth caterpillar	All of the viral insecticides used for the control of forests pests are produced and used by the US forest services.
Tussock moth	TM biocontrol	Tussock moth catterpillar	
Codling moth granulosis virus (GV)		Codling moth caterpillars	Commercially produced and marketed briefly, no longer registered or available.

18.3 TYPES OF INDUSTRIAL-SCALE MICROBIAL INSECTICIDE PRODUCTION

The production of microbial pesticides can be carried out in one of the following ways:

18.3.1 Submerged Fermentation

The growth of bacilli can be done in various ways which require the preparation of nutrient broth. In one shake-flask initial growth occurs in the flask which contains nutrient broth. In another flask best molasses (1%), corn steep liqueur (0.85%), pharma media (1.4%) and $CaCO_3$ (0.1%) are used. The second media contains starch (6.8%), sucrose (0.64%), casein (9.94%), corn steep liquor (4.7%), yeast extract (0.6%) and phosphate buffer (0.6%). The third media contains soya bean meal (15%), dextrose (5%), corn starch (5%), MgSO4 (0.3%), $FeSO_4$ (0.02%) and $CaCO_3$ (1%). The above media preparation can be used for both the strains of bacillus called *Bacillus thuringiensis var insaraelensis* and *Bacillus sphericus.* They do not require carbohydrates to grow (Bailey, 2010).

At the end of the fermentation process the components have to be extracted; for that one can use centrifugation, vacuum filtration and precipitation (with the help of $CaCO_3$), which has the highest potency (Chandler et al., 2008).

18.3.2 Surface Culture

This technique is used for fungi and spore formers. First the organisms are cultured in a shake flask, from where the organisms are transferred to flat bins with perforated bottoms. There is also the use of a semi-solid media (composition: kieselguhr, soy bean, dextrose and mineral salts) which increases the surface area and hence aeration. This semi-solid media is used inside the bins into which the broth is transferred. In the next step hot air is passed through the perforations to dry the material (Chandler et al., 2008). The last step of processing includes grinding, and assaying the compound. An example is *Hirsutella thompsonii.*

18.3.2.1 In Vivo Culture

These methods are used for producing caterpillar viruses, mosquito protozoa and *Bacillus papillae*. These methods requires labour and precautions. Once the organism to be cultured has been obtained, it is lyophilized and stored at low temperature. The viruses are fed to larvae via food; upon the death of larvae they are crushed, centrifuged to remove large particles, and the rest are dried. The number of viruses obtained from different larvae varies. The steps involved are: rearing caterpillars, infecting them, and extraction of virus particles. This technique is commonly used for treating cotton moths (Chandler et al., 2008).

18.4 POTENCY OF MICROBIAL PESTICIDE

There has to be a standard reference set up for preparation analysis. The standard will differ from microbe to microbe. There are some parameters to be considered such as: age and type of insect, food of the insect, temperature etc. An LD_{50} value is the dose of insecticide that will kill 50% of a total insect population. The standards are prepared and deposited at the Institute Pasteur in Paris (Kumar et al., 2019).

18.5 FORMULATION OF BIOINSECTICIDE

Formulation means to prepare any material according to its formula. The formulation of bioinsecticides is important because the product formed may be effective in the laboratory but during field testing it may be valueless (Gan-Mor and Matthews, 2003). A pesticide formulation is a mixture of three things: the active principle, an inert material and a synergist (a type of active ingredient which is added to enhance the ability of pesticide). The inert material determines the adequate presentation of larvicide to the target insect; this is also referred to as a carrier or extender, and it can be a solid or liquid into which the active principle is diluted. The active principle is what kills or repels an insect. Formulation can be divided into two categories:

18.5.1 Dusts

Dusts are semi-solid preparations and have one or more inert material, and contain celite, chalk, kaolin, bentonite, starch and lactose. Dusts are always used dry. When the inert material attracts an insect then it is known as bait. Dusts are used for *Bacillus thuringiensis*. The bait for *Bacillus thuringiensis* is ground corn meal and for protozoa can be cotton seed oil, or honey etc. Wetting agents can be added to dusts to suspend particles easily in water (Hynes and Boyetchko, 2006).

Advantages: I. They give greater stability to the preparations; II. They can reach the underside of low-lying crops such as cabbage. III. Good alternative for wet spray.

Disadvantages: I. They can be washed off by heavy rains, and can be inhaled by the person spraying. II. Difficult to apply equitable allocation.

18.5.2 Liquids

These are made from an oil-soluble liquid active ingredient along with one or more petroleum-based solvents, and a mixing agent. Mixing agents allow particles to form emulsions with water. In liquid formulations both crystals and spores are stable in water (Chandler et al., 2008). Apart from the composition, other agents are added such as emulsifiers to stabilize emulsions when used e.g. for *Bacillus thuringiensis,* and virus tween 80, triton B 1956, span 20 are used.

Spreader or wetting agent: When the surface is slippery then these can be applied to act as surface-tension reducers e.g. for agricultural *Bacillus thuringiensis,* alkyl phenols tween 20, triton X114 are used. For viruses, triton X100 and arlacel 'C' are used.

Stickers or adhesive: As the name suggests, these are added to hold the insect on the surface e.g. for bacteria and viruses, skimmed milk, dried blood, corn syrup, casein, molasses, polyvinyl chloride latexes are used.

Protectants: They protect the active agent from the effects of ultraviolet light, oxidation, heat, desiccation etc.

Dyes: When combined with protein, these help in protecting virus preparations from ultraviolet light e.g. brewer's yeast plus charcoal, albumin plus charcoal, skimmed milk plus charcoal.

18.6 SAFETY TESTING ASSOCIATED WITH BIOPESTICIDES

There are some tests which need to be carried out to prove that the micro-insecticides produce no harm to humans (Chandler et al., 2008). Animals are the major source for safety testing. Common animals used for this purpose are rats, mice, monkeys, rabbits, and fish.

- Tests conducted in Russia, United States and Japan on agricultural entomopathogens have shown them to be non-toxic to man, some animals and plants.
  Bacteria: *B. papillae, B. thuringiensis, B. moritai.*
  5 viruses: *Heliothus, Orgyia, Lymantria, Autographa, Dendrolimus.*
  2 protozoa: *Beauveria bassiana, Hirsutella thompsonii.*
- Tests by WHO carried out in France and United States have also shown that the use of entomopathogens is safe. These are listed below:
  B. sphericus strain SS11-1, B. sphericus strain 1593-4, B. sphericus strain 1404-9, B. thuringiensis var israelensis (serotype H14), Metarrhuzium anisophilli.
- The OCED pesticide assessment and testing project is also responsible for testing the safety of micro-insecticides.

WHO have provided criteria and stages for the development of bio-insecticides:

The stages are as follows: screening, evaluation, safety and impact of entomopathogens (Gan-Mor and Matthews, 2003).

Criteria to be followed:

- Materials can be prepared anywhere, such as on cottage-industry level, or small-scale level, but final production should be carried out by industry, which can formulate the product.
- Based on already-existing information about safety it will take less than five years to produce a bioinsecticide.
- It will take more than five years without previous listing.
- The cost of developing a biological insecticide is far less than developing a chemical insecticide, by between 20% and 50%.

Biopesticide market trends: The global biopesticides market size is projected to grow at a compound annual growth rate of 14.7% from an estimated value of USD 4.3 billion in 2020 to reach USD 8.5 billion by 2025. Fruits and vegetables are expected to develop the market for biopesticides as the fastest-growing crop group. Biopesticides will provide a sustainable alternative for managing such insect pests, with the latest desert locust assault on key crops across regions. Farmers have also been introducing biopesticides in conjunction with traditional crop defence chemicals in order to achieve a longer pest control period. The need for biopesticides is also motivated by a rise in safe farming areas such as greenhouses and the need for residue-free food crops. Leading companies include BASF SE (Germany), Bayer AG (Germany), Biobest Group NV (Belgium), Certis USA L.L.C (US), Novozymes A/S (Denmark).

18.7 BIOPLASTICS

The use of synthetic resources has increased in this economic area over the last couple of decades and become more regular in it. Polymers used as raw resources can very easily be dumped, and serious harm takes place when they return to the environment. As a result of these actions, globalization has been expanding by reducing the waste produced. PHAs are biopolyesters, processed by microbes inside cells as energy storage materials. PHAs are renewable plastics that have significant environmental and social impact. They are a promising raw resource for bioplastics. PHAs have a wide variety of uses in different sectors due to biocompatibility, and biodegradability. Apart from PHAs, another notable material is polyhydroxy butyrate (PHB), which has many applications for use as resource polymer polypropylene (PP), which produces an effective bio-based, biodegradable approach. PHB's degradable properties have enhanced awareness of biodegradable plastics, whose production can minimize unwanted pollution and waste. This chapter highlights some conditions of these emerging biopolymers including their composition, structure, classifications of PHA and PHB, substrate used, developments in their production, along with the microbes known for degrading bioplastics from various microbial communities, and biological applications. PLA (poly-lactic acid) has also recently been used for bioplastics.

Plastics are broadly used polymers in day to day life, mainly for packing purposes. Commercially they are used in various industries including automobiles and medicines. Plastics, also known as 'synthetic plastics', are one of the major inventions in history, and are widely used everywhere in the world for different purposes; from making bottles to automobiles parts, and from making polymer carrier bags to compound surgical and medical instruments. Plastic use extends to electronic appliances, plastic bottles, carpets, tyres, holders, paints, shopping bags, containers etc. According to a survey, in 2015, the global output of petroleum plastics increased to 300 million tonnes. About 20% of urban solid waste volume is occupied by plastics. The daily used plastics are made from petroleum material, which is an undegradable source. Plastic materials are very convenient due to their flexible structure and dynamic nature. Their structure can easily be manipulated chemically and exhibits a wide spectrum of lengths and shapes. Higher molecular weight and tight carbon–carbon bonding between molecules make them non-biodegradable and they can accumulate in the environment for a long period. Their undegradable nature is harmful to the environment. One solution to undegradable plastics is burning them but this is a dangerous method because it releases toxic chemical compounds such as hydrogen chloride and hydrogen cyanide into the environment. This overconsumption of petroleum plastics demands sustainable alternative renewable sources. Adverse effects on the environment, emission of carbon dioxide, and long-period accumulation are the foremost drawbacks of the use of non-biodegradable plastics on daily basis. Synthetic polymers are derived from petroleum, which is a non-renewable source in nature. The production of plastic in recent years has decreased due to increased cost of oil and limited availability of petrochemical resources. The adverse effect of plastics on the ecosystem is responsible for the limited production of the plastics in recent years. In such situations, alternatives to synthetic plastics are needed. To sustain the environment, bioplastics have gained attention due to their biodegradability.

Bioplastics are a replacement for synthetic plastics. They are significantly advantageous to the ecosystem and decoupling of the plastic production from fossil fuels. Bioplastics belong to the family of polyalkanoates (PHAs). PHAs includes polymeric esters such as: polyhydroxybutyrate (PHB), polyhydroxybutyrate co-hydroxy valerates (PHBV), polyhydroxybutyrate co-hydroxy-hexanoate (PHBHx), and polyhydroxybutyrate co-hydroxy octanoate (PHBO), polyhydroxyalkanoate (PHA), poly-lactic acid (PLA) and aliphatic plastic such as polybutylene succinate (PBS), which can be used as a substratum by microbes (Tokiwa et al 2009, Mekonnen et al 2013). Bioplastics are biodegradable plastics that degrade completely through microbial metabolism in a short period, undermaintained conditions. PHAs composed of linear polyesters consist of monomeric hydroxy acid (HA) bound together with an ester linkage and accumulated as a source of energy and nutrients by many microbes. As at2018, it was projected that the production of bioplastic had increased up to 205–300%. However, the main difficulties with commercial bioplastics are their production costs, poor recycling facilities, and inefficient waste handling technology. The major part of the cost is spent on media sterilization and maintenance of bioreactors. Hence, there is a need to find cost-effective and worthwhile feedstock for PHA production.

Aggregation of synthetic plastic material in the environment and its non-biodegradable nature encourages industries to produce a degradable sustainable plastic. Biodegradation means biological activity or biological action of microorganisms. It involves three steps: i) Biodeterioration is the amendment of mechanical, physical, and chemical properties on or inside the polymer by the action of biological catalysts. ii) Biofragmentation involves the conversion of polymers to simple form such as monomers and oligomers by the activity of microbes, and iii) Assimilation, in which microbes are supplemented by required energy, nutrients, and carbon sources from the fragmentation of the polymers (Lucas et al. 2008, Emadian, 2016). The main aspects that influence plastic's biodegradation are its chemical structure, polymer chain, attached functional groups, crystallinity, and polymer complexity. Polymers with less complex structures are more sensitive to biodegradation through the action of microbes. The environment where plastic is dumped also plays an important role in degradation. Environmental factors involving pH, temperature, moisture, and availability of oxygen can enhance the degradation process (Massardier-Nageotte et al. 2006, Emadian, 2016).

Biopolymers (bioplastic) are safer than synthetic plastics and are well degraded naturally without emission of toxins and are also biocompatible to human systems (Wani et al. 2016). Enhanced production of bioplastics could minimize greenhouse gas emissions, waste generation of plastic and also reduce plastics usage. Bioplastic is a promising discovery that reduces pollution and is eco-friendly. The production and recovery cost of plastic is high compared to the traditional production methods of synthetic plastics. Hence there is a need to discover polymers which are profitable in addition to being completely biodegradable. Bioplastics are an exciting new area that offers challenges to research fields, together with genetic engineering for the development of cost-effective bioplastics.

18.8 VARIOUS BIOPLASTICS

Where humans are present, plastics are typically found in urban areas but often accumulate in all environments. In the deep ocean, they aggregate in wide areas of partly degraded product and causes the death of many aquatic organisms. In recent years, plastic pollution recovery methods have increased. Around the globe, there is a continuous increase in population, which causes a greater requirement for the production of plastic and consequently also an increase in the amount of waste plastics. Particularly in urban regions, incineration of plastic pollution is implemented. During this, some environmental limitations can be encountered, which include large amounts of coal, slag usually contains toxic and hazardous substances, and high CO_2 emissions (Rajesh Banu et al., 2019). Due to harmful effects to the environment, the community's global awareness of reducing plastic waste has evolved and achieved particular interest in significant environmental initiatives. On the other hand, to solve this problem, the research community is trying to establish technical replacements. Regarding renewable resources, the last few years are linked with growing interest in the use and production of biopolymers in scientific and public areas. The main features are biodegradability and desirable physicochemical characteristics of traditional synthetic polymers.

It is essential to recognize that biodegradable plastic materials are not only produced from biopolymers, but all sustainable plastics are biodegradable although often made from fossil fuel resources. To characterize both biodegradable plastics and renewable plastics, many commentators use the common term bioplastic. However, the first relates to end-of-life alternatives, whereas the second relates to plastics produced from sustainable sources of raw products. The second category includes polyethylene and bio-based polyamide (Yates and Barlow, 2013). Biodegradable plastic products derived from renewable sources, although natural or modified starch, PLAs, collagen, polyvinyl alcohols (PVA), polyglycolate (PGA), PHAs, and polysaccharides, can therefore be classified in both categories. Biopolymers operate as a safe way of accessing non-biodegradable and non-renewable polymeric materials and are considered as sustainable in packaging manufacturing (Mahdavi et al., 2014). Due to the initial substrate used as a source of carbon, the cost of production is a concern closely connected to bioplastics, which can account for 40% of total value. The development of biopolymers from localized and renewable energy sources, such as maize, dairy effluents, and agro-industrial wastes has been found via research and become commercially significant. Manufacturing demand accounts for around 40% of the production of plastics, half of which is intended for food packaging. In general, there are few disposal strategies for synthetic plastic packaging; these cause destruction of natural resources associated with conventional non-biodegradable plastic packaging, and are a growing worldwide issue regarding harm to the environment. To reduce impact on the environment it is necessary to seek alternative processing, uses, and disposal replacements for the polymeric materials. Biodegradable properties in the use of plastic materials can help as an approach to this issue. As a result, the amount of bioplastic currently used in the application of bottles and wrappings is much greater than for average plastic specification.

18.8.1 Polyhydroxyalkanoate (PHA)

PHA compositions can be developed conveniently using mixed cultures, focusing on a broad range of various feedstocks, with other units of monomer such as 3-hydroxy-2-methyl butyrate, 3-hydroxyhexanoate, 3-hydroxy valerate, 3-hydroxy-2-methyl valerate, and 4-hydroxybutyrate. These components are common in P(3HB)-based copolymers (Karthikeyan et al., 2015),e.g. P(3HB-co-3HV) copolymers with HV percent varying between 17 to 85% reported in acetate/propionate feed focused on propionic acid fractions (Lemos et al. 2006). In comparison, pure cultures require massive quantities of co-substrate to generate polymers that have comparatively small percentages of monomers other than P(3HB) (Srikanth et al., 2015). P(3HB-co-3HV) derived from concrete mixtures has similar quantities of vegetable oils and propionic acid with only 2–8 mo % HV. This relates to the fact that mixed cultures contain a variety of species and are likely to use a variety of development pathways for PHA (Karthikeyan et al., 2015). These have already attained comparable cellular PHA content and development levels equal to those of culture medium; using mixed crops, P(3HB) is produced in less than 8 h at 89%w/w of dried biomass. Despite the various organisms contained in a mixed culture conditions, the compositional distribution of the polymer formed, and the impact of these mixture formulations on chemical-mechanical properties, is one of the major questions (Pagliano et al., 2017).

The carboxyl group and two distinct monomer hydroxyl groups form an ester linkage consisting of saturated or unsaturated fatty-acid monomer unit polyhydroxyalkanoates. Polyhydroxyalkanoates are linear polyesters (Vieira et al., 2011). Polyhydroxyalkanoate comprises 2 to 6 fatty acids with various classes of ether, halogen, ester, alkyl, epoxy, cyano, acid groups, and alkyl. The nomenclature of various types of PHAs is based on the number of units $(CH_2)(n)$ and the category of side group (R) connected to the PHA (Endres and Raths 2011). PHAs are graded according to their monomer unit composition consist of two large categories – i) Medium-chain length (mcl) polyhydroxyalkanoate of 6–14 carbon chain length; and ii) Short-chain length (scl) polyhydroxyalkanoates of 3–5 carbon chain length. Copolymers or mixtures of two such kinds of PHA are commonly called scl-co-mcl polymers, as compared to normal scl- and mcl-PHAs (Pagliano et al., 2017). These comprise long-chain lengths (LCL-PHAs) with 16 or more carbon atoms, and 3–14 carbons; around 150 various units of polymerized monomer to provide PHAs are reported (Mahdavi et al., 2014).

Monomer units that are oxidized at a place different from the 3rd carbon include 4-hydroxybutyrate, 4-hydroxy valerate, and 5-hydroxy valerate. Short-chain length PHAs primarily comprise 3-hydroxybutyrate and 3-hydroxy valerate. 3-hydroxyheptanoate, 3-hydroxyhexanoate, 3-hydroxyoctanoate, and 3-hydroxydecanoate are included in medium-chain length PHAs. There are various metabolic processes concerned in the formation of scl and mcl-PHAs (Pagliano et al., 2017). Microbe aggregation of various categories of PHA is determined by the metabolic history of the organism, role of PHA biosynthesis, and genetic composition (Karthikeyan et al., 2015).

18.8.2 Polyhydroxybutyrate (PHB)

Polyhydroxybutyrate (PHB) was first discovered by the French scientist Lemoigne, and is described as a lipid found in the *Bacillus megaterium* (35-biopl and envi). PHB is the most commonly used PHA, and is a homopolymer of 3-hydroxybutyric acid, having methyl group attached at the side chain and high molecular weight compound tangled in carbon energy storage by certain bacteria. As it is biodegradable and synthesized by bacteria, it is an alternative option to synthetic plastic; it is also economically safe. According to research, granules of PHB are situated in subcellular and cytoplasm partitions such as Gram-positive bacteria vacuoles. PHB particles are also found in some Gram-negative species such as cyanobacteria, photosynthetic bacteria, and archaebacteria (Pagliano et al., 2017). PHB granules have mainly been synthesized in environments where cells becomes extremely rich in sources of carbon and there are absences of other nutrition such as oxygen, nitrogen, magnesium, and phosphorus. A carbon-rich environment enhances the construction of PHB molecules into cells, then acts as an energy stock during stressed conditions of cells (Mukherjee and Kao, 2011).

Biosynthesis of PHB: Microorganisms react rapidly to changing environments. They have an innate ability to collect vital nutrients that assist survival throughout various stress conditions. Under unfavourable conditions, the microorganisms accumulate PHB granules as a carbon source. Accumulated carbon is then converted to hydroxyalkanoate (HA) compounds. Polymerization of HA compounds leads to the formation of PHB, which is insoluble in water and is stored in cells. Synthesis of PHB involves different pathways that have been clarified in certain species of *Azobacter beijerinckii, R. eutropha, and Zoogloea ramigera.* PHB is the most common scl-PHA synthesized by microbes via a three-step pathway. The pathway starts with the combination of two moles of acetyl-CoA, which gives rise to acetoacetyl-CoA. The reduction reaction of acetoacetyl-CoA results in 3-hydroxybutyryl-CoA formation followed by a formation of one mole of PHB through polymerization (Yates and Barlow, 2013).

Acetoacetyl-co-reductase, PHA synthase, and β-ketothiolase are the three major enzymes involved in the PHB synthesis process. These three enzymes have different roles in the pathway. β-ketothiolase catalyses the first stage of the process and is a family member included in thiolytic substrate cleavage into acetyl-CoA and acyl-CoA. It catalyses the condensation of acetyl-CoA to give rise to acetoacetyl-CoA. The second stage in the process is catalysed through the enzyme acetoacetyl-CoA-reductase, which is a soluble protein and results in the formation of 3-hydroxybutyryl-CoA in the stereo-specific reduction reaction. The last reaction in the pathway leads to the formation of PHB, and this step indicates the type of synthesized PHA (Mukherjee and Kao, 2011). This reaction is carried out by the PHB synthase enzyme that makes ester linkage polymerization of 3-hydroxylbyutryl. PHB synthase is a significant enzyme and is a soluble enzyme. After PHB biosynthesis, it becomes linked with PHB granules and is stored in the cells. According to research, the genes involved in the PHB biosynthetic pathway are highly mixed. *Ralstonia eutropha* and other PHB-producing microorganisms consist of three genes *pha A, pha B and pha C.* These genes are crucial for synthesis and accumulation and are found on the single operon, *phb*CAB (Karthikeyan et al., 2015).

PHB extraction from microorganisms is a challenging downstream process. PHB extraction from cells involves three major steps: polymer extraction, harvesting of cells pre-treatment, and post-treatment purification (Mukherjee and Kao, 2011). The main step in PHB abstraction is polymer extraction from the cell biomass. Polymer extraction is carried out through various processes, such as chemical digestion surfactant, solvent extraction, sodium hypochlorite, surfactant-chelate, enzymatic digestion, bead mills, gamma radiation, SDS sonication, and two-phase aqueous systems (Karthikeyan et al., 2015). PHB purity is dependent on the techniques used for the extraction, and it varies from 86% to 96%. The yield of PHB is dependent on the strain used and the various solvents used in the extraction process. PHB is soluble in organic solvents such as chloroform, 1,2-dichloroethane, and 1,2-propylene carbonate, and hence solvent extraction is considered a rapid and easy process for PHB extraction and is a routinely used method at laboratory level. PHB extraction from cells consists of two-step-solvent extraction; increase cell permeability to allow PHB availability, and then dissolve PHB in different organic solvents (Yates and Barlow, 2013). The use of organic solvents in PHB extraction gives high purity of PHB but is harmful to the environment; when used at large scale it releases toxic chemicals and volatile solvents without treatment. High cost and complexity in recovery when viscosity of PHB is more than 5% are the main disadvantages (Srikanth et al., 2015). The prime aim of surfactants is a lipid bilayer. Disturbance of the cell membrane is the primary objective in chemical digestive surfactant systems because surfactant incorporation creates micelle development, and carbonosomes are revealed; PHBs and cell debris are distributed into the lysis solvent stream (Du et al., 2016). Chemical digestion surfactant with sodium hypochlorite processes can be used to improve PHB efficiency. This gives a high level of purity with good quality (Mukherjee and Kao, 2011). Also, chemical surfactants for digestion followed by chelating agents are used to boost PHB consistency.

Enzyme digestion is an eco-friendly approach for the recovery of PHB, with a gentle operating environment. Proteolytic enzymes such as protease and glycosidase are increasing the purity of PHB with fewer environmental drawbacks (Rajesh Banu et al., 2019). Chemical and solvent methods are costly and non-sustainable as compared with biological methods such as enzymatic methods, which are eco-friendly. Bacteriophages are the latest green biological solution that has been developed to reduce the cost of the process and eliminate environmental disadvantages (Mahdavi et al., 2014). Bioextraction is the latest process for extracting PHB from the cells of producers – without the use of any environmentally inappropriate substances –that provides an affordable extraction procedure (Srikanth et al., 2015). PHB accumulates in the cell as carbon storage of the cell during stressed conditions. PHB molecules are surrounded by various proteins that play a crucial role in depolymerization and synthesis. In the extraction process of PHB, cell wall disruption is a common step. Cell wall disruption is carried out by several methods, and involves the use of a variety of solvents and chemicals. Another approach is the enzymatic disruption of cell walls during biotechnological-based approaches (Pagliano et al., 2017). Bioextraction methods of PHB involve bacteriophage-related system, mealworm digestive systems, and predatory systems. These methods are gaining more attention due to their strong potential in extraction (Rajesh Banu et al., 2019).

18.8.3 Naturally Occurring Microorganisms to Produce Biopolymers

PHA is metabolized by some microbes combining bodies to create a source of carbon, and electrons sinks (Srikanth et al., 2015). Such granules behave as extremely refractive inclusion bodies under an electron microscope. At the beginning of the twentieth century, *Bacillus megaterium* was the first microbe identified as producing PHA. Over 200 distinct microbes belong to the Gram-positive and Gram-negative classes recorded as intracellularly aggregating PHA under specified environmental conditions by the end of the 20th century. Maximum density of PHB granules from clay soil strains is reported by a few investigators (Mukherjee and Kao, 2011). However, according to species type, the size and number of these granules in cells differ. It has been observed that *Ralstonia eutropha* accumulates PHA as high as 80% of cell biomass. *Pseudomonas oleovorans, Cupriavidus necator, Alcaligenes latus,* and *Bacillus megaterium* exist naturally. For PHA production, various carbon streams, including industrial waste and plant oils, have been established (Karthikeyan et al., 2015). *C. necator* is among the most economical microbiota and is recognized as producing PHA from various microorganisms. Methyl bacterium sp. *Pseudomonas putida*, *Azotobacter chroococcum,* and *Rhizobium meliloti* are PHB producers that have been identified (Yates and Barlow, 2013).

18.8.4 Biopolymer Development in Genetically Engineered Microbes

In comparison to chemically manufactured petroleum-based plastic materials, PHAs are an acceptable replacement. The production of PHA in natural sources is affected by some factors, mainly because of the mobilizing enzymes in the form of in vivo depolymerases and esterases (Mukherjee and Kao, 2011). However, several native microbes expand slowly, and the cost of production of PHB by microbial fermentation is quite high. A difference in improving the development of PHB in native species is to utilize a mutant without PHA degradation or to make explicit the genes heterologically in fast-growing species that lack PHB biosynthesis and degrading genes. Thus to reduce the cost of production, there is a need not just to improve efficient fermentation methods but also to select new genetically engineered microbes.

In various heterologous systems, the PHA biosynthetic pathway gene has been expressed. Using affordable carbon sources and fast processing of downstream, microbes that grow faster and accumulate large volumes of PHB are best suited for metabolic engineering. For PHB mass production, *R. eutropha*, and *E. coli* have been identified to be of economic significance among the numerous heterologous processes (Mukherjee and Kao, 2011). There are various reasons recombinant *E. coli* produces a better level of PHA, so there is no need to supply the restrictive number of nutrients of PHA development. For production of the *E. coli* recombinant expression system, genes of the PHB biosynthetic route from many microbes are cloned into *E. coli*.

Apart from PHA and PHB, PLA is becoming a potential environmentally friendly material for bioplastics production; PLA is a thermoplastic aliphatic polyester, biodegradable polymer. It is derived from renewable resources by means of a fermentation process using sugar from corn, followed by either ring-opening polymerization or by condensation polymerization of lactic acid. PLA is considered both as biodegradable and as biocompatible in contact with living tissues (e.g. for biomedical applications such as implants, drug encapsulation etc.). PLA can be degraded by abiotic degradation (i.e. simple hydrolysis of the ester bond without requiring the presence of enzymes to catalyse it).

18.8.5 Nutritional Requirements For PHA Production

The use of bioplastic-containing PHA can offer countless advantages to both industry and the environment. Existing technologies of PHA production have certain disadvantages. Genetically modified or selected strains require an aseptic, sterilized medium to avoid contamination. To achieve highly sterilized media for PHA production requires expenditure for sterilization of the medium, and equipment. For production of pure PHAs, strains need costly nutrients including pure mineral salts and pure carbon and energy sources. As compared to the synthetic plastic, bioplastic has a high cost. Carbon source costs are an important element that contribute 50% of the total cost of production. Researchers, therefore, focus on minimizing the production costs, employing various waste materials as carbon sources (Vieira et al., 2011).

The composition and quality of PHAs are dependent upon the different microbial strains used and the types of carbon sources available. Biochemical and physiological characteristics of the microorganisms are responsible for PHB composition as well as quality. *Pseudomonas putida* can synthesize when developed on appropriate substrates; PHAs contain various functional groups such as esters, phenoxy, halogens, alkyls, olefin, and phenyl. *Pseudomonas* species strain DR2 has been used for the production of PHAs by using substrates such as glycerol, glucose, acetate, citrate, corn oil, and palmitate as a source of carbon. Of these, corn oil shows the maximum PHA production rate. A recent study on the mutant strain of *P.aeruginosa* suggested that it can produce 50.3 %(w/w) PHA with soybean oil as substrate and 40.7 %(w/w) PHA with *n*-hexadecane (Yates and Barlow, 2013). Hence the choice of carbon source is a vital factor that affects the yield of PHA. Whey is the co-product of production of cheese and casein. Half of the whey is turned into a functional form for humans, animals, and the rest of the components are discarded in the environment. Worldwide, wheat is cultivated in large amounts. Bran is the outer layer of the wheat grain; it is hard and is made up of the aleurone and pericarp. It contains carbohydrates, proteins, and other minerals, and its waste materials can create problems. Several experiments have reported that the use of sources of carbon for microbial production and PHA production carried out by growing *Halomonas boliviensis* LC1 result in biomass production of 3.19 g/L and PHB production of 1.08 g/L (Mahdavi et al., 2014). In the market, starch is an available and inexpensive material. Starch has a complicated nature, which restricts the hydrolysis capacity of several strains of bacteria and generates α-amylase. *H. mediterranei* with enzymatic extruded starch gives biomass yield of 1.14 g/L with PHAs accumulation of 43%(w/w) of

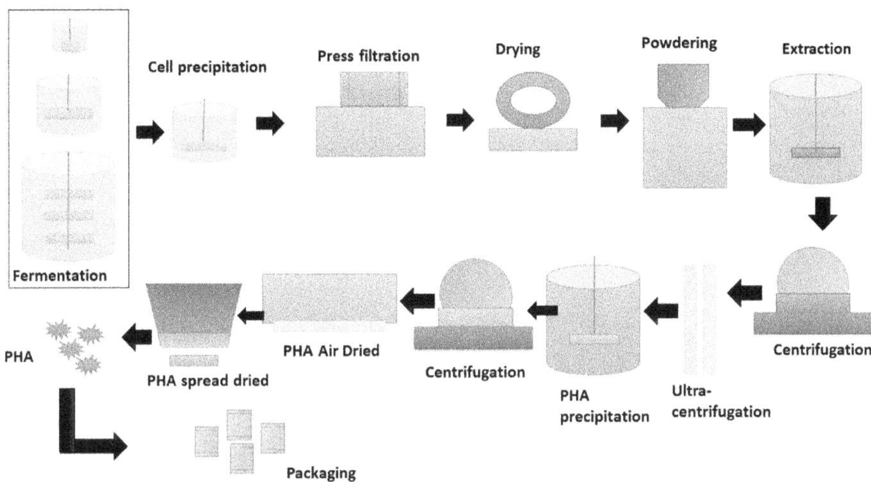

FIGURE 18.2 Industrial production of polyhydroxyalkanoates (PHAs).

CDW (Karthikeyan et al., 2015). Sulfur and unsulfured molasses are two types of molasses that are by-products of sugar cane extraction. Experiments state that up to 6.0 g/L of PHAs are produced when sugar cane molasses are used as carbon sources (Vieira et al., 2011). Waste frying oils are suitable substrates for the PHAs production. The main source of frying oil wastes are fast-food industries that produce waste frying vegetable oils that need proper disposal. Many strains of bacteria can produce PHAs from these wastes, but plant oil yields low PHA production (Du et al., 2016). *P. aeruginosa* 45A2 exhibits different PHA production on various substrates: 29.4% from waste frying oils, 16.8% from glucose, 66.1% from waste fatty acids, and 54.6% from oleic acid (Fernandez et al 2005). Along with a nutrient source, temperature also affects PHAs yield with variation in temperature of incubation, phosphorus, and nitrogen limitation. Higher temperatures lead to production of a lower amount of PHAs. Agricultural wastes, food processing, or biofuel production wastes containing palm oil, fats, glycerol, as well as cooking oil can be used for low-cost production of PHAs (Vieira et al., 2011). N, P, S, Fe, and microelements can also supply components of organic wastes. Figure 18.2 depicted the steps involved in the production of PHAs.

After fermentation of PHAs, cells are removed using press filtration. The DSP involves drying, powdering, extraction, centrifugation, and ultracentrifugation.

18.9 BIODEGRADATION OF BIOPLASTICS

To study bioplastics biodegradability under various environmental conditions, such as manure, soil, aquatic, and other marine ecosystems several experiments have been performed. Due to high microbial diversity, these environmental circumstances, mainly soil and compost, are considered (Pagliano et al., 2017). The main purpose of using bioplastics is to make them eco-friendly (Figure 18.3). Many plastic wastes,

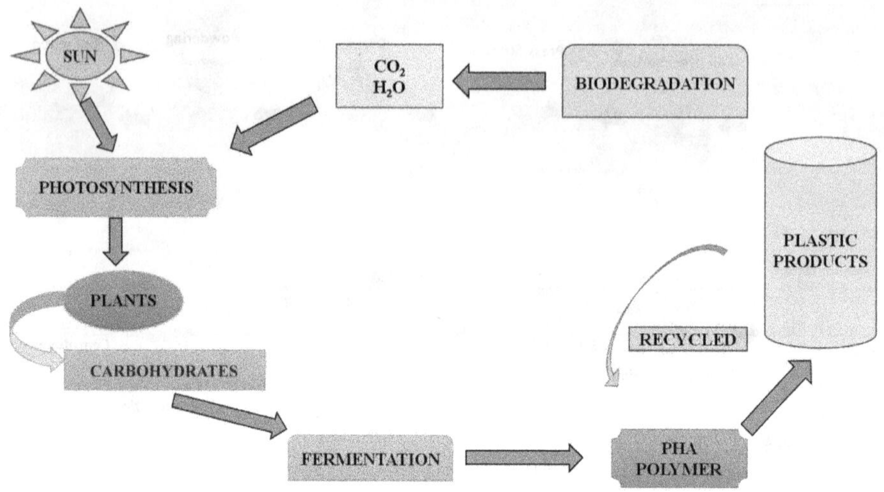

FIGURE 18.3 Bioplastics cycle: their production and biodegradation.

however, are discarded in landfills due to biodegradation of bioplastics in landfills that have not yet been widely investigated.

Different sources of biodegradable bioplastics are discussed below:

1. *Compost*: A large quantity of plastic waste discarded in landfills ultimately contributes to greenhouse gas production and leachate. Hence other means of handling solid waste such as composting or recycling are considered to be more suitable for plastic materials recovery (Du et al., 2016). A significant contribution in bioplastic biodegradation is the different conditions between home composting and commercial composting. In domestic composting conditions, biodegradation was fairly slow for 11 months in the study of biodegradation of PLA bioplastic. This can be due to lower temperatures than a large-scale study could carry out (Pagliano et al., 2017).

To increase the biodegradability of bioplastics in a compost environment several experiments were carried out. For example, by adding materials comprising a high amount of protein to increase the amount of soluble sugar in the biocomposites, bioplastics are biodegradable. Biofuel by-products were used in the mixture as composite bioplastics to enhance biodegradability of PBS, including bioplastics. This demonstrated that the level of a meal-based filler increases the biodegradation rate in comparison with pure bioplastic PBS associated with high content of soluble sugars in meal fillers (Rajesh Banu et al., 2019). The existence of maize in PLA/corn bioplastic tended to increase compost biodegradation, because maize was a strongly biodegradable material. Thus microbes more effectively destroyed the substance and the PLA fraction (Rajesh Banu et al., 2019). Poultry feather fibres (PFF) in combination with PLA pots showed a significantly higher degradation rate than those of pure PLA, which can be linked to the other materials used in the PLA pot-manufacturing process for extrusion and moulding which inhibited biodegradation (Mukherjee and Kao, 2011). The biodegradation of acrylic acid-grafted biocomposites polyhydroxyalkanoates/ rice husk (PHA-g-AA/RH), and polyhydroxyalkanoate/rice husk (PHA/RH) directly related to their RH content. The existence of such substances in a PHA matrix enhances the performance of the biocomposite along with tensile strength. Although the compressive strength of PHA-g-AA/RH was higher than that of PHA/RH, its biodegradation was significantly lower because of the sensitivity of the former composite materials to water absorption (Pagliano et al., 2017).

A significantly lower level of biodegradation than pure PLA or PHB biodegradation appeared with blending PHB or PLA bioplastics with poly(butylene adipateco-terephthalate) (PBAT). Fourier Transform Infrared (FTIR) outcomes showed that changes in the composition of these mixtures during the composting cycle led to the start of biodegradation from the PHB or PLA portion of the mixture. However, the development of a PBAT rich 3D network resulted in less disintegration of a mix (Vieira et al., 2011). In some analyses, it confirmed that PHA films with various arrangements had biodegraded. The biodegradation of CA bioplastics is reported to be 44% and 35% in low-cost fabric liners and cotton liners after 14 days of composting in a new analysis (Srikanth et al., 2015).

Considering that waste plastic is still popularly discarded in soil, it is also worth exploring the changes and effects in this specific context. A vast biodiversity of microbes is contained in soil environments that make plastic biodegradation more possible compared with other environments such as air or water. To increase the biodegradability of bioplastics, in a recent study blends of other biodegradable materials were examined. Biodegradation of PHB/PPW-FR – i.e. potato peel waste fermentation residue – biocomposite was documented to be more efficient than sole PHB because PPW-FR fibres reduce PHB biocomposite crystallinity (Yates and Barlow, 2013). The biodegradation rate of PLA biocomposite increases due to the introduction of empty fruit bunch (EFB) fibres found in another analysis (Pagliano et al., 2017). The biodegradation of bioplastics can depend on the soil environment (Table 18.3).

18.10 APPLICATION OF PHAS

PHAs are non-toxic to the environment. Biopolymers are biocompatible and biodegradable, and these characteristics make them successful with synthetic plastics. They find applications in major fields such as agriculture, environment, packaging, medical, and electronic fields.

Due to the unique chemical and physical properties of bioplastics, their use over the decades increased in the medical field (Pagliano et al., 2017). Members of the biodegradable plastics are biocompatible, and they do not cause any allergic reaction to humans. The main applications of plastic in the medical field include use as surgical implants and implant matrices for controlled drug delivery (Vieira et al., 2011). One can use PHAs in tissue engineering technological development to utilize the qualities of biodegradability and biocompatibility. Research suggests that PHAs are used as a heart valve framework in sheep (Karthikeyan et al., 2015). High-tensile strength properties of PHAs are used in tissue repair structures, bone, and ligament

TABLE 18.3
Bioplastic Biodegradation in Different Conditions of Environment

Sr. No.	Type of environment	Name of bioplastic	Biodegradable ability (%)	Biodegradability method	Period of biodegradation (days)	References
1	Compost	PLA	13	CO_2 Produce	60	Ahn et al. (2011)
2	Compost	PLA	70	CO_2 Produce	28	Tabasi and Ajji (2015)
3	Compost	PLA	60	Loss in Weight	30	Mihai et al. (2014)
4	Soil	PLA	10	Loss in Weight	98	Wu (2012b)
5	Synthetic components contain compost	Corn/PLA	79.8	Loss in Weight	90	Sarasa et al. (2009)
6	Soil	PHB	64	Loss in Weight	180	Jain and Tiwari (2015)
7	Compost/Soil	PHA	42-50	CO_2 Produce	15	Arcos-Hernandez et al. (2012)
8	Soil	PHA	35	Loss in Weight	60	Wu (2014)
9	Soil	Rice Husk/PHA (40%/60%)	>90	Loss in Weight	60	Wu (2014)

repair devices. Poly-4-hydroxybutyrate has also been used in the fabrication of surgical materials.

Biomaterials are also used in agricultural applications such as greenhouse coverings, fumigation, and mulching, etc. Biomaterials are also used as agricultural mulches and as planting containers (Pagliano et al., 2017).

PHAs can be used to make packaging films, bottles, bags, foils, and coatings of paper. PHAs have a good ability to absorb oils, and hence they have been used as an oil indicator in cosmetic oil-blotting films (Srikanth et al., 2015).

Piezoelectric properties of PHAs help in making different electronics such as loudspeakers, headphones, keyboard pressure sensors, and microphone shock wave sensors (Karthikeyan et al., 2015).

PHAs can be used as solid substrates to denitrify water. The insolubility of PHAs helps in their use as a source of power reduction for denitrification and also in their action as matrices for microbial growth. PHAs provide good results for the removal of nitrogen (Yates and Barlow, 2013). A few companies have already started using bioplastics for commercial purposes (Figure 18.4).

PLA and its production as bioplastics materials: PLA (polylactic acid) is a thermoplastic aliphatic polyester, and a biodegradable polymer. It is derived from renewable resources by means of a fermentation process using sugar from corn, followed by either ring-opening polymerization or by condensation polymerization of lactic acid. PLA is considered both as biodegradable and as biocompatible in contact with living tissues (e.g. for biomedical applications such as implants, drug encapsulation etc.). PLA can be degraded by abiotic degradation (i.e. simple hydrolysis of the ester bond without requiring the presence of enzymes to catalyse it). Therefore, it is considered as an environmentally friendly material. Figure 18.5 depicts unit operations involved in the industrial production of PLA.

Uses: The global bioplastic market size was estimated at 8.3 billion USD in 2019 and is expected to have a compound annual growth rate of 16.1 billion USD from 2020 to 2027. With a share of 62.4% in 2019, packaging applications have dominated the industry. The category of non-biodegradable plastics led the industry in 2019 with

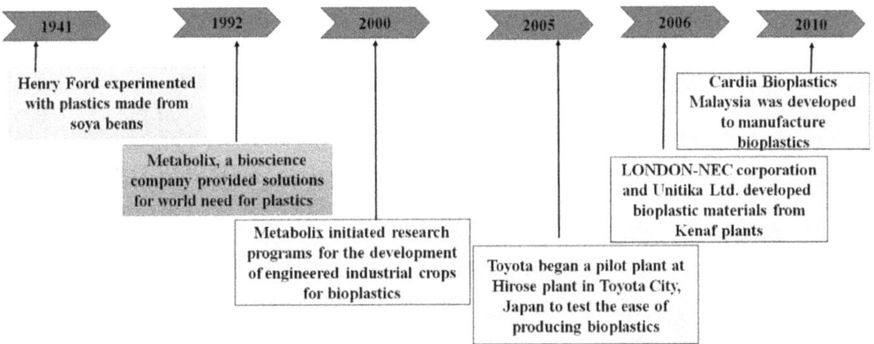

FIGURE 18.4 Chronological development of bioplastic production for commercial purposes.

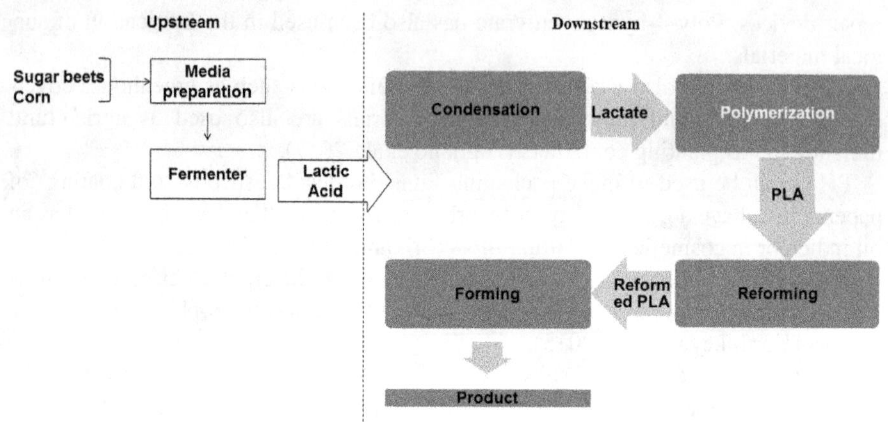

FIGURE 18.5 Industrial production of polylactic acid (PLA).

a share of 54.96 % and will retain its leading position over the projected years. High demand for electronic equipment, automotive housing, and consumer goods, such as bags, cups, films, and bottles, is due to this development. The key factor responsible for the growth of the segment is the expanded use of bioplastics to manufacture compost bags, agricultural foils, horticultural products, nursery products, toys and textiles. Europe led the global market in 2019 with a share of 44.3%. Asia Pacific is likely to register a steady growth rate due to the availability of skilled labour at low cost. Key players are Teijin Ltd. and Toray Industries.

18.11 CONCLUSIONS

Biopesticides play an important role in overcoming the detrimental effects of chemical pesticides. One advantage of using biopesticides is their biodegradability. Several microbial strains are used for the production of biopesticides. Increasingly depleted fossil fuel sources and adverse effects on the environment by synthetic plastics lead researchers to discover and develop new alternative materials to replace synthetic plastics. In addition to this, disposal of plastics wastes meaningfully contribute to the release of liquid and gaseous pollutants in the environment causing threats to humans and nature. Meanwhile, new bio-based plastics are being produced from renewable sources such as agricultural wastes, and crops. Bio-based plastic, known as 'bioplastic,' has become an important field, having the ability to help save the environment. PHA is the most common and promising form of bioplastic, showing properties similar to conventional plastics; however, commercial production of PHAs is quite expensive. Bioextraction is the most relevant and promising approach, as well being as an eco-friendly method that shows good results in the production of PHAs. The development of a new recombinant bacterial system through genetic engineering has the potential to increase production rates of PHAs on a large scale. At present, research is focused on discovering processes and new recombinants for the production of bioplastics on a commercial level.

REFERENCES

Bailey KL, Canadian innovations in microbial biopesticides. Can. J. Plant Pathol. 32, 113–121, 2010.

Berini F, Katz C, Gruzdev N, Casartelli M, Tettamanti G, Marinelli F, Microbial and viral chitinases: Attractive biopesticides for integrated pest management. Biotechnol. Adv., Prospects in Biotechnology 36, 818–838, 2018.

Burges HD, Formulation of Microbial Biopesticides: Beneficial microorganisms, nematodes and seed treatments. Springer Science & Business Media, 2012.

Butt TM, Harris JG, Powel KA, Microbial Biopesticides, in: Hall FR, Menn JJ (Eds.), Biopesticides: Use and Delivery, Methods in Biotechnology. Humana Press, Totowa, NJ, pp. 23–44, 1999.

Chandler D, Davidson G, Grant WP, Greaves J, Tatchell GM, Microbial biopesticides for integrated crop management: an assessment of environmental and regulatory sustainability. Trends Food Sci. Technol., Towards Sustainable Food Chains: Harnessing the Social and Natural Sciences. 19, 275–283, 2008.

Du Y, Li S, Zhang Y, Rempel C, Liu Q, Treatments of protein for biopolymer production in view of processability and physical properties: A review. J. Appl. Polym. Sci. 133, 2016.

Gan-Mor S, Matthews GA, Recent Developments in Sprayers for Application of Biopesticides—an Overview. Biosyst. Eng. 84, 119–125, 2003.

Hynes RK, Boyetchko SM, Research initiatives in the art and science of biopesticide formulations. Soil Biol. Biochem. 38, 845–849, 2006.

Karthikeyan OP, ChidambarampadmavathyK, CirésS, HeimannK, Review of Sustainable Methane Mitigation and Biopolymer Production. Crit. Rev. Environ. Sci. Technol. 45, 1579–1610, 2015.

Kumar KK, Sridhar J, Murali-Baskaran RK, Senthil-Nathan S, Kaushal P, Dara SK, Arthurs S, Microbial biopesticides for insect pest management in India: Current status and future prospects. J. Invertebr. Pathol., Global status of Microbial Control Programs and Practices 165, 74–81, 2019.

Mahdavi SA, Jafari SM, Ghorbani M, Assadpoor E, Spray-Drying Microencapsulation of Anthocyanins by Natural Biopolymers: A Review. Dry. Technol. 32, 509–518, 2014.

Mukherjee T, Kao N, PLA Based Biopolymer Reinforced with Natural Fibre: A Review.J. Polym. Environ. 19, 714, 2011.

Pagliano G, Ventorino V, Panico A, Pepe O, Integrated systems for biopolymers and bioenergy production from organic waste and by-products: a review of microbial processes. Biotechnol. Biofuels 10, 113, 2017.

Rajesh Banu J, Kavitha S, Yukesh Kannah R, Poornima Devi T, Gunasekaran M, Kim S-H, Kumar G, A review on biopolymer production via lignin valorization. Bioresour. Technol. 290, 121790, 2019.

Srikanth R, Reddy CHSSS, Siddartha G, Ramaiah MJ, Uppuluri KB, Review on production, characterization and applications of microbial levan. Carbohydr. Polym. 120, 102–114,2015.

Vieira MGA, da Silva MA, dos Santos LO, Beppu MM, Natural-based plasticizers and biopolymer films: A review. Eur. Polym. J. 47, 254–263, 2011.

Yates MR, Barlow CY, Life cycle assessments of biodegradable, commercial biopolymers - A critical review. Resour. Conserv. Recycl. 78, 54–66, 2013.

19 Microbial Metal Leaching

19.1 INTRODUCTION

Microbial metal leaching is the commercial extraction of rich metals from ores and mines by using microorganisms with the least effect on the environment. It is a simple and effective technology for metal extraction from low grade ores and mineral concentrates by the use of mostly mesophilic and moderately thermophilic microorganisms. Industrially it is a more convenient method thanthe conventional method. Microbes play an important role in bioleaching. This method is preferred because of its lower cost and useful by-product production. In industry this technique ismainly used to boost the process of copper and uranium extraction. 20% of the world's copper extraction is done with this technique. Mainly *Acidophilic – Sulfobacillus*, *Rhodococcus*, *Ferromicrobium* – and *Chemolithotrophic* microorganisms are used for bioleaching. Conventional pyrometallurgical techniques affect the environment by production of harmful gases, and depletion of ores (Asghari et al., 2013). In the microbial leaching process the oxidation of insoluble metal sulphides to soluble sulphates takes place and it also increases mineral extraction by isolating thermophile microorganisms. Extraction of gold is doubled by the use of diverse microbes during gold mining.Nowadays about 40 plants are in industrial use for copper, gold, zinc, cobalt, and uranium.

19. 2 PRINCIPLES OF MICROBIAL METAL LEACHING

The first acidophilic iron and sulphur-oxidizing bacteria was discovered and isolated in the late 1950s. In the 1960s the first industrial copper heap leaching operation started. The first industrial gold bioleaching plant started in the 1980s. At Bingham Canyon, Utah mine, the oxidation of pyrite, copper minerals and released copper into solution by culturing microorganisms from the leach stream was reported (Potysz et al., 2018). Microorganisms usually undergo chemosynthetic metabolism and this requires acidic pH for metal leaching (Figure 19.1). During the leaching process, microbes derive carbon dioxide and oxygen from the atmosphere. *T. thiooxidans, T. caldus, T. ferrooxidans* are included under new a genus *Acidithiobacillus*. *L. ferrooxidans* is not easily recognized in metal sulphide bio-oxidation because of its much slower growth rate than *A. ferrooxidans* in common ferrous iron-rich media.

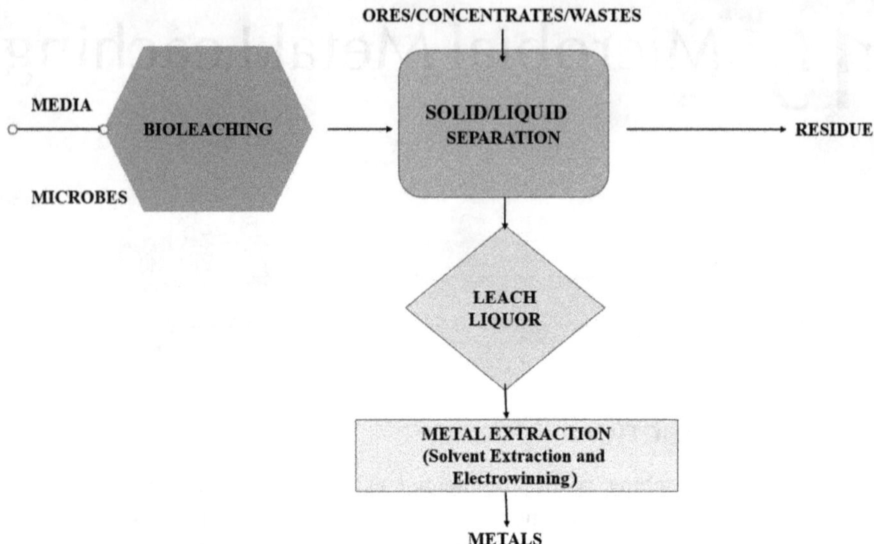

FIGURE 19.1 Schematic representation of the process of bioleaching.

On the other hand, this showed in some studies in 1978 that it can oxidize pyrite faster than *A. ferrooxidans*. It was also found that *L. ferrooxidans* had high affinity to ferrous iron (Km 0.25 mM as compared to 1.34 mM in *A. ferrooxidans)* and low sensitivity to inhibition by ferric iron (Km 42.8 mM compared to 3.10 mM in *A. ferrooxidans.)* (Asghari et al., 2013).

For the bioleaching process, mostly chemolithic microbes were isolated (Olson et al., 2003). The main aim is to use its mechanism for oxidation of inorganic substances, mainly sulphur-containing compounds. These bacteria live in harsh environments; they are mainly autotrophic and get carbon sources from inorganic substances rather than organic carbon sources. Most of bacteria are thermophiles, and some bacteria can even survive up to 50 °C. Almost all are lithotropic and can survive in highly soluble metallic environments and oxidize large amounts of substrate with high availability of CO_2 (Rohwerder et al., 2003).

19.2.1 *Thiobacillus ferrooxidans*

Thiobacillus ferrooxidans bacteria are mostly used for bioleaching. These bacteria were discovered from acid coal mine drainage. They are mesophilic and survive at40–48 °C and the optimum temperature is 70–75 °C. They can tolerate pH up to 1-6 and the optimum pH is 2. They can oxidize elemental sulphur, and soluble and insoluble sulphides. In the case of iron, the ferrous bivalent form is oxidized to the ferric trivalent form. Moreover, *T. thiooxidans* and *T. ferroxidans* are morphologically the same but in anaerobic conditions*T. ferrooxidans* use Fe as an electron acceptor (Gericke et al., 2009). A new species, acidophilic *Thiobacillus,* has been discovered; this is mostly used to mobilize copper from chalcopyrite. *T. prosperus* represents a new group of halotolerant metal-mobilizing bacteria. *T. cupurins* are facultative chemolithotrophic; they oxidize metal sulphide but not ferrous ions (Pathak et al., 2009).

19.2.2 Leptospirillum

However, after the discovery of *Leptospirillum ferrooxidans*, it became dominant over *Thermophilic ferroxidans* because it tolerates lower pH and a high concentration of molybdenum, uranium, and silver. Specifically, it is more sensitive to copper so it cannot oxidize sulphur. Consequently mobilization is doneusing *T. ferrooxidans* and *Thioxidans*. *L. ferrooxidans* is an obligatory chemolithotrophic ferrous iron-oxidizing bacterium. This bacterium was first isolated from mine water *T. thiooxidans, T. ferrooxidans,* and *L. ferrooxidans* are mesophilic bacteria which grow best at a temperature of 25–35 °C (Cui and Zhang, 2008). *Leptospirillum* is a superior iron-oxidizing bacterium in continuously stirred tank reactors. The predominance of *Leptospirillum* species over *A. ferrooxidans is* also relevant to column bioleaching of copper sulphide ore. Moreover, in the absence of iron *L. ferrooxidans* is superior to *A. ferrooxidans* and viceversa. Additionally, *Leptospirillum* divides into 2 groups. However, there is no significant physiological physical difference observed between them. Group 2 can grow at 45 °C. Group 3 is unable to grow at 45 °C. Industrially, isolates at the Fairview mine bio-oxidation tanks are members of Group 2. Group 3 includes slime in extreme acid mine drainage at an iron mountain and this investigation was carried outvia molecular analyses (Olson et al., 2003).

19.2.3 Ferroplasma

This is a new genus observed by Edward et al. in a low pH, high dissolved solids environment at an iron mountain. *Ferroplasma* is an iron-oxidizing archaeon. *Ferroplasma* also contributesa little in bio-oxidation in some areas. It comprises significant microflora at particular conditions. In three-reactor trains of polymetallic sulphide, bio-oxidation occurs at 45 °C. *Ferroplasma* is dominant in a second and third reactor; it occurs with increasingly acidic, high dissolved solid, organic carbon availability, and these conditions influence the growth of *ferroplasma.* It is assumed that it lies under the boundary of mesophilic and moderately thermophilic. The first reactor contains *Acidithiobacilluscaldus* (sulphur oxidation) and *Leptospirillum* (iron oxidation); from the first reactor, the extremely acidophilic, moderately thermophilic strain was isolated (Pathak et al., 2009).

19.2.4 Thermophilic Bacteria

These are moderately thermophilic bacteria and grow at a range of 50 °C. They grow on pyrite, pentlandite, and chalcopyrite. *Acidophilus* includes mainly *sulfolobus* and *ascidians. Sulfolobus* belongs to the Archaea domain and its characteristics are a round-shaped cell, rigid cell walls and it about 0.8–0.9 mm in diameter (Pathak et al., 2009). It is an aerobic, facultative chemolithotrophic bacterium, oxidizing ferrous iron, elemental sulphur, and sulphide minerals. Only in the presence of yeast extract can ferrous iron be utilized and used as an energy source. An extremely thermophilic bacterium growing above 60 °C was discovered. *Sulfobacillus thermosulfidoxidans* is a spore-forming facultative autotrophic bacterium and it is a member of the genus sulfobacillus; it also oxidizes ferrous iron, and elemental, sulphide minerals.

19.2.5 Acidianus

Obligate *acidophilus* especially eukaryotic bacteria form colonies with prokaryotic bacteria at low temperatures < 35 °C. These area very important source in the bioleaching process. This genus is mostly used for extraction of zinc, copper, molybdenum, aluminium, nickel, and iron etc. These bacteria ideally are used because they tolerate and survive in a highly acidic and toxic environment (Rohwerder et al., 2003). They are isolated in hot regions. *Acidianus brierleyi*, associated with the genus *sulfolobus,* is a chemolithoautotrophic, facultative aerobic extremely acidophilic archon growing on ferrous iron, elemental sulphur, and metal sulphides. Anaerobically, elemental sulphur is used as an electron acceptor and reduced to H_2S (Gericke, 2012).

Thiobacillus ferrooxidans has the ability for oxidation of ferrous iron aerobically and anaerobically. Sulphur found in elemental form in hot regions is oxidized to form sulphuric acid. *Thiobacillus* bacteria oxidize sulphides and sulphur-containing bacteria as a substrate. They occur in sulphur ore deposits. They produce acid, which increases the acidity of the environment and is favourable to other acidic bacteria (Pathak et al., 2009). *Thermothrix thiopara* are present in neutral conditions at temperatures ranges from 50–70 °C as well as oxidizing sulfhydryl ions, sulphite ions, elemental sulphur, and thiosulfate ions to sulphate ions (Tichy et al., 1998). It is found in volcanic fissures, hot springs, and other thermal regions. Chalcopyrite, pentlandite, covellite, and pyrite are divalent iron showing maximum growth.

19.3 LEACHING TECHNIQUES

This is done on a technical scale (Figure 19.1). It is a simple operative technique for processing of sulphide ores and also used for retrieval of copper and uranium. The activity of bacteria and composition of ore determine the profit and productivity of the industry. Before the use of technology, it is very important to know the optimum conditions for each type of ore (Rohwerder et al., 2003).

19.3.1 Percolator Leaching

The first experiments were carried out in airlift percolators. These were made of simple glass tubes provided in the bottom part with a sieve plate and filled with ore particles. Ore packing was irrigated or flooded with a nutrient sieve plate and filled with ore particles. This was inoculated with bacteria. For recirculation, the leach liquor trickling through the column was pumped up by compressed sterile air to the top of the aeration system. Determination of leaching process liquid samples was made at intervals and was determined based on pH, microbiological investigation, and chemical analysis (Potysz et al., 2018). This includes *in situ* leaching, vat leaching, and heap leaching. *In situ* leaching involves pumping of air inside, forced under pressure into mines and ore deposits making it permeable through explosive charging. Consequently, in wells metal leach solutions were obtained which are drilled and recovered. Various bodies come under *in situ* leaching are as follows:

1) Deep deposits below water tables.
2) Surface deposits above water tables.
3) Surface deposits below water tables.

19.3.2 Submerged Leaching

This technique is used to replace percolator leaching when the oxygen supply is not sufficient and the surface ratio is unfavourable. This technique containsfine-grained material (particle size <100 micrometres) which is suspended in leaching liquid and kept in motion by shaking and stirring. This suspension technique is carried out in Erlenmeyer flasks or in a more sophisticated manner in a bioreactor. Economically, shortened time and metal extraction are important steps determined by higher rate aeration, and more accurate monitoring and control of parameter favour growth and activity of bacteria. Along with mechanically stirred systems, airlift reactors have been proved suitable for the treatment of ore concentration, industrial waste products, and biodesulphurization of coal (Olson et al., 2003).

19.3.3 Column Leaching

This is a laboratory method. It uses glass, plastic, lined concrete, and steel material, and it depends on size or capacities from several kilograms to a few tons. The column is regularly recharged from the top with leaching solution and collected at the bottom, usually filtered, and next it is analysed for contaminants. This leaching process operates under the principle of percolator leaching and is used as a model for heap or dump leach processes. These techniques determine the type and concentration of contaminants likely to be mobilized from a wetted solid sample and moved into the environment. This method is more expensive, time-consuming, and labour-intensive than batch leaching. The advantage is it can allow observation of long-term chemical reactions between solid samples and leachates. It also containsa special instrument for measuring temperature, pH, humidity, oxygen, or carbon dioxide. It makes clear the concept of what to expect in heap and dump leaching and optimum requirements for leaching conditions (Rohwerder et al., 2003).

19.3.4 Industrial Leaching

Naturally occurring bioleaching process is very slow. For commercial extraction of metal by bioleaching the process is optimized by controlling pH, temperature, humidity, O_2 and CO_2 concentrations. In industry, the technique is used for extraction of low-grade ore which normally contains 0.5%(w/w). Initially, material is piled in a heap (uncrushed waste ore), water is allowed to trickle, and the seepage is collected. However, sulphur oxidation requires time; hence leachate is recirculated. There are three main procedures in use: dump leaching, heap leaching, and underground leaching (Gericke et al., 2009).

Dump leaching: This is the oldest process. A pile of uncrushed waste ores and rocks is used for leaching. The size of the dumps varies and the amount of ore can be 100–1,000 tons. The procedure starts with sprinkled lixiviant, which may be water, acidified water, or acidic ferric sulphate. Moreover, the recirculation process helps bacteria and ferric iron to regenerate by leachate oxidation. However, economically copper recovery is too low (Rohwerder et al., 2003).

Heap leaching: For the formulation of ore firstly size is reduced to fine-grained ores to enhance the interaction between mineral and lixiviate (Figure 19.2). A large basin can accommodate 12,000 tons of sample. A heap of crushed ore is placed on an impermeable base which prevents loss of lixiviate and water pollution. Lixiviate is a chemical solution used to accelerate the dissolution of ore in the mining process. When the heap is arranged on a linear impermeable membrane, lixiviate percolates and consequently it reaches the bottom and leachesmetal (Tichy et al., 1998). During settlement of the fine-grained ore, pipelines are arrange strategically to give a deeper region to a heap and to provide a sufficient amount of O_2. Generally, cyanide or sulphuric acid is preferred. Recovery is done by solvent extraction and electrowinning. Both of these methods use lixiviate from the top of the heap or dump to leach metals. Due to gravity flow, it flows down. Because of sulphuric acid, it lowers the pH of ore and enhances or makes favourable conditions for *acidophilus spp*. Furthermore, acid

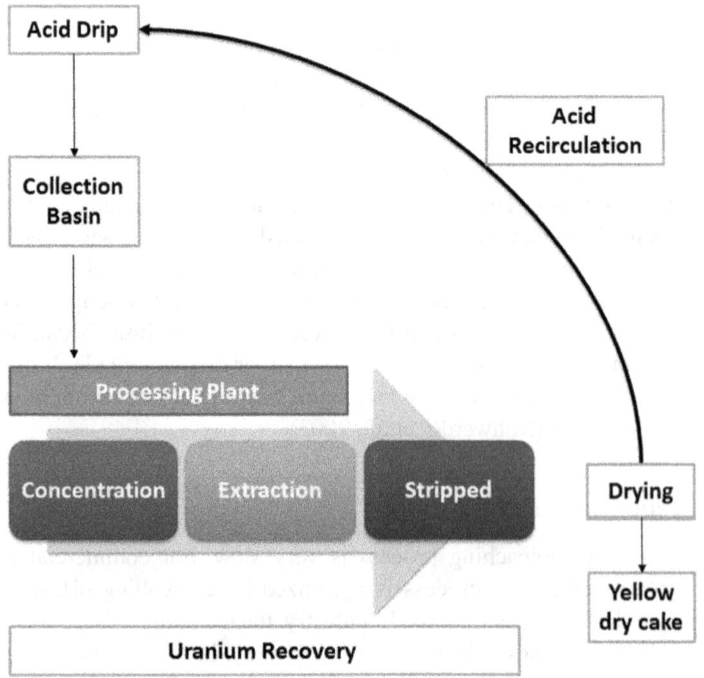

FIGURE 19.2 Schematics of heap leaching.

is collected from the bottom and transferred to the recovery centre and is recovered by using electrolysis and solvent extraction methods. Bioreactors can enhance the recovery of metal but are costly. Nowadays this method is used for oxides of ores and precious metals are actively extracted (Asghari et al., 2013).

Underground leaching: Stan rock uranium mine at Elliot Lake in Ontario, Canada, is the best example to explain underground leaching. This is mainly done in unrestrained mines. Galleries are flooded, or unmined ore or mine waste work on the principle of washing under pressure or side-tunnel sprinkling. Conventional methods cannot work on ore deposits because they are too low grade or too small to be leached *in situ*. Adequate bacteria are implanted into bore holes to fracture the ore body. After some time, leachate is pumped out from neighbouring wells or collected in drifts. Impermeability of ore body and permeability of rock gangue is important to prevent seepage of pregnant leachate (Olson et al., 2003).

Tank leaching: This is based on a submerged leaching mechanism, and produces a high yield. Therefore in early investigations, flasks are replaced by a bioreactor. It is more effective for the extraction of zinc; up to 80% from zinc sulphide. Nowadays this technique is used for the extraction of gold (Gericke et al., 2009).

19.4 FACTORS AFFECTING THE BIOMINING/BIOLEACHING OF METALS

Optimum pH and optimum temperature are the most important factors. These decide the efficiency of bioleaching. Only in optimum conditions can maximum yield be achieved. It is an inherent property of microbes to respond to temperature and pH (Gericke et al., 2009).

Effect of pH: pH depends on the different tolerance capacities of microbes. Generally *T. ferrooxidans and Leptospirillum ferrooxidans* are used for bioleaching. The significance of these microbes is oxidation potential Fe^{3+}/Fe^{2+}. Optimum pH is below 2.0–2.5. pH values below 2.0 are unfavourable for *T. ferrooxidans* but it can be adopted by increasing the concentration of acid. Also, *L. ferrooxidans* has a more acidic resistant ability and can grow at a pH of 1–2. Biooxidation requires a pH of 2.5–3 for maximum results. The rate of biooxidation decreases significantly if the pH is not in this range since the activity of acidophilic bacteria is reduced.

Nutrients: Bacteria used for bioleaching are mostly chemolithoautotrophic. They obtain their nutrients from inorganic compounds. Inorganic compounds are available from the environment and material to be leached. Iron, sulphur, ammonia, phosphorus, and magnesium salts are the additional supplements require for optimum growth (Asghari et al., 2013).

O_2 and CO_2: Growth and activity also depend on O_2 and CO_2. O_2 can be provided by aeration, stirring, or shaking. CO_2 is only a carbon source in the process but there is no need for addition.

Temperatures: For bio-oxidation of iron and sulphide *T. ferrooxidans* can grow at 28–30 °C. Moreover, it is observed that temperatures less than 4 °C can be toleratedby this bacterium in the presence of nickel, zinc, cobalt, or copper (Tichy et al., 1998). *Leptospirillum* can survive up to 20–45 °C. Generally, thermophilic bacteria can tolerate up to 50–80 °C. Gold mining is doneata temperature of 40 °C and the pH range is 1–6.

Heavy metals: Thiobacillus species can tolerate high metal sulphide concentrations such as 50 g/L Ni, 55 g/L Cu, and 1/2 g/L Zn. By gradually increasing metal concentration, it is easier for bacteria to adopt the concentrate environments (Gericke, 2012).

Surfactants and organic extractants: These decrease the activity of bacteria, surface tension and mass transfer of O_2. Solvent extraction methods are significant for concentration and recovery of metals. Solvent extraction and bacterial leaching are coupled to become enriched in an aqueous phase, and this needs to be removed before a barren solution is recirculated to the leaching operation (Rohwerder et al., 2003).

Choice of bacteria: This is the most important factor that determines the success of bioleaching. Suitable bacteria that can survive at high temperatures, acid concentrations, high concentrations of heavy metals, and remain active under such circumstances, are to be selected to ensure successful bioleaching.

Crystal lattice energy: This determines the mechanical stability and degree of solubility of the sulphides. Sulphide ores with lower crystal lattice energy have higher solubility, and hence are easily extracted into solution by the action of bacteria.

Aeration: The bacteria used in biomining are aerobic and thus require an abundant supply of oxygen for survival and growth. Oxygen is provided by aerators and pipes. Mechanical agitation is also an effective method to provide a continuous air supply uniformly and also to mix the contents.

Solid–liquid ratio: The ratio of ore/sulphide to the leach solution (water + acid solution + bacteria inoculum) should be maintained at optimum level to ensure that biooxidation proceeds at maximum speed. The leach solution containing leached minerals should be removed periodically and replaced with new solution.

Surface area: The rate of oxidation by the bacteria depends on the particle size of the ore. The rate increases with reduction in size of the ore and vice versa.

Ore composition: Composition of ore such as concentration of sulphides, amount of mineral present, and the extent of contamination, has a direct effect on the rate of bio-oxidation being selected. The rate of bio-oxidation is reduced significantly if the temperature is above or below the optimum temperature.

19.5 MECHANISM OF BIOLEACHING OF METALS

Bacteria help in regenerating major ore oxidizers, mostly ferric ions. This reaction takes place in the cell membrane of bacteria. In the first step, disulphide is spontaneously oxidized to thiosulfate by ferric iron (Fe^{3+}), which in turn is reduced to give ferrous iron (Fe2+):

$$FeS_2 + 6Fe^{3+} + 3H_2O \longrightarrow 7Fe^{2+} + S_2O_3^{2-} + 6H^+$$

In the second step microorganisms catalyse the oxidation of ferrous iron and sulphur, to produce ferric iron and sulphuric acid:

$$Fe^{2+} + 1/4O_2 + H^+ \rightarrow Fe^{3+} + 1/2\ H_2O$$

$$S + 3/2O_2 + H_2O \rightarrow H_2SO_4$$

Thiosulfate is also oxidized by bacteria to give sulphate:

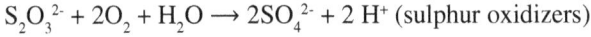

$$S_2O_3^{2-} + 2O_2 + H_2O \longrightarrow 2SO_4^{2-} + 2\,H^+ \text{ (sulphur oxidizers)}$$

The ferric iron produced in reaction (2) oxidizes more sulphide as in reaction (1), closing the cycle and given the net reaction

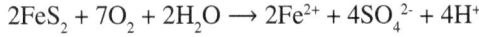

$$2FeS_2 + 7O_2 + 2H_2O \longrightarrow 2Fe^{2+} + 4SO_4^{2-} + 4H^+$$

Bioleaching mostly depends on *T. ferrooxidans, L. ferrooxidans, T. Thiooxidans.* These convert heavily soluble metal sulphides into water-soluble metal sulphates. Two methods have been invented: direct and indirect from bacterial leaching (Gericke, 2012).

19.5.1 Direct Bacterial Leaching

There is physical contact between the bacterial cell and mineral sulphides occur. Enzymes catalyse the reaction by oxidation of sulphates (Figure 19.3).

$$4FeS_2 + 14O_2 + 4H_2O \text{ -------> } 4FeSO_4 + 4H_2SO_4$$

$$4FeSO_4 + O_2 + 2H_2SO_4 \text{ ------> } 2Fe(SO_4)_3 + 2H_2O$$

Summary of oxidation of pyrite:

$$4FeS_2 + 15O_2 + 2H_2O \text{ -------> } 2Fe_2(SO_4)_3 + 2H_2SO_4.$$

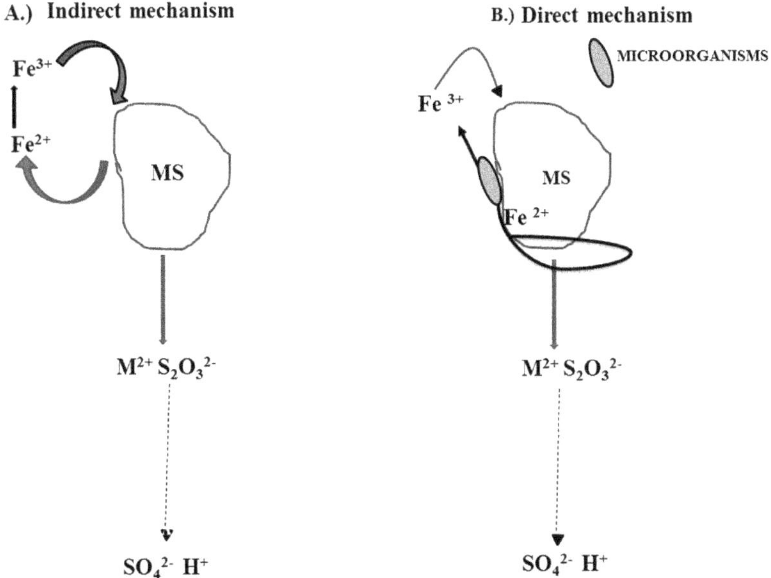

FIGURE 19.3 Mechanisms involved in bioleaching.

Torm discovered that some non-iron metal sulphides can be oxidized by *T. ferrooxidans*. Furthermore, there is some evidence that bacteria do not attach to the whole mineral; they prefer a specific site of crystal imperfection and solubilization is carried out by electrochemical interaction. Much has been discovered about the mechanism of attachment of bacteria and solubilization of minerals (Olson et al., 2003).

Indirect bacterial leaching: this technique works by production of lixiviant which oxidizes the sulphidemineral. In the example below metal solubilization is done by ferric iron as a lixiviant (Potysz et al., 2018).

$$MeS + Fe_2(SO_4)_3 \text{ -------> } MeSO_4 + 2FeSO_4 + S$$

To keep more iron in the environment the pH should be below 5. Also, reoxidation is done by *T. ferrooxidans* and *L. ferrooxidans* by conversion of ferrous iron to ferric iron. These bacteria play the role of catalytic activity. However, in the absence of bacteria, oxidation becomes slower. This was proved by experiments done in the range of 2–3 pH. Bacterial oxidation of ferrous iron is about 10^5–10^6 times faster than chemical oxidation. The sulphur produced in the above reaction is simultaneously oxidized to sulphuric acid by *T. ferrooxidans* but oxidation done by *T. thiooxidans* and *T. ferrooxidans* is much faster (Asghari et al., 2013).

$$2S + 3O_2 + 2H_2O \text{ -------> } 2H_2SO_4$$

The role of *T. thiooxidans* is just to create an acidic environment to make favourable conditions for *T. ferrooxidans* and *L. ferrooxidans.* According to the newest publication of Sand and co-workers, there is uncertainty about the existence of indirect leaching mechanisms (Olson et al., 2003).

19.5.2 Bioleaching of Gold

Microbial leaching of refractory-processed metal ores to enhance gold and silver recovery is one of the promising applications. The gold ores used here are calaverite ($AuTe_2$), sylvanite *Ag.* $AuTe_2$, petzite (Ag_3AuTe_2), and arsenopyrite/pyrite. It is difficult to extract gold from low-grade sulphidic ores. Gold ores need to be pretreated by roasting or by pressure oxidation to free the gold before cyanide leaching. This pretreatment is costly. After this 70–95% of the gold in the ore can be recovered by the cyanide leaching process. The sodium cyanide leaching process converts gold to a soluble cyanide complex. Silver is also obtained by bioleaching of arsenopyrite but it is more readily solubilized than gold during microbial leaching of iron sulphide. Bioleaching occurs in reactors or heap leaching processes using *Thiobacillus ferrooxidans.* After leaching, the porous ore of exposed gold is converted for cyanide leaching (Rohwerder et al., 2003).

$$4Au + 8NaCN + 2H_2O + O_2 \text{ -------> } 4Na\,[Au(CN)_2] + 4NaOH$$

$$2Na[Au(CN)_2] + Zn \text{ -------> } Na2[Zn(CN)_4] + 2Au$$

19.5.3 Bioleaching of Uranium

The uranium ores used are uraninite or pitchblende (UO_2), and brannerite (UTi_2O_6). Uranium ores occur in low-grade ores and are insoluble. They can be converted to leachable form by oxidation with ferric ions. Fe^{2+} is reoxidized by *Acidithiobacillus ferroxidase*. The ferrous ions produced during uranium oxidation are converted back to Fe^{3+} by chemical oxidants, such as chlorate, manganese dioxide, or hydrogen peroxide, and this ion in turn acts as an oxidant to convert UO_2 chemically to the leachable UO_2SO_4 (Asghari et al., 2013).

$$UO_2 + 2Fe^{3+} + SO_4^{2-} \text{------>} UO_2SO_4 + 2Fe^{2+}$$

$$U^{4+} + 2Fe^{3+} \text{------>} U^6 + 2Fe^{2+}$$

19.6 ADVANTAGES AND BOTTLENECKS OF MICROBIAL METAL LEACHING

Microbial mining or leaching is a simple and inexpensive technology. It is an eco-friendly process which can be carried out at ambient temperature and pressure. In addition, no poisonous sulphur dioxide is emitted during microbial leaching as in smeltersand other chemical processes. It is used to extract refined and expensive metals, which is not possible by other chemical processes. This type of technology is ideal for low-grade sulphide ores. Itis usually employed for collecting metals from waste and drainages. However, this process has several bottlenecks: it takes about 6–24 months or longer; has a very low yield of mineral, requires a large open area for treatment, and may have no process control with a high risk of contamination. This process has inconsistent yield because bacteria cannot grow uniformly. Key players in the global bioleaching market are Rio Algom Ltd, BHP, Teck Resources Ltd, and Nyrstar NV. The leading countries are US, Canada, Mexico, Germany, UK and France.

19.7 CONCLUSIONS

Microbial metal leaching is a better alternative to conventional methods. Microbes such as bacteria and fungi convert metal compounds into their water-soluble forms and are biocatalysts of these leaching processes. Due to the nature of numerous toxicants, the application of waste sludge on land as a fertilizer is related to environmental issues. Under low pH conditions established in the sludge, the bioleaching cycle will eliminate the poisonous compounds and pathogens present in the sludge and thereby make the sludge useful as a fertilizer. The bioleaching process is based on the action of the genus *Acidithiobacillus* which directly and indirectly transforms reduced iron and sulphur compounds into soluble sulphate. Most bioleaching experiments have currently been performed in batch mode and under laboratory conditions. More in-depth experiments for large-scale deployment in real-field environments bioleaching are needed in a continuous mode of operation coupled with digestion of the sludge so that a large amount of sludge can be handled in a single step.

REFERENCES

Asghari I, Mousavi SM, Amiri F, Tavassoli S, Bioleaching of spent refinery catalysts: A review. J. Ind. Eng. Chem. 19, 1069–1081, 2013.

Cui J, Zhang L, Metallurgical recovery of metals from electronic waste: A review. J. Hazard. Mater. 158, 228–256, 2008.

Gericke M, Review of the role of microbiology in the design and operation of heap bioleaching processes. J. South. Afr. Inst. Min. Metall. 112, 1005–1012, 2012.

Gericke M, Neale JW, van Staden PJ, A Mintek perspective of the past 25 years in minerals bioleaching. J. South. Afr. Inst. Min. Metall. 109, 567–585, 2009.

Olson GJ, Brierley JA, Brierley CL, Bioleaching review part B: Appl. Microbiol. Biotechnol. 63, 249–257, 2003.

Pathak A, Dastidar MG, Sreekrishnan TR, Bioleaching of heavy metals from sewage sludge: A review. J. Environ. Manage. 90, 2343–2353, 2009.

Potysz A, van Hullebusch ED, Kierczak J, Perspectives regarding the use of metallurgical slags as secondary metal resources – A review of bioleaching approaches. J. Environ. Manage. 219, 138–152, 2018.

Rohwerder T, Gehrke T, Kinzler K, S and W, Bioleaching review part A: Appl. Microbiol. Biotechnol. 63, 239–248, 2003.

Tichy R, Rulkens WH, Grotenhuis JTC, Nydl V, Cuypers C, Fajtl J, Bioleaching of metals from soils or sediments. Water Sci. Technol. 37, 119–127, 1998.

20 Aerobic Effluent Treatment Processes, Biohythane Processes and Biofertilizers

The Central Pollution Control Board of India carried out a survey of the industries responsible for water pollution problems in India and they identified mostly chemical and biochemical industries. The question arises as to how industrial effluents pollute our water streams. When wastewater mixes with water streams, microorganisms present in the water will use the biodegradable organic matter for their growth and metabolism. This causes depletion of the dissolved oxygen (DO) concentration of the water and many toxic metabolites which are not only detrimental to the growth of aquatic plants and animals but also create problems in our drinking water sources. Therefore all chemical and biochemical industries must have wastewater treatment plants in order to reduce the biodegradable organic matter present in wastewater to keep our environment safe. More than 70% of these wastewater treatment plants are controlled through biological means. These biological treatment processes are classified as aerobic and anaerobic. Carbon and energy analyses of aerobic and anaerobic microbial degradation processes are shown in the Table 20.1 (Das and Sen Gupta, 1990).

20.1 CHARACTERISTICS OF WASTEWATER

The characteristic of wastewater are determined with respect to physical, chemical or biological parameters as listed below.

- Physical Characteristics: temperature, turbidity, colour, odour etc.
- Chemical Characteristics: Chemical Oxygen Demand (COD), total organic carbon (TOC), nitrogen, total solid, soluble solid etc.
- Biological Characteristics: Biochemical Oxygen Demand (BOD), oxygen required for nitrification, most probable number (MPN) etc.

The temperature of wastewater should be maintained in such a way that growth of the aquatic plants and animals is not affected. Turbidity of the wastewater is a measure of contamination. This is determined by the degree by which the water loses its transparency due to the presence of suspended particulates. The colour of the effluent discharged by different textile industries varies. Odours of wastewater are mostly due to the presence of different organic acids (Metcalf and Eddy, 1991).

TABLE 20.1
Carbon and energy analysis of aerobic and anaerobic microbial degradation processes

	Aerobic condition	Anaerobic condition
Carbon balance	About 50% converted to biomass and 50% into CO_2	About 95% decomposed into biogas (CO_2 and CH_4) and 5% incorporated into biomass.
Energy balance	Approximately 60% stored in the form of new cells and 40% is lost as process heat	Almost 90% of the energy in organic material can be recovered in biogas, 5-7 % is used for the growth of the cell and 3-5 % is wasted as heat.

Chemical oxygen demand (COD) is a measure of water and wastewater quality. The COD is the amount of oxygen consumed to chemically oxidize organic materials present in the water. The COD test is often used to monitor the efficiency of wastewater treatment plants. The basis for the COD test is that nearly all organic compounds can be fully oxidized to carbon dioxide and water in the presence of a strong oxidizing agent under acidic conditions. The stoichiometry of this oxidation process is shown in Eq. 20.1 (AOAC, 1980).

$$C_nH_aO_bN_c + \left(n + \frac{a}{4} - \frac{b}{2} - \frac{3}{4}c\right)O_2 \rightarrow nCO_2 + \left(\frac{a}{2} - \frac{3}{2}c\right)H_2O + NH_3 \quad (20.1)$$

Acidified potassium dichromate (using sulfuric acid) is a strong oxidizing agent. The reaction of potassium dichromate with organic compounds is represented as follows:

$$C_nH_aO_bN_c + d\,Cr_2O_7^{2-} + (8d + c)\,H^+ \rightarrow CO_2 + \frac{a + 8d - c}{2}H_2O + c\,NH_4^+ + 2d\,Cr^{3-} \quad (20.2)$$

During oxidation of the organic substances present in the wastewater sample, potassium dichromate is reduced (since in all redox reactions, one reagent is oxidized and the other is reduced), forming Cr^{3+}. Determination of the amount of Cr^{3+} produced in the wastewater sample is an indirect measure of the organic content of the wastewater sample. The total time of estimation of the COD value of wastewater samples is around 2.5 h. This parameter is mostly used to find the pollutants present in wastewater samples. Other chemical parameters are: total organic carbon (TOC), nitrogen, total solid, soluble solid etc. Total solid present in the wastewater is classified as volatile solid and non-volatile solid. Total solid is estimated by drying the wastewater sample. This dry solid is burned in the presence of oxygen at 600 °C in a muffle furnace. The remaining material is known as ash or non-volatile material. The volatile solid is calculated as shown in Eq. 20.3.

$$\text{Volatile solid} = (\text{total solid} - \text{non-volatile solid or ash}) \tag{20.3}$$

Volatile solids mostly comprise organic materials. Biochemical oxygen demand (BOD) is the amount of dissolved oxygen required by aerobic microorganisms to break down organic materials present in the water or wastewater sample at certain temperature over a specific period of time. BOD indicates the biodegradable organic matter present in the water or wastewater sample. BOD is an experimental parameter which is determined by measuring the dissolved oxygen concentration of the water sample before and after the incubation of the aerobic microbial culture for a specific time. Total BOD (known as ultimate BOD, BOD_u) of the sample requires 20 d, which is a quite long time. Therefore the BOD_5 parameter is used for characterizing wastewater samples. BOD_5 is the total amount of oxygen consumed by microorganisms during the first five days of biodegradation. Usually BOD_5 is approximately equal to 70% of BOD_u. COD of wastewater samples is generally greater than BOD_u because COD includes the total amount of oxygen required for the oxidation of both biodegradable and non-biodegradable materials present in wastewater. Therefore most of industries prepare calibration curves between the COD and BOD of their wastewater samples as shown in Figure 20.1 (Das and Varanasi, 2019; Das and Sen Gupta, 1990).

Most Probable Number (MPN) is a statistical method which is used to estimate the concentration of viable coliform bacteria in a water or wastewater sample by means of replicate wastewater sample in ten-fold dilutions. MPN is commonly used to determine the quality testing of water i.e. to ensure whether the water is safe or not in terms of bacteria contamination. A higher value of MPN is an indicator of the drinking water supply line within the sanitary system, because every human being discharges millions of Coli forms of bacteria every day that are a source of pathogens.

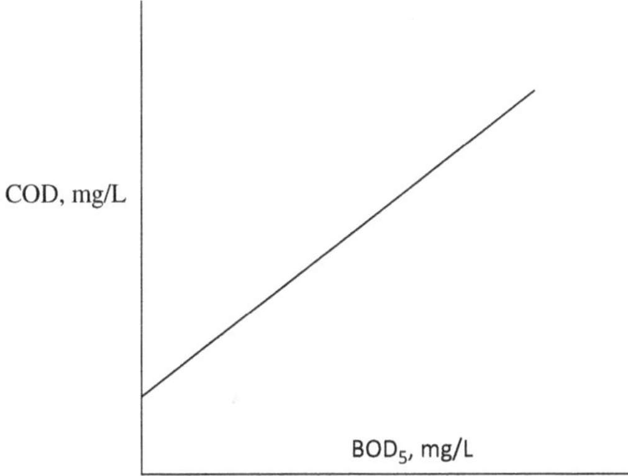

FIGURE 20.1 Correlation between COD and BOD_5 of the industrial wastewater sample.

TABLE 20.2
Raw water quality for drinking and bathing

	WHO standard	International Standard (IS)
pH	6.9	6.9
TDS	≤1500 mg/L	-
Iron	≤ 50 mg/L	-
Nitrate as NO_3	≤ 45 mg/L	-
Fluoride	≤ 1.5 mg/L	≤ 1.5 mg/L
BOD	≤ 6 mg/L	≤ 3 mg/L
COD	≤ 10 mg/L	-
DO	-	% solubility 40
Phenolic substances	≤ 0.002 mg/L	≤ 0.001 mg/L
Cyanide	≤ 0.2 mg/L	≤ 0.1 mg/L
Chromium	≤ 0.05 mg/L	≤ 0.05 mg/L
Lead	≤ 0.05 mg/L	≤ 0.1 mg/L
Arsenic	≤ 0.05 mg/L	≤ 0.2 mg/L
Chlorides	-	≤ 600 mg/L
MPN count	≤ 5000-10000/L drinking source. Normal 5000/L for bathing	Max permissible 20000/L

The raw water quality for drinking and bathing is the same as shown in Table 20.2.

20.2 AEROBIC EFFLUENT TREATMENT PROCESS

The aerobic process is largely practiced for the treatment of low-strength industrial effluent or sewage water. The rate of degradation of biodegradable organic materials is very fast in the case of aerobic wastewater treatment processes as compared to that of anaerobic processes. Bacteria are insoluble material which can convert soluble organics into cell mass and carbon dioxide. In the aerobic process, 50% carbon present in the wastewater is converted to biomass and 50% to carbon dioxide, whereas in the anaerobic process, 95% decomposes into biogas which comprises methane and carbon dioxide and only 5% is converted to biomass (Table 20.1). Cell mass can be separated by the sedimentation process, and soluble organic matter present in the wastewater is reduced significantly. Pollution control boards of different countries have formulated the rules and regulations as Minimum National Standards (MINAS) for the disposal standards of wastewater generated by different industries. For example, brewing industry treated effluent should have BOD_5 of 30 mg/L and total suspended solids of 100 mg/L as the disposal standard in the water course in India (Metcalf and Eddy, 1991; Buch, 1971).

20.2.1 Lagooning Process

Different aerobic wastewater treatment processes are available. The oldest wastewater treatment process is the lagooning technique. This is a natural treatment technique

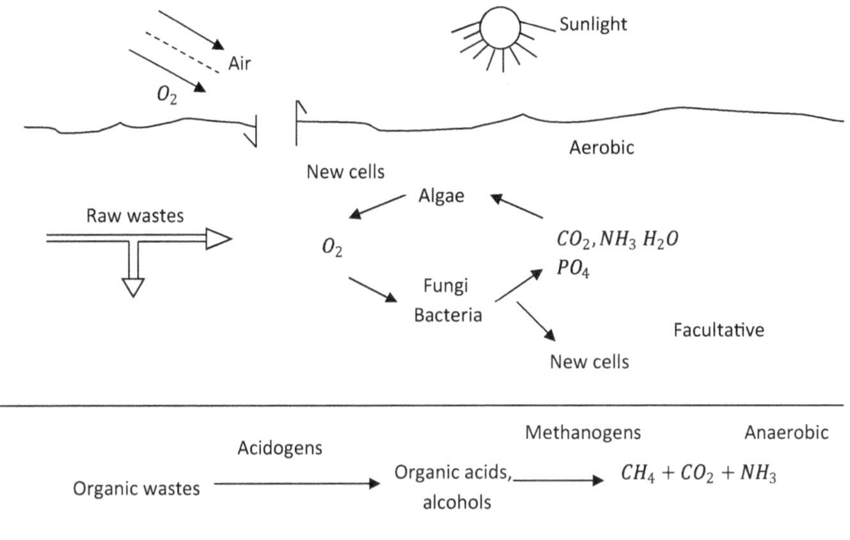

FIGURE 20.2 Lagooning process.

where wastewater is accumulated in ponds or basins, known as biological or stabilization ponds (Figure 20.2). A series of biological, biochemical and physical processes takes place. The wastewater is kept in shallow ponds and aerobic organisms grow and multiply for the degradation of biodegradable organic materials. The residence time in ponds from varies from 2 to 10 d depending on the organic load in the wastewater. Usually in the lagooning process, low water height is used for the growth of the aerobic organisms because oxygen transfer takes place via natural diffusion processes.

20.2.2 Activated Sludge Processes

Activated sludge contains different types of bacteria and other microorganisms. These microorganisms perform the following activities under aerobic conditions:

- Oxidization of organic compounds
- Promotion of coagulation and flocculation and conversion of dissolved, colloid and suspended solids into settleable solids

Activated sludge processes (ASP) are basically a CSTR process with cell mass recycling. The following operations are carried out in an activated sludge process:

- Sewage water is passed through the primary sedimentation tank (PST). The retention time is about 1.5 h.
- Settled sewage sludge from the PST is mixed with the required amount of activated-sludge in the aeration tank. The mixture of activated sludge and wastewater in the aeration tank is known as mixed liquor suspended solids

(MLSS). However, mixed liquor volatile suspended solids (MLVSS) are considered to be cell mass.

- The hydraulic retention time (HRT) is about 6–8 h. 8 m^3 of air is required for the treatment of 1 m^3 of wastewater. The volume of sludge returned to the aeration basin varies from 20 to 30% of wastewater.
- The aerated MLSS results in the formation of floc particles, ranging in size from 50 to 200 μm. MLSS is removed in the secondary sedimentation tank (SST) by gravity settling, More than 99% of suspended solids can be removed in the SST.
- A portion of the settled sludge is returned to the aeration tank (and is called cell mass recycling) to keep the population of microbes in the vessel constant in order to keep degradation of the organic compound constant under steady state condition. The excess sludge produced in the ASP is taken out from the process, known as sludge bleeding.
- Clear water is discharged into the water course.

Zoogloea ramigera is the key organism in ASP. Typical characteristics of activated sludge bacteria are the formation of activated sludge flocs. The *Zoogloea* has Greek-language origin and translates as living glue; this glue is due to polysaccharide gels. *Zoogloea ramigera* is an aerobic and chemoorganotrophic Gram-negative, straight and slightly curved, rod shaped and non-spore forming bacterium (Metcalf and Eddy, 1991; Green and Sheelef, 1980).

A schematic diagram of the ASP is shown in Fig. 20.2. The detailed mathematical analysis of ASP has been discussed in the Chapter 5. The volume of the bioreactor of ASP can be determined by using Eq. 5.80 as follows:

$$V = \frac{Y_{x/S}\left(S_0 - S\right)\theta_c F_0}{x\left(\theta_c \mu_d + 1\right)}$$

In the case of chemostats, $D = \mu$, under steady state conditions and sterile feed. On the other hand, in the case of chemostats with cell mass recycling, $D > \mu$.

This may be justified from the following mathematical analysis of the ASP.

The names of the different operational parameters shown in Figure 20.3 are given below:

F_0 = feed flow rate, S_0 = initial substrate concentration, x_0 = cell mass concentration in the feed, F_a = overall feed flow rate, S = steady-state biodegradable substrate concentration x = steady-state MLVSS concentration, F_r = recycle feed flow rate, x_u = settled MLVSS concentration in the cell separator, F_0 = wastewater feed flow rate, x_e = effluent MLVSS concentration, V = volume of the bioreactor, and F_w = rate of sludge bleeding.

At steady state, the cell mass balance across the bioreactor can be written as follows:

Rate of cell mass input + Rate of cell mass formation = Rate of cell mass output + Rate of cell mass accumulation + Rate of cell death

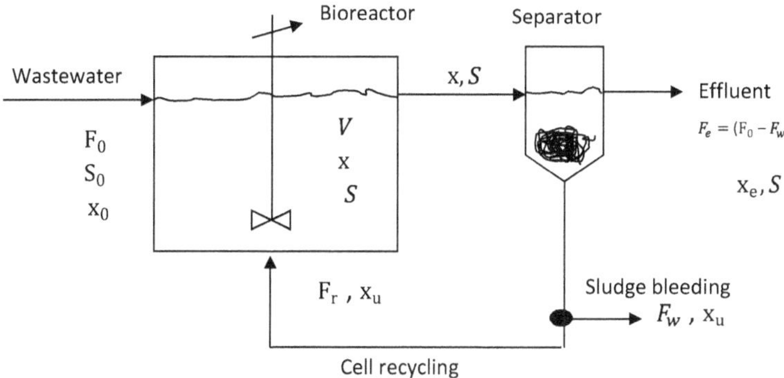

FIGURE 20.3 Activated Sludge Process (ASP).

$$\left(F_0\ x_0 + F_r\ x_u\right) + V\frac{dx}{dt} = F_a\ x + 0 + 0 \tag{20.4}$$

Now, $\alpha = \frac{F_r}{F_0}$; where α is the recycle ratio

$$\text{Thus, } F_r = \alpha\ F_0 \tag{20.5}$$

$$\text{Also, } F_a = F_0 + F_r \tag{20.6}$$

$$\text{Therefore, } F_a = F_0 + \alpha F_0 \tag{20.7}$$

$$\text{Hence, } F_a = F_0\left(1+\alpha\right)F_0 \tag{20.8}$$

Eq. 20.4 can be written as

$$\left(F_0\ x_0 + \alpha\ F_0\ x_u\right) + V\frac{dx}{dt} = F_0\left(1+\alpha\right)x + 0 + 0 \tag{20.9}$$

In chemostats with MLVSS recycling, assuming no cells present in the feed, $x_0 = 0$; Eq. 20.9 becomes

$$\alpha\ F_0 x_u + V\ \mu\ x = F_0\left(1+\ \alpha\right)\ x \tag{20.10}$$

(since $\frac{dx}{dt} = \mu\ x$)

Dividing Eq. 20.10 by V, we get

$$\alpha \frac{F_0}{V} x_u + \mu x = \frac{F_0}{V}(1+\alpha)\, x \qquad (20.11)$$

$$\alpha D x_u + \mu x = D(1+\alpha)\, x \qquad (20.12)$$

(Since $\frac{F_0}{V} = D$)

Now, assuming $C = \frac{x_u}{x}$, where C is the concentration ratio.

Therefore, $x_u = Cx$ and inserting into Eq. 20.12, we get

$$\alpha D C x + \mu x = D(1+\alpha)\, x \qquad (20.13)$$

By rearranging we get; $\mu = D\,[1+\alpha(1-C)]$ (20.14)

In Eq. 20.14, $\alpha < 0$ and $C \gg 1$, $\frac{\mu}{D} < 0$ i.e. $D > \mu$ (20.15)

We have seen in Chapter 5 that in a chemostat D_{max} is very close to $D_{wash\ out}$. Therefore, in the case of chemostats with cell recycling (ASP), the process can be easily operated at D_{max} because cell washout problems will not arise. The correlation between $\frac{\mu}{D}$ vs. α is shown in Figure 20.4 at different F_e/F_0 ratios. At any value of α with $F_e/F_0 > 0$, $\frac{\mu}{D}$ is less than 1 i.e. $D > \mu$ (Aiba et al., 1973; Das and Das, 2019).

At steady state, the substrate mass balance across the bioreactor in ASP can be given as

Rate of substrate input + Rate of substrate formation = Rate of substrate output + Rate of substrate consumption + Rate of substrate accumulation

$$\left(F_0 S_0 + F_r S\right) + 0 = F_a S + V\frac{dS}{dt} + 0 \qquad (20.16)$$

Eq. 6.16 can be written as:

$$F_0 S_0 + \alpha F_0 S = F_0 (1+\alpha) S + V\left(\frac{dS}{dx}\frac{dx}{dt}\right) \qquad (20.17)$$

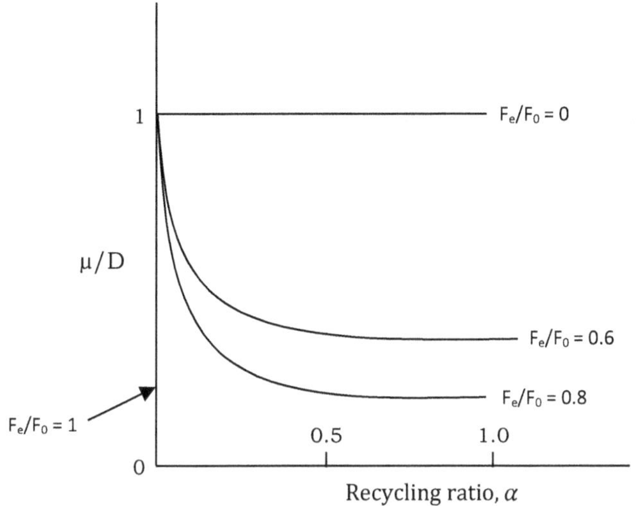

FIGURE 20.4 Plot of μ/D vs. recycling ratio (α) at different F_e/F_0 ratios.

$$F_0S_0 + \alpha F_0S - F_0S - \alpha F_0S = V\left(\frac{1}{Y_{x/S}}\mu x\right) \tag{20.18}$$

(Since $\frac{dx}{dS} = Y_{X/S}$; $\frac{dx}{dt} = \mu x$)

$$F_0\left(S_0 - S\right) = V\left(\frac{1}{Y_{x/S}}\mu x\right) \tag{20.19}$$

$$D\left(S_0 - S\right) = \frac{1}{Y_{x/S}}\mu x \tag{20.20}$$

$$\left(\text{Since } D = \frac{F_0}{V}\right)$$

$$x = \frac{D\left(S_0 - S\right)}{\mu}Y_{x/S} \tag{20.21}$$

Using the value of μ from Eq. 20.14, we get

$$x = \frac{\left(S_0 - S\right)}{1+\alpha\left(1-C\right)}Y_{x/S} \tag{20.22}$$

Thus, the biomass increases by a factor of $\frac{1}{1+\alpha(1-C)}$ as compared to chemostats without recycling.

The substrate concentration 'S' can be obtained by applying Monod kinetics to Eq. 20.14

$$\frac{\mu_{max} S}{K_S + S} = D\,[1+\alpha(1-C)] \tag{20.23}$$

By rearranging,

$$\frac{\mu_{max}}{D\left[1+\alpha(1-C)\right]} = \frac{K_S + S}{S} \tag{20.24}$$

$$S = \frac{K_S\ D\,[1+\alpha(1-C)]}{\mu_{max} - D\,[1+\alpha(1-C)]} \tag{20.25}$$

Using the value of S in Eq. 20.22, we get

$$x = \frac{Y_{x/S}}{\left[1+\alpha(1-C)\right]}\left[S_0 - \frac{K_S\left[1+\alpha(1-C)\right]D}{\mu_{max} - \left[1+\alpha(1-C)\right]D}\right] \tag{20.26}$$

Figure 20.5 shows that $D_{washout}(with\,recycling) \gg D_{washout}(without\,recycling)$. Typical values of the kinetic parameters are shown in Table 20.3.

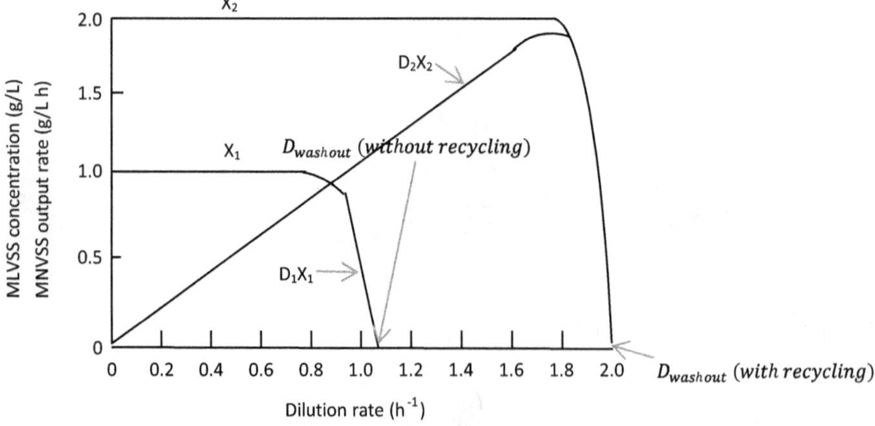

FIGURE 20.5 Rate of cell mass formation or cell mass productivity without recycling (D_1X_1) and with recycling (D_2X_2) at different dilution rates.

TABLE 20.3
Typical values of the kinetic parameters of ASP at 20°C

Coefficient	Range	Typical
μ_{max} (d^{-1})	1.5-4	1.6
K_s $(g\ m^{-3})$	10-50	25
$Y_{\frac{x}{S}}$ (g biomass g^{-1} BOD_u)	0.3-0.8	0.5
Hydraulic retention time (θ) (d)	1.02-1.15	1.05
Mean cell residence time (θ_c) (d)	3-20	5
Rate of substrate consumption/Total amount of cell mass (*F*/*M*) (d^{-1})	0.05-0.5	0.3
MLVSS/MLSS	0.75-0.95	0.80
$\alpha = F_r/F_0$	0.2-1.0	0.4

The settling characteristics of the sludge play a very important role in the separation of sludge in the separator. The settling characteristics of the sludge are determined by the sludge volume index (SVI). The SVI is the volume in mL occupied by 1 g of an insoluble solid suspension after 30 min settling. Settling characteristics of activated sludge and other biological suspensions are determined by SVI.

Desirable SVI for activated sludge process: 100 – 200 mL/g

Problem 20.1 Design an activated sludge process (ASP) for the treatment of 0.20 m^3/s of settled wastewater having 500 mg/L of BOD_5. BOD_5 of the effluent is 30 mg/L. The temperature is assumed to be 20 °C and the following conditions are applicable. Influent MLVSS to the reactor is negligible. The following data is available:

$$\text{MLVSS/MLSS} = 0.8,\ x_u = 10{,}000 \text{ mg/L of suspended solids, } x\ (\text{MLVSS}) = 4200 \text{ mg/L,}$$

$\theta_c = 10$ d, $BOD_5 = 0.68\ BOD_u$, the effluent contains 25 mg/L of biological suspended solids, of which 70% is biodegradable, $\mu_{death} = 0.06\ d^{-1}$ and $Y_{x/S} = 0.5$

The waste contains sufficient nitrogen, phosphorous, and other trace nutrients for cell growth. The reactor is considered a continuous stirred tank reactor.

SOLUTION

1. Estimation of soluble BOD_5 of the effluent.

   Effluent BOD_5 = BOD_5 of the soluble effluent + BOD_5 of effluent suspended solids

Determination of BOD_5 of Effluent S.S

(i) Biodegradable portion of effluent S.S = 0.7 (25 mg/L) = 17.5 mg/L
(ii) BOD_u of effluent S.S. = (17.5 mg/L) × (1.42 mg/mg) = 24.85 mg/L
[1g biodegradable organic solid requires 1.42 g of oxygen for complete oxidation]
(iii) BOD_5 of effluent S.S = 24.85 mg/L × 0.68 = 16.9 mg/L

Therefore, BOD_5 of soluble effluent = (30 – 16.9) mg/L = 13.1 mg/L

2. Determination of the treatment efficiency of the process

a) Biological conversion efficiency of the process $= \frac{(500-30)\,mg/L}{500\,mg/L} \times 100 = 94\%$

b) The overall plant efficiency $= \frac{(500-13.1)\,mg/L}{500\,mg/L} \times 100 = 97.4\%$

3. Volume of the bioreactor

From Eq. 5.80, we get

$$V = \frac{Y_{x/S}(S_0 - S)\theta_c F_0}{x(\theta_c \mu_d + 1)} = \frac{0.5\times(500-13.1)\frac{mg}{L} \times 10d\times 17,280\ m^3/d}{4200\ mg/L\times(10d\times 0.06d^{-1}+1)} = 6,260\ m^3$$

4. Determination of rate of sludge bleeding

a. Observed yield,

$$Y_{obs} = \frac{Y_{x/S}}{(1+\mu_d\theta_c)} = \frac{0.5}{(1+10\ d\times 0.06\ d^{-1})} = 0.3125$$

b. Rate of cell mass (MLVSS) formation

$$= Y_{obs} F_0 (S_0 - S) = \frac{0.3125\times 17,280\,m^3/d \times (500-13.1)\,mg/L}{1000\ mg/g}\ \text{kg/d}$$

= 2,629 kg/d

c. Rate of sludge (MLSS) formation

$$= \frac{2,629\,kg/d}{0.8} = 3,286\,\text{kg/d}$$

= amount of sludge to be wasted per day

5. Determination of volumetric sludge-wasting rate

$$F_w = \frac{V\,x}{\theta_c\,x_u} = \frac{6,260\,m^3 \times 4,200\,mg/L}{10\,d \times 10,000}\ 262.9\,m^3/d$$

6. Computation of recycling ratio

Bioreactor cell mass concentration $(x) = 4,200$ mg/L = 4.2 g/L = $4.2 kg\,m^{-3}$

Return cell mass concentration $(x_u) = 10,000$ mg/L = 10 g/L = $10 kg\,m^{-3}$

Therefore, $4.2\ kg\,m^{-3}\ (F_0 + F_r) = 10\ kg\,m^{-3}\ F_r + 2,629\,kg/d$

$F_r = 12,059\ m^3/d$

$$\therefore \alpha = Recycling\,ratio = \frac{F_r}{F_0} = \frac{12,059\,m^3/d}{17,280\,m^3/d} = 0.7$$

7. Hydraulic retention time (HRT)

$$= \theta = \frac{V}{F_0} = \frac{6,260\ m^3}{17,280\ m^3/d} = 0.36\ d$$

8. Determination of oxygen requirement
   Assumption: Nitrification is neglected

   a. Mass of BOD_u utilized

$$= \frac{F_0\left(S_0 - S\right)}{0.68} = \frac{17,280\,m^3/d \times \left(500 - 13.1\right)mg/L}{0.68 \times 1000\,mg/g} = 12,373\frac{kg}{d}$$

   [since $BOD_5 = 0.68\,BOD_u$]

   b. Oxygen required $= 12,373\frac{kg}{d} - 1.42 \times 2629\frac{kg}{d} = 8,640\frac{kg}{d}$

9. F/M ratio and volumetric loading factor

   a. $\text{F/M} = \frac{F_0\left(S_0 - S\right)}{\text{V x}} = \frac{17,280\,m^3/d \times \left(500 - 13.1\right)\ mg/L}{6,260\,m^3\ \times 4,200\ \ mg/L}\ 0.32\,d^{-1}$

   b. Volumetric loading $= \frac{F_0 S_0}{V} = \frac{17,280\frac{m^3}{d} \times 0.5\,kg\,/\,m^3}{6,260\ m^3} = k\ 1.38\,kg/m^3.d$

10. Computation of air requirement
    a. Theoretical air requirement

$$= \frac{8{,}640 \dfrac{kg}{d}}{1.201 \dfrac{kg}{m^3} \times 0.21} = 34{,}257\ m^3/d$$

(Assuming density of air = 1.201 $\dfrac{kg}{m^3}$ and air contains 21% oxygen)

    b. Actual air requirement

$$= \frac{34{,}257\, m^3/d}{0.08} = 428{,}214\, m^3/d$$

(Assuming Oxygen-transfer efficiency = 8%)

    c. Design air requirement

$$= 2 \times 428{,}214 \frac{m^3}{d} = 856{,}428 \frac{m^3}{d}$$

(assuming factor of safety = 2)

11. a. Air required per unit volume of effluent

$$= \frac{428{,}214 \dfrac{m^3}{d}}{17{,}280 \dfrac{m^3}{d}} = 24.8 \frac{m^3}{m^3}$$

    b. Air required per kg. of BOD_5 removal

$$= \frac{428{,}214\, m^3/d}{17{,}280 \dfrac{m^3}{d} \times 0.4869\, kg/m^3} = 50.3\, m^3/kg$$

20.3 ANAEROBIC DIGESTION PROCESS

All the chemical and biochemical industries must have wastewater treatment plants which incur about 10% of total installation cost. Biodegradable organic matter can be treated with aerobic and anaerobic treatment processes. Low-strength industrial effluent is usually treated with aerobic processes, whereas high-strength wastewater is found to suitable for the anaerobic process. The main advantage of the aerobic process is the higher rate of degradation of the substrate as compared to the anaerobic process. However, a major drawback of the aerobic process is the conversion of 50% of the carbon to cell mass, which creates a disposal problem. On the other hand, in the anaerobic digestion process biodegradable organic matters will be converted to hydrogen and/or methane, which have heating values. Development of this anaerobic digestion process makes it possible to recover some of the expenditure invested in the wastewater treatment plant using hydrogen and methane as fuels. For example it has been found that 50 % of the energy required for distillation of alcohol can be fulfilled by anaerobic digestion of distillery effluent.

The biomethanation process is very old. This has both small-scale and large-scale applicability. There are millions of biogas plants located in the villages of China as well as India. High-strength biodegradable organic wastes discharge by the chemical and biochemical industries can be degraded by using anaerobic treatment processes for the dual purposes of reduction of pollution load and generation of gaseous energy. The digested materials may be taken to be an organic fertilizer, which increases not only the water retention property of the soil but also its fertility (Das, 1985; Das and Sen Gupta, 1990).

Compressed natural gas (CNG) is largely used both by developed and developing countries to avoid environmental pollution problems due to formation of fog by the gasoline used in the automobile sector. CNG mostly comprises methane, which is the simplest hydrocarbon. Therefore the possibility of formation of particulate matter in the air is less.

20.3.1 Major Steps Involved in Anaerobic Digestion

Methane is produced in nature by bacterial decay of vegetables and animal wastes in the absence of molecular oxygen. This process is known as anaerobic digestion (AD). AD is a series of biological processes in which microorganisms are involved in the breakdown of biodegradable organic matter in the absence of oxygen (Figure 20.6). The first step is the hydrolysis of organic compounds where bigger molecules such as cellulose, starch, proteins etc. are converted to smaller soluble organic materials such as glucose; in the second step these are converted to volatile fatty acids such as acetic acid, butyric acid, propionic acid etc. In the third step, the higher volatile fatty acids will be converted to acetic acid and hydrogen. These reactions are controlled by the mixed microbial consortium known as acidogens. Acetic acid is converted to methane and carbon dioxide by another group of mixed microflora known as methanogens. Acidogens are micro-aerophilic in nature whereas methanogens are obligatory anaerobes. Therefore the major constraints of the process are absence of oxygen – i.e. the system should be free from air – and proper environmental conditions for growth of acidogens and methanogens. Acidogens usually prefer acidic pH (5.5), whereas methanogens require alkaline pH (7.5). Therefore a two-stage process is found to be suitable for the biomethanation process (Figure 20.7) where it is possible to maintain desired conditions for the microorganisms. In the first reactor acidogens are active, whereas the second reactor is dominated by methanogens. In the first reactor the gas comprises hydrogen and carbon dioxide and in the second reactor the gas

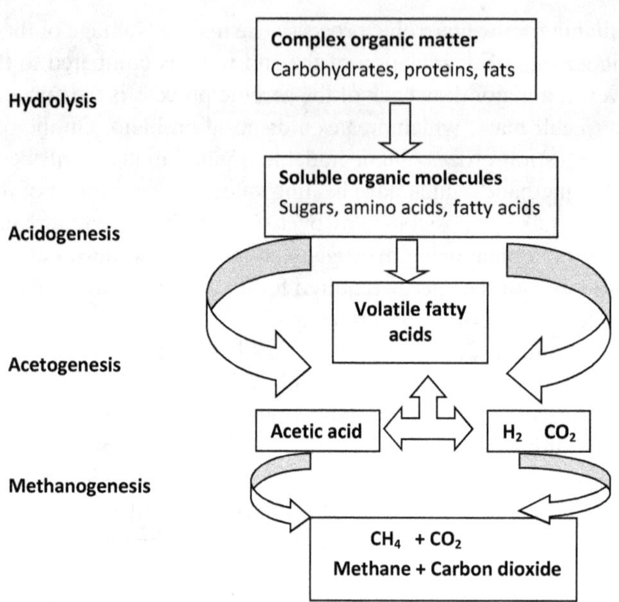

FIGURE 20.6 Major steps involved in the anaerobic digestion process of organic wastes.

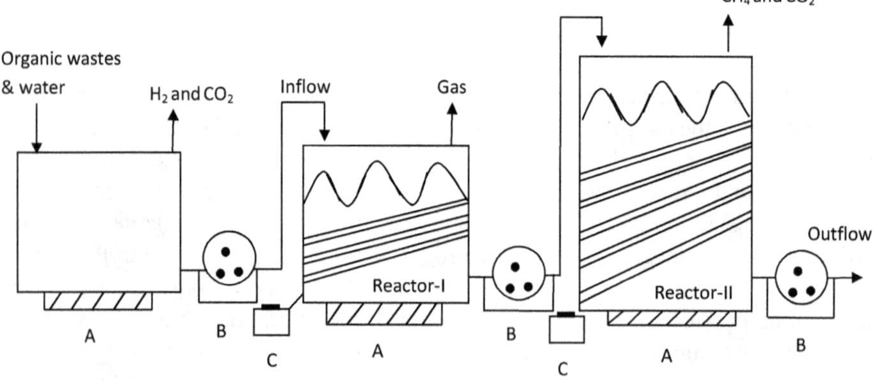

FIGURE 20.7 Two-stage biomethanation process.

contains mostly methane and carbon dioxide. Methane can be used for transportation or cooking fuel and generation of electricity. The digested material is used as an organic fertilizer.

20.3.2 Organic Wastes Suitable for Anaerobic Digestion

The following organic wastes are found to be suitable for AD:

Poultry wastes

- Animal dung (cow dung, pig dung etc.)
- Plant wastes (husk, grass, weeds etc.)
- Human excreta
- Industrial wastes (distillery effluent, wastewater from food processing industries etc.)
- Domestic wastes (vegetable peels, waste food material etc.)

20.3.3 Two-Stage Anaerobic Digestion or Biomethanation Process

These organic wastes contain different organic compounds such as polysaccharides (e.g. cellulose, hemicellulose, starch etc.), proteins (polymer of amino acids), lipids etc. These compounds undergo hydrolysis in order to produce soluble sugars, peptides, amino acids etc. in the presence of the hydrolysing enzymes present in

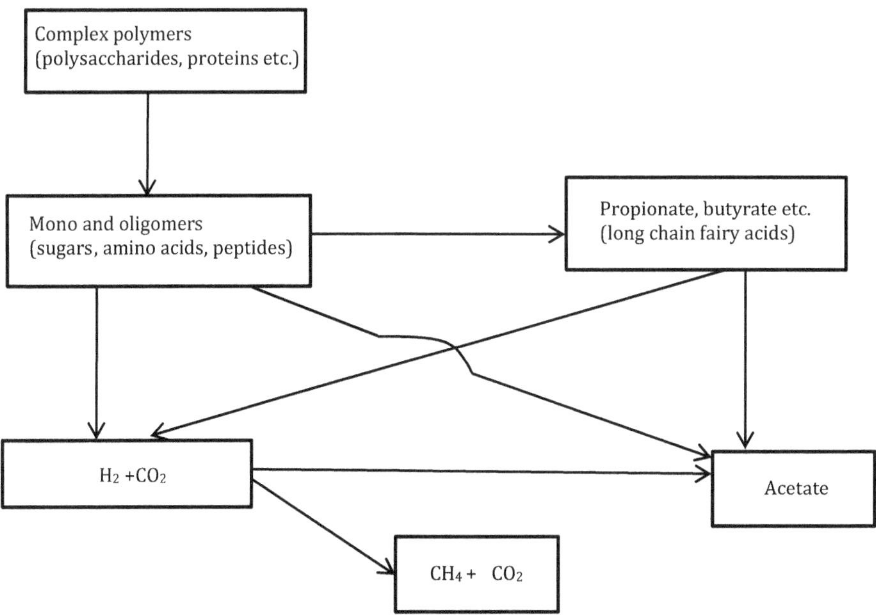

FIGURE 20.8 Mode of degradation of the components present in organic wastes in AD.

acidogens. This is followed by the formation of long- and short-chain volatile fatty acids (VFAs) such as acetic acid, propionic acid, butyric acid etc. The VFAs are degraded to methane and carbon dioxide by methanogens (Fig. 20.8). This process is largely applicable throughout the world. Acidogens and methanogens are present in AD. Acidogens are fast-growing microorganisms as compared to methanogens. Therefore acidogens can easily be separated from methanogens by controlling the dilution rate, because $D_{washout\ (acidogens)} \gg D_{washout\ (methanogens)}$. This has been discussed in detail in Chapter 5. Methanogen cultures can be developed using VFA as feed in alkaline pH. A temperature range of 35–40 °C is suitable for AD. However, this process can be operated at 60 °C using thermotolerant cultures (Das and Varanasi, 2019; Das, 1985). Methanogens have the following characteristics:

- They belong to archaebacteria
- They are distinguished from eubacteria by virtue of many contrasting characteristics, such as the presence of isoprenoid-rich membrane lipids
- They are linked with glycerol, and the absence of a muramic acid-based peptidoglycan cell wall and distinct ribosomal RNA
- They are slow-growing organisms as compared to acidogens
- They prefer alkaline pH

On the basis of their metabolic characteristics, methanogens can be categorized into three groups:

CO_2-reducing – The CO_2-reducing methanogens convert CO_2 or bicarbonate to methane, for which they require two electrons (Rouviere and Wolfe, 1989).
Methylotrophic – can reduce CO_2 to CH_4 by oxidizing methyl group containing substrates such as methanol, trimethylamine, and dimethyl sulfide (Hippe et al., 1979; Mathrani and Boone, 1985).
Acetoclastic pathway – utilize acetate for methane production

Methanogens can be classified into five different genera as mentioned below

- *Methanobacteriales*
- *Methanococcales*,
- *Methanomicrobiales*
- *Methanosarcinales*
- *Methanopyrales*

The biochemical characteristics of methanogens differ from other bacteria as follows:

- The biosynthesis of methane takes place in restricted ranges of oxidizable substrates
- Contains an unusual range of cell-wall components
- Produces biphytanyl glycerol ethers as well as high amounts of squalene

- Co-enzyme M and growth factors present in the cells are considered to be unusual coenzymes
- The synthesis of rRNA is differed from that of the other bacteria
- Genome size (DNA) is about 1/3 that of *Escherichia coli*

The conventional anaerobic digestion process is shown in Figure 20.9. The quantity of methane gas can be calculated as follows:

$$V_{CH_4} = (0.35\text{m}^3/\text{kg})\left[\frac{EQS_0}{1000} - 1.42\text{Px}\right] \tag{20.27}$$

Where V_{CH_4} = volume of methane produced, m³/d

E = efficiency of waste utilization (normally ranges from 0.6 to 0.9 under satisfactory operating conditions)

Q = flow rate, m³/d

= ultimate BOD_u of influent, g/m³

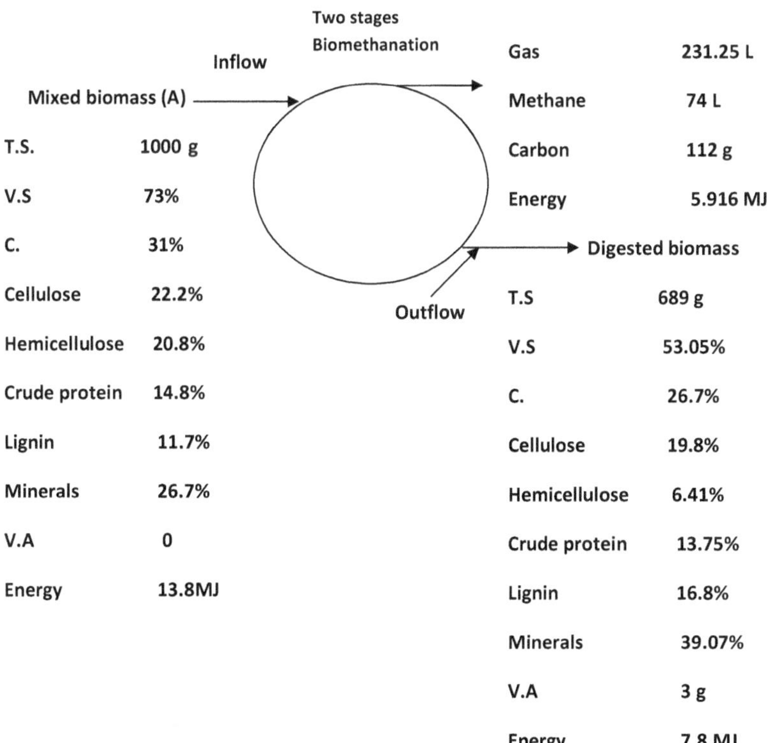

FIGURE 20.9 Materials and energy analysis of the batch biomethanation process using mixed organic residues (Das, 1985).

1.42 = Conversion factor for cell mass to BOD_u
P_x = net rate of cell mass production, kg/d
0.35 = theoretical conversion factor for the amount of methane produced from the conversion of 1 kg of BOD_u. The calculations are shown below:

$$\underset{180}{C_6H_{12}O_6} = \underset{132}{3CO_2} + \underset{48}{3CH_4} \tag{20.28}$$

$$\underset{48}{3CH_4} + \underset{192}{6O_2} = 3CO_2 + 6H_2$$

$$\frac{kg\,CH_4}{kg\,BOD_u} = \frac{48/180}{192/180} = 0.25 \tag{20.29}$$

Therefore, 1 kg BOD_u = 0.25 kg CH_4 = 0.35 m^3 CH_4
(at NTP)

For a continuous flow-stirred tank digester without recycling, the mass of biological solids synthesized daily P_X can be estimated as follows.

$$Px = \frac{YQ(ES_0)}{(1+\mu_d\theta_c)1000} \tag{20.30}$$

where, Y = yield coefficient, g/g, Q = flow rate, m^3/d, μ_d = specific death rate of the microorganisms, d^{-1}, θ_c = mean cell residence time, d

In the conventional anaerobic digestion process both acidogens and methanogens are present in the same reactor but they have different physiological characteristics. Therefore the advanced anaerobic digestion process is conducted in two separate stages: one is acid-forming and another is methane-forming. The performance of the batch and two-stage fermentation processes using cow dung and mixed organic residues (comprising cow dung, water hyacinth, microalgae in 1:1:1 proportion) is shown in the Table 20.4. The experiments were conducted at 37 °C for 15 d.

It was observed that mixed residues produce more methane than cow dung mostly due to the availability of the desired nutrients and substrates. Microorganisms require not only carbon and nitrogen sources but also minerals and vitamins. Mixed residues fulfil these requirements. Material and energy analysis of the batch biomethanation process using mixed residues is shown in Figure 20.9.

Distillery effluent is found to be suitable for the biomethanation process. The performance of the two-stage biomethanation process using immobilized whole cell is shown in Table 20.5 (Switzenbaun, 1982; Switzenbaun and Jewel, 1980; Das et al. 1983).

The largest AD plant in Europe is located in Denmark. The biogas plant in Vrå University, Denmark annually processes 300,000 tons of biomass, of which approximately 250,000 tons is manure that would otherwise have ended up as untreated

TABLE 20.4
Comparison of batch and two-stage biomethanation processes using solid organic wastes (Das D,1985)

Parameters	Cow dung		Mixed biomass (A)	
	Batch process	Two-stage process	Batch process	Two-stage process
Volatile solids (V.S.) reduction,%	31	36	35	50
Biogas production, m^3 kg^{-1} dry residue	0.102	0.12	0.16	0.231
Methane production, m^3 kg^{-1} dry residue	0.061	0.08	0.12	0.174
Methane yield, m^3 kg^{-1}V.S. digested	0.246	0.278	0.468	0.475
Energy recovery as methane, %	16.0	20.5	29.56	43.5
Increase in NPK value, %	33	41	34	57

manure on fields. In India, there are about 400 distilleries (Das, 1988). Most of these distilleries have AD for the treatment of their effluents.

Biomethanation of organic waste satisfies all three principles – security of supply, sustainability and competitiveness (Das and Varanasi, 2019). There are several applications for the methane produced by the AD process.

- Electricity generation – Methane present in biogas is used for electricity production in gas engines.
- Heat generation – As with natural gas, biomethane can be used for household applications such as cooking and heating.
- Automobiles – Methane can be used as CNG and LNG for vehicles
- Material use – Methane is used as a chemical feed stock in chemical processes e.g. fertilizer production (Haber–Bosch process) and iron-ore reduction processes.

Problem 20.1 Determine the capacity of anaerobic digester required to treat 10,000 m^3/d distillery effluent from an alcohol-producing industry. Find the volumetric loading and estimate the percent gaseous energy recovery and the amount of gas production. Assume that the distillery effluent contains 60,000 mg COD/L and has a specific gravity of 1.0. Other pertinent design assumptions are as follows:

1. The hydraulic regime of the reactor is continuous flow stirred tank.
2. Hydraulic retention time, $\theta = 5$ d
3. Mean cell residence time, $\theta_c = 10$ d
4. E = 0.80
5. The waste contains adequate nitrogen and phosphorus for biological growth.
6. $Y_{x/S} = 0.05$ and $\mu_d = 0.03\ d^{-1}$
7. Constants are for a temperature of 35 ^{0}C
8. Heat value of methane: 38.7 MJ/m^3 and distillery effluent: 18 MJ/Kg COD.

TABLE 20.5
Performance of two-stage biomethanation process using distillery effluent at Indian Institute of Technology Delhi (Das et al. 1983; Das and Biswas, 1992)

Reactors Parameters	Reactor-I	Reactor-II
Loading (Kg COD/m³.d)	24.26	7.28
HRT (d)	1.5	6.0
Biogas Productivity (m³/m³.d)	1.33	6.66
Biogas (m³/m³ distillery effluents)	2	44
Gasification Rate (L methane/Kg COD)	-	329
COD Reduction	6%	81%
Acid Concentration (g/L in terms of acetic acid)	24	7

SOLUTION

Given data:

COD of the effluent = 60,000 mg/L = 60 g/L = 60 Kg/m³

Flow rate of the effluent = 19,000 m³/d = 791.7 m³/h

Assuming BOD/COD ratio = 0.8

Total amount of BOD loading = 60 Kg/m³ × 791.7 m³/h × 0.8 = 38,000 kg/h

Hydraulic retention time = 5 d = 120 h = $\frac{V}{F}$

Volume of the reactor = 120 h × 791.7 m³/h = 95,004 m³

Volumetric loading = $\frac{38{,}000 \text{ kg BOD/h}}{95{,}004 \text{ m}^3}$ = 0.4 kg BOD/h m³

$$\text{Methane production} = 0.35\left(E\,F\,S_0 - 1.42\,\frac{dx}{dt}\right) \tag{20.31}$$

$$\text{Again, } \frac{dx}{dt} = \frac{Y_{x/S}\,E\,F\,S_0}{\left(1+\mu_d\,\theta_c\right)} = \frac{0.05\times0.8\times791.7\frac{m^3}{h}\times60\frac{kg}{m^3}}{(1+0.03d^{-1}\times10d)} = 1461.6 \text{ kg/h}$$

From Eq. 20.31, we get, methane production = 0.35 (0.8 × 0.8 × 791.7 $\frac{m^3}{h}\times60\frac{kg}{m^3}$ – 1.42 × 1461.6 kg/h) = 0.35 × 28,753 = 10,063 m³/h

Heating value of distillery effluent = 60 Kg/m³ × 791.7 m³/h × 18 MJ/Kg = 855,036 MJ/h

Energy recovered as methane = 10,063 m³/h × 38.7 MJ/m³ = 389,457 MJ/h

Therefore, gaseous energy recovery = $\frac{389{,}457 \text{ MJ/h}}{855{,}036 \text{ MJ/h}}\times100$ = 45.5%

Assuming gas content 60%v/v of methane, the rate of gas production =
$$\frac{10,063\ m^3/h}{0.6} = 16,771\ m^3/h$$

20.3.4 Landfill Gas Generation

The Landfill gas generation process is used for the degradation of municipal solid wastes (MSW) to methane and carbon dioxide by anaerobic digestion (AD). This is successfully implemented both in the USA and European countries. AD is governed by two groups of microflora: acidogens and methanogens as mentioned previously. The age of landfills varies from 5 to 10 years. Due to rapid industrialization and urbanization MSW production is increasing drastically. Landfill processes are usually designed to fill up the low land areas. Most of the low land areas near metropolitan cities are filled up. Therefore, the hauling distance for the disposal of MSW increases. This increases the cost of MSW disposal. It has already been found that the two-stage process increases the conversion efficiency of the MSW significantly within a short time. Therefore, one particular landfill area may be identified near metropolitan cities and reused again and again because digestion time can be reduced to 40–50 d. The digested materials may be used as an organic fertilizer. The advanced landfill gas-generation process is depicted in Figure 20.10. After weighing the track, MSW is unloaded for screening purposes. This is followed by shredding of the bigger particles to a smaller size so that microorganisms can interact with more surfaces of MSW. This will increase the rate of biodegradation of solids to volatile fatty acids (VFAs) by the acidogens. These VFAs can be degraded to methane and carbon dioxide by methanogens in a separate reactor. Digested solids from landfill can be used as fertilizer (Das and Sen Gupta, 1990).

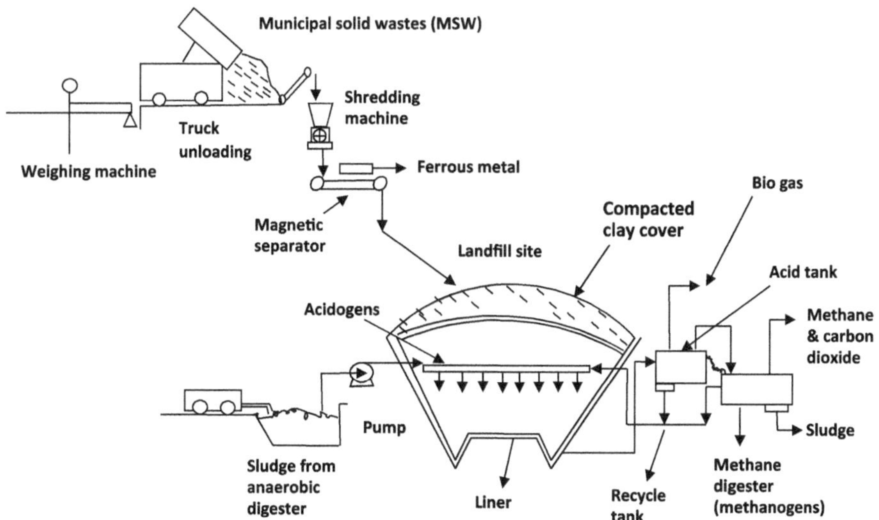

FIGURE 20.10 Schematic diagram of the advanced landfill gas generation process.

20.4 BIOHYDROGEN PRODUCTION BY DARK FERMENTATION PROCESS

Hydrogen is considered as a unique energy source due to the following:

- No greenhouse gases
- No ozone depleting chemicals
- No acid rain ingredients
- No pollutants
- Clean environment for biodiversity
- High utilization efficiency conserving resources
- Abundant energy for economic development

20.4.1 Special Features of Biohydrogen Production Process

The special features of the biohydrogen production process are given below.

- Essentiality: Constraints in the conventional production techniques due to higher energy consumption.
- Versatility: Different renewable resources are used as potential substrates (particularly through bioremediation of organic wastes).
- Naturality: On combustion, hydrogen produces water i.e. zero pollution and a cost-effective process.

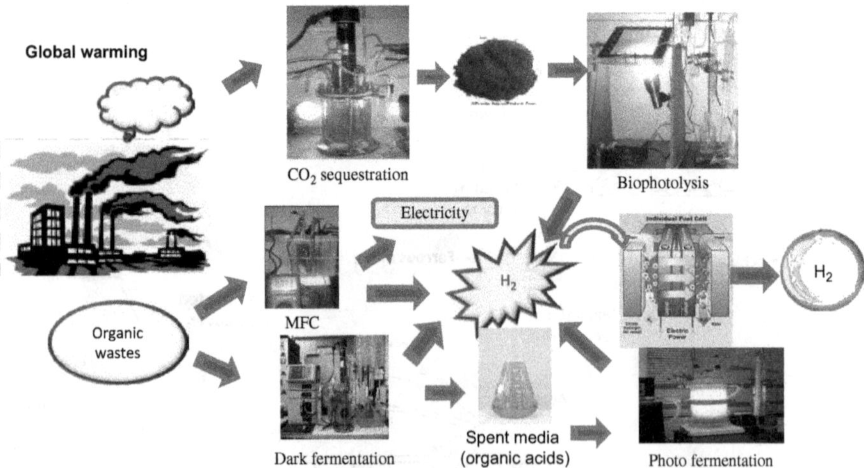

FIGURE 20.11 Various biohydrogen production processes.

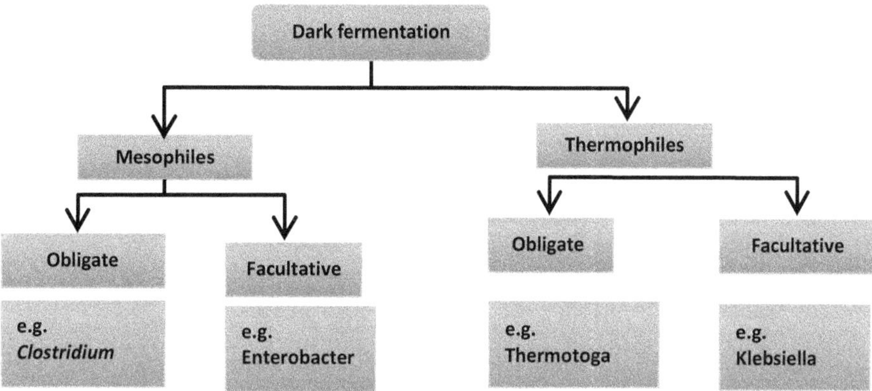

FIGURE 20.12 Classification of the microorganisms participating in the dark fermentation process.

20.4.2 Various Biohydrogen Production Processes

Biohydrogen can be produced by various processes as shown in Fig. 20.11 (Das et al., 2014). These processes are as follows:

- Biophotolysis
- Dark fermentation
- Photofermentation
- Microbial Fuel Cell (MFC)

Biophotolysis processes are carried out by microalgae and cyanobacteria. They can produce hydrogen both from the hydrolysis of water and organic compounds. These processes require light energy sources for growth and metabolism. The key enzyme responsible for hydrogen production is FeFe-hydrogenase, which is inhibited by the molecular oxygen present in the process. Therefore, the rate of hydrogen production is very low. The photofermentation process utilizes soluble organic compounds such as acetic acid, and it also requires a light energy source. However, hydrogen production drastically reduces due to red pigment formation which has a shading effect on the light. MFC can produce electricity directly from biodegradable organics present in wastewater. However, it has scaling-up problems. The dark fermentation process is found to be most suitable for hydrogen production. (Das et al., 2014; Das and Varanasi, 2019)

20.4.3 Biohydrogen Production by Dark Fermentation Process

Microorganisms involved in the dark fermentation process can be classified as mesophiles (20–45 °C) and thermophiles (> 45 °C) (Fig. 20.12). Facultative organisms are microaerophilic in nature but obligate anaerobes cannot tolerate trace amounts of oxygen concentration in the fermentation medium. These microorganisms can utilize different organic materials as stated below.

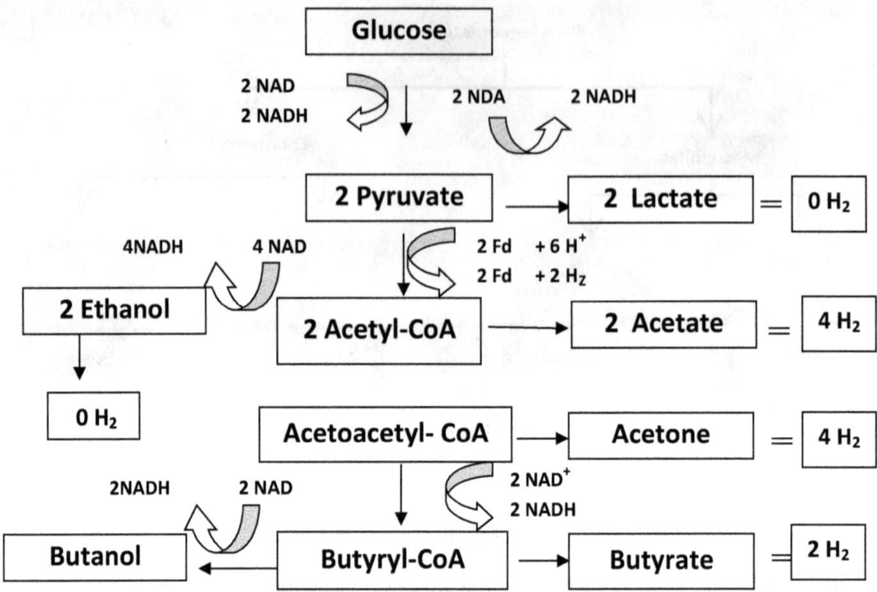

FIGURE 20.13 Mixed-acid pathways for biohydrogen production.

- Carbohydrates: glucose, sucrose, maltose, cellulose, starch, cane molasses.
- Cellulosic biomass: energy crops (corn, wheat straw), paper pulp.
- Wastes: municipal solid waste (MSW), food wastes, industrial wastes, wastewater.

When pyruvate is oxidized to acetate as the sole metabolic end product, four moles of hydrogen per mole of glucose are formed (Das et al., 2014; 2015). In the case of oxidation of pyruvate to butyrate this produces 2 moles of hydrogen per mole of glucose (Fig. 20.13). The overall stoichiometry of the process is given below:

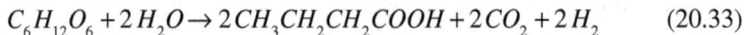

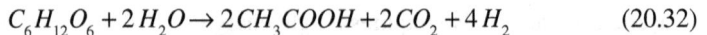

$$C_6H_{12}O_6 + 2H_2O \rightarrow 2CH_3COOH + 2CO_2 + 4H_2 \quad (20.32)$$

$$C_6H_{12}O_6 + 2H_2O \rightarrow 2CH_3CH_2CH_2COOH + 2CO_2 + 2H_2 \quad (20.33)$$

In the case of a few facultative anaerobic enteric bacteria such as *E. coli*, the biochemical reaction involves conversion of pyruvate to acetyl CoA and formate which is followed by hydrogen and carbon dioxide production.

$$CH_3COCOOH + CoA \rightarrow CH_3 - CoA + HCOOH \quad (20.34)$$

Subsequently, formate is cleaved by formate hydrogen lyase (FHL) to produce carbon dioxide and hydrogen.

FIGURE 20.14 10 m^3 biohydrogen pilot plant at Indian Institute of Technology Kharagpur, India.

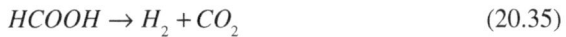

$$HCOOH \rightarrow H_2 + CO_2 \quad (20.35)$$

20.4.4 10,000 L Biohydrogen Pilot Plant at IIT Kharagpur

It has been observed that organic wastes as such are not suitable for high-rate biohydrogen production. Therefore, it is necessary to add some micro-nutrients (minerals and vitamins) which facilitate the growth of microbial cells. Addition of de-oiled ground nut cake in starch-based wastewater as well as cane molasses based distillery effluents increases hydrogen production drastically. A 10 m^3 biohydrogen pilot plant has been in operation at the Indian Institute of Technology Kharagpur using cane molasses and groundnut de-oiled cake (Fig. 20.14). 67 m^3 of hydrogen was produced per day (Balachandar, 2020; Das and Varanasi, 2019). A live demonstration of the process is shown on the website: www.bioh2iitkgp.ac.in.

20.5 BIOHYTHANE PROCESS

It has been observed that biohydrogen production from organic wastes using a dark fermentation process is very promising. However, economic viability of the process and overall energy recovery of the process are major bottlenecks for commercialization. It has been observed that the biohydrogen conversion process produces significant amounts of volatile fatty acids as a by-product. This is a very good substrate for biomethanation processes where these volatile fatty acids can be converted to methane and carbon dioxide. Biohydrogen followed by a biomethanation process is known as a biohythane process (Figure 20.15). This process has great potential for the

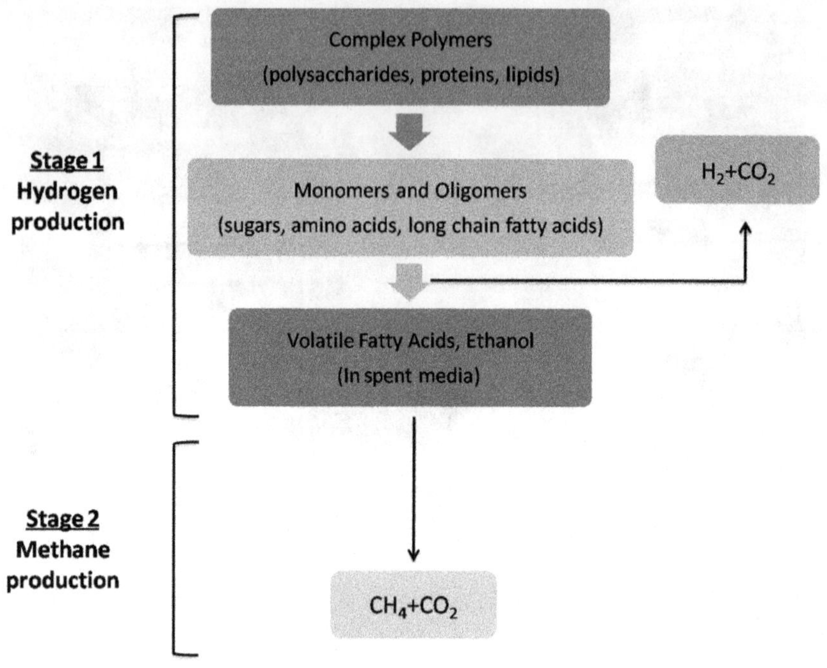

FIGURE 20.15 Biohythane process from organic wastes.

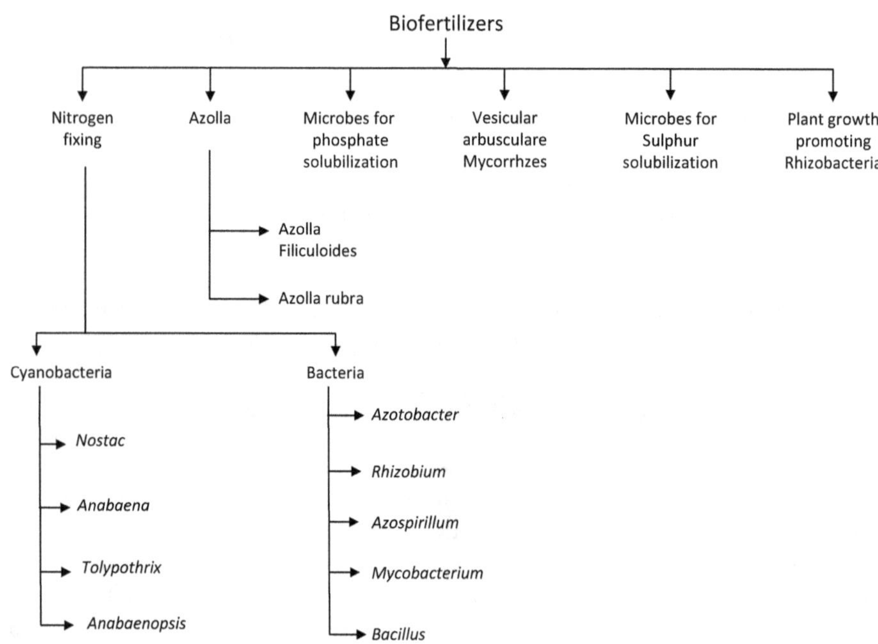

FIGURE 20.16 Microorganisms involved in biofertilizer production.

effective utilization of organic wastes. (Das and Roy, 2017; Das and Varanasi, 2019; Balachandar et al., 2020)

A comparative study of the biomethanation, biohydrogen production and biohythane processes shows that the total energy recovery is highest in the case of the biohythane process (91%) as shown below.

Biomethanation: $C_6H_{12}O_6 + 2H_2O \rightarrow 3CO_2 + 3CH_4$

Maximum energy recovery = 3 × 801/2826.6 × 100 % = 85 %

Biohydrogen: $C_6H_{12}O_6 + 2H_2O \rightarrow 2CH_3COOH + 2CO_2 + 4H_2$

Maximum energy recovery = 4 × 242/2826.6 × 100 % = 34.2 %

Biohythane: $C_6H_{12}O_6 + 2H_2O \rightarrow 3CO_2 + 3CH_4$

$2CH_3COOH + 2H_2O \rightarrow 4CO_2 + 2CH_4 + 2H_2O$

Maximum energy recovery = (4 × 242 + 2 × 801)/2826.6 × 100 % = 91 %

20.6 BIOFERTILIZERS

In the last century, chemical fertilizers were used extensively in agriculture to increase the productivity of agricultural crops. However, chemical fertilizers have several detrimental effects such as

- Water basin pollution due to leaching
- Destruction of microorganisms and friendly insects
- Crops are more susceptible to the attack of diseases
- Reduction of soil fertility

Biofertilizers comprise large populations of a specific group of beneficial microorganisms (mycorrhizal fungi, blue-green algae, and bacteria) to enhance the productivity of soil. The fertility of the soil improves either by fixing atmospheric nitrogen or by solubilizing soil phosphorus or by stimulating plant growth through synthesis of growth-promoting substances. These are responsible for the improvement of water-retention properties of the soil because plants take on the nutrients through the diffusion process. Biofertilizers are based on renewable energy sources, are cost effective, and eco-friendly. There are several benefits of the biofertilizers as listed below.

- 25% chemical nitrogen and phosphorus can be replaced
- 20–30% increase in crop yield
- Stimulate plant growth
- Retain the biological activity of the soil
- Soil fertility is restored
- Provide protection against some soil-borne diseases and drought

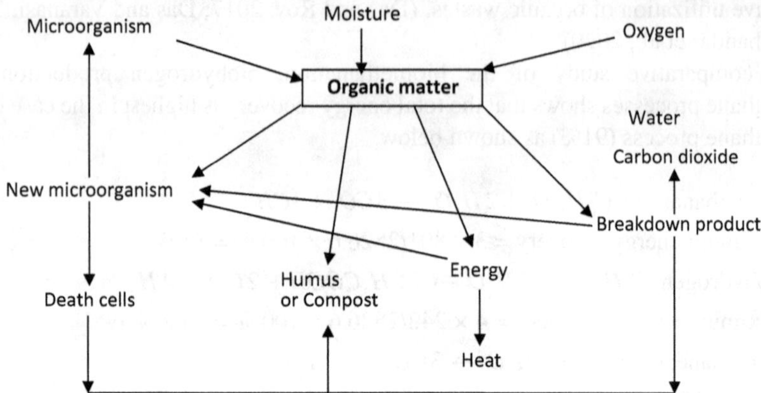

FIGURE 20.17 Interaction between organic waste, microorganisms, moisture and oxygen during aerobic composting.

Farmers can prepare biofertilizers very easily. The microorganisms involved in the biofertilizer production are shown in Figure 20.16 (Das and Sen Gupta, 1990).

The following types of biofertilizer are available to farmers in India:

- Nitrogen-fixing biofertilizers e.g. *Rhizobium, Bradyrhizobium, Azospirillum* and *Azotobacter*
- Phosphorus-solubilizing biofertilizers (PSB) e.g. *Bacillus, Pseudomonas* and *Aspergillus*
- Phosphate-mobilizing biofertilizers e.g. *Mycorrhiza*
- Plant growth promoting biofertilizers e.g. *Pseudomonas* sp

Biofertilizers (*Rhizobium* microorganisms) have the characteristics of fixing atmospheric nitrogen in the soil and root nodules of legume crops so that plants can utilize these. They solubilize insoluble forms of phosphate such as tricalcium, iron and aluminium phosphates. They also produce hormones and anti-metabolites which promote root growth. They decompose organic matter and help in the mineralization of soil, which increases the availability of nutrients and improves plant yield by 10 to 25% without adversely affecting the soil and environment. Biofertilizers are produced by two processes:

- Aerobic composting
- Anaerobic composting

20.6.1 Aerobic Composting

Aerobic composting is a rapid but partial decomposition of wet solid organic matter – primarily garbage – which contains plant nutrients (nitrogen, phosphorus and potassium) by the use of aerobic organisms under controlled conditions. The result is

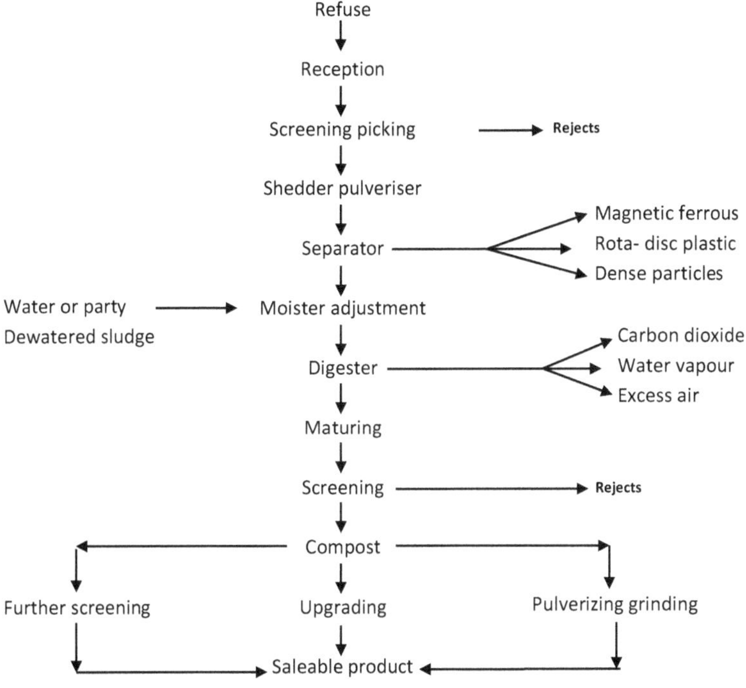

FIGURE 20.18 Flow diagram of the aerobic composting process.

sanitary, nuisance-free, humus-like material that can be used as a soil conditioner and fertilizer. Figure 20.17 shows the interaction between the organic waste, microorganisms, moisture and oxygen during aerobic composting. Aerobic composting can be done by either an open or closed system.

Aerobic composting takes place in three different phases:

- Mesophilic phase: Mesophilic microorganisms are responsible for the degradation of organics at moderate temperatures.
- Thermophilic phase: Various thermophilic bacteria help in the decomposition of organic waste at high temperatures.
- Maturation phase: As the supply of high-energy compounds dwindles, the temperature starts to decrease, and the mesophiles once again predominate.

The flow diagram of the aerobic composting process is shown in Figure 20.18.

20.6.2 Anaerobic Composting

Anaerobic composting is a biological organic waste-treatment process where degradation of organic wastes such as organic sludge, food wastes, and municipal solid wastes etc. take place. The treated solid organic waste is easy to dispose of in low volume due to degradation of the organic wastes. Microbial action during the process

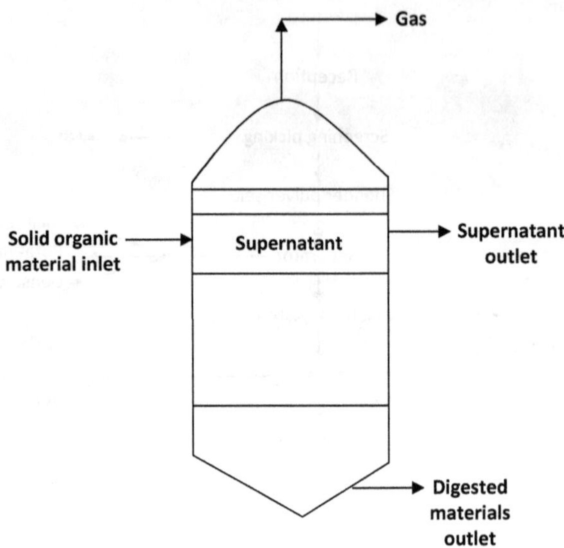

FIGURE 20.19 Anaerobic digestion process.

also helps in biogas generation (methane), and a clean energy source will be generated in the anaerobic degradation process. Nutrients remain in effluent as a source of biofertilizer.

In the anaerobic composting of domestic and farm wastes, there is no destruction of the nutrients, but it makes them more available to plants in useful forms such as ammonia. Anaerobic digester slurry also acts as a soil conditioner which helps to improve the physical properties of the soil. The quality of unproductive soils is improved by using digester slurry, which helps in the growth of plants. Digested organic wastes do not have a bad smell (Figure 20.19). Therefore, the process is environmentally friendly. This process is largely used by farmers by putting vegetable wastes under soil for some time (Das and Sengupta, 1990).

20.7 CONCLUSIONS

More than 70% of wastewater treatment processes are controlled by biological means. Wastewater has been characterized on the basis of parameters such as COD, BOD, TS, MPN etc. The COD estimation procedure is much less time-consuming than BOD. Therefore mainly COD is used to find the strength of wastewater. However, BOD indicates the presence of biodegradable matter in waste, and therefore most industries prepare calibration curves between COD and BOD of their wastes; this helps to find the degradation efficiency of biodegradable organic matter. Biological wastewater treatment processes are broadly controlled by two types of microorganism: aerobes and anaerobes. Aerobic wastewater treatment processes are found to be suitable for treating low-strength wastewater whereas high strength wastewater is treated with anaerobic digestion processes. The main problem of the aerobic treatment process is that 50% of the carbon present in wastewater is converted to cell mass which is biodegradable in nature. Therefore there is a disposal problem. However, in the case of anaerobic digestion processes 95% of carbon present in the wastewater is converted to methane and carbon dioxide. However, aerobic processes are much faster than anaerobic processes. In addition, methane has a calorific value which can be used as gaseous fuel. The biomethanation process is controlled by two groups of microflora: acidogens and methanogens, which differ from each other with respect to substrate, pH etc. Therefore, a two-stage biomethanation process is very effective for degradation of organic matter present in wastewater. Both aerobic and anaerobic processes are successfully used for the production of biofertilizer, which not only increases water retention properties of soil but also its nutritional quality. Organic wastes can be converted to hydrogen and methane by biohythane processes, which have great promise in the future for greater energy recovery.

REFERENCES

Balachandar G, Varanasi JL, Singh V, Singh H, and Das D, Biological hydrogen production via Dark fermentation: A holistic approach from Lab-scale to Pilot-scale, *International Journal of Hydrogen Energy,* 45: 5202–5215, 2020.

Busch AW, Aerobic Biological Treatment of Wastewaters, Oligo dynamic press, Houston, 1971.

Das D, Ph.D. Thesis "Optimization of methane production from agricultural residues", Indian Institute of Technology Delhi, 1985.

Das D, "Scale-up studies of the two stage Biomethanation Process for the treatment of Cane Molasses based Distillery Wastes", IFCON'88, Mysore, 1988.

Das D, Algal Biorefinery: An Integrated Approach, Capital Pub. Co, New Delhi and Springer, Switzerland, 2015.

Das D and Biswas AK, Short Term Course on "Application of immobilized techniques in biotechnology", Indian Institute of Technology Kharagpur, India, 1992.

Das D and Das D, Biochemical Engineering: An Introductory Text Book, Jenny Stanford, Singapore, 2019

Das D, Khanna N, and Nag Dasgupta C, Biohydrogen Production: Fundamentals and Technology Advances, CRC Press, Florida, 2014

Das D and Roy S, Biohythane: Fuel for the future, Pan Stanford Publishing, Singapore, 2017.

Das D and Sen Gupta P, Short Term Course on "Biotechnology for pollution abatement", Indian Institute of Technology Kharagpur, India, 1990.

Das D et.al, "Treatment of Distillery Wastes by a Two Phase Biomethanation Process", Symposium papers "Energy from Biomass and Wastes VII", Florida, USA p. 601–626. 1983.

Das D and Varanasi JL, Fundamentals of biofuel production processes, CRC Press, Boca Raton, FL, 2019.

Green M, and Sheelef G, Sludge Viability in a biological reactor, Water Research, 15, 953, 1980.

Metcalf and Eddy, Wastewater Engineering: Treatment, Disposal and Reuse (Revised by Tchobamoglous G and Burton FL), TATA McGraw-Hill, 1991.

Official Methods of Analysis of the A.O.A.C. (1980), 13th ed., Horwitz W (ed.), A.O.A.C., Washington, D.C., 1980.

Switzenbaum MS, A comparison of the Anaerobic filter and the Anaerobic Expanded/Fluidized Bed Processes, IAWPR Specialized Seminar on Anaerobic Treatment, Copenhagen, Denmark, p.399–413, 1982.

Switzenbaum MS and Jewel WJ, Anaerobic Attached Film Expanded Bed Reactor Treatment, Journal WPCF 52, 1933–1965, 1980.

Index

For Product Safety Concerns and Information please contact our EU representative GPSR@taylorandfrancis.com
Taylor & Francis Verlag GmbH, Kaufingerstraße 24, 80331 München, Germany

www.ingramcontent.com/pod-product-compliance
Lightning Source LLC
Chambersburg PA
CBHW060423220326
41598CB00021BA/2269

* 9 7 8 0 3 6 7 7 6 9 5 6 7 *